Volume 2

Functional Foods

Biochemical and Processing Aspects

FUNCTIONAL FOODS AND NUTRACEUTICALS SERIES

Series Editor
G. Mazza, Ph.D.

Senior Research Scientist and Head
Food Research Program
Pacific Agri-Food Research Centre
Agriculture and Agri-Food Canada
Summerland, British Columbia

Functional Foods: Biochemical and Processing Aspects
Volume 1
Edited by G. Mazza, Ph.D.

Herbs, Botanicals, and Teas
Edited by G. Mazza, Ph.D. and B.D. Oomah, Ph.D.

Functional Foods: Biochemical and Processing Aspects
Volume 2
Edited by John Shi, Ph.D., G. Mazza, Ph.D., and Marc Le Maguer, Ph.D.

Methods of Analysis for Functional Foods and Nutraceuticals
Edited by W. Jeffrey Hurst, Ph.D.

Handbook of Functional Dairy Products
Edited by Collete Short and John O'Brien

Handbook of Fermented Functional Foods
Edited by Edward R. Farnworth, Ph.D.

Handbook of Functional Lipids
Edited by Casimir C. Akoh, Ph.D.

Dictionary of Nutraceuticals and Functional Foods
N. A. Michael Eskin, Ph.D. and Snait Tamir, Ph.D.

Volume 2

Functional Foods

Biochemical and Processing Aspects

EDITED BY

John Shi, Ph.D.
Guelph Food Research Center
Agriculture and Agri-Food Canada

G. Mazza, Ph.D.
Pacific Agri-Food Research Centre
Agriculture and Agri-Foods Canada

Marc Le Maguer, Ph.D.
Department of Food Science
University of Guelph

CRC PRESS

Boca Raton   London   New York   Washington, D.C.

Library of Congress Control Number: LC98-085171

Visit the CRC Press Web site at www.crcpress.com

© 2002 by CRC Press LLC

No claim to original U.S. Government works
International Standard Book Number 1-5667-6902-7
Library of Congress Control Number LC98-085171
Printed in the United States of America   2 3 4 5 6 7 8 9 0
Printed on acid-free paper

Foreword

The science of functional foods and nutraceuticals is at the confluence of two major factors in our society — food and health. The link between diet and disease has now been quite widely accepted, not only at the institutional level by organizations such as Health Canada, the U.S. Surgeon General and the Japan Ministry of Health and Welfare, but also by a large portion of the populace. In recent years, there appears to have been a growing desire by individuals to play a greater role in their own health and well-being rather than rely strictly on conventional medical practice. Much of this drive has been attributed to the aging of the "baby boomer" generation and their efforts to hold the inevitable effects of time on the body at bay. As a result, there has been a burgeoning market for a wide range of dietary supplements and nutraceutical products that are perceived by the consuming public to be beneficial in the maintenance of their health and in the prevention of disease.

This book continues a series of timely publications on functional foods under the editorship of Dr. G. Mazza and explores new sources of nutraceuticals and functional food ingredients as a logical step between foods and therapeutic drugs. Specifically, the book addresses the biochemical and processing aspects associated with the production of functional foods and nutraceutical products. By definition, a nutraceutical is a product isolated or purified from a biological material that is generally sold in medicinal form and is not usually associated with food. A nutraceutical is demonstrated to have a physiological benefit or provide protection against chronic disease. A functional food is similar in appearance to conventional foods, is consumed as part of a usual diet, and has demonstrated physiological benefits or reduces the risk of chronic disease beyond basic nutritional functions.* Based on these definitions, the book addresses the key factors associated with functional foods and nutraceuticals: the biochemistry of various bioactives, their physiological effects and the engineering and process technology associated with the isolation and purification of the desired compounds from the complex biological matrices in which they are found.

In the first book in the series, the biochemical and processing aspects of functional foods were examined in detail. This book now moves the topic forward in its study of the biochemistry and processing aspects of tocopherols and tocotrienols from oil and cereal grains, isoflavones from soybeans, flavonoids from berries, lycopene from tomatoes, limonene from citrus, phenolic diterpenes from rosemary and sage, organosulfur constituents from garlic, pectin from fruits, bioactives from *Echinacea* and omega-3 fatty acids from fish products. Many of these biologically active compounds appear regularly in articles in the popular press and take on almost a "folk medicine" aura. These products will benefit from the sound and thorough

---

* Nutraceuticals/Functional Foods and Health Claims on Foods. Policy Paper, Therapeutic Products Programme and the Food Directorate, Health Canada. November 1998.

approach taken in this book that will help them achieve recognized status in the lexicon of the medical and nutrition professionals.

While each product-specific chapter addresses the biochemical and processing of the subject compound and material from which it is being extracted, Chapter 11 addresses the engineering and technology associated with the extraction of the nutraceutical compounds from plant material from the technology side. Given that essentially all of the target biologically active compounds exist at very low levels in the source material, a suitable technique to extract and separate the compounds is vital to the economic feasibility and will help govern whether the product can be brought to the market at a realistic price.

The book ends with a more general chapter, albeit one that is critical to the success of the sector. Safety of the products is paramount and needs to be carefully addressed if the functional foods and nutraceutical products are to take their legitimate place in the kit of tools that can be used to maintain the health and well-being of the population. Anecdotal evidence of either safety or efficacy will not suffice to ensure the long-term impact that nutraceuticals and functional foods can bring to the quality of life in the general population. Market forces will continue to be a major driver for the industry, as many players ranging from small entrepreneurs to multinational food and pharmaceutical companies seek to enter the marketplace.

The contributors, coeditors and the series editor are to be congratulated on development of a very useful and timely book that serves to advance the science of functional foods and nutraceuticals. A thorough and complete understanding of the many aspects of functional foods and nutraceuticals, the biochemistry, physiological effects, impact of processing and the technologies required for the production, will all contribute to moving this exciting area ahead.

Brian Morrissey, Ph.D.
Former Assistant Deputy Minister
and Research Director of Agriculture and Agri-Food Canada
Ottawa, Canada

Series Preface

The titles in the Functional Foods and Nutraceuticals Series offer food, nutrition and health professionals a comprehensive treatment of the emerging science and technology of functional foods and nutraceuticals. The first two books in the series, *Functional Foods: Biochemical and Processing Aspects, Volume 1* and *Herbs Botanicals and Teas*, have received worldwide acceptance by practitioners in these fields. This latest book and upcoming volumes present the state-of-the-science and technology of all aspects of functional foods and nutraceuticals, from chemistry and pharmacology to process engineering and clinical trials.

With over 2100 scientific references, *Functional Foods: Biochemical and Processing Aspects, Volume 2* provides readers with a comprehensive and up-to-date scientific publication. This volume discusses the occurrence, chemistry, bioavailability, health effects, processing and engineering aspects of tocopherols and tocotrienols from oil and cereal grain, isoflavones from soybeans and soy foods, flavonoids from berries and grapes, lycopene from tomatoes, limonene from citrus, phenolic diterpenes from rosemary and sage, organosulfur constitutes from garlic, phytochemicals from *Echinacea*, pectin from fruit, omega-3 fatty acids from fish products and safety aspects of botanicals.

The book also presents a detailed chapter on solid–liquid extraction technologies for manufacturing nutraceuticals and dietary supplements, and several chapters provide information on physical and chemical properties of bioactives. This information, which is seldom available, is essential for the reliable and economical scaleup of laboratory-based extraction and purification techniques for secondary plant metabolites.

A critical issue in the development of functional foods and nutraceuticals is product safety and efficacy. In Chapter 12, Safety of Botanical Dietary Supplements, a variety of safety issues related to dietary supplements are discussed. These include the complex issue of herb–drug interactions and inherent toxicity of some plants, such as comfrey, which has been used for inflammatory disorders including arthritis, thrombophlebitis and gout, and as a treatment for diarrhea. Recently, however, the U.S. Food and Drug Administration (FDA) has requested the withdrawal of dietary supplements containing the herb comfrey due to the potential danger of liver damage and its possible role as a cancer-causing agent. With respect to herb–drug interactions, the recent case reports that implicate St. John's wort herb–drug interactions with cyclosporine, warfarin, theophylline and ethinyl estrodiol are reviewed.

The present volume also addresses the issue of processing and its effects on the bioavailability of bioactives. It shows that processing of a functional food may have profound effects on specific health benefits it claims to deliver. As illustrated in Chapter 4, the physiological effects of lycopene are altered significantly during processing, primarily due to its isomerization and oxidation. Worth noting, however, is the fact that the *cis*-isomers, which are produced in the processing of tomatoes, are better absorbed by the human body than the naturally occurring all-*trans* form.

Upcoming volumes of the series include methods of analysis for functional foods and nutraceuticals, and functional dairy products.

I hope that this work will be useful to all those interested in functional foods and nutraceuticals, especially food scientists and technologists, nutritionists, phytochemists, physiologists, food process engineers and public health professionals.

G. Mazza
Series Editor

Preface

During the past decade, functional foods and nutraceuticals have emerged as a major consumer-driven trend, serving the desire of aging populations to exercise greater control over health, delay aging, prevent disease and enhance well-being and performance. This trend is expected to continue, and the need for and interest in scientific information on all aspects of functional foods will continue to be vital to the advancement of this emerging sector.

The Functional Foods and Nutraceuticals Series, launched in 1998, was developed to provide a timely, comprehensive treatment of the emerging science and technology of functional foods and nutraceuticals that are shown to play a role in preventing or delaying the onset of diseases, especially chronic diseases. *Functional Foods: Biochemical and Processing Aspects, Volume 1*, the first volume of the series, is a bestseller devoted to functional food products from oats, wheat, rice, flaxseed, mustard, fruits, vegetables, fish and dairy products. In Volume 2, the focus is on presenting the latest developments in chemistry, biochemistry, pharmacology, epidemiology and engineering of tocopherols and tocotrienols from oil and cereal grain, isoflavones from soybeans and soy foods, flavonoids from berries and grapes, lycopene from tomatoes, limonene from citrus, phenolic diterpenes from rosemary and sage, organosulfur constitutes from garlic, phytochemicals from *Echinacea*, pectin from fruit and omega-3 fatty acids and docosahexanoic acid from fish products. Also covered is solid–liquid extraction technologies for manufacturing nutraceuticals and dietary supplements.

All chapters, especially the last two, contain information that has not been published previously. The chapter on solid–liquid extraction technologies, for example, presents and discusses both theoretical and practical aspects of these technologies, from fundamental concepts of equilibrium and mass transfer to equipment selection and design. Similarly, the chapter on safety of botanicals reviews safety issues of botanicals associated with misidentification of plant species, misuse of products, product adulteration and botanical/drug interactions.

The contributing authors are international experts on the subjects covered, and we are grateful to every one of them for their thoughtful and well-written contributions. We hope that this book will be of interest to a wide spectrum of food scientists and technologists, nutritionists, biochemists, engineers and entrepreneurs worldwide, and that it will serve to further stimulate the development of functional foods and nutraceuticals and contribute to providing consumers worldwide with products that prevent diseases and help to maintain a healthier life. We believe the scientific community will benefit from the overall summary of each area presented.

John Shi
G. Mazza
Marc Le Maguer

Acknowledgments

The editors would like to express sincere gratitude and appreciation to all those who reviewed chapters: Dr. Cathy Y. W. Ang (National Center for Toxicological Research/FDA); Dr. Joseph M. Betz (American Herbal Products Association); Professor Robert Shewfelt (University of Georgia); Professor Bor S. Luh (University of California at Davis); Professor Patricia Murphy (Iowa State University); Professor Bhimu S. Patil (Citrus Center, Texas A&M University); Drs. Steve Nagy and Steven Pao (Florida Department of Citrus Research Center); Dr. Keshun Lui (Monsanto Company); Dr. Marijan A. Boskovic (Kraft Foods); Dr. Matt Bernart (Om-Chi Herbs Company); Professors John DeMan, Yukio Kakuda and Gauri Mittal (University of Guelph, Canada); Drs. Akhtar Humayoun, John K. G. Kramer, Albert Liptay, Peter J. Wood and Chris J. Young (Agriculture and Agri-Food Canada). Support from Dr. Brian Morrissey (former Assistant Deputy Minister and Research Director of Agriculture and Agri-Food Canada) and Dr. Gordon Timbers (Food Research Coordinator Agriculture and Agri-Food Canada) is appreciated.

In addition, John Shi wishes to acknowledge the encouragement and help of Greg Poushinsky and Dr. Gary Whitfield (Agriculture and Agri-Food Canada), Dr. Asbjørn Gildberg (Norwegian Institute of Fishery and Aquaculture, Norway), Dr. Albert Ibarz (University of Lleida, Spain), Dr. Pedro Fito (Polytechnic University of Valencia, Spain), Dr. Sam Chang (North Dakota State University), and Dr. Samuel L. Wang (Ameripec, Inc.). The editors would also like to thank Dr. Eleanor Riemer (CRC Press) for her assistance in the preparation and organization of the manuscript.

Contributors

Karen Arnott
Department of Human Biology
  and Nutritional Science
University of Guelph
Guelph, Ontario Canada

Joseph M. Betz, Ph.D.
Vice President for Scientific and
  Technical Affairs
American Herbal Products Association
  (APHA)
Silver Spring, Maryland U.S.A.

Mike Bryan
Chemical Specialist
Food Research Center
Agriculture and Agri-Food Canada
Guelph, Ontario Canada

Teresa Cháfer, Ph.D.
Assistant Professor
Department of Food Technology
Polytechnic University of Valencia
Valencia, Spain

Sam K. C. Chang, Ph.D.
Professor
Department of Cereal Food
  and Sciences
North Dakota State University
Fargo, North Dakota U.S.A.

Amparo Chiralt, Ph.D.
Professor
Department of Food Technology
Polytechnic University of Valencia
Valencia, Spain

Julie Conquer, Ph.D.
Director
Human Nutraceutical Research Unit
University of Guelph
Guelph, Ontario Canada

Jean-Paul Davis
Department of Human Biology
  and Nutritional Science
University of Guelph
Guelph, Ontario Canada

Pedro Fito, Ph.D.
Professor
Department of Food Technology
Polytechnic University of Valencia
Valencia, Spain

Tam Garland, Ph.D.
Professor
Department of Veterinary Physiology
  and Pharmacology
College of Veterinary Medicine
Texas A&M University
College Station, Texas U.S.A.

Dennis D. Gertenbach, Ph.D.
Senior Project Manager
Hazen Research, Inc.
Golden, Colorado U.S.A.

Clifford Hall III, Ph.D.
Assistant Professor
Department of Cereal and Food
  Sciences
North Dakota State University
Fargo, North Dakota U.S.A.

Bruce J. Holub, Ph.D.
Professor
Department of Human Biology
  and Nutritional Science
University of Guelph
Guelph, Ontario Canada

Afaf Kamal-Eldin, Ph.D.
Associate Professor
Department of Food Science
SLU
Swedish University of Agricultural
  Sciences
Uppsala, Sweden

Anna-Maija Lampi, Ph.D.
Research Scientist
Department of Applied Chemistry
  and Microbiology
University of Helsinki
Helsinki, Finland

Marc Le Maguer, Ph.D.
Professor
Department of Food Science
University of Guelph
Guelph, Ontario Canada

Javier Martínez-Monzó, Ph.D.
Assistant Professor
Department of Food Technology
Polytechnic University of Valencia
Valencia, Spain

Arun Nagpurkar
Department of Human Biology
  and Nutritional Science
University of Guelph
Guelph, Ontario Canada

Jordi Pagán, Ph.D.
Professor
Department of Food Technology
University of Lleida
Lleida, Spain

Samuel W. Page, Ph.D.
Scientific Director
Institute for Food Safety
  and Applied Nutrition
U.S. Food and Drug Administration
Washington, D.C. U.S.A.

Jason Peschell
Department of Human Biology
  and Nutritional Science
University of Guelph
Guelph, Ontario Canada

Vieno Piironen, Ph.D.
Professor
Department of Applied Chemistry
  and Microbiology
University of Helsinki
Helsinki, Finland

Jurgen G. Schwarz, Ph.D.
Associate Professor
Department of Cereal and Food
  Sciences
North Dakota State University
Fargo, North Dakota U.S.A.

Karin Schwarz, Ph.D.
Professor
Department for Human Nutrition
  and Food Science
Christian-Albrecht University in Kiel
Kiel, Germany

John Shi, Ph.D.
Research Scientist
Food Research Center
Agriculture and Agri-Food Canada
Guelph, Ontario Canada

Grete Skrede, Ph.D.
Senior Research Scientist
MATFORSK
Norwegian Food Research Institute
Osloveien, Norway

Qi Wang, Ph.D.
Research Scientist
Food Research Center
Agriculture and Agri-Food Canada
Guelph, Ontario Canada

Ronald E. Wrolstad, Ph.D.
Professor
Department of Food Science
  and Technology
Oregon State University
Corvallis, Oregon U.S.A.

Contents

1 Tocopherols and Tocotrienols from Oil and Cereal Grains

Anna-Maija Lampi, Afaf Kamal-Eldin and Vieno Piironen

CONTENTS

1-5667-6902-7/02/$0.00+$1.50
© 2002 by CRC Press LLC

1

1.1   INTRODUCTION

Because tocopherols and tocotrienols exhibit the biological activity of α-tocopherol, they belong to the group of compounds called vitamin E (VERIS, 1998). It was first discovered as a factor present in, for example, cereal grains to prevent reproductive failure in the rat, nutritional muscular dystrophy in the quinea pig and rabbit, and later other vitamin E deficiency symptoms. There is no specific reaction for which vitamin E is a cofactor. Instead, its role is to prevent a range of oxidation reactions of polyunsaturated lipids *in vivo* and to function as a biological antioxidant (Machlin, 1991; Combs, 1998; Traber, 1999). In the human body, vitamin E is the most important lipid-soluble antioxidant that provides an effective protective network against oxidative stress together with other antioxidants, such as vitamin C. For healthy people, it is relatively easy to get enough tocopherols and tocotrienols from the diet to prevent vitamin E deficiency, but higher daily intakes may provide other beneficial effects and may be needed when the diet contains large amounts of polyunsaturated fats (Horwitt, 1991).

Both tocopherols and tocotrienols are important antioxidants in foods, feeds and their raw materials, where they scavenge lipid radicals similarly as they do *in vivo* (Schuler, 1990; Combs, 1998). Tocopherols and tocotrienols improve storage and processing stability of many fat-containing materials and also are added as antioxidants to some foods and feeds during manufacturing. Protection needed in foods and feeds is, however, different from that of a living tissue, because these materials are subjected to a large variety of chemical and physical environments. Moreover, antioxidant activities of different tocopherols and tocotrienols vary depending on the conditions. Despite the fact that all E-vitamers are believed to be absorbed to the same extent in the human body, the bioavailability and bioactivity of α-tocopherol is much higher than that of the others.

This chapter deals with tocopherols and tocotrienols. It begins with a review of their chemical and physical properties and extends to their nutritional and health effects. The main focus is on tocopherols and tocotrienols in oil and cereal grains

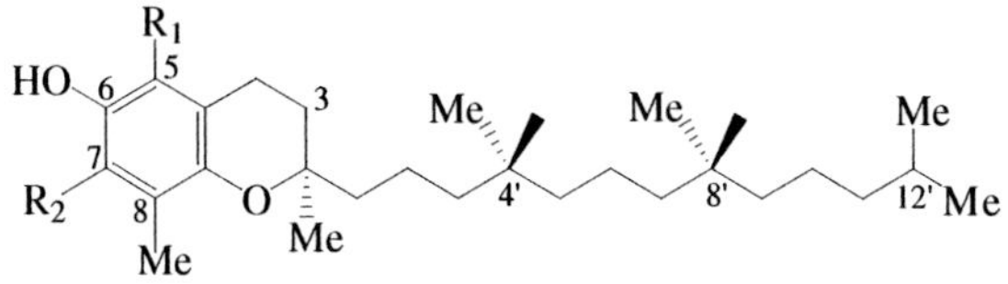

Tocopherols

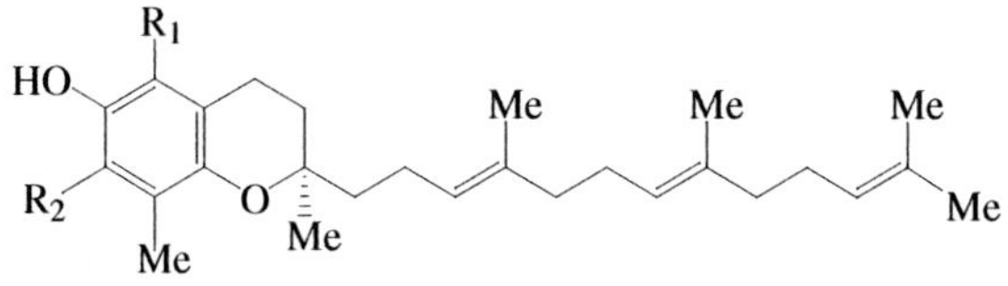

Tocotrienols

Substitution in the chromanol ring of tocopherols and tocotrienols

	R_1	R_2
α	Me	Me
β	Me	H
γ	H	Me
δ	H	H

FIGURE 1.1 Structures of tocopherols and tocotrienols.

that are the richest sources of vitamin E in our diet and provide valuable raw materials for the production of tocopherol and tocotrienol concentrates. Regulations of adding these compounds as vitamin E and technological aspects of their utilization as antioxidants are discussed together with their stability during processing and storage of oil and cereal grains and foods. Finally, to be able to study tocopherol and tocotrienol contents in oil and cereal grain-based materials, different analytical methods are introduced and evaluated.

1.2 CHEMICAL AND PHYSICAL PROPERTIES OF TOCOPHEROLS AND TOCOTRIENOLS

The term "vitamin E" is used as a generic description for all tocol and tocotrienol derivatives exhibiting the biological activity of α-tocopherol (IUPAC-IUB, 1982). There are, at least, eight E vitamers (four tocopherols and four tocotrienols) consisting of a chromanol ring and a hydrophobic side chain linked to the ring at carbon atom 2. The four members of each subfamily, i.e., α-, β-, γ- and δ-, differ from each other by the number and position of methyl groups on the phenolic part of the chromanol ring. Tocopherols have phytyl side chains with three chiral centers of asymmetry at C2, C4′ and C8′, while tocotrienols have isoprenyl side chains with only one chiral center at C2 but three double bonds at C-3′, C-7′ and C-11′ (Figure 1.1). In natural forms, all tocopherols have the 2R, 4′R, 8′R (d- or RRR) configuration and all tocotrienols the 2R, 3'-*trans*, 7'-*trans* configuration (Diplock, 1985). Synthesis may lead to stereoisomerism of tocopherols or tocotrienols. Synthetic α-tocopherol is an equal mixture of all eight stereoisomeric forms and is generally referred to as all-*rac*-α-tocopherol. There is no report describing the existence of geometrical isomers of tocotrienols. Knowledge about the isomeric configuration of tocopherols/tocotrienols is important for the evaluation of their biological biopotency (*vide infra*).

Tocopherols and tocotrienols form pale yellow and viscous liquids at room temperature that are soluble in lipids and lipophilic solvents but are insoluble in water. In agreement with the substitution pattern, the hydrophobic properties of tocopherols decrease in the order of α- > β- > γ- > δ-tocopherol (Sliwiok and Kocjan, 1992). The melting point of RRR-α-tocopherol is *ca* 3°C, while that of RRR-γ-tocopherol is at −3° to −2°C (Schudel et al., 1972). The ultraviolet absorption spectra of tocopherols and tocotrienols in ethanol show maxima in the range of 290–298 nm (Schudel et al., 1972; Diplock, 1985). The pure tocopherols isolated from natural sources exhibit relatively small optical rotations; tocopherols are dextrorotatory in ethanol, while they are levorotatory in benzene. The free phenolic hydroxyl group of tocopherol molecules mainly determines their chemical properties (Schudel et al., 1972).

Tocopherols and tocotrienols are easily oxidized by oxidizing agents, especially in the presence of heat, light and/or alkali. They are readily oxidized and darken when exposed to light (Schudel et al., 1972). Tocopherols and tocotrienols are, however, rather stable in nonoxidizable materials and oxidize very slowly by atmospheric oxygen in the dark (Schudel et al., 1972; Lehman and Slover, 1976). In the absence of oxygen, they are also resistant to alkali and heat treatment up to 200°C. In the presence of oxygen, the stability of α-tocopherol decreases by half with each increase of 10°C above 40°C in lipid material (Lips, 1957). Tocopherol esters are more stable to oxidation than free tocopherols (Schudel et al., 1972). When tocopherols are added to food and feed, they are often esterified to acetate or succinate derivatives to prevent oxidation by air during processing and storage. The esters are readily hydrolyzed in the intestine by enzymes, thus providing similar bioactivity as tocopherols.

The predominant reaction responsible for tocopherol antioxidant action in foods and biological systems is hydrogen atom donation, where a tocopheroxyl radical is formed. Tocopherols and tocotrienols scavenge lipid peroxyl radicals and thus retard propagation of the lipid oxidation chain reaction (Porter et al., 1995; Kamal-Eldin and Appelqvist, 1996). First, a tocopherol or tocotrienol (TocOH) molecule reacts with a peroxyl radical (ROO•) to form a tocopheroxyl or tocotrienoxyl radical (TocO•) and a lipid hydroperoxide (ROOH), which is a relatively stable product that decomposes only when present at high levels or by metal or other catalysts. The TocO• then scavenges another ROO• through radical-radical coupling to form a variety of stable secondary oxidation products.

$$ROO• + TocOH \rightarrow ROOH + TocO•$$

$$ROO• + TocO• \rightarrow \text{stable products}$$

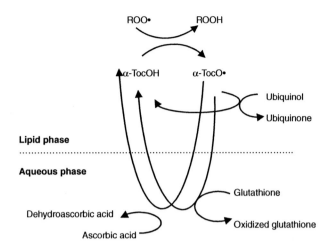

FIGURE 1.2 Interactions between α-tocopherol and other antioxidant defense systems in cells.

The slowness of reaction between a α-TocO• and oxygen is one main reason why α-tocopherol appears to have been selected as nature's major lipid-soluble chain-breaking antioxidant (Burton et al., 1985). Tocopherols may also retard lipid oxidation by stabilizing lipid hydroperoxides. The mechanism of this secondary antioxidant activity has been recently explained (Gottstein and Grosch, 1990; Hopia et al., 1996; Mäkinen et al., 2000). Besides scavenging peroxyl radicals, tocopherols and tocotrienols are also excellent quenchers for singlet oxygen (Kamal-Eldin and Appelqvist, 1996) and nitrogen oxide radicals (Christen et al., 1997). In biological systems, α-tocopherol works as part of an antioxidant network including ascorbic acid (vitamin C) and other water-soluble antioxidants, as shown in Figure 1.2.

1.3 VITAMIN E ACTIVITY AND NUTRITIONAL AND HEALTH EFFECTS OF TOCOPHEROLS AND TOCOTRIENOLS

1.3.1 VITAMIN E BIOAVAILABILITY AND BIOACTIVITY

All tocopherols and tocotrienols have vitamin E activity (Table 1.1). The biopotency of natural RRR-α-tocopherol is the highest at 1.49 IU mg^{-1}, while those of β-, γ- and δ-tocopherols are 50%, 10% and 3% of that of α-tocopherol (Kayden and Traber, 1993; Stone and Papas, 1997; Eitenmiller and Landen, 1999; Traber, 1999). The biological activity of each tocopherol stereoisomer is mainly determined by the chirality of the number 2 carbon, where the phytyl tail is attached to the chromanol ring. Stereoisomers having a 2R configuration have higher activities than those with a 2S configuration (Weiser et al., 1996; Stone and Papas, 1997). Thus, the synthetic all-*rac*-α-tocopherol has a relative activity of 74% compared with RRR-α-tocopherol. Tocotrienols have lower vitamin E activity than the respective tocopherols. Although all four geometrical isomers of α-tocotrienols are bioavailable, the preferred isomer

TABLE 1.1
Vitamin Activity of Different E-Vitamers

Compound	Activity (IU/mg)	Relative Activity (%)
Natural Vitamers		
RRR-α-tocopherol	1.49	100
RRR-β-tocopherol	0.75	50
RRR-γ-tocopherol	0.15	10
RRR-δ-tocopherol	0.05	3
RRR-α-tocotrienol	0.75	30
RRR-β-tocotrienol	0.08	5
RRR-γ-tocotrienol	not known	not known
RRR-δ-tocotrienol	not known	not known
Synthetic Vitamers		
RRR-α-tocopherol	1.49	100
SRR-α-tocopherol	0.46	31
RRS-α-tocopherol	1.34	90
RSR-α-tocopherol	0.85	57
SRS-α-tocopherol	0.55	37
RSS-α-tocopherol	1.09	73
SSR-α-tocopherol	0.31	21
SSS-α-tocopherol	0.89	60
all-rac-α-tocopherol	1.10	74
RRR-α-tocopheryl acetate	1.36	91
all-rac-α-tocopheryl acetate	1.00	67
RRR-α-tocopheryl succinate	1.21	81
all-rac-α-tocopheryl succinate	0.89	60

Data modified from VERIS *Vitamin E and Carotenoid Abstracts*, LaGrange, IL (1998).

is the natural 3′-*trans*, 7′-*trans* isomer (Drotleff and Ternes, 1999). Esterification of tocopherols does not change their vitamin E activity on a molar basis. Due to the higher bioavailability, α-tocopherol is the predominant form of vitamin E in human and animal plasma and tissues.

Tocopherols and tocotrienols are believed to be absorbed to the same extent, but a tocopherol transfer protein with a specific affinity for α-tocopherol is considered responsible for the discrimination between the different E-vitamers and the selective incorporation of α-tocopherol into nascent, very-low-density lipoprotein (Hosomi et al., 1997). Other forms of vitamin E, which are discriminated during this process, are most likely excreted via the bile after shortening and carboxylation of their phytyl tails (Traber et al., 1998; Swanson et al., 1999). Recently, the biological importance of γ-tocopherol has gained more attention. Because it is preferentially secreted from the liver into the small intestine via the bile duct, it may be superior to α-tocopherol in increasing the antioxidant status of digesta (Stone and Papas, 1997). When tissues are saturated with α-tocopherol, this E-vitamer is excreted via the bile to the same extent as the other vitamers (Kayden and Traber, 1993).

The bioactivity of E-vitamers is defined as their ability to prevent or reverse specific vitamin E deficiency symptoms, such as fetal resorption in rats and muscular dystrophy and encephalomalacia in chicken (Machlin, 1991). The bioactivities of different E-vitamers parallel their bioavailability, indicating that bioactivity is more a function of the amounts available in the body than the chemical activities of these vitamers. Because diets are reasonably abundant in E-vitamers, vitamin E deficiency is rare in humans and is almost limited to malnourished people, people with fat malabsorption or those with a defect in the hepatic tocopherol binding protein (Dutta-Roy et al., 1994).

In humans, plasma or serum α-tocopherol concentrations of <11.6, 11.6–16.2 and >16.2 μmol/L are considered to indicate deficient, low and normal levels, respectively (Sauberlich et al., 1974; Morrissey and Sheehy, 1999).

1.3.2 VITAMIN E, DIABETES AND AGING

Oxidative stress was found to be implicated in aging and in the pathogenesis and complications of diabetes and to cause low plasma levels of vitamins A, E and carotenoids (Astley et al., 1999; Polidori et al., 2000). Despite low levels of α-tocopherol in people with type 1 diabetes, plasma low-density lipoprotein (LDL) and lymphocyte DNA were more resistant to induced oxidation than in control subjects (Astley et al., 1999). Diabetics and elderly people are, however, known to develop cataracts and cardiovascular problems as a result of enhanced free radical production and/or binding of glucose to proteins and LDL, and vitamin E was suggested to inhibit these pathogenic alterations (Ceriello et al., 1991; Gazis et al., 1997). It was also suggested that intake of vitamin E improves insulin action in diabetic and nondiabetic subjects (Paolisso et al., 1993).

1.3.3 VITAMIN E AND CORONARY HEART DISEASE

Vitamin E is suggested to have protective effects against the progression of coronary heart diseases. The antioxidant hypothesis suggests that the inhibition of LDL oxidation is the main mechanism by which vitamin E exerts its protective effects against the progression of coronary heart disease (CHD) (Stampfer and Rimm, 1995). Other mechanisms, distinctive of vitamin E's antioxidant properties, were also suggested to explain the protection against CHD: inhibition of platelet adherence to proteins such as fibrinogen, fibronectin and collagen (Steiner, 1993); control of smooth muscle cell proliferation through binding to a "receptor protein" and acting as a sensor and information transducer of the cell's oxidation state (Azzi et al., 1995); and inhibition of the secretion of interleukin-1 beta and the adhesion monocyte to endothelium (Devaraj and Jialal, 1996). E-vitamers, especially tocotrienols, may also protect against CHD by lowering LDL cholesterol via inhibition of the HMG-CoA reductase enzyme involved in cholesterol biosynthesis (Hood, 1998; Theriault et al., 1999).

1.3.4 VITAMIN E AND CANCER

Although the evidence for vitamin E protection against cancer is weaker than for cardiovascular disease, vitamin E forms may also act as chemoprotective agents that block or suppress mutation, promotion and proliferation in different types of cancer, including hormonal and alimentary-canal-related cancers (Patterson et al., 1997; Stone and Papas, 1997; Kline et al., 1998; Shklar and Oh, 2000). The E-vitamers can protect against carcinogenesis and tumor growth through antioxidant properties and/or immunomodulatory functions (Das, 1994). The anticarcinogenic effect of vitamin E was related to its ability to scavenge mutagenic, superoxide and/or nitrogen dioxide radicals (Stone and Papas, 1997), to inhibit peroxidation of DNA and proteins (Summerfield and Tappel, 1984) and/or to induce apoptosis by inhibition of DNA synthesis in cancer cells (Carroll et al., 1995; Yu et al., 1999). The report published by the American Institute for Cancer Research (1997) nevertheless concluded that no evidence exists for the association between vitamin E intake and reduced risk of cancer. It should be stressed here that research strongly indicates that protection against degenerative diseases cannot be attributed to a single dietary factor, such as vitamin E, but that a wide range of environmental, dietary and lifestyle factors determine our resistance or vulnerability to such diseases (Chan, 1998).

1.3.5 INTAKES, REQUIREMENTS AND RECOMMENDATIONS

Recommendations for vitamin E intakes are based on estimates for the minimum levels needed to avoid deficiency symptoms. Dietary requirements of vitamin E are related to intake of polyunsaturated fatty acids (Horwitt, 1991; Weber et al., 1997). Daily intakes by adults in U.S. populations were estimated to be between 7–11 mg of RRR-α-tocopherol equivalents (NRC, 1989). The average value in Germany was 12.4 mg (18.6 IU) for adults, and in the U.K., the average value was 11.7 mg (17.6 IU) for men and 8.6 mg (12.9 IU) for women (VERIS, 1998). According to the tenth edition of the Recommended Dietary Allowances (RDA) (NRC, 1989), the daily RDA for vitamin E is 10 mg (15 IU) and 8 mg (12 IU) α-tocopherol equivalents for men and women, respectively. Pregnancy and lactation increase the value to 10–12 mg. The Canadian Recommended Nutrient Intakes (RNI, 1983) are slightly lower, being 6–10 and 5–7 mg α-tocopherol equivalents/day for men and women, respectively. In Europe, the recommended daily allowance of vitamin E is 10 mg (EEC, 1989, 1990). Comparison of these intake figures and recommendations indicates that the average vitamin E intakes in Western societies meet the RDA estimated for prevention of vitamin E deficiency.

The optimal dietary requirements of vitamin E for humans are not yet known, especially with the emergence of new paradigms regarding adequate levels of dietary micronutrients (Chalem, 1999). Recommendations in the United States and Canada have been reevaluated, and a new concept of Dietary Reference Intake (DRI, 2000) was issued for vitamin E and other antioxidants. The DRI recommendation should prevent specific deficiency disorders, support health in general ways and minimize the risk of toxicity, which carries more tasks than the previous recommendations (DRI, 2000). Accordingly, the recommendations for intakes were set to higher levels than previously. Estimated Average Requirements (EAR) for adults, both men and women, were set to 12 mg α-tocopherol/day, RDA to 15 mg/day and Tolerable Upper Intake Level (UL) to 1000 mg/day. Moreover, the EAR and RDA are based only on the 2R-stereoisomeric forms of α-tocopherol, because the other vitamers were considered not to contribute to meeting the vitamin E requirement. This approach represents a change from earlier recommendations and the concept of vitamin activity of different vitamers (Table 1.1). Vitamin UL was set on the adverse effect of osmotic diarrhea. It is based on intake from supplements only.

1.3.6 SUPPLEMENTATION

Although healthy people who have access to well-balanced diets may not need vitamin E supplements, certain groups of people fearing risk of oxidative stress (e.g., smokers, diabetics, athletics) may consider taking supplements. The doses of vitamin E required for protection against CHD are questionable, because in intervention studies, only high doses of vitamin E (>400 IU/day) were found to reduce the risk of CHD (Hense et al., 1993; Steiner, 1993; Stampfer and Rimm, 1995). Daily intakes corresponding to 15–25 mg RRR-α-tocopherol were found in the Mediterranean diet corresponding to low risk of CHD (Dutta-Roy et al., 1994). Traber and Sies (1996) estimated 15–30 mg RRR-α-tocopherol equivalents as required to maintain the plasma level of >30 μmol/L required for protection against CHD. Other workers claimed that daily intakes of 87–100 mg RRR-α-tocopherol equivalents should be recommended for protection against CHD (Horwitt, 1991; Weber et al., 1997).

1.4 OCCURRENCE OF TOCOPHEROLS AND TOCOTRIENOLS IN OIL AND CEREAL GRAINS

1.4.1 BIOSYNTHESIS AND COMPARTMENTALIZATION IN PLANTS AND GRAINS

Tocopherols and/or tocotrienols have been found in all photosynthetic organisms, and they are always located in membranes of the cell. In higher plants, tocopherols are mainly synthesized in chloroplasts and plastids, and tocotrienols outside chloroplasts. The chromanol ring is built up from a shikimic acid pathway intermediate homogentisate and the phytyl side chain from an isoprenoid pathway intermediate phytyl pyrophosphate (Figure 1.3). Phytyl transferase combines these two units to form 2-methyl-6-phytyl benzoquinol that is methylated and cyclized by methyl transferases and cyclases. S-adenosylmethionine serves as the methyl donor. To produce δ-tocopherol, 2-methyl-6-phytyl benzoquinol undergoes cyclization, and to produce β- and γ-tocopherols, it is also methylated. The final product of the biosynthetic pathway, α-tocopherol, is fully methylated. The biosynthesis of tocotrienols follows the same reactions, except geranylgeranyl pyrophosphate is used as the source for the unsaturated side chain. Thus, the availability of homogentizate, S-adenosylmethionine and phytol or geranylgeranol is essential for the synthesis of tocopherols and tocotrienols (e.g., Hess, 1993; Bramley et al., 2000). Many of the biosynthetic reactions are known to regulate the total amount of tocopherols and tocotrienols. The activity of different methyltransferases is important in determining the relative proportions of different vitamers in seeds/grains, which is an interesting finding for plant breeding purposes (see Bramley et al., 2000).

In general, total amounts of tocopherols and tocotrienols in plants are greatest in mature leaves and other light-exposed tissues and smallest in roots and other

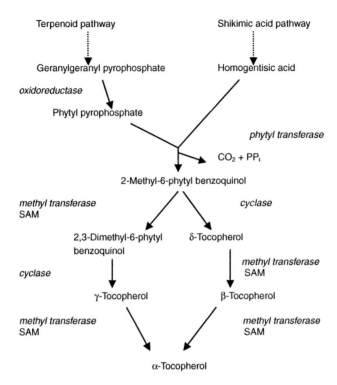

FIGURE 1.3 General biosynthetic pathway of tocopherols in chloroplasts and proplasts of plant cells. Biosynthesis of tocotrienols is similar except for reduction of geranylgeranyl pyrophosphate to phytyl pyrophosphate, which is omitted. SAM = S-adenosylmethionine. [Data modified from Hess (1993).]

tissues grown under diminished light. In photosynthetic tissues, α-vitamers dominate, while in nonphotosynthetic tissues, the ratio of γ- over α-vitamers is higher (Hess, 1993). The vitamin E content of plants varies depending on, e.g., the intensity of sunlight and soil state and other growing conditions (Crawley, 1993).

1.4.2 OILS AND CEREALS AS IMPORTANT SOURCES OF TOCOPHEROLS AND TOCOTRIENOLS

The major sources of vitamin E in the Western diet are fats and oils, cereal and vegetable products. In the Unites States, these three groups of food items contributed to 20.2%, 14.6% and 15.1% (Murphy et al., 1990) and in Finland, to 41%, 18% and 3% (Heinonen and Piironen, 1991) of the total vitamin E intake from the diet, respectively.

1.4.3 OIL GRAINS

The descending order of total tocopherols and tocotrienols in commercial oils is as follows: corn, soybean, palm (800–1100 μg/g) > cottonseed, sunflower, rapeseed

TABLE 1.2
Tocopherol (T) and Tocotrienol (T3) Contents of Refined Vegetable Oils (µg/g FW)

Oil	α-T	α-T3	β-T	β-T3	γ-T	γ-T3	δ-T	δ-T3	Reference
Soybean	95	—	13	—	689	—	239	—	1
	120	—	10	—	610	—	260	—	2
	179	—	28	4	604	—	371	—	3
Corn	257	15	9	—	752	20	32	—	1
	120	—	10	—	400	—	20	—	2
	272	54	—	11	566	62	25	—	3
Cottonseed	403	—	—	9	383	—	4	—	3
Sunflower	622	—	23	—	27	—	—	—	1
	610	—	10	—	30	—	10	—	2
	564	—	24	—	4	—	—	—	3
Safflower	449	—	12	—	26	—	6	—	1
Rapeseed (LEAR)	189	—	—	—	486	—	12	—	1
	200	—	—	—	430	—	10	—	2
Linseed	5	—	—	—	573	—	7	—	1
Wheat germ	1507	36	312	—	527	18	—	—	1
Olive oil	119	—	—	—	13	—	—	—	1
	90	—	—	4	5	—	—	—	3

References: 1. Syväoja et al., *J. Am. Oil Chem. Soc.* 63: 328–329 (1986); 2. Warner and Mounts, *J. Am. Oil Chem. Soc.* 67: 827–831 (1990); and 3. Van Niekerk and Burger, *J. Am. Oil Chem. Soc.* 62: 531–538 (1985).

(550–800 µg/g) > peanut, olive (150–350 µg/g) > coconut, palm kernel (≤50 µg/g) (Tan, 1989). The contents in some breeding lines significantly exceed those mentioned above. Most oils contain only tocopherols, while palm, palm kernel and rice bran oil contain significant amounts of tocotrienols (Eitenmiller, 1997). The richest source of tocopherols, wheat germ oil, is mainly consumed as a tocopherol supplement and not as ordinary edible oil (Table 1.2).

Soybean oil is the most important source of vitamin E in the Western world. It contains, on average, 1000 µg/g of tocopherols (Eitenmiller, 1997; Kamal-Eldin and Andersson, 1997). The total tocopherol contents of 14 breeding lines with commodity-type fatty acid composition were 1406–2195 µg/g, while those of the major isomers γ-, δ- and α-tocopherols were 850–1171, 254–477 and 44–158 µg/g, respectively (Dolde et al., 1999).

Sunflower oil is a rich source of vitamin E. It contains, on average, 400–700 µg/g total tocopherols (Van Niekerk and Burger, 1985; Eitenmiller, 1997; Kamal-Eldin and Andersson, 1997). Total tocopherol concentration of 66 experimental breeding lines with commodity-type fatty acid compositions was 982 ± 27 µg/g, of which >99% was α-tocopherol. α-Tocopherol was present in some lines as a minor component. The tocopherol compositions of breeding lines with adjusted fatty acid compositions were similar (Dolde et al., 1999).

Palm oil is an exception among oil crops, because more than 85% of its vitamin E content consists of γ-tocotrienol, α-tocotrienol and α-tocopherol (Tan, 1989).

Other tocotrienols (i.e., β-, δ-) are also present (Syväoja et al., 1986; Ong and Choo, 1997). Total vitamin E content in crude palm oil is between 600–1200 μg/g, while contents in palm oil fractions may be significantly smaller. Palm kernel oil has only <100 μg/g vitamers (Eitenmiller, 1997; Ong and Choo, 1997). Tocotrienols and tocopherols are concentrated up to eightfold in the palm fatty acid distillate compared with the crude palm oil, which enables extraction of vitamin E compounds from this by-product.

Oils from genetically modified rapeseed/canola with a wide range of fatty acid compositions contained 478–677 μg/g of total tocopherols, of which >50% was γ- and about 30% was α-tocopherol. There was no relationship between the fatty acid and tocopherol compositions (Abidi et al., 1999). Genetically modified canola oils had similar tocopherol compositions with a range of 504–687 μg/g of total tocopherols (Dolde et al., 1999).

A study on flaxseed showed that this seed contains an average of 93 μg/g tocopherols, of which >96% was γ-vitamer (Oomah et al., 1997). The level of tocopherols was cultivar specific and was regulated by environmental conditions. The hulls contained about 26% of total tocopherols and relatively more α- and δ-tocopherols than the embryo (Oomah and Mazza, 1998). Flaxseeds also contain an analogous compound, plastochromanol-8, consisting of a γ-chromanol and eight isoprene units in the side chain (Olejnik et al., 1997). The level of plastochromanol-8 in seeds was 17–74 μg/g and that of total tocopherols was 74–184 μg/g, of which 98–100% was γ-tocopherol (Velasco and Goffman, 2000). Tocopherol contents in linseed oil range from 400 to 600 μg/g (Syväoja et al., 1986; Kamal-Eldin and Andersson, 1997; Oomah and Mazza, 1998; Schöne et al., 1998).

For olive oil, total tocopherol level is *ca* 100–150 μg/g and α-tocopherol is the dominant (>90%) tocopherol present (Van Niekerk and Burger, 1985; Kamal-Eldin and Andersson, 1997). α-Tocopherol values of virgin olive oils from Greece ranged between 98 and 370 μg/g, while the sum of other vitamers was <50 μg/g. Small-sized Greek olives had slightly higher contents of α-tocopherol than medium-sized olives, at 239 and 198 μg/g, respectively (Psomiadou et al., 2000).

1.4.4 CEREAL GRAINS

Cereal grains are mainly recognized for their proteins and carbohydrates and for being valuable sources of energy and other nutrients, but they also contain substantial amounts of various lipids. Lipids are located in the cereal grain in membranes and spherosomes. They are unevenly distributed in grain fractions. Nonstarch lipid contents of cereal grains are for oat groats 5–9%, maize kernels 3.9–5.8%, barley 3.3–4.6%, rye 2.0–3.5%, millet 4.0–5.5% and rice 0.8–3.1% (Morrison, 1978; Youngs, 1986; Chung and Ohm, 2000). Cereals are considered only moderate sources of vitamin E, providing 6–23 mg of α-tocopherol equivalents/kg (Bramley et al., 2000). Because cereals contain great amounts of tocotrienols (Table 1.3) with very low vitamin E activity, they are more valuable as sources of tocotrienols and tocopherols than vitamin E.

Wheat grains contain total tocopherol and tocotrienol of >40 μg/g (e.g., Barnes, 1983; White and Xing, 1997), where β-tocotrienol (>50%) and α-tocopherol

TABLE 1.3

Tocopherol (T) and Tocotrienol (T3) Contents of Cereal Grains (μg/g FW)

Cereal	α-T	α-T3	β-T	β-T3	γ-T	γ-T3	δ-T	δ-T3	Total T+T3	Reference
Wheat	10	4	5	21	—	—	<1	—	40.4	1
	12	4	6	26	—	—	—	—	47.8	2
Rye	10	14	3	11	—	—	—	—	38.0	1
	10	12	2	7	<0.2	—	—	—	32.4	2
Barley	3	16	<1	6	1	6	<1	—	32.0	1
	4	20	<0.2	4	1	8	—	0.6	36.9	2
	10	33	0.7	7	0.7	5	0.7	0.9	58.5	3
Oats	9	25	0.6	3	—	<0.2	—	—	38.0	2
	8	15	0.9	0.9	0.9	—	—	0.2	25.9	3
Rice, brown	6	4	1	<0.1	1	7	<1	—	19.0	1
	8	4	0.7	<0.2	4	10	0.5	0.7	27.0	2

References: 1. Piironen et al., *Cereal Chem.* 63: 78–81 (1986); 2. Balz et al., *Fat Sci. Technol.* 94: 209–213 (1992); and 3. Peterson and Qureshi, *Cereal Chem.* 70: 157–162 (1993).

(*ca* 20%) are the major E-vitamers, and α-tocotrienol and β-tocopherol are the minor E-vitamers (Table 1.3). Wheat germ is an especially rich source of α- and β-tocopherols and oil produced from it provides more vitamin E activity than most other oils (Piironen et al., 1986; Syväoja et al., 1986; Balz et al., 1992). In wheat, as in other cereals, tocopherols are concentrated in the germ, and tocotrienols are concentrated in the bran and endosperm (Table 1.4).

Barley grains contain all eight tocopherols and tocotrienols (Table 1.3), but there is a large variation in the total contents of tocopherols and tocotrienols. The major vitamers are α-tocotrienol (contributing to >50%), α-tocopherol, β-tocotrienol and γ-tocotrienol, and they are generally correlated positively with each other (Peterson and Qureshi, 1993). The total vitamer content in whole-grain barley of 30 genotypes

TABLE 1.4

Distribution of Tocopherols (T) and Tocotrienols (T3) in Milling Fractions of Swedish Winter Wheat (μg/g dry matter)

Milling Fraction	α-T	α-T3	β-T	β-T3
Wholemeal	12	5	5	30
White flour	6	2	2	21
Bran	4	20	2	94
Germ	130	5	50	25

Data modified from Jägerstad and Håkansson (1988) and Wennermark and Jägerstad (1992).

grown in three locations in the United States ranged from 42 to 80 µg/g with an average of 58.5 µg/g and suggested that the variation was mainly between genotypes and to a lesser extent between locations (Peterson and Qureshi, 1993). The tocopherol and tocotrienol contents of these 30 genotypes as well as that of two hull-less barley cultivars, 71.0 ± 9.5 µg/g (Wang et al., 1993), were greater than the average level of 30 µg/g reported earlier (Barnes, 1983; Piironen et al., 1986; Balz et al., 1992). Within a barley grain, the germ contains high levels of α-tocopherol, while the hull and endosperm contain all vitamers and are especially rich in tocotrienols (Barnes, 1983; White and Xing, 1997). Unlike most cereals, the barley contains significant amounts of tocotrienol, namely, the β-vitamer, in the germ (Peterson, 1994). α-Tocotrienol seems to accumulate in the starchy endosperm (Wang et al., 1993). When studying the distribution of tocopherols and tocotrienols in hull-less barley fractions, pearling flour had the greatest total concentration of 205 µg/g, which was almost three times as much as in the whole grain. Since the yield of pearling flour is ca 20% of the total grain weight, it is a potential ingredient for health-promoting food products (Wang et al., 1993; Peterson, 1994).

Oat grains and dehulled grains (called groats) contain on average 2–11% of lipids, most of which are located in endosperm and bran, as reviewed by Zhou et al. (1999). The major E-vitamers of oat groats are α-tocotrienol and α-tocopherol, contributing to 57% and 31% of the total, respectively (Peterson and Qureshi, 1993). β-, γ- and δ-Vitamers occur only as minor components (Table 1.3). Oat groats of 12 genotypes contained 25.9 µg/g of tocopherols and tocotrienols with a range of 19–30 µg/g (Peterson and Qureshi, 1993), which is at a similar level to that found in other studies (Barnes, 1983; Piironen et al., 1986; Peterson and Wood, 1997). There were significant differences in the total vitamin contents between the 12 genotypes and the three growing locations (Peterson and Qureshi, 1993). Within the oat grain, the germ was richest in total E-vitamers, because the contents in the hulls, germ and endosperm were 3, 130 and 40 µg/g, respectively (Peterson, 1995). In 25 high-oil oat selections with 6.9–18.1% of lipids, the total tocopherol and tocotrienol contents ranged from 25 to 67 µg/g with a significant positive correlation to lipids, especially for tocotrienols (Peterson and Wood, 1997). In these high-oil oats, more lipids including tocotrienols were distributed in the endosperm than in control oats, while the germs were similar. Analysis of hand-dissected fractions indicates that tocotrienols, mainly α-tocotrienol, were located in the endosperm, while tocopherols, mainly α-tocopherol and γ-tocopherol, were present in the germ (Peterson, 1995; Peterson and Wood, 1997).

In rye grain and its products, α-tocopherol and α- and β-tocotrienols are the three dominant vitamers (Table 1.3). The total amount of tocopherols and toco-trienols in whole grain lies in the range of 30–50 µg/g (e.g., Barnes, 1983; White and Xing, 1997).

Two other cereal grains; namely, corn and rice, are important in terms of being utilized as sources of tocopherols and tocotrienols. In corn, there is a large variation, 28–102 µg/g, in the total amount and profile of tocopherols and tocotrienols (White and Xing, 1997). In general, γ-tocopherol is the predominant vitamer contributing up to 80% of the total amount in most cultivars, while α-tocopherol levels are greater or equal to those of γ-tocopherol in some varieties and genotypes (Weber, 1984;

Kurilich and Juvik, 1999). The germ and tip-cap contained 63–91% of total toco-pherols and tocotrienols of hand-dissected corn kernels, while the endosperm con-tained 9–37% of them. As with other cereals, tocotrienols (25–35% of total) are concentrated in the endosperm and are almost lacking in the germ, which is rich in α- and γ-tocopherols (Weber, 1987).

In rice grain, γ-tocotrienol, γ-tocopherol and α-tocopherol are the major vitamers (White and Xing, 1997), although some studies report α-tocotrienol levels higher than those of γ-vitamers (Table 1.3). Comparison of tocopherol and tocotrienol results from different studies is difficult, because rice samples may have been processed to different extents, being rough (unshelled), brown (hull was removed) or milled (bran was removed and sometimes polished) (Moldenhauer et al., 1998). A coproduct from milling, rice bran, is an exceptionally rich source of tocotrienols, and rice bran oil is produced from it. There is evident variation in the tocopherol and tocotrienol compositions of rice bran oils, which can be explained both by differences in rice varieties and/or by processing. For example, total vitamer contents of five brands varied from 88 to 1610 μg/g, and the proportion of tocotrienols varied from 51 to 82% (Rogers et al., 1993).

Tocopherol and tocotrienol profiles of buckwheat and millet are different compared to other cereal grains. In both cereals, γ-tocopherol is the dominant vitamer, and tocotrienols occur in minor amounts. Buckwheat contains 54–62 μg/g and millet 22–32 μg/g of vitamers, of which ca 90% and 70% is γ-tocopherol (Piironen et al., 1986; Balz et al., 1992). Grain amaranth of several cultivars and growing environments showed a large variation in the content of total E-vitamers, 5–29 μg/g, and tocotrienols, mainly β- and γ-, representing 39–74% of this total (Lehmann et al., 1994).

1.4.5 Effects of Growing Conditions

Tocopherol and tocotrienol contents depend to a large extent on the geographical and climatic conditions, the maturity of seeds during harvesting and the variety of the plant (Kamal-Eldin and Andersson, 1997; White and Xing, 1997). The temperature is the most effective environmental factor controlling tocopherol and tocotrienol content. In general, the levels are higher in oilseeds and cereal grains grown at high temperatures than at low temperatures, indicating that the biosynthesis is promoted by increased temperature (Marquard, 1990; White and Xing, 1997). However, an experiment performed by Dolde et al. (1999) showed the opposite for soybeans, in which tocopherol contents were higher at low rather than high temperatures. Despite the variation due to growing temperature, the tocopherol composition is mainly determined by the crop (Dolde et al., 1999).

Tocopherol and tocotrienol contents and compositions may be modified by traditional and modern plant breeding techniques (Marquard, 1990). Genetic modifi-cation has been used to change the profile and content of canola tocopherols (Abidi et al., 1999). In general, plant breeding of grains is focused on the fatty acid composition to produce tailor-made oils with special characteristics such as good stability or high linolenic content. Only recently, the importance of tocopherols and other minor lipid components has been recognized. It is suggested that genetic modifications for altered tocopherol and tocotrienol contents and for altered fatty acid composition of oil grains can be done independently in many oil grains, while there may be some relationship in soybean (Dolde et al., 1999).

1.5 PRODUCTION OF TOCOPHEROL AND TOCOTRIENOL CONCENTRATES/PRODUCTS

Tocopherols and tocotrienols used for supplementation of foods and feeds or production of nutraceuticals, e.g., pills and powders, are obtained either by extraction from natural sources or by chemical synthesis (Schuler, 1990; O'Leary, 1993). Extraction of tocopherols and tocotrienols from natural sources is economically feasible only if tocopherol/tocotrienol-rich raw materials are available at large quantities and are of high quality. In production of both natural and synthetic compounds, there are several extraction and purification processes that have to be optimized and controlled. Moreover, the environmental impacts of these processes should be acknowledged. Technological innovations in these areas are constantly needed.

Natural tocopherols and tocotrienols have the 2R, 4′R, 8′R and 2R, 3′-*trans*, 7′-*trans* configurations, respectively, while synthetic products are racemic mixtures. Natural and synthetic products have similar potentials as antioxidants, but natural products have higher vitamin E activity than synthetic ones. Production capacities of natural and synthetic tocopherols and tocotrienols are enlarging (ANON, 1998). In 1991, the world production of vitamin E compounds was 6800 tons (O'Leary, 1993).

1.5.1 Extraction from Natural Sources

The major sources of natural tocopherols and tocotrienols are the deodorization distillates from edible oil processing. Soybean oil has the highest level of tocopherols in distillates (10–14%), followed by corn oil (7–10%), cottonseed oil (6–10%), sunflower oil (5–8%) and rapeseed oil (4–7%) (Walsh et al., 1998). Refining processes of oils rich in both tocopherols and tocotrienols, such as palm oil, are especially tailored for manufacturing these value-added by-products (Ong and Choo, 1997; Moldenhauer et al., 1998). Utilization of rice bran and corn fiber and other by-products that are rich in E-vitamers is also expanding. In general, in Europe, refining processes are focused on manufacturing oils with maximum tocopherol contents, while in North America, high levels of tocopherols in the deodorization distillates are favored and utilized in special products (De Greyt et al., 1999; Dijkstra, 2000).

Deodorizer distillate, a by-product of the vegetable oil refining process, consists mainly of fatty acids, sterols, tocopherols, sterol esters, acyl glycerol species and several breakdown products. Tocopherols and sterols are the two valuable groups of compounds for which the distillate is sold and used (Schuler, 1990; Ramamurthi and McCurdy, 1993; Barnicki et al., 1996), but depending on the oil material and the refining conditions used, the distillates may contain 27–75% of free fatty acids. Isolation of tocopherols and tocotrienols from the distillate requires three main steps: removal of free fatty acids and their esters, separation of tocopherols/tocotrienols from other compounds and purification of the product. Fatty acids are removed by vacuum distillation after being chemically or enzymatically methylated, by saponification, fractional solvent extraction or supercritical fluid extraction (Schuler, 1990; Ramamurthi and McCurdy, 1993). Small consumption of chemicals and solvents has made vacuum distillation of fatty acid methyl esters the most commonly used method. Chemical methylation is often used, although enzymatic methylation under gentle conditions would retain higher tocopherol yields than the chemical process. When studying lipase-catalyzed esterification of soybean and canola oil distillates, the loss of tocopherols during esterification was <5%, and the overall recovery after vacuum distillation was >90% (Ramamurthi and McCurdy, 1993). Finally, to produce pure tocopherols and tocotrienols, they have to be purified from plant sterols and squalene, which are also concentrated in the distillates. Sterols may be removed from the distillates by crystallization or may be separated by chromatography (Diack and Saska, 1994). Sterols may also be converted to steryl esters, which are bigger molecules than vitamers and thus can be separated by fractional distillation (Shimada et al., 2000). Tocopherol and tocotrienol distillates need to be stored at ambient temperatures, under nitrogen and in noncorrosive tanks with no direct light (Walsh et al., 1998).

Palm fatty acid distillate (PFAD) is a valuable by-product of palm oil processing, from which the production of tocopherols and tocotrienols is technically feasible (Ong and Choo, 1997). PFAD is wildly abundant, as palm oil is the second largest oil produced in the world. It contains 4000–8000 µg/g total vitamin E compounds, which is up to 10 times as much as crude palm oil contains (Tan, 1989; Ong and Choo, 1997). The tocopherol and tocotrienol profiles of PFAD and crude palm oil are similar, consisting of about 85% of tocotrienols, i.e., γ-tocotrienol > α-tocotrienol > δ-tocotrienol (Tan, 1989). Because of cholesterol-lowering properties (Lane et al., 1999), the high level of tocotrienols makes PFAD an important source, because most other natural sources are devoid of tocotrienols. Fatty acids and esters, sterols and squalene are removed from PFAD as described above. Finally, the vitamin-E-rich product is purified and deodorized to yield a tocopherol/tocotrienol-rich product of 95–99% purity.

Oil extracted from rice bran and corn fiber is especially rich in tocopherols, tocotrienols and sterols. Extraction of these valuable compounds is remarkably enhanced by heat pretreatment (Lane et al., 1999). For example, the level of extracted γ-tocopherol increased from 0.3% to 3.6% by subjecting corn fiber to temperatures of 100–175°C (Moreau et al., 1999). Tocopherol- and tocotrienol-rich oils may be used as such as important sources of vitamin E, or the vitamers can be purified from them following the same protocols as above.

1.5.2 CHEMICAL SYNTHESIS

Synthesis of α-tocopherol is based on acid-catalyzed condensation of 2,3,5-trimethyl hydroquinone with isophytol. The 2,3,5-trimethyl hydroquinone is prepared by hydrogenation of the quinone that is an oxidation product of trimethylphenol. The isopropyl component is a product of isoprenoid chain extension reaction (O'Leary, 1993). With synthetic isophytol, a racemic mixture of eight stereoisomers is formed. High yields (>90%) and selectivity are obtained with new technology, including strong solid catalysts (Schager and Bonrath, 1999). Other tocopherols are produced using the same method (see Figure 1.3), i.e., γ-tocopherol is built up from 2,3-dimethylquinone and isophytol. Because α-tocopherol has the highest vitamin E activity, protocols to produce it from non-α-tocopherol vitamers by hydroxymethylation or permethylation have been developed (Breuninger, 1999; Bruggemann et al., 1999).

After synthesis, tocopherols are purified by vacuum distillation. In order to improve stability of tocopherols, they are converted to acetate and other esters before final purification by vacuum distillation (Schuler, 1990; O'Leary, 1993).

1.6 UTILIZATION OF TOCOPHEROLS AND TOCOTRIENOLS AS FOOD ADDITIVES AND INGREDIENTS

In general, vitamins are added to a food for vitaminization, restoration, standardization and fortification. Vitaminization makes foods that usually do not contain vitamins carriers of vitamins. Restoration is used to compensate for vitamin losses during processing of food. Fortification is used to ensure vitamin adequacy for populations. Standardization describes fortification of a product to a standard within its class (O'Brien and Roberton, 1993; Combs, 1998). Tocopherols and tocotrienols may also be added to foods and feeds as an antioxidant.

1.6.1 REGULATIONS AND LEGISLATION REGARDING ADDITION AS VITAMIN

Mixed tocopherol concentrates containing α- and other tocopherols have a Generally Recognized as Safe (GRAS) status. Their use is limited to an antioxidant and as a nutrient supplement (Six, 1994). For all Western and Middle European countries, which have harmonized their relevant legislation within the European Union (EU), the directive (EC, 1995) allows addition of tocopherol mixture (E 306), α- (E 307), γ- (E 308), δ-tocopherols (E 309) and tocopherol-rich extracts as additives to most foodstuffs following the *quantum satis* principle. It is not permitted to add them in unprocessed foodstuffs (e.g., most dairy products and nonemulsified oils and fats), and there are special regulations about their contents in foods for infants and young children.

The Food and Drug Adminstration of the United States (FDA) approves 1 mg α-tocopherol (1.5 IU vitamin E) to be added to 100 kcal of food. Guidelines for the use of vitamin supplements emphasize that healthy individuals can and should obtain adequate amounts of all nutrients from a well-balanced diet. Only certain circumstances warrant the use of vitamin supplements, i.e., by individuals with very low caloric intakes and patients with diseases or medications that may interfere with the utilization of vitamins (Combs, 1998).

According to the Nutrition Labeling and Education Act (NLEA, 1993), it is optional to give information about the contents of tocopherols and tocotrienols in food labeling in the United States. If given, the contents should be compared with the Reference Daily Intakes (RDI) for vitamin E of 20 mg α-tocopherol equivalents (30 IU)/day. In Europe, only significant quantities of vitamins can be declared. This means that either 100 g of the food or the whole package containing one portion should contain at least 15% of the RDA. However, if there is a nutrient content claim on a package, nutrient information is required.

There is worldwide confusion about the regulations on natural products and dialogue between countries for harmonization is needed. This is important for both the consumers and the developers and producers of new products. Some general regulations exist in the United States, Canada and Europe, where health claims permitted are generic and not product-specific. Products that claim benefits in the prevention or treatment of disease or illness are considered drugs or medicinal products and not foods (Stephen, 1998). This trend may, however, change with the renewed interest in functional foods and alternative medicine.

1.6.2 ANTIOXIDANT ACTIVITY IN FOODS

As previously mentioned, tocopherols/tocotrienols may be added to foods for the stabilization of their polyunsaturated fatty acids. Chemical conditions and the physical state of the food to be stabilized have a major influence on lipid oxidation and antioxidant reactions. The number of possible reaction routes and stabilizing and destabilizing factors is enormous (e.g., Pokorný, 1987; Warner, 1997; Frankel, 1998). In foods, tocopherols can be regenerated from their tocopheroxyl radicals by reducing compounds such as ascorbic acid. Regeneration is possible only if the lipid-soluble tocopheroxyl radical and the water-soluble ascorbic acid are close enough to be able to react with each other, e.g., in emulsion. Thus, the interfacial region between lipid and water phases is important for the antioxidant action of tocopherols and tocotrienols (Frankel, 1998).

Kinetic studies have confirmed that α-tocopherol is the most efficient antioxidant toward peroxyl radicals (Burton et al., 1985). The order of reaction rates is α- > β- $\approx$ γ- > δ-vitamer. Because α-tocopherol is the most reactive vitamer, it is also the unstable one and is consumed first among all vitamin E compounds. This disputes the role of α-tocopherol as the most effective antioxidant in foods, which it confidently is *in vivo*. There are scarce data on antioxidant activity of tocotrienols. Tocotrienols are as efficient as tocopherols in quenching peroxyl radicals in solutions (Packer, 1995), whereas in membranes, tocotrienols seem to be more active (Packer, 1995; Qureshi et al., 2000). Depending on the environment and conditions, α-tocopherol may act as a better or a worse antioxidant than the other vitamers.

The superior reactivity of α-tocopherol causes another problem, because above a certain concentration, it loses efficiency in stabilizing polyunsaturated fatty acids (Huang et al., 1995; Fuster et al., 1998; Lampi et al., 1999; Mäkinen et al., 2000). The concept that α-tocopherol is a prooxidant is, however, not accepted, because samples containing tocopherols are more stable than samples void of them. Because the addition of α-tocopherol to vegetable oils does not generally improve their oxidative stability, the natural tocopherol level in these oils seems to be high enough not that added tocopherols are ineffective antioxidants. In vegetable oils, the antioxidant activity of α-tocopherol seems to level off at 100–500 µg/g (e.g., Parkhurst et al., 1968; Huang et al., 1994). However, less is known about the reactions in the more complex food systems, where it is possible that interactions with matrix and catalysts may direct oxidation reactions differently. It has been observed that α-tocopherol may have a considerable prooxidant activity in aqueous systems in the absence of chelators and reducing compounds (Cillard et al., 1980; Cillard and Cillard, 1986). Thus the optimal level of total α-tocopherol in vegetable oils seems to be in the range of 50–500 μg/g.

Other tocopherols and tocotrienols have not been shown to act as prooxidants. For example, antioxidant activity of δ-tocopherol increased parallel to concentration up to 2000 μg/g in a simple oil model (Huang et al., 1995). Adding tocopherols as mixtures is an efficient means to improve oxidative stability of oils (Parkhurst et al., 1968; Huang et al., 1995), because in mixtures, they protect and regenerate each other (Niki et al., 1986; Lampi et al., 1999). Thus, addition of γ- or other tocopherols and tocotrienols together with α-tocopherol may enhance the antioxidant activity of α-tocopherol.

1.6.3 UTILIZATION OF TOCOPHEROLS AND TOCOTRIENOLS IN FOODS AND FEEDS

Because they lack the phenolic hydrogen, esters are more stable than free vitamers, but they are not antioxidants *in vitro*. Tocopheryl esters hydrolyze slowly under, e.g., acidic aqueous systems, thus providing antioxidant activity for emulsions, soft drinks and some dairy products (Schuler, 1990).

Mixed tocopherols can be added dispersed in dry raw materials or suspended in water or mixed with ethanol as a carrier (O'Brien and Robertson, 1993; Six, 1994). Typical usage levels are 150–450 μg/g. Partly, this large range of different addition levels is due to the large range of tocopherol and tocotrienol contents of the raw materials (Cort et al., 1983) and their susceptibility to oxidation. Often, tocopherols and tocotrienols are added mixed with other compounds such as ascorbic acid or chelating agents that enhance the effect of tocopherols as natural antioxidants (Six, 1994).

1.6.4 FOOD ITEMS

Because vegetable oils contain optimum content of tocopherols and tocotrienols for antioxidant activity, they are not usually supplemented. Fats and oils used for frying may need some more antioxidant protection and tocopherol mixtures together with citric acid, and/or pentapolyphosphate may be added.

It is not common to add vitamin E to cereals. Restoration of vitamin E losses during heating processes, such as extrusion, roller-drying and making bread, may be feasible. In the case of cereals and cereal products, tocopherols and tocotrienols are added as vitamins and not as antioxidants.

The trend toward multivitamin drinks has increased in developed countries. Vitamins, including vitamin E, are added to fruit-juice-based products to be used by consumers on their way toward "optimal health" (O'Brien and Roberton, 1993). Many dieters eat yogurt as a meal replacement, and the addition of a wide range of vitamins is justified (O'Brien and Robertson, 1993).

Infant formulas are highly regulated products, because they often provide the only source of nutrients for infants. The need for fortifying infant formulas with vitamin E is evident, because they usually contain elevated levels of polyunsaturated fatty acids. Thus, tocopherols are utilized as antioxidants, and their usage is regulated both in ratio to energy and polyunsaturated fatty acids. There is some variation in the recommendations for supplementation of vitamin E between different authorities. The National Research Council has the lowest recommendation of 0.5 mg/g polyunsaturated fatty acids (NRC, 1989) followed by the American Academy of Pediatrics recommending 0.7 mg/g (AAP, 1985) and the European Society of Pediatric Gastroenterology and Nutrition recommending 0.9 mg/g (ESPGAN, 1991). For example, the minimum ESPGAN recommendation (equal to 0.6 mg/100 kcal) reaches up to the maximum allowable level of 10 mg/100 kcal. Because the range of recommended levels is broad, the manufacturers prefer to add vitamin E more to compensate for the possible loss during processing and storage. A recent European study showed that 85% of 30 infant formulas had higher levels of vitamin E than stated on the label. The mean value was 2.4 ± 1.5 mg/g polyunsaturated fatty acids and 2.0 ± 1.2 mg/100 kcal (Gonzáles-Corbella et al., 1999).

1.6.5 ANIMAL FEEDING

The use of purified vitamins and vitamin premixes as supplements to animal feeds has increased the economy of animal feeding. Vitamin premixes are composed of required vitamins and a suitable carrier and are given to animals together with feedstuffs providing energy and protein. Vitamin E is generally added to premixes for most domestic and laboratory animals (Combs, 1998). They are added as esters, because they are more stable than free tocopherols. Free tocopherols are known to stabilize other vitamins such as vitamin A in premixes. As an example, practical diets for chicks contain 5500 µg α-tocopherol equivalents/g (Combs, 1998).

1.7 STABILITY OF TOCOPHEROLS AND TOCOTRIENOLS DURING PROCESSING AND STORAGE OF OIL AND CEREAL GRAINS AND THEIR PRODUCTS

As mentioned before, tocopherols and tocotrienols are easily oxidized in the presence of light and metals and at high temperature or at alkaline pH, and they are sensitive to ionizing radiation, but tocopheryl esters are much more stable than tocopherols.

1.7.1 STORAGE STABILITY IN OILSEEDS AND CEREAL GRAINS

Tocopherols and tocotrienols in intact grains are relatively stable under proper storage conditions. For example, tocopherols in rapeseeds were stable during storage at 40°C for 24 weeks in low availability of oxygen, but some degradation occurred in open containers (Goffman and Möllers, 2000). Corn grains stored in the dark at room temperature lost only 5% of their total tocopherols in six months (Weber, 1987).

1.7.2 PRODUCTION OF VEGETABLE OILS

Refining of vegetable oils is a process during which phospholipids, free fatty acids, oxidized lipids, most prooxidants and impurities in crude oils are reduced as much as possible in order to produce oils with good oxidation stability and sensory quality

(Young et al., 1994). Unfortunately, tocopherols and tocotrienols that have a positive effect on the end product are also lost to some extent. The refining processes are divided into chemical and physical ones. In the chemical process, free fatty acids are neutralized using an alkaline reagent, and the soaps thus formed are removed by washing. In the physical process, free fatty acids are distilled off. Other important unit processes in refining are degumming, bleaching and deodorizing. In physical refining, the distillation stage in deodorization is especially important, because it should remove more free fatty acids than in chemical refining. The source of oil to be refined and the quality of raw material determine which processing conditions are needed. Because chemical and physical parameters in each stage may vary considerably, there are differences in losses of tocopherols and tocotrienols.

When crude soybean oil was refined in a commercial refinery, the process removed 31.8% of total tocopherols (Jung et al., 1989). The effects of degumming, alkaline refining and bleaching were small, 5.4%, 2.0% and 4.8%, respectively, compared to deodorization, during which 19.7% of initial tocopherols were lost. Similarly, degumming and alkaline refining had only minor effects on the total tocopherol content of rapeseed and linseed oils, whereas deodorization of rapeseed oil decreased it by 12% (Schöne et al., 1998). Greater losses of tocopherols were found in refining rapeseed press oil at laboratory scale; namely, 20% and 40%, on degumming stage and on bleaching stage, respectively (Prior et al., 1991). Fewer tocopherols were lost during bleaching, using bleaching medium with a lot of acid sites for purifying alkali-refined vegetable oils, than when standard active clays were used. For example, for rapeseed oil, the decreases were 7% and 44% for acid and standard clay bleaching, respectively (Boki et al., 1992). Great losses of tocopherols and tocotrienols occur during refining of rice bran oil, because it contains up to 20% of free fatty acids and a high amount of unsaponifiable matter and thus needs a thorough process. Total loss of tocopherols and tocotrienols of physically refined rice bran oil after a combined degumming-dewaxing pretreatment compared to crude oil was >75% and was markedly increased by increasing the degumming-dewaxing temperature from 10°C to 35°C (De and Bhattacharyya, 1998).

The relative composition of different vitamers remained constant during refining of soybean oil (Jung et al., 1989) and rapeseed oil (Schöne et al., 1998), but in one study, α-tocopherol was selectively adsorbed on the bleaching clay (Prior et al., 1991).

1.7.3 PROCESSING OF CEREAL AND OIL GRAINS

When flour for baking bread is manufactured, the germ and some of the outer layers are removed, and the levels of tocopherols and tocotrienols together with many other vitamins are reduced. By removing germ fraction, significant amounts of tocopherols are lost, because the germ, in general, is especially rich in E-vitamers. Bran fractions contain more tocotrienols than the germ and endosperm fractions (Barnes, 1983) (Table 1.4).

Any processing of grains subjects tocopherols and tocotrienols to oxidation. When linseeds and rapeseeds to be used for poultry feeding were crushed and stored either at room temperature or in a cold room, a 50% reduction in the tocopherol content occurred in 30 days (Gopalakrishnan et al., 1996).

Tocopherols and tocotrienols are sensitive to irradiation. To reduce the effects of irradiation, it should be performed at very low temperatures and in the absence of oxygen (Ottaway, 1993). They are also sensitive to extrusion cooking, steam-flaking, autoclaving and drum-drying (Jägerstad and Håkansson, 1988; O'Brien and Robertson, 1993). Stability of vitamin E during drum-drying could be markedly improved by using wholemeal wheat flour that was freshly milled (Wennermark et al., 1994).

1.7.4 PRODUCTION OF FOOD AND COOKING

Cooking of porridges of rolled oats and rye meal implies only minor effects on their tocopherol and tocotrienol contents. About 4–9% of tocopherols and 0–5% of toco-trienols were lost in rolled oats and rye meal porridges, when the cereals were of good quality (Piironen et al., 1987). Malting had no effects on tocopherol and tocotrienol concentrations of barley and brewers' coproduct of barley rich in these vitamers (Peterson, 1994).

Baking of whole wheat bread (200°C, 30 min) destroyed 25% of α-tocopherol and tocotrienols and 12% of β-tocopherol and tocotrienols. In baking of rye bread (200°C, 50–60 min), the losses were greater at 45% and 35%, respectively, which can be explained by both longer dough-making and baking times (Piironen et al., 1987). Jägerstad and Håkansson (1988) confirmed that vitamin E was consumed during the dough-making and baking steps but not during the fermenting step of bread making. Industrial-scale bread making (220–270°C, 22–25 min) caused a reduction in α-tocopherol content of 56–65% in French bread and of 28–40% in rye bread compared with the flours and other ingredients (Wennermark and Jägerstad, 1992). Yet, both tocopherols and tocotrienols were equally stable, and α-vitamers were less stable than β-vitamers during these food preparation procedures. Breads fortified with α-tocopherol acetate retained nearly the same percentage of 67% within the large range of 200–1600 IU/pound added to the ingredients (Ranhotra et al., 2000).

In deep-fat frying, the temperature reaches >170°C, and tocopherols are unstable. The order of stability of different tocopherols remains similar to that at lower temper-atures, i.e., $\alpha < \gamma < \delta$ (Gordon and Kourimska, 1995; Lampi and Kamal-Eldin, 1998). However, an opposite order of stability has also been found. During deep-fat frying, the relative stabilities of natural tocopherols and tocotrienols in soybean oil were α-tocopherol > δ-tocopherol > β-tocopherol > γ-tocopherol, in corn oil were α-tocopherol > γ-tocopherol > δ-tocopherol > γ-tocotrienol, and in palm oil were α-tocopherol > δ-tocotrienol > α-tocotrienol > γ-tocotrienol, respectively (Simonne and Eitenmiller, 1998). The authors assumed that tocotrienols were less stable than tocopherols, because they acted more effectively as antioxidants.

Tocopherol losses during microwave cooking are mainly caused by the effect of high temperature and not by microwaves as such. When sunflower oil was subjected to microwaves discontinuously for 120 min at two constant temperatures; namely, 170°C and <40°C, tocopherol losses were 72% and 21%, respectively (Albi et al., 1997). During continuous microwave cooking of soybean oil, the temperature increased and reached 170°C after 8 min and 210°C after 25 min. This caused greater losses of tocopherols, i.e., 5% and 10–30% after 8 and 25 min of heating, respectively (Yoshida and Takagi, 1999). During microwave cooking, α-tocopherol is less stable to heating than β-, γ- and δ-tocopherols (Yoshida et al., 1991; Yoshida and Takagi, 1999), which is similar to conventional heating at frying temperatures (Barrera-Arellano et al., 1999). Other studies have confirmed that α-tocopherol (Ha and Igarashi, 1990; Fuster et al., 1998; Lampi et al., 1999) and α-tocotrienol (Lehman and Slover, 1976) are consumed before other tocopherols and tocotrienols, also at temperatures $\leq 100^\circ$C.

At high temperatures such as during deep-fat frying and microwave heating, the rate of tocopherol consumption in the oil was greater in highly unsaturated oils than in less unsaturated oils (Yuki and Ishikawa, 1976; Yoshida et al., 1991; Simonne and Eitenmiller, 1998), which is opposite to that at moderate temperatures $<100^\circ$C. It is suggested that at low temperatures, less unsaturated oils do not oxidize, while they participate in thermooxidation at high temperatures.

1.7.5 STABILITY DURING STORAGE OF FOOD

Both food matrix and storage conditions have an effect on the stability of tocopherols and tocotrienols in food. In wheat and rye flours, storage at room temperature caused significant losses in tocopherol and tocotrienol contents during two months, while in cookies, they were stable for a year (Piironen et al., 1988). Thus, vitamers are prone to oxidation in flours exposed to air and protected in cookie matrix with a low surface-to-volume ratio. Similar to food preparation procedures, α-vitamers were less stable than β-vitamers during storage. In flours after 2 and 12 months of storage at room temperature, losses of α-tocopherol and α-tocotrienol were 20% and 80%, respectively, and those of β-tocopherol and β-tocotrienol were 10% and 60%, respectively (Piironen et al., 1988). The same difference in stability was found in another study, in which wholemeal flour, white flour, bran and germ of wheat were stored under similar conditions. Despite major differences in the profile of tocopherols and tocotrienols in these milling fractions, 40–45% and 25–35% of α-vitamers and β-vitamers were lost during one year in all of them (Jägerstad and Håkansson, 1988).

Storage stability of α-tocopherol in crisps was mainly dependent on exposure to oxygen, the storage temperature and the oil used for frying. When crisps (deep-fried in soybean oil) were stored at room temperature for 25 weeks and at 60°C for 11 days, α-tocopherol content decreased by 12% and 34%, respectively. The loss during storage was smaller in products deep-fried in more saturated high-oleic sunflower oil and palm olein compared with conventional sunflower oil (Martin-Polvillo et al., 1996). Because the life of fresh bread is limited, tocopherols and tocotrienols are stable as long as the bread is consumed before the end of the shelf life (Piironen et al., 1988; Ranhotra et al., 2000).

1.8 ANALYSIS OF TOCOPHEROLS AND TOCOTRIENOLS IN OIL AND CEREAL GRAINS AND THEIR PRODUCTS

To be able to evaluate vitamin E activity of foods and for quality control purposes, it is important to analyze all eight tocopherols and tocotrienols separately. Specific analysis of all eight vitamers is possible using either capillary gas chromatography (GC) or high-performance liquid chromatography (HPLC). Because the concentration of vitamers is small compared to other components in a food matrix, sample preparation including extraction and purification is important before GC analysis. Fewer purification steps are needed to remove neutral lipids from a sample with HPLC compared to GC applications (De Greyt et al., 1998). Another advantage of HPLC over GC analysis is that the vitamers can be analyzed without derivatization, whereas, in general, tocopherols and tocotrienols are derivatized, e.g., to trimethyl-silyl ethers, prior to GC analysis. Recently, HPLC has become the predominant method for tocopherol and tocotrienol analyses (Abidi, 2000; Bramley et al., 2000; Kamal-Eldin et al., 2000). There are some special applications in which GC is used, e.g., analysis of tocopherol steroisomers (Piironen et al., 1991; Weiser et al., 1996) and identification of vitamers by GC-mass spectroscopy (Frega et al., 1998). Both GC analysis and HPLC analysis are used when both tocopherols and sterols of fats and oils are analyzed (Slover et al., 1983; Warner and Mounts, 1990; Lechner et al., 1999; Parcerisa et al., 2000). Nelis et al. (2000) and Abidi (2000) recently published comprehensive reviews of the analysis of tocopherols and tocotrienols, including numerous chromatographic and spectroscopic techniques.

1.8.1 SAMPLE PREPARATION

Because tocopherols and tocotrienols and their esters are lipid-soluble compounds, they are soluble in organic solvents and can be extracted with them. Simple extraction of vitamers from cereal products and other foods containing high amounts of fiber and other polysaccharides may be less effective because of the complex matrix. Inclusion of the vitamers in the matrix by various interactions may hinder the penetration of the solvent into the sample, thus lowering its extraction power. To improve the extraction of lipids from cereals, dynamic extraction with hot solvents (Balz et al., 1992; Zhou et al., 1999), and modern techniques such as supercritical fluid extraction, may be used.

Saponification prior to or after lipid extraction may be used to improve extraction of vitamers from a food matrix. Heating and exposure to light as well as excessive amounts of alkali should be avoided during saponification, which should be carried out under inert atmosphere and with added antioxidants. Large amounts of lipids interfere with the extraction of E-vitamers with nonpolar solvents after saponification, because soap may solubilize them. This effect of lipids was more pronounced in the recovery of δ- than α-tocopherol (Ueda and Igarashi, 1987a,b). With these precautions, the cold saponification step is recommended for the vitamer analysis of foods (Kramer et al., 1997). It should, however, be acknowledged that losses of tocopherols and tocotrienols occur during analysis, and that the vitamers have different stabilities (e.g., Katsanidis and Addis, 1999). Reasonable recoveries of tocopherols from cereal products and other food samples have been obtained using cold saponification combined with extraction into organic solvents, i.e., 99%, 95%, 99% and 80% for α-, β-, γ- and δ-tocopherols, respectively (Piironen et al., 1984). Recently, an extraction procedure for tocopherols in vegetables was optimized and validated. The procedure included saponification at 70°C for about 50 min, and the recoveries of added α- and γ-tocopherols were 90–102% (Lee et al., 2000). This means that under optimized potassium hydroxide and ethanol contents, saponification can also be performed at this temperature.

Replacing saponification, triacylglycerols and other saponifiable lipids can be removed by chromatographic techniques. For example, saponifiable lipids were removed from vegetable oils by solid-phase extraction with silica cartridges with 98% recoveries of tocopherols (Lechner et al., 1999), and α-tocopheryl acetate was extracted from emulsified nutritional supplements by solid-phase extraction with octadecyl silica cartridges with >90% recoveries (Iwase, 2000).

1.8.2 HPLC SEPARATION

Both normal-phase and reversed-phase HPLC have been applied in vitamin E analysis. Reversed-phase HPLC is unable to completely separate all tocopherols and tocotrienols. Because β- and γ-vitamers have very similar structures, their separation cannot be obtained with reversed-phase HPLC. It is, however, applicable when only tocopherols or α-tocopheryl esters are analyzed (Gimeno et al., 2000; Iwase, 2000). There are reversed-phase methods to analyze tocopherols together with other lipid constituents from biological and food samples such as carotenoids (Epler et al., 1993; Salo-Väänänen et al., 2000), ubiquinols and ubiquinones (Podda et al., 1996) or sterols (Warner and Mounts, 1990).

There are many applications of normal-phase chromatography to separate all eight vitamers using silica or bonded silica phases such as diol or amino (Table 1.5; Figure 1.4). The separation of β- and γ-tocopherols and tocotrienols is the most difficult task, but it can be achieved with normal-phase HPLC. In addition to its

---

TABLE 1.5
Normal-Phase HPLC Conditions Used to Separate Eight Tocopherol and Tocotrienol Isomers from Food and Feed Samples

Column	Mobile Phase	Reference
Silica 150 × 4.6 mm, 5 μm	Isooctane:tetrahydrofuran, 97.5:2.5 (v/v), 1.5 ml/min	Cort et al., 1983
Silica 250 × 4 mm, 5 μm	Hexane:diisopropyl ether, 93:7 (v/v), 2.1–2.3 ml/min, 29–37°C	Piironen et al., 1984
Silica 250 × 4 mm, ≈5 μm	Hexane:2-propanol, 99.5:0.5 (v/v), 0.7–1.5 ml/min	AOCS, 1990
Diol 250 × 4 mm, 5 μm	Hexane:tert-butylmethylether, 96:4 (v/v), 1.3 ml/min	Balz et al., 1992
Diol 250 × 4.6 mm, 5 μm	Hexane:2-propanol, 99:1 (v/v) 1 ml/min	Kramer et al., 1997, 1999
Silica 250 × 4.6 mm, 5 μm	Hexane:2-propanol, 99:1 (v/v), 1.3 ml/min	Katsanidis and Addis, 1999
Silica 250 × 4.6 mm, 4–5 μm	Hexane:dioxane, 96:4 and 95:5 (v/v) 1.5–2 ml/min	Kamal-Eldin et al., 2000

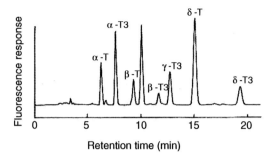

FIGURE 1.4 Normal-phase HPLC chromatogram of tocopherols (T) and tocotrienols (T3) of oat extract. Column: Genesis Silica (Jones Chromatography; 250 × 4.6 mm, d_p = 4 μm); eluent: *n*-hexane:1,4-dioxane (96:4, v/v) 1.5 ml/min at 30°C; fluorescence detection; λ_{ex} = 294 nm, λ_{em} = 326 nm.

higher separation ability, it is superior to reversed-phase chromatography because neutral lipids and other nonpolar compounds do not interfere with the analysis. The reproducibility of normal-phase columns, which used to reduce the use of these columns, has recently improved (Kramer et al., 1999; Kamal-Eldin et al., 2000).

An advanced HPLC system has been developed to separate *cis/trans* isomers of tocotrienols using a chiral permethylated β-cyclodextrin column and an acetonitrile/water eluent mixture (Drotleff and Ternes, 1999).

Detection of tocopherols and tocotrienols after HPLC separation is based on their ability to absorb ultraviolet light and create fluorescence. Tocopherols and tocotrienols show typical UV spectra with maximum absorption at 290–300 nm (Table 1.6). If the samples contain sufficient amounts of analytes, e.g., vegetable oils and supplemented products, a UV detector is sensitive enough. When higher sensitivity and better selectivity is needed, a fluorescence detector is the commonly used detector. With a fluorescence detector, it is possible to analyze tocopherols

TABLE 1.6
Characterizing Tocopherols and Tocotrienols by Ultraviolet Light

Vitamer	λ_{max} (nm), in EtOH	$E^{1\%}$, 1 cm
α-tocopherol	292	75.8
β-tocopherol	296	89.4
γ-tocopherol	298	91.4
δ-tocopherol	298	87.3
α-tocotrienol	292	91.0
β-tocotrienol	295	87.5
γ-tocotrienol	298	103.0
δ-tocotrienol	292	83.0

Data modified from Podda et al. (1996).

from as little as single-seed samples (3–7 mg) of rapeseed (Goffman et al., 1999). Usually, the excitation wavelength is set at 292 nm (285–297 nm) and the emission wavelength at 320 nm (310–324 nm) (Eitenmiller and Landen, 1999). Different tocopherols and tocotrienols have different fluorescence responses (Chase et al., 1994; Kramer et al., 1997), which means that calibration curves should be prepared individually. However, all tocopherols and their respective tocotrienols have similar fluorescence response, e.g., α-tocopherol and α-tocotrienol. Thus, tocotrienols are quantitated with their respective tocopherols (Piironen et al., 1984; AOCS, 1990; Kramer et al., 1999).

Electrochemical detection has also been applied to detect tocopherols and tocotrienols (Ueda and Igarashi, 1987a; Podda et al., 1996). It has the advantage of about 20-fold greater sensitivity relative to fluorescence detection (Nelis et al., 2000).

1.8.3 QUALITY CONTROL OF TOCOPHEROL AND TOCOTRIENOL ANALYSIS

For identification of tocopherols and tocotrienols after chromatographic separation, pure tocopherols and tocotrienol mixtures are commercially available. Their purity can be checked by both HPLC and GC and the concentration of stock solutions by UV spectroscopy (Table 1.6). It is common to use lipid extracts from cereals and rice bran and their mixtures with pure compounds and vegetable oils to tentatively identify the vitamers and to study the performance of the HPLC separation.

Quantitation of tocopherols and tocotrienols is based on either internal or external standard calibration procedures. When analyzing oils and biological samples, it is easy to find a suitable internal standard among the other tocopherols that are not present in the sample, e.g., δ-tocopherol. However, with complex materials, it is more difficult to find a relevant internal standard that would be similar enough to pass the sample preparation step similar to the vitamers, and would be different enough to be separated by HPLC. Compounds used as internal standards include 2,2,5,7,8-pentamethyl-6-chromanol (Ueda and Igarashi, 1987b), 3-octadecycloxy-1,2-propanediol (De Greyt et al., 1998) and 5,8-dimethyl tocol (Epler et al., 1993; Chase et al., 1994; Lechner et al., 1999).

1.8.4 OFFICIAL METHODS AND REFERENCE MATERIALS

Determination of four tocopherols and four tocotrienols in vegetable oils and fats by the official American Oil Chemists' Society method is based on separation by normal-phase HPLC and fluorescence detection (AOCS, 1990). Oil samples are dissolved in hexane, whereas margarines and other fats containing vitamer esters need a cold saponification step to liberate the vitamers. The American Association of Cereal Chemists has a method to analyze vitamin E in various foods. This method (AACC, 1997) is applicable to a vitamin E range of $1 \times 10^{-4} - 100\%$, and it includes hot saponification and separation by reversed-phase HPLC. Results are calculated as α-tocopherol acetate. The Royal Society of Chemistry has approved a method to analyze vitamin E in animal feedstuffs by normal-phase HPLC after the vitamers have been liberated by hot saponification (Analytical Methods Committee, 1990).

TABLE 1.7
Certified Reference Materials for Vitamin E Analysis in Foods

Material	Certified Constituent	Source
Margarine, CRM 122	α-tocopherol	Community Bureau of References—BCR, European Commission, EU
Coconut oil, fortified, SRM 1563	Tocopheryl acetate	National Institute of Standards and Technology, U.S.
Baby food composite, SRM 2383	Tocopherols	National Institute of Standards and Technology, U.S.
Infant formula, SRM 1846	α-tocopherol	National Institute of Standards and Technology, U.S.

The National Institute of Standards and Technology (NIST) and the Community Bureau of Reference (BCR) have developed several food matrix standard reference materials for vitamin E analysis (Sharpless et al., 2000). Most of the materials contain either tocopherol acetate or α-tocopherol as the only certified constituent, but some materials contain all tocopherols (Table 1.7). It has been proposed to add tocopherols in baking chocolate for reference material (Sharpless et al., 2000).

REFERENCES

AAP. 1985. American Academy of Pediatrics. Committee on Nutrition, Nutritional Needs of Low-birth-weight Infants. *Pediatrics* 75:976–986.

AACC. 1997. Method 86-06. Vitamin A and Vitamin E in Food by HPLC, in *Approved Methods of the American Association of Cereal Chemists.* 9th ed. reprint. St. Paul, MN: Am. Assoc. Cereal Chem., 8 pp.

Abidi, S.L. 2000. Chromatographic Analysis of Tocol-Derived Lipid Antioxidants. *J. Chromatogr. A* 881:197–216.

Abidi, S.L., List, G.R., and Rennick, K.A. 1999. Effect of Genetic Modification on the Distribution of Minor Constituents in Canola Oil. *J. Am. Oil Chem. Soc.* 76:463–467.

Albi, T., Lanzón, A., Guinda, A., León, M., and Pérez-Camino, M.C. 1997. Microwave and Conventional Heating Effects on Thermoxidative Degradation of Edible Fats. *J. Agric. Food Chem.* 45:3795–3798.

American Institute for Cancer Research. 1997. *Food, Nutrition and the Prevention of Cancer: A Global Perspective.* World Cancer Research Fund. Washington D.C: American Institute for Cancer Research.

Analytical Methods Committee. 1990. Determination of Vitamin E in Animal Feedingstuffs by High-Performance Liquid Chromatography. *Analyst.* 116:421–430.

ANON. 1998. News: Vitamin E Producers Expanding their Capacity. *Inform.* 9:553–554.

AOCS. 1990. Official Method Ce 8-89. Determination of Tocopherols and Tocotrienols in Vegetable Oils and Fats by HPLC, in *Official Methods and Recommended Practices of the* Am. Oil Chem. Soc.: *Sampling and Analysis of Commercial Fats and Oils.* 4th ed. Champaign, IL: Am. Oil Chem. Soc. Press, 5 pp.

Astley, S., Langrish-Smith, A., Southon, S., and Sampson, M. 1999. Vitamin E Supplementation and Oxidative Damage to DNA and Plasma LDL in Type 1 Diabetes. *Diabetes.* 22:1626–1631.

Azzi, A., Boscoboinik, D., Marilley, D., Ozer, N.K., Stauble, B., and Tasinato, A. 1995. Vitamin E: A Sensor and an Information Transducer of the Cell Oxidation State. *Am. J. Clin. Nutr.* 62:1337–1346.

Balz, M., Schulte, E., and Thier, H.-P. 1992. Trennung von Tocopherolen und Tocotrienolen durch HPLC. *Fat Sci. Technol.* 94:209–213.

Barnes, P.J. 1983. Non-Saponifiable Lipids in Cereals, in *Lipids in Cereal Technology*. Barnes, P.J. (ed.). London, UK: Academic Press, pp. 33–55.

Barnicki, S.D., Sumner, C.E. Jr., and Williams, H.C. 1996. "Process for the Production of Tocopherol Concentrates." United States Patent 5,512,691.

Barrera-Arellano, D., Ruiz-Méndez, V., Márquez-Ruiz, G., and Dobarganes, C. 1999. Loss of Tocopherols and Formation of Degradation Compounds in Triacylglycerol Model Systems Heated at High Temperature. *J. Sci. Food Agric.* 79:1923–1928.

Boki, K., Kubo, M., Wada, T., and Tamura, T. 1992. Bleaching of Alkali-Refined Vegetable Oils with Clay Minerals. *J. Am. Oil Chem. Soc.* 69:232–236.

Bramley, P.M., Elmadfa, I., Kafatos, A., Kelly, F.J., Manios, Y., Roxborough, H.E., Schuch, W., Sheehy, P.J.A., and Wagner, K.-H. 2000. Vitamin E. *J. Sci. Food Agric.* 80:913–938.

Breuninger, M. 1999. "Process for Permethylating Non-α-Tocopherols to produce α-Tocopherol." United States Patent 5,593,748.

Bruggemann, K., Herguijuela, J.R., Netscher, T., and Riegl, J. 1999. "Hydroxymethylation of Tocopherols." United States Patent 5,892,058.

Burton, G.G., Doba, T., Gabe, E.J., Hughes, L., Lee, F.L., Prasad, L., and Ingold, K.U. 1985. Autoxidation of Biological Molecules. 4. Maximizing the Antioxidant Activity of Phenols. *J. Am. Chem. Soc.* 107:7053–7065.

Carroll, K.K., Guthrie, N., Nesaretnam, K., Gapor, A., and Chambers, A.F. 1995. Anticancer Properties of Tocotrienols from Palm Oil, in *Nutrition, Lipids, Health and Disease*. Ong, A.S.N., Niki, E. and Packer, L. (eds.) Champaign, IL: Am. Oil Chem. Soc. Press, pp. 117–121.

Ceriello, A., Giugliano, D., Quatrarto, A., Donzella, C., Dipalo, G., and Lefebvre, P.J. 1991. Vitamin E Reduction of Protein Glycosylation in Diabetes. New Prospect for Prevention of Diabetic Complications? *Diabetes Care.* 14:68–72.

Challem, J.J. 1999. Toward a New Definition of Essential Nutrients: Is it Now Time for a Third "Vitamin" Paradigm? *Med. Hypotheses.* 52:417–422.

Chan, A.C. 1998. Vitamin E and Atherosclerosis. *J. Nutr.* 128:1593–1596.

Chase, G.W. Jr., Akoh, C.C., and Eitenmiller, R.R. 1994. Analysis of Tocopherols in Vegetable Oils by High-Performance Liquid Chromatography: Comparison of Fluorescence and Evaporative Light-Scattering Detection. *J. Am. Oil Chem. Soc.* 71:877–880.

Christen, S., Woodall, A.A., Shigenaga, M.K., Southwell-Keely, P.T., Duncan, M.W., and Ames, B.N. 1997. γ-Tocopherol Traps Mutagenic Electrophiles such as Nox and Complements α-Tocopherol: Physiological Implications. *Proc. Nat. Acad. Sci.* 94:3217–3222.

Chung, O.K. and Ohm, J.-B. 2000. Cereal Lipids, in *Handbook of Cereal Science and Technology*, 2nd ed. Kulp, K. and Ponte, J.G. Jr. (eds) New York, NY: Marcel Dekker, Inc., pp. 417–477.

Cillard, J. and Cillard, P. 1986. Inhibitors of the Poroxidant Activity of α-Tocopherol. *J. Am. Oil Chem. Soc.* 63:1165–1169.

Cillard, J., Cillard, P., and Cormier, M. 1980. Effect of Experimental Factors on the Prooxidant behavior of α-Tocopherol. *J. Am. Oil Chem. Soc.* 57:255–261.

Combs, G.F. Jr. 1998. *The Vitamins. Fundamental Aspects in Nutrition and Health.* 2nd ed. London: Academic Press Ltd., pp. 189–223, 459–491, 511–535.

Cort, W.M., Vicente, T.S., Waysek, E.H., and Williams, B.D. 1983. Vitamin E Content of Feedsuffs Determined by High-Performance Liquid Chromatographic Fluorescence. *J. Agric. Food Chem.* 31:1330–1333.

Crawley, H. 1993. Natural Occurrence of Vitamins in Food, in *The Technology of Vitamins in Food.* Ottaway, P.B. (ed.) Glasgow, UK: Blackie Academic & Professional, pp. 19–41.

Das, S. 1994. Vitamin E in the Genesis and Prevention of Cancer. A Review. *Acta Oncol.* 33:615–619.

De, B.K. and Bhattacharyya, D.K. 1998. Physical Refining of Rice Bran Oil in Relation to Degumming and Dewaxing. *J. Am. Oil Chem. Soc.* 75:1683–1686.

De Greyt, W.F., Kellens, M.J., and Huyghebaert, A.D. 1999. Effect of Physical Refining on Selected Minor Components in Vegetable Oils. *Fett/Lipid.* 101:428–432.

De Greyt, W.F., Petrauskaite, V., Kellens M.J., and Huyghebaert, A.D. 1998. Analysis of Toco-pherols by Gas-Liquid and High-Performance Liquid Chromatography: A Comparative Study. *Fett/Lipid.* 100:503–507.

Devaraj, S., Li, D., and Jialal. I. 1996. The Effects of α-Tocopherol Supplementation on Monocyte Function: Decreased Lipid Oxidation, Interleukin-1β Secretion, and Monocyte Adhesion to Endothelium. *J. Clin. Invest.* 98:756–763.

Diack, M. and Saska, M. 1994. Separation of Vitamin E and γ-Oryzanols from Rice Bran by Normal-Phase Chromatography. *J. Am. Oil Chem. Soc.* 71:1211–1217.

Dijkstra, A.J. 2000. Edible Oils in Europe. *Inform.* 11:386–394.

Diplock, A.T. 1985. Vitamin E, in *Fat-Soluble Vitamins, Their Biochemistry and Applications.* London, UK: William Heinemann Ltd., pp.154–224.

Dolde, D., Vlahakis, C., and Hazebroek, J., 1999. Tocopherols in Breeding Lines and Effects of Planting Location, Fatty Acid Composition, and Temperature during Development. *J. Am. Oil Chem. Soc.* 76:349–355.

DRI. 2000. Vitamin E, in *Dietary Reference Intakes for Vitamin C, Vitamin E, Selenium, and Carotenoids.* Washington, D.C.: National Academy Press, pp. 186–283.

Drotleff, A.M. and Ternes, W. 1999. *Cis/trans* Isomers of Tocotrienols—Occurence and Bioavailability. *Eur. Food Res. Technol.* 210:1–8.

Dutta-Roy, A.K., Gordon, M.J., Campbell, F.M., Duthie, G.G., and James, W.P. 1994. Vitamin E Requirements, Transport and Metabolism: Role of α-Tocopherol Binding Proteins. *J. Nutr. Biochem.* 5:562–570.

EC. 1995. European Parliament and Council Directive No 95/2/EC on Food Additives other than Colours and Sweeteners. OJ L 061, 18/03/95, pp. 1–40.

EEC. 1989. Commission of the European Communities. Council Directive No 89/398/EEC on the Approximation of the Laws of the Member States Relating to Foodstuffs Intended for Particular Nutritional Uses. OJ L 186, 30/06/89, pp. 27–32.

EEC. 1990. Commission of the European Communities. Council Directive No 90/496/EEC on Nutrition Labelling for Foodstuffs. OJ L 276, 06/10/90, pp. 40–44.

Eitenmiller, R.R. 1997. Vitamin E Content of Fats and Oils—Nutritional Implications. *Food Technol.* 51:78–81.

Eitenmiller, R.R., and Landen, W.O. Jr. 1999. Vitamin E, Tocopherols and Tocotrienols, in *Vitamin Analysis for the Health and Food Sciences.* Boca Raton, FL: CRC Press, pp. 109–148.

Epler, K.S., Ziegler, R.G., and Craft, N.E. 1993. Liquid Chromatographic Method for the Determination of Carotenoids, Retinoids and Tocopherols in Human Serum and in Food. *J. Chromatogr.* 619:37–48.

ESPGAN. 1991. European Society of Paediatric Gastroenterol and Nutrition: Committee on Nutrition Comment on the Content and Composition of Lipids in Infant Formulas. *Acta Paediatrica Scand.* 80:887–896.

Frankel, E.N. 1998. Antioxidants, in *Lipid Oxidation*. Dundee, UK: The Oily Press Ltd., pp. 129–160.

Frega, N., Mozzon, M., and Bocci, F. 1998. Identification and Estimation of Tocotrienols in the Annatto Lipid Fraction by Gas Chromatography-Mass Spectrometry. *J. Am. Oil. Chem. Soc.* 75:1723–1727.

Fuster, M.D., Lampi, A.-M., Hopia, A., and Kamal-Eldin, A. 1998. Effects of α- and γ-Tocopherols on the Autoxidation of Purified Sunflower Triacylglycerols. *Lipids.* 33:715–722.

Gazis, A., Page, S., and Cockcroft, J. 1997. Vitamin E and Cardiovascular Protection in Diabetes (editorial). *Br. Med. J.* 314:1845–1846.

Gimeno, E., Castellote, A.I., Lamuela-Raventós, R.M., de la Torre, M.C., and López-Sabater, M.C. 2000. Rapid Determination of Vitamin E in Vegetable Oils by Reversed-Phase High-Performance Liquid Chromatography. *J. Chromatogr. A.* 881:251–254.

Goffman, F.D. and Möllers, C. 2000. Changes in Tocopherol and Plastochromanol-8 Contents in Seeds and Oil of Oilseed Rape (*Brassica napus* L.) during Storage as Influenced by Temperature and Air Oxygen. *J. Agric. Food Chem.* 48:1605–1609.

Goffman, F.D., Velasco, L., and Thies, W. 1999. Quantitative Determination of Tocopherols in Single Seeds of Rapeseed (*Brassica napus* L.). *Fett/Lipid.* 101:142–145.

Gonzáles-Corbella, M.J., Tortras-Biosca, M., Castellote-Bargalló A.I., and López-Sabater, M.C. 1999. Retinol and α-Tocopherol in Infant Formulas Produced in the EEC. *Food Chem.* 66:221–226.

Gopalakrishnan, N., Cherian, G., and Sim, J.S. 1996. Chemical Changes in the Lipids of Canola and Flax Seeds during Storage. *Fett/Lipid.* 98:168–171.

Gordon, M.H. and Kourimska, L. 1995. Effect of Antioxidants on Losses of Tocopherols during Deep-Fat Frying. *Food Chem.* 52:175–177.

Gottstein, T. and Grosch, W. 1990. Model Study of Different Antioxidant Properties of α-and γ-Tocopherol in Fats. *Fat Sci. Technol.* 92:139–144.

Ha, K.-H. and Igarashi, O. 1990. The Oxidation Products from Two Kinds of Tocopherols Co-Existing in Autoxidation System of Methyl Linoleate. *J. Nutr. Sci. Vitaminol.* 36:411–421.

Heinonen, M. and Piironen, V. 1991. The Tocopherol, Tocotrienol, and Vitamin E Content of the Average Finnish Diet. *Int. J. Vitam. Nutr. Res.* 61:27–32.

Hense, H.W., Stender, M., Bors, W., and Keil, W. 1993. Lack of an Association between Serum Vitamin E and Mycocardial Infarction in a Population with High Vitamin E Levels. *Atherosclerosis.* 103:21–28.

Hess, J.L. 1993. Vitamin E, α–Tocopherol, in *Antioxidants in Higher Plants*. Alscher, R.G., and Hess, J.L. (eds.) Boca Raton, FL: CRC Press, pp. 111–134.

Hood, R.L. 1998. Tocotrienols in Metabolism, in *Phytochemicals—A New Paradigm*. Bidlack, W.R., Omaye, S.T., Meskin, M.S., and Jahner, D. (eds.) Lancaster, PA: Technomic Publishing Co., Inc., pp. 33–51.

Hopia, A.I., Huang, S.-W., and Frankel, E.N. 1996. Effect of α-Tocopherol and Trolox on the Decomposition of Methyl Linoleate Hydroperoxides. *Lipids.* 31:357–365.

Horwitt, M.K. 1991. Data Supporting Supplementation of Humans with Vitamin E. *J. Nutr.* 121:424–429.

Hosomi, A., Arita, M., Sato, Y., Kiyose, C., Ueda, T., Igarashi, O., Arai, H., and Inoue, K. 1997. Affinity for Alpha-Tocopherol Transfer Protein as a Determinant of the Biological Activities of Vitamin E Analogs. *FEBS Lett.* 409:105–108.

Huang, S.-W., Frankel, E.N., and German, J.B. 1994. Antioxidant Activity of α- and γ-Tocopherols in Bulk Oils and in Oil-in-Water Emulsions. *J. Agric. Food Chem.* 42:2108–2114.

Huang, S.-W., Frankel, E.N., and German J.B. 1995. Effects of Individual Tocopherols and Tocopherol Mixtures on the Oxidative Stability of Corn Oil Triglycerides. *J. Agric. Food Chem.* 43:2345–2350.

IUPAC-IUB. 1982. Nomenclature of Tocopherols and Related Compounds. International Union of Pure and Applied Chemistry and International Union of Biochemistry and Molecular Biology (IUPAC-IUB). *Pure Appl. Chem.* 54:1507–1510.

Iwase H. 2000. Determination of Tocopherol Acetate in Emulsified Nutritional Supplements by Solid-Phase Extraction and High-Performance Liquid Chromatography with Fluorescence Detection. *J. Chromatogr. A.* 881:243–249.

Jägerstad, M. and Håkansson, B. 1988. Vitamin Changes during Processing of Cereals, in *Cereal Science and Technology in Sweden—Proceedings from an International Symposium*, June 13–16, 1988. Asp, N.G. (ed.) Lund: BTJ Tryck, pp. 207–219.

Jung, M.Y., Yoon, S.H., and Min, D.B. 1989. Effects of Processing Steps on the Contents of Minor Compounds and Oxidation of Soybean Oil. *J. Am. Oil Chem. Soc.* 66:118–120.

Kamal-Eldin, A. and Andersson R. 1997. A Multivariate Study of the Correlation between Tocopherol Content and Fatty Acid Composition in Vegetable Oils. *J. Am. Oil Chem. Soc.* 74:375–380.

Kamal-Eldin, A. and Appelqvist, L.-Å. 1996. The Chemistry and Antioxidant Properties of Tocopherols and Tocotrienols. *Lipids.* 31:671–701.

Kamal-Eldin, A., Görgen, S., Petterson, J., and Lampi A.-M. 2000. Normal-Phase High-Performance Liquid Chromatography of Tocopherols and Tocotrienols. Comparison of Different Chromatographic Columns. *J. Chromatogr. A.* 881:217–227.

Katsanidis, E. and Addis, P.B. 1999. Novel HPLC Analysis of Tocopherols, Tocotrienols, and Cholesterol in Tissue. *Free Radic. Biol. Med.* 27:1137–1140.

Kayden, H.J. and Traber, M.G. 1993. Absorption, Lipoprotein Transport, and Regulation of Plasma Concentrations of Vitamin E in Humans. *J. Lipid Res.* 34:343–358.

Kline, K., Yu, W., and Sanders, B.G., 1998. Vitamin E: Mechanisms of Action as Tumor Cell Growth Inhibitors, in *Cancer and Nutrition.* Prasad, K.N. and Cole, W.C. (eds.) Amsterdam, The Netherlands: IOS, pp. 37–53.

Kramer, J.K.G., Blais, L., Fouchard, R.C., Melnyk, R.A., and Kallury, K.M.R. 1997. A Rapid Method for the Determination of Vitamin E Forms in Tissues and Diet by High-Performance Liquid Chromatography Using a Normal-Phase Diol Column. *Lipids.* 32:323–330.

Kramer, J.K.G., Fouchard, R.C., and Kallury, K.M.R. 1999. Determination of Vitamin E Forms in Tissues and Diets by High-Performance Liquid Chromatography Using Normal-Phase Diol Column, in *Analysis of Vitamin E Forms.* New York, NY: Academic Press Ltd., pp. 318–329.

Kurilich, A.C. and Juvik, J.A. 1999. Quantification of Carotenoid and Tocopherol Antioxidants in *Zea mays. J. Am. Agric. Chem. Soc.* 47:1948–1955.

Lampi, A.-M. and Kamal-Eldin, A. 1998. Effect of α- and γ-Tocopherols on Thermal Polymerization of Purified High-Oleic Sunflower Triacylglycerols. *J. Am. Oil Chem. Soc.* 75:1699–1703.

Lampi, A.-M., Kataja, L., Kamal-Eldin, A., and Piironen, V. 1999. Antioxidant Activities of α- and γ-Tocopherols in the Oxidation of Rapeseed Oil Triacylglycerols. *J. Am. Oil Chem. Soc.* 76:749–755.

Lane, R.H., Qureshi, A.A., and Salser, W.A. 1999. "Tocotrienols and Tocotrienol-Like Compounds and Methods for Their Use." United States Patent 5,919,818.

Lechner, M., Reiter, B., and Lorbeer, E. 1999. Determination of Tocopherols and Sterols in Vegetable Oils by Solid-Phase Extraction and Subsequent Capillary Gas Chromatographic Analysis. *J. Chromatogr. A.* 857:231–238.

Lee, J., Ye, L., Landen, W.O. Jr., and Eitenmiller, R.R. 2000. Optimization of an Extraction Procedure for the Quantification of Vitamin E in Tomato and Broccoli Using Response Surface Methodology. *J. Food Comp. Anal.* 13:45–57.

Lehman, J. and Slover, H.T. 1976. Relative Autoxidative and Photolytic Stabilities of Tocols and Tocotrienols. *Lipids.* 11:853–857.

Lehmann, J.W., Putnam, D.H., and Qureshi, A.A. 1994. Vitamin E Isomers in Grain Amaranths (*Amaranthus* spp.). *Lipids.* 29:177–181.

Lips, H.J. 1957. Stability of d-α-Tocopherol Alone, in Solvents, and in Methyl Esters of Fatty Acids. *J. Am. Oil Chem. Soc.* 34:513–515.

Machlin, L.J. 1991. Vitamin E, in *Handbook of Vitamins*, Machlin L.J. (ed.) New York, NY: Marcel Dekker, pp. 99–144.

Mäkinen, M., Kamal-Eldin, A., Lampi, A.-M., and Hopia, A. 2000. Effects of α- and γ-Tocopherols on Formation of Hydroperoxides and Two Decomposition Products from Methyl Linoleate. *J. Am. Oil Chem. Soc.* 77:801–806.

Marquard, V. 1990. Investigations on the Influence of Genotype and Location on the Tocopherol Content of the Oil from Different Oil Crops. *Fat Sci. Technol.* 92:452–455.

Martín-Polvillo, M., Márquez-Ruis, G., Jorge, N., Ruiz-Méndez, M.V., and Dobarganes, M.C. 1996. Evolution of Oxidation during Storage of Crisps and French Fries Prepared with Sunflower Oil and High Oleic Sunflower Oil. *Grasas y Aceites.* 47:54–58.

Moldenhauer, K.A., Champagne, E.T., McCaskill, D.R., and Guraya, H. 1998. Functional Products from Rice, in *Functional Foods. Biochemical & Processing Aspects.* Mazza, G. (ed.) Lancaster, PA: Technomic Publishing Co., Inc., pp. 71–89.

Moreau, R.A., Hicks, K.B., and Powell, M.J. 1999. Effect of Heat Pretreatment on the Yield and Composition of Oil Extracted from Corn Fiber. *J. Agric. Food Chem.* 47:2869–2871.

Morrison, W.R. 1978. Cereal Lipids, in *Advances in Cereal Sciece and Technology*, vol. 2. Pomeranz, Y. (ed.) St Paul, MN: Am. Assoc. Cereal Chem., pp. 221–348.

Morrissey, P.A. and Sheehy, P.J.A. 1999. Optimal Nutrition: Vitamin E. *Proc. Nutr. Soc.* 58:459–468.

Murphy, S.P., Subar, A.F., and Block, G. 1990. Vitamin E Intakes and Sources in the United States. *Am. J. Clin. Nutr.* 52:361–367.

NRC. 1989. *Vitamin E.* National Research Council. Recommended Dietary Allowances. 10th ed., Washington, D.C.: National Academy Press, pp. 99–107.

Nelis, H.J., D'Haese, E., and Vermis, K. 2000. Vitamin E, in *Modern Chromatographic Analysis of Vitamins.* 3rd ed. De Leenheer, A.P., Lambert, W.E., and Van Bocxlaer, J.F. (eds.) New York, NY: Marcel Dekker, pp. 143–228.

Niki, E., Tsuchiya, J., Yoshikawa, Y., Yamamoto, Y., and Kamiya, Y. 1986. Oxidation of Lipids. XIII. Antioxidant Activities of α-, β-, γ- and δ-Tocopherols. *Bull. Chem. Soc. Jpn.* 59:487–501.

NLEA. 1993. Nutrition Labeling and Education Act of 1990. *Fed. Reg.* 58:2080.

O'Brien, A. and Roberton, D. 1993. Vitamin Fortification of Foods—Specific Applications, in *The Technology of Vitamins in Food.* Ottaway, P.B. (ed.) Glasgow, UK: Blackie Academic & Professional, pp. 114–130.

O'Leary, M.J. 1993. Industrial Production, in *The Technology of Vitamins in Food.* Ottaway, P.B. (ed.) Glasgow, UK: Blackie Academic & Professional, pp. 63–89.

Olejnik, D., Gogolewski, M., and Nogala-Kalucka, M. 1997. Isolation and some Properties of Plastochromanol-8. *Nahrung.* 41:101–104.

Ong, A.S.H. and Choo, Y.M. 1997. Carotenoids and Tocols from Palm Oil, in *Natural Antioxidants—Chemistry, Health Effects, and Applications.* Shahidi, F. (ed.) Champaign, IL: Am. Oil Chem. Soc. Press, pp. 133–1149.

Oomah, B.D., Kenaschuk, E.O., and Mazza, G. 1997. Tocopherols in Flaxseed. *J. Agric. Food Chem.* 45:2076–2080.

Oomah, B.D. and Mazza, G. 1998. Flaxseed Products for Disease Prevention, in *Functional Foods—Biochemical and Processing Aspects.* Mazza, G. (ed.) Lancaster, PA: Technomic Publishing Co., Inc., pp. 91–138.

Ottaway, P.B. 1993. Stability of Vitamins in Food, in *The Technology of Vitamins in Food.* Ottaway, P.B. (ed.) Glasgow, UK: Blackie Academic & Professional, pp. 90–113.

Packer, L. 1995. Nutrition and Biochemistry of the Lipophilic Antioxidants Vitamin E and Carotenoids, in *Nutrition, Lipids, Health, and Disease.* Ong, A.S.H., Niki, E., and Packer, L. (eds.) Champaign, IL: Am. Oil Chem. Soc. Press, pp. 8–35.

Paolisso, G., D'Amore, A., Giugliano, D., Ceriello, A., Varricchio, M., and D'Onofrio, F. 1993. Pharmacologic Doses of Vitamin E Improve Insulin Action in Healthy Subjects and Non-Insulin-Dependent Diabetic Patients. *Am. J. Clin. Nutr.* 57:650–656.

Parcerisa, J., Casals, I., Boatella, J., Codony, R., and Rafecas, M. 2000. Analysis of Olive and Hazelnut Oil Mixtures by High-Performance Liquid Chromatography — Atmospheric Pressure Chemical Ionisation Mass Spectrometry of Triacyglycerols and Gas-Liquid Chromatography of Non-Saponifiable Compounds (Tocopherols and Sterols). *J. Chromatogr. A.* 881:149–158.

Parkhurst, R.M., Skinner, W.A., and Sturm, P.A. 1968. The Effect of Various Concentrations of Tocopherols and Tocopherol Mixtures on the Oxidative Stability of a Sample of Lard. *J. Am. Oil Chem. Soc.* 45:641–642.

Patterson, R.E., White, E., Kristal, A.R., Neuhouser, M.L., and Potter, J.D. 1997. Vitamin Supplements and Cancer Risk: the Epidemiologic Evidence. *Cancer Causes Control.* 8:786–802.

Peterson, D.M. 1994. Barley Tocols: Effects of Milling, Malting, and Mashing. *Cereal Chem.* 71:42–44.

Peterson, D.M. 1995. Oat Tocols: Concentration and Stability in Oat Products and Distribution within the Kernel. *Cereal Chem.* 72:21–24.

Peterson, D.M. and Qureshi, A.A. 1993. Genotype and Environment Effects on Tocols of Barley and Oats. *Cereal Chem.* 70:157–162.

Peterson, D.M. and Wood, D.F. 1997. Composition and Structure of High-Oil Oat. *J. Cereal Sci.* 26:121–128.

Piironen, V., Liljeroos, A., and Koivistoinen, P. 1991. Transfer of α-Tocopherol Stereoisomers from Feeds to Eggs. *J. Agric. Food Chem.* 39:99–101.

Piironen, V., Syväoja, E.-L., Varo, P., Salminen, K., and Koivistoinen, P. 1986. Tocopherols and Tocotrienols in Cereal Products from Finland. *Cereal Chem.* 63:78–81.

Piironen, V., Varo, P., and Koivistoinen, P. 1987. Stability of Tocopherols and Tocotrienols in Food Preparation Procedures. *J. Food Comp. Anal.* 1:53–58.

Piironen, V., Varo, P., and Koivistoinen, P. 1988. Stability of Tocopherols and Tocotrienols during Storage of Foods. *J. Food Comp. Anal.* 1:124–129.

Piironen, V., Varo, P., Syväoja, E.-L., Salminen, K., and Koivistoinen, P. 1984. High-Performance Liquid Chromatographic Determination of Tocopherols and Tocotrienols and its Application to Diets and Plasma of Finnish Men. *Intern. J. Vitam. Nutr. Res.* 53:35–40.

Podda, M., Weber, C., Traber, M.G., and Packer, L. 1996. Simultaneous Determination of Tissue Tocopherols, Tocotrienols, Ubiquinols, and Ubiquinones. *J. Lipid Res.* 37:893–901.

Pokorný, J. 1987. Major Factors Affecting the Autoxidation of Lipids, in *Autoxidation of Unsaturated Lipids.* Chan, H.W.-S. (ed.) London, UK: Academic Press Ltd., pp. 140–233.

Polidori, M.C., Mecocci, P., Stahl, W., Parente, B., Cecchetti, R., Cherubini, A., Cao, P., Sies, H., and Senin, U. 2000. Plasma Levels of Lipophilic Antioxidants in Very Old Patients with Type 2 Diabetes. *Diabetes Metab. Res. Rev.* 16:15–19.

Porter, N.A., Caldwell, S.E., and Mills, K.A. 1995. Mechanisms of Free Radical Oxidation of Unsaturated Lipids. *Lipids.* 30:277–290.

Prior, E.M., Vadke, V.S., and Sosulski, F.W. 1991. Effect of Heat Treatments on Canola Press Oils. I. Non-Triglyceride Components. *J. Am. Oil Chem. Soc.* 68:401–406.

Psomiadou, E., Tsimidou, M., and Boskou, D. 2000. α-Tocopherol Content of Greek Virgin Olive Oils. *J. Agric. Food Chem.* 48:1770–1775.

Qureshi, A.A., Mo, H., Packer, L., and Peterson, D.M. 2000. Isolation and Identification of Novel Tocotrienols from Rice Bran with Hypocholesterolemic, Antioxidant and Antitumor Properties. *J. Agric. Food Chem.* 48:3130–3140.

Ramamurthi, B. and McCurdy, A.R. 1993. Enzymatic Pretreatment of Deodorizer Distillate for Concentration of Sterols and Tocopherols. *J. Am. Oil Chem. Soc.* 70:287–295.

Ranhotra, G.S., Gelroth, J.A., and Okot-Kotber, B.M. 2000. Stability and Dietary Contribution of Vitamin E Added to Bread. *Cereal Chem.* 77:159–162.

RNI. 1983. *Recommended Nutrient Intakes for Canadians.* Committee for the Revisions of the Dietary Standard for Canada. Bureau of Nutritional Sciences, Food Directorate. Department of National Health and Welfare, Ottawa, Ontario, Canada.

Rogers, E.J., Rice, S.M., Nicolosi, R.J., Carpenter, D.R., McCelland, C.A., and Romanczyk, L.J. Jr. 1993. Identification and Quantitation of γ-Oryzanol Components and Simultaneous Assessment of Tocols in Rice Bran Oil. *J. Am. Oil Chem. Soc.* 70:301–307.

Salo-Väänänen, P., Ollilainen, V., Mattila, P., Lehikoinen, K., Salmela-Mölsä, E., and Piironen, V. 2000. Simultaneous HPLC Analysis of Fat-Soluble Vitamins in Selected Animal Products after Small-Scale Extraction. *Food Chem.* 71:535–543.

Sauberlich, H.E., Dowdy, R.P., and Skala, J.H., 1974. Laboratory Tests for the Assessment of Nutritional Status. Cleveland OH: CRC Press, pp. 74–80.

Schager, F. and Bonrath, W. 1999. Synthesis of D,L-α-Tocopherol Using Strong Solid Acids as Catalysts. *J. Catal.* 182:282–284.

Schöne, F., Fritsche, J., Bargholz, J., Leiterer, M., Jahreis, G., and Matthäus, B. 1998. Zu den Veränderungen von Rapsöl und Leinöl während der Bearbeitung. *Fett/Lipid.* 100:539–545.

Schudel, P., Mayer, H., and Isler, O. 1972. Tocopherols—Chemistry, in *The Vitamins.* 2nd ed. Vol. V. Sebrell, W.H. Jr. and Harris, R.S. (eds.) New York, NY: Academic Press, pp. 168–218.

Schuler, P. 1990. Natural Antioxidants Exploited Commercially, in *Food Antioxidants.* Hudson, B.J.F. (ed.) Barking, UK: Elsevier Applied Publishers, Ltd., pp. 99–170.

Sharpless, K.E., Margolis, S., and Brown Thomas, J. 2000. Determination of Vitamins in Food-Matrix Standard Reference Materials. *J. Chromatogr. A.* 881:171–181.

Shimada, Y., Nakai, S., Suenaga, M., Sugihara, A., Kitano, M., and Tominaga, Y. 2000. Facile Purification of Tocopherols from Soybean Oil Deodorizer Distillate in High Yield Using Lipase. *J. Am. Oil Chem. Soc.* 77:1009–1013.

Shklar, G. and Oh, S.K. 2000. Experimental Basis for Cancer Prevention by Vitamin E. *Cancer Invest.* 18:214–222.

Simonne, A.H. and Eitenmiller, R.R. 1998. Retention of Vitamin E and Added Retinyl Palmitate in Selected Vegetable Oils during Deep-Fat Frying and in Fried Breaded Products. *J. Agric. Food Chem.* 46:5273–5277.

Six, P. 1994. Current Research in Natural Fat Antioxidants. *Inform.* 5:679–688.

Sliwiok, J. and Kocjan, B. 1992. Chromatographische Untersuchungen der hydrophoben Eigenschaften von Tocopherolen. *Fat Sci. Technol.* 94:157–159.

Slover, H.T., Thompson, R.H. Jr., and Merola, G.V. 1983. Determination of Tocopherols and Sterols by Capillary Gas Chromatography. *J. Am. Oil Chem. Soc.* 60:1524–1528.

Stampfer, M.J. and Rimm, E.B. 1995. Epidemiologic Evidence for Vitamin E in Prevention of Cardiovascular Disease. *Am. J. Clin Nutr.* 62:1365–1369.

Steiner, M. 1993. Vitamin E: More Than an Antioxidant. *Clin. Cardiol.* 16:116–118.

Stephen, A.M. 1998. Regulatory Aspects of Functional Products, in *Functional Foods— Biochemical & Processing Aspects.* Mazza, G. (ed.) Lancaster, PA: Technomic Publishing Co., Inc., pp. 403–437.

Stone, W.L. and Papas, A.M. 1997. Tocopherols and the Etiology of Colon Cancer. *J. Natl. Cancer Inst.* 89:1006–1014.

Summerfield, F.W. and Tappel, A.A. 1984. Effect of Dietary Polyunsaturated Fats and Vitamin E on Aging and Peroxidation Damage to DNS. *Arch. Biochem. Biophys.* 233:408–416.

Swanson, J.E., Ben, R.N., Burton, G.W., and Parker, R.S. 1999. Urinary Excretion of 2,7,8-trimethyl-2-(β-carboxyethyl)-6-hydroxychroman is a Major Route of Elimination of γ-Tocopherol in Humans. *J. Lipid Res.* 40:665–671.

Syväoja, E.-L., Piironen, V., Varo, P., Koivistoinen, P., and Salminen, K. 1986. Tocopherols and Tocotrienols in Finnish Foods: Oils and Fats. *J. Am. Oil Chem. Soc.* 63:328–329.

Tan, B. 1989. Palm Carotenoids, Tocopherols and Tocotrienols. *J. Am. Oil Chem. Soc.* 66:770–776.

Theriault, A., Chao, J.T., Wang, Q., Gapor, A., and Adeli, K. 1999. Tocotrienol: A Review of its Therapeutic Potential. *Clin. Biochem.* 32:309–319.

Traber, M.G. 1999. Vitamin E, in *Modern Nutrition in Health and Disease*, 9th ed. Baltimore, MD: Williams & Wilkins, pp. 347–362.

Traber, M.G. and Sies, H. 1996. Vitamin E in Humans: Demand and Delivery. *Ann. Rev. Nutr.* 16:321–347.

Traber, M.G., Elsner, A., and Brigelius-Flohe, R. 1998. Synthetic as compared with Natural Vitamin E is Preferentially Excreted as α-CEHC in Human Urine: Studies Using Deuterated α-Tocopheryl Acetates. *FEBS Lett.* 437:145–148.

Ueda, T. and Igarashi, O. 1987a. Effect of Coexisting Fat on the Extraction of Tocopherols from Tissues after Saponification as a Pretreatment for HPLC Determination. *J. Micronutr. Anal.* 3:15–25.

Ueda, T. and Igarashi, O. 1987b. New Solvents System for Extraction of Tocopherols from Biological Specimens for HPLC Determination and the Evaluation of 2,2,5,7,8-Pentamethyl-6-Chromanol as an Internal Standard. *J. Micronutr. Anal.* 3:185–198.

Van Niekerk, P.J. and Burger, A.E.C. 1985. The Estimation of the Composition of Edible Oil Mixtures. *J. Am. Oil Chem. Soc.* 62:531–538.

Velasco, L. and Goffman, F.D. 2000. Tocopherol, Plastochromanol and Fatty Acid Patterns in the Genus *Linum*. *Plant Syst. Evol.* 221:77–88.

VERIS. 1998. *Vitamin E and Carotenoid Abstracts*, LaGrange, IL: VERIS.

Walsh, L., Winters, R.L., and Gonzalez, R.G. 1998. Optimizing Deodorizer Distillate Tocopherol Yields. *Inform.* 9:78–83.

Wang, L., Xue, Q., Newman, R.K., and Newman, C.W. 1993. Enrichment of Tocopherols, Tocotrienols, and Oil in Barley Fractions by Milling and Pearling. *Cereal Chem.* 70:499–501.

Warner, K. 1997. Measuring Tocopherol Efficacy in Fats and Oils, in *Antioxidant Methodology.* Aruoma, O.I. and Cuppett, S.L. (eds.) Champaign, IL: Am. Oil Chem. Soc. Press, pp. 223–233.

Warner, K. and Mounts, T.L. 1990. Analysis of Tocopherols and Phytosterols in Vegetable Oils by HPLC with Evaporative Light-Scattering Detection. *J. Am. Oil Chem. Soc.* 67:827–831.

Weber, E.J. 1984. High Performance Liquid Chromatography of Tocols in Corn Grain. *J. Am. Oil Chem. Soc.* 61:1231–1234.

Weber, E.J. 1987. Carotenoids and Tocols of Corn Grain Determined by HPLC. *J. Am. Oil Chem. Soc.* 64:1129–1134.

Weber, P., Bendich, A., and Machlin, L.J. 1997. Vitamin E and Human Health: Rationale for Determining Recommended Intake Levels. *Nutrition.* 13:450–460.

Weiser, H., Riss, G., and Kormann, A.W. 1996. Biodiscrimination of the Eight α-Tocopherol Stereoisomers Results in Preferential Accumulation of the Four 2R Forms in Tissues and Plasma of Rats. *J. Nutr.* 126:2539–2549.

Wennermark, B. and Jägerstad, M. 1992. Breadmaking and Storage of Various Wheat Fractions Affect Vitamin E. *J. Food Sci.* 57:1205–1209.

Wennermark, B., Ahlmen, H., and Jägerstad, M. 1994. Improved Vitamin E Retention by Using Freshly Milled Whole-Meal Wheat Flour during Drum-Drying. *J. Agric. Food Chem.* 42:1348–1351.

White, P.J. and Xing, Y. 1997. Antioxidants from Cereals and Legumes, in *Natural Antioxidants—Chemistry, Health Effects, and Application.* Shahidi, F. (ed.) Champaign, IL: Am. Oil Chem. Soc. Press, pp. 25–63.

Yoshida, H., Hirooka, N., and Kajimoto, G. 1991. Microwave Heating Effects on Relative Stabilites of Tocopherols in Oils. *J. Food Sci.* 56:1042–1046.

Yoshida, H. and Takagi, S. 1999. Antioxidative Effects of Sesamol and Tocopherols at Various Concentrations in Oils during Microwave Heating. *J. Sci. Food Agric.* 79:220–226.

Young, F.V.K., Poot, C., Biernothe, E., Krog, N., Davidson, N.G.J., and Gunstone, F.D. 1994. Processing of Fats and Oils, in *The Lipid Handbook*, 2nd ed. Gunstone, F.D., Harwood, J.L., and Padley, F.B. (eds.) London, UK: Chapman & Hall, pp. 249–276.

Youngs, V.L. 1986. Oat Lipids and Lipid-Related Enzymes, in *Oats—Chemistry and Technology.* Webster, F.H. (ed.) St Paul, MN: Am. Assoc. Cereal Chem., pp. 205–226.

Yuki, E. and Ishikawa, Y. 1976. Tocopherol Contents of Nine Vegetable Frying Oils, and Their Changes under Simulated Deep-Fat Frying Conditions. *J. Am. Oil Chem. Soc.* 53:673–676.

Yu, W., Simmons-Menchaca, M., Gapor, A., Sanders, B.G., and Kline K. 1999. Induction of Apoptosis in Human Breast Cancer Cells by Tocopherols and Tocotrienols. *Nutr. Cancer.* 33:26–32.

Zhou, M., Robards, K., Glennie-Holmes, M., and Helliwell, S. 1999. Oat Lipids. *J. Am. Oil Chem. Soc.* 76:159–169.

2 Isoflavones from Soybeans and Soy Foods

Sam K. C. Chang

CONTENTS

1-5667-6902-7/02/$0.00+$1.50
© 2002 by CRC Press LLC

2.1 INTRODUCTION

Epidemiological studies (Aldercreutz et al., 1991; Aldercreutz, 1998) indicate that consumption of tofu and other soy foods may be associated with the low incidence of breast cancer in Japanese women. This discovery has led numerous researchers in recent years to search for the biochemical components in soybean that are responsible for the cancer risk-lowering effect. Aside from the potential cancer prevention effect (Wu et al., 1996; Cline and Hughes, 1998; Griffiths et al., 1998; Messina and Bennink, 1998; Stephens, 1999), isoflavones also have been found to have other potential health benefits, including heart disease prevention (Anthony et al., 1998), bone mass density increase to prevent osteoporosis (Anderson and Carner, 1997) and the reduction of postmenopausal syndromes in women (Knight et al., 1996).

Although isoflavones are present in several species of legumes, most of the research has been on soybean and soybean foods, due to the widespread use of soybean in traditional and modern foods (Liu, 1997; Messina, 1997). Recently, the Food and Drug Administration (FDA) approved a health claim for the relationship between the consumption of soy protein and heart disease (*Federal Register*, 1999). This approval further encourages soy isoflavone research. There may be health claims that could be made in the future to associate the intake of the isoflavonoids and the lowering of the risk of certain chronic diseases.

Excellent reviews (Knight et al., 1996; Wu et al., 1996; Anderson and Carner, 1997; Anthony et al., 1998; Cline and Hughes, 1998; Griffiths et al., 1998; Messina and Bennink, 1998; Stephens, 1999) are available on the effects of isoflavones and phytoestrogens on various types of chronic diseases, however, none have emphasized the chemistry of isoflavones and the processing effect. Many innovative patent processes have been invented recently for concentrating isoflavones or for producing isoflavone-rich proteins in soybean processing. It is timely to review these processes.

2.2 OCCURRENCE, STRUCTURE AND BIOSYNTHESIS

2.2.1 OCCURRENCE

In the plant kingdom, isoflavonoids occur primarily in the subfamily Papilion-oideae of the Leguminosae. It is still unknown why certain legumes contain isoflavones, whereas some do not. The biological functions of isoflavones in the life cycle of the plant are not well understood. There are hundreds of isoflavones in legumes (Dewick, 1994). Kaufman et al. (1997) reported the genistein and daidzein contents in vegetative parts of many legumes, including varieties such as psoralea (*Psoralea corylifolia*), kudzu (*Pueraria lobata*), fava bean (broad bean, *Vicia faba*), lupine (*Lupinus leteus*), soybean (*Glycine max*), clover (*Trifolium* spp.), peas (*Pisum sativum*), chickpea (*Cicer arietinum*), mung bean (*Phaseolus aureus*) and lima bean (*Phaseolus lunatus*). Although isoflavones are present in various other food legume seeds, the concentrations in all except chickpea are very small as compared with that in soybean. In soybean, isoflavones exist in the entire plant, including in the seeds, leaves, stems, seedlings and roots. In soybean seeds, hypocotyls have a high concentration of isoflavones (Kudou et al., 1991), although the major quantity of the isoflavones are in the cotyledons. Twelve forms of isoflavones in soybean seeds and foods have received the most attention. Unless pointed out otherwise, the discussion in this chapter will focus on isoflavones in soybean seeds and foods made from soybean seeds.

2.2.2 CHEMICAL STRUCTURES

The major difference between flavonoids and isoflavonoids is in their basic skeleton structures. Flavonoids contain a 2-phenylchroman, whereas isoflavonoids contain a 3-phenylchroman.

2-Phenylchroman   3-Phenylchroman
(Flavonoids)   (Isoflavonoids)

The structural variations of isoflavones are numerous, depending upon their substituents, oxidation levels and extra heterocyclic rings attached on the basic skeleton (Dewick, 1994). The most commonly studied of the 12 forms in soy are genistein, daidzein, and glycitein and their glucosides, 6″-O-acetyl and 6″-O-malonyl glucosides. Aside from these 12 forms, succinyl glucosides of genistein, daidzein and glycitein also have been reported (Kanaoka et al., 1998). There are also two methylated derivatives of genistein and daidzein, which are formononetin and bio-chanin A in clover, garbanzo and alfalfa, respectively. Equol and O-desmethylango-lensin are not present in seeds or foods but are metabolites of daidzein found in the urine of animals. The chemical structures of the 12 forms of isoflavones related to genistein, daidzein and glycitein as well as formononetin, biochanin A and equol are as follows:

R_1	R_2	Aglucone
H	H	Daidzein
OH	H	Genistein
H	OCH_3	Glycitein

R_1	R_2	Glucoside
H	H	Daidzin
OH	H	Genistin
H	OCH_3	Glycitin

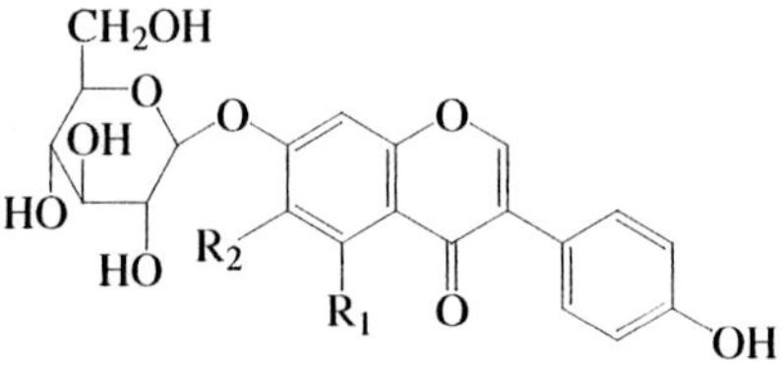

R_1	R_2	Malonyl glucoside
H	H	6″-O-malonyl daidzin
OH	H	6″-O-malonyl genistin
H	OCH_3	6″-O-malonyl glycitin

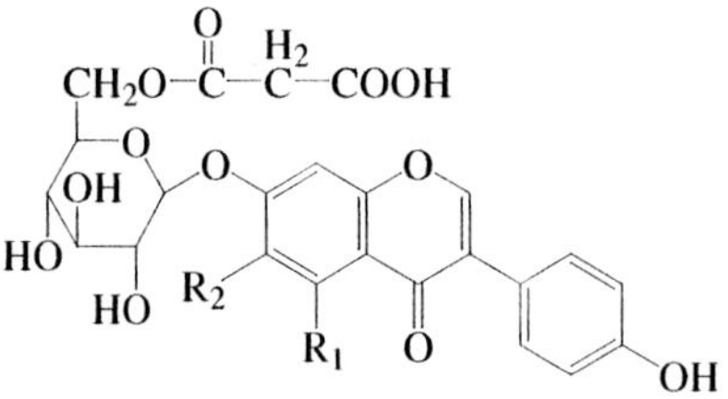

R_1	R_2	Acetyl glucoside
H	H	6″-O-acetyl daidzin
OH	H	6″-O-acetyl genistin
H	OCH_3	6″-O-acetyl glycitin

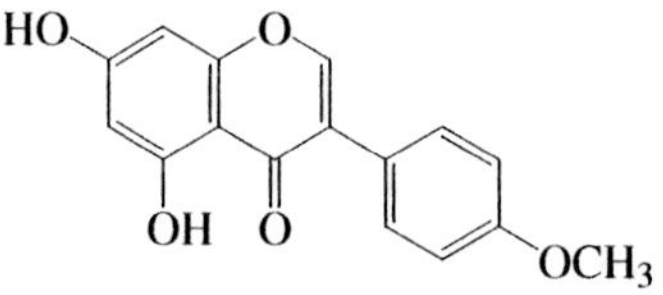

Biochanin A

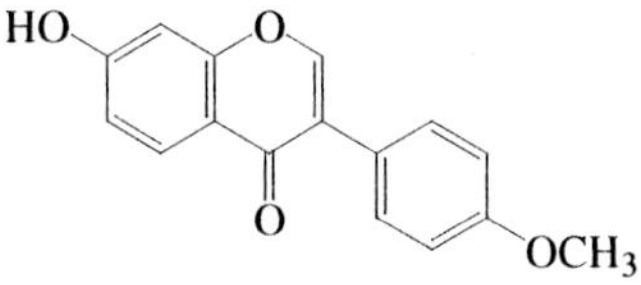

Formononetin 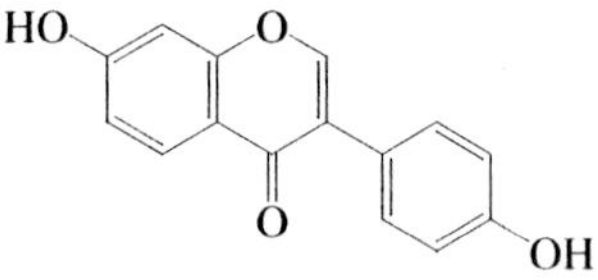   Equol

Aside from genistein, daidzein, glycitein and their derivatives, various parts other than the seeds of soybean plant also have been found to contain isoflavonoids, including coumestrol, coumestrin (coumestrol 3-O-glucoside), pterocarpans, including 6a-hydroxypterocarpan, glyceollidin I, glyceocarpin (glyceollidin II), glyceolin I, glyceolin II, glyceofuran, and 9-O-methyl glyceofuran (Dewick, 1994). Among these compounds, coumestrol's weak estrogenic activity has been studied. Coumestrol has the following structure:

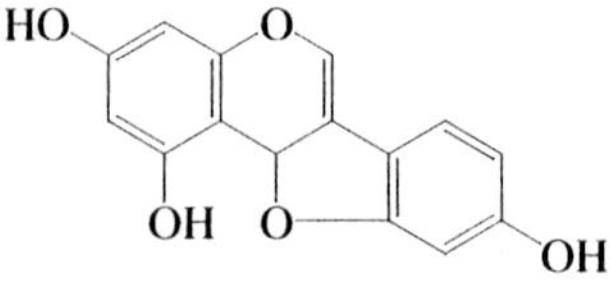

Coumestrol

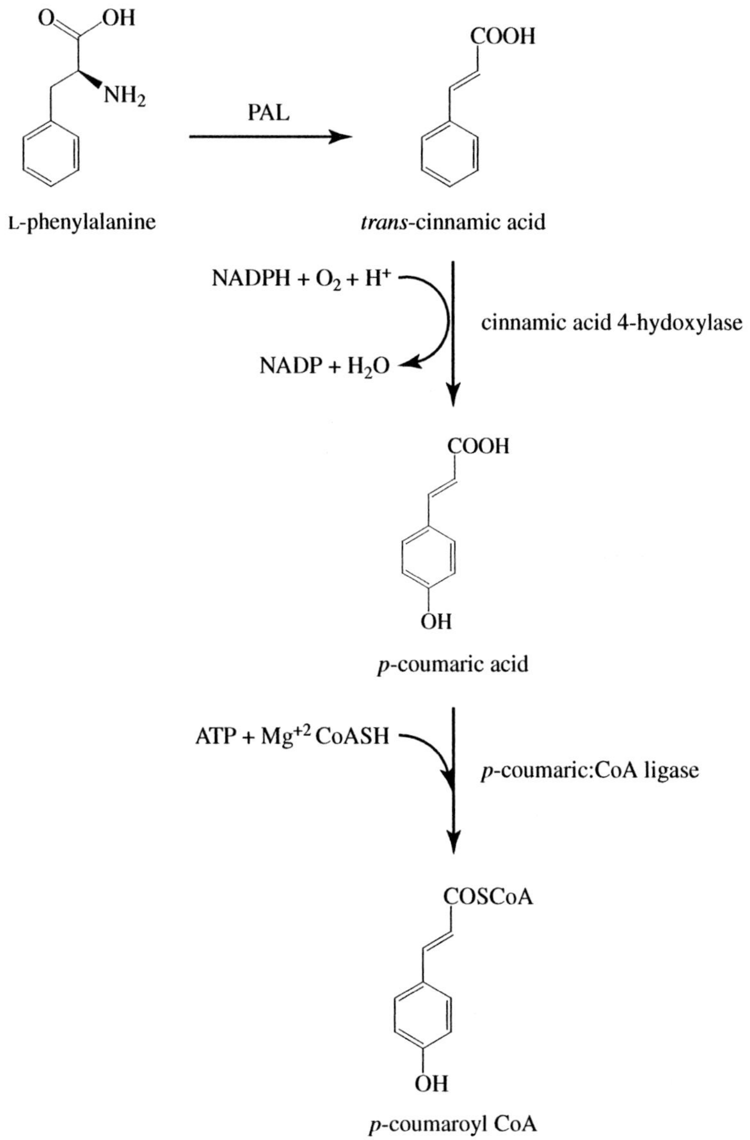

L-phenylalanine → *trans*-cinnamic acid (PAL)

NADPH + O$_2$ + H$^+$ → NADP + H$_2$O, cinnamic acid 4-hydoxylase

p-coumaric acid

ATP + Mg^{+2} CoASH → *p*-coumaric:CoA ligase

p-coumaroyl CoA

2.2.3 BIOSYNTHESIS

Biosynthesis of flavonoids starts with the conversion of phenylalanine or tyrosine to cinnamic acid by phenylalanine ammonia lyase (PAL) (Hahlbrock and Grisebach, 1975). Subsequent reactions are catalyzed by cinnamic acid 4-hydrolase to form 4-hydroxyl cinnamic acid (*p*-coumaric acid). The *p*-coumaric acid is then catalyzed by *p*-coumarate CoA ligase to form *p*-coumaroyl CoA.

In the intermediate steps of the synthesis, three malonyl CoA molecules react with one *p*-coumaroyl CoA to form chalcone, which may be isomerized to flavanone

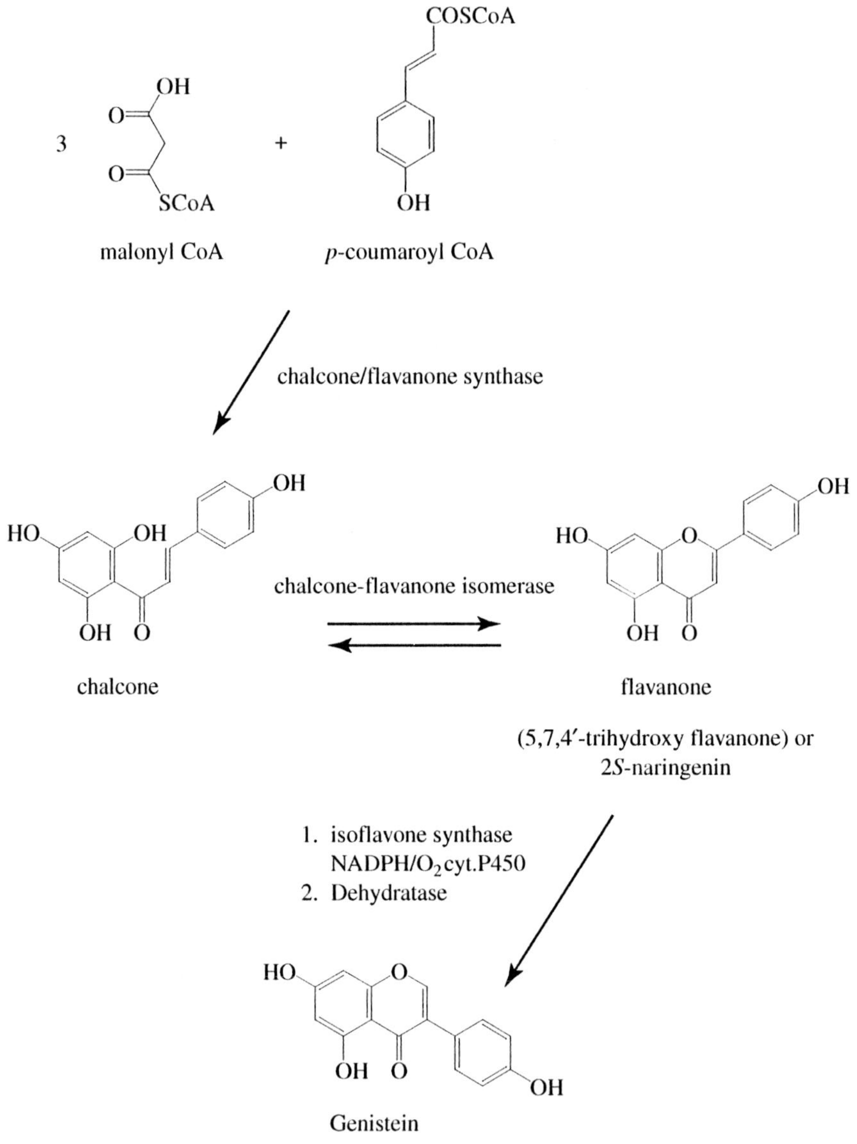

malonyl CoA   p-coumaroyl CoA

chalcone/flavanone synthase

chalcone   chalcone-flavanone isomerase   flavanone

(5,7,4′-trihydroxy flavanone) or
2S-naringenin

1. isoflavone synthase
   NADPH/O$_2$cyt.P450
2. Dehydratase

Genistein

(5,7,4′-trihydroxyl flavanone (2S-naringenin) by chalcone-flavanone isomerase (Heller and Forkmann, 1988). Naringenin may be converted to genistein through 2,3-aryl rearrangement. Similarly, daidzein is formed through aryl migration by isoflavone synthase enzymes on the substrate (2S)-liquiritigenin (Dewick, 1994).

2.3 EXTRACTION AND ANALYTICAL METHODS

Methanol has been used for the extraction of isoflavones since 1941 (Walter). The total recovery of genistin from defatted soy flakes was 0.1% (Walter, 1941). Methanol extraction also extracts sugars, phosphotides and saponins. Addition of acetone to the methanol extract precipitates some of these impurities. Genistin crystallizes by adding water to the acetone extract concentrate. The crystals are dissolved in a hot alcohol solution. Further crystallization is accomplished by cooling the hot ethanol extract to room temperature. Maximum absorption of genistin is at 262.5 nm. Absorption of genistein and genistin in the range of 220–380 nm gives similar spectrum shape and maximum absorption wavelength.

Since 1940, many researchers have extracted isoflavones with either methanol or ethanol (Ohta et al., 1979). There also have been several modifications of the extraction solvent composition to improve recoveries or to improve chromatographic separation. Reinli and Block (1996) compared analytical methods published between 1976 and 1995. Analysis of isoflavones was accomplished with gas chromatography (GC) in the early days and, more commonly, by reversed-phase high-performance liquid chromatography (HPLC) in recent years. Because some isoflavones standards are not available, photodiode array UV scanning (Wang et al., 1990; Franke et al., 1994, 1995; Song et al., 1998) and mass spectroscopy (Barnes et al., 1994; 1998) have been used to aid in the identification and quantification following GC or HPLC separation. Barnes et al. (1998) found a high correlation between the results obtained by HPLC-mass spectroscopy (MS) and HPLC-UV detection methods. Table 2.1 gives a comparison of several analytical procedures published recently.

Several studies convert isoflavones glucosides and glucoside conjugates to aglycones by acid or enzyme hydrolysis to simplify the analysis. Acid hydrolysis produces a higher yield than enzyme hydrolysis (Franke et al., 1995). One of the reasons for the conversion of glucosides to aglycones is that all glucosides are converted to aglycones either in food processing or by bacteria in the intestine and are absorbed effectively (Song et al., 1998). Only recently, analysis of all 12 forms of isoflavones can be done using a UV detector, because the standards of these forms are commercially available (LC laboratories, Woburn, MA, U.S.). Conversion of isoflavones glucosides to aglycones can be accomplished by refluxing in 1–3 N HCl for 2–4 hr (Walter, 1941; Wang et al., 1990; Franke et al., 1994, 1995). An overnight storage of the hydrolyzate at –8 to –10°C could improve resolution by GC (Fenner, 1996). Recently, an improved GC method for the analysis of daidzein and genistein was described by Parthasarathi and Fenner (1999).

Isoflavones are responsible for the slightly bitter flavor of soy isolate. The flavor of soy isolate can be improved by extracting soy flakes with an azeotropic mixture of hexane and alcohol (Eldridge et al., 1971). After extraction, the isolate has a bland taste. Acetonitrile containing HCl and water is more effective than methanol for extracting isoflavones from soybean products (Murphy, 1981). Isoflavones can be hydrolyzed by acid hydrolysis and followed by thin-layer chromatography (TLC) and GC analysis. GC analysis is performed on silylated isoflavones (Naim et al., 1974).

Murphy (1981) reported the analysis of genistin, daidzin and their aglycones and coumestrol in toasted, defatted soy flakes by HPLC without the use of external standards. Twelve extraction combinations with methanol, chloroform-methanol, acetonitrile and acetone, with or without water, and HCl were compared. The addition of water or HCl to the extraction solvents improved recoveries. Acetonitrile-HCl-water was found to co-extract less impurity. Based on the studies, three best

TABLE 2.1
Comparison of Analytical Procedures Used in Recent Publications

Reference	Method and Detection	Compounds Analyzed	Extraction Solvent	Sample Hydrolysis	Precision CV	Recovery	Detection Limit	Internal Standard
Mazur et al., 1998	GC-MS	Phytoestrogens	Diethyl ether	Enzyme	3–9%	99.5%	.02–.03 mg/kg	NA
Chandra and Nair, 1996	HPLC, UV 254 nm	Phytoestrogens	Supercritical CO_2-ethanol	None	NA[1]	93%	NA	NA
Kaufman et al., 1997	HPLC, UV 280 nm and MS	Genistein and daidzein	Acetonitrile-water	None	NA	NA	NA	NA
Song et al., 1998; Murphy et al., 1999	HPLC, UV photodiode array	12 forms of isoflavones	Acetonitrile-HCl-water	Acid	5–9%	81–99%	NA	2,4,4'-trihydroxy-benzoin (THB)
Liggins et al., 1998	GC-MS	Genistein and daidzein	Methanol-water	Acid or enzyme	NA	NA	20 pg/g	NA
Aussenac et al., 1998	Capillary electrophoresis, UV 260 nm	Aglycones and glucosides, conjugates	Methanol-water	None	NA	NA	NA	NA
Barnes et al., 1998; Coward et al., 1998	HPLC (C8)-MS, UV 262 nm	Genistein daidzein and glucosides	Acetonitrile-HCl-water or methanol-water	Acid	2–5%	NA	NA	Fluorescein
Franke et al., 1994, 1998	HPLC-UV 260 nm, photodiode array	Phytoestrogens	Methanol	Acid or none	3–8%	94–104%	1 ppm	Flavone
Tsukamoto et al., 1995	HPLC-UV 260 nm	Glucosides and conjugates	Ethanol-water-acetic acid	NA	NA	NA	NA	NA
Wang et al., 1998	HPLC-UV 254 nm	9 isoflavones	Methanol-water	NA	NA	NA	NA	NA

[1] NA means not available.

extraction systems, including acetonitrile-water, acetonitrile-HCl and acetone-HCl, were reported to yield different recoveries of genistein, daidzein and coumestrol. Farmakalidis and Murphy (1985) analyzed 6″-O-acetylgenistin and 6″-O-acetyl daidzein in toasted, defatted flakes and found that acetone with water was better than 80% methanol for extraction.

Eldridge (1982a,b) reported analysis of soybean and soy products using HPLC with an external standard method. They found 80% ethanol extraction for 4 hr was suitable for extraction of isoflavones from soybean, soy protein concentrate and protein isolate. They also found that daidzin and genistin accounted for 50–75% of the total isoflavones. Eldridge and Kwolek (1983) reported analysis using selected external standards and n-butyrophenone as an internal standard.

Jones et al. (1989) developed an HPLC-UV (262 nm) method (C18 column) for analysis of phytoestrogens and coumestrol. Ethyl benzoate was used as an internal standard. The recovery rate was from 90% to 105% for daidzin and daidzein. The detection limit was approximately 20 ng based on a signal-to-noise ratio of 3:1.

Pettersson and Kiessling (1984) used C18 Sep-pak to clean the extract from animal feed and used an isocratic elution to separate five phytoestrogens, including daidzein, genistein, formononetin, biochanin A and coumestrol. The method gave a sensitivity of 2.5 ppm for coumestrol and 10 ppm for isoflavones. Differences from normal extraction and after-acid hydrolysis were used to estimate glucosides. Extraction of phytoestrogens from silage with acetonitrile-HCl gave less yield than that with 75% ethanol. Ha et al. (1992) used C18 HPLC to analyze extract from defatted soybean. A linear gradient with methanol and detection with UV 280 nm were used for elution and quantification, respectively.

Wang and Murphy (1994a) analyzed isoflavones in commercial soy foods by HPLC using an external standard method. Song et al. (1998) extracted isoflavones from soybean and soy foods using an acetonitrile-HCl-water mixture and separated 12 forms of isoflavones by HPLC and quantitated them using isoflavone standards and an internal standard 2,4,4′-trihydroxy benzoin (THB). HPLC was performed on a YMC-Pak ODS-AM-303 column and eluted with a linear gradient of 0.1% glacial acetic acid and 0.1% glacial acetic acid in acetonitrile. A photodiode array detector was used to monitor spectrum of peaks from 200 to 350 nm. They reported that the method took approximately 60 min. For malonyl and acetyl glucosides without pure standards, they used the standard curves of aglycones and adjusted the standard curves on the basis of differences in molecular weight. Glucosides are approximately twice the weight of aglycones. The authors stated that glucosides should be normalized to aglycones mass rather than a simple sum of all forms for reporting the isoflavone concentrations.

In a study of retail soy foods, Murphy et al. (1999) optimized extraction of isoflavones and found that the amount of water included in the mixture of acetonitrile-0.1 N HCl-water affected the yield of isoflavones in various foods matrices. Two grams of samples were extracted with 10 ml acetonitrile, 2 ml of 0.1 N HCl and an amount of water. The optimal amount of water ranged from 5 to 10 ml, depending upon the type of food matrix. The standard deviation of the optimized extractions ranged from 0.6% to 4.8% for total daidzein and 0.2–3.5% for genistein analysis. HPLC with a C18 column was used to separate the isoflavones, which were monitored by a photodiode array detector (Murphy et al., 1999). The recovery rates were in the range of 81–92%, 87–95%, 94–99% and 94–98% for daidzein, genistein, glycitein and internal standard THB, respectively. The coefficient of variations from the same day or day-to-day analysis of individual isoflavones and total isoflavones ranged from 2% to 4.5% and 3.7% to 11%, respectively. Genistin analysis had the highest variation.

Chandra and Nair (1996) used supercritical carbon dioxide fluid extraction for extracting daidzein, genistein, biochanin A and formononetin from soy products. The separation was done by HPLC with a C18 column, followed by UV detection at 254 nm. Ninety-three percent of the isoflavones were extracted by using the super-critical fluid carbon dioxide with 20% ethanol at 50°C and 600 atmospheric pressure.

Franke et al. (1994, 1995, 1998) developed a rapid HPLC method for measuring phytoestrogens (daidzein, genistein, coumestrol, formononetin, biochanin A, equol and O-desmethyl angolensin) from legumes and human urine. Extraction was done with approximately 80% methanol with or without acid hydrolysis. The extract was applied to a NovaPak C18 HPLC with eluents consisting of a gradient of acetonitrile and acetic acid/water. The chromatographic run was completed in 20 min for food samples and 24 min for urine samples. The effluent was monitored at 260 nm for all isoflavones except 280 nm for equol and 342 for coumestrol. Flavone was used as an internal standard. The range of detection limits was 5–25 nm (less than 1 ppm) for most phytoestrogens, whereas equol and O-desmethyl angolensin had detection limits of 600–700 nm. The rapid HPLC method had good precision and accuracy. The recoveries of all phytoestrogens in urine after a solid reversed-phase extraction were excellent, ranging from 94% to 104%.

Barnes et al. (1994, 1998) and Coward et al. (1998a) analyzed isoflavones and their glucoside conjugates in soy foods and in human physiological fluid and tissue samples, and in cooked food samples using HPLC-MS. Extraction of isoflavones with 80% aqueous methanol at room temperature was as efficient as extraction at 60–80°C. High-temperature extraction caused decomposition of acetyl and malonyl glucosides to the glucoside forms, therefore changing the composition of isoflavones. They compared 80% aqueous methanol extraction with 80% acetonitrile-0.1% HCl. An Aquopore C8 column was used for the HPLC separation using fluorescein as the internal standard. Detection and quantification were at 262 nm. A heated nebu-lizer-atmospheric pressure chemical ionization apparatus was interfaced with the HPLC to generate positive ion mass spectrum, which gave structural information for each conjugate. They found no differences in the results between these two extraction methods. The correlation between these two methods was high.

Wang et al. (1990) used acetonitrile-water to extract four isoflavones and coumestrol from soybean and processed soy products. Reversed-phase C18 HPLC followed by UV or fluorescence detection were used to separate and quantitate isoflavones in 20 min. Peaks were identified by retention time and standard addition. The method was sensitive to 2 ppm with UV detection at 254 nm for isoflavones and was sensitive to 0.5 ppm for coumestrol detection (using the fluorescent detector with excitation at 365 nm and emission at 418 nm). Recoveries of isoflavones and coumestrol ranged from 75% to 110% in spiked samples. Coefficients of variation for three soy products were below 9%. They compared three extraction solvents in the volume ratios of 80:0.5:19.5 of acetonitrile:HCl:water, methanol:HCl:water and ethanol:HCl:water and found that extraction efficiencies were similar. However, they stated that acetonitrile-HCl-water was preferred, because it resulted in a fast settling of sample particles and impurities, and cleaner separation of the peaks in the chromatograms. They also compared hydrolysis of glucosides into aglycones by various hydrolysis procedures and concluded that 1 N HCl at 98–100°C for 2 h was the most suitable for total isoflavone analysis.

Liggins et al. (1998) used 80% aqueous methanol to extract isoflavones. They found that partitioning genistein and daidzein into hexane used for lipid extraction was dependent upon concentration of isoflavone, and they concluded that lipid extraction of the 80% ethanol extract was not needed. Also, they tested enzymes from three sources for hydrolyzing glucosides and concluded that cellulase from *Aspergillus niger* was the most suitable as compared to 3 N HCl hydrolysis. A GC-MS method was used to validate analysis over a wide range of concentrations as might be present in processed foods containing soy products. The aglycones resulting from the hydrolysis were derivatized with *N-tert*-(butyl dimethyl silyl) *N*-methyl trifluoro-actamide. The derivatives were injected into GC-MS for quantification and compared to that derivatized from synthetic daidzein and genistein standards. A detection limit as low as 20 pg/g soy food was reported.

Mazur et al. (1998) reported a method for the extraction of phytoestrogens, including formononetin, biochanin A, daidzein, genistein and coumestrol, with diethyl ether after enzymatic hydrolysis. Separation and quantification were on silylated derivatives using isotope dilution GC-MS equipped with selective ion mode. The method was several times more sensitive than HPLC-UV methods. Several foods that contained very low amounts of phytoestrogens, which could not be detected by HPLC-UV methods, were detected by this method. The coefficient of variation was between 3% and 9% for concentrations in the range of 0.05–0.13 mg/kg. The recovery was 99.5%, and the detection limit was 0.02–0.03 mg/kg (ppm). The isoflavone contents of four soybean cultivars were 18–121 µg, 0–15 µg, 12,500–56,000 µg, 26,000–84,000 µg and 0–185 µg/100 g for formononetin, biochanin A, daidzein, genistein and coumestrol, respectively. Soybeans also were found to contain lignan (13–273 µg/100 g).

Aussenac et al. (1998) used capillary zone electrophoresis with UV detection at 260 nm to analyze isoflavones in soybean seeds of various varieties grown in various locations. Methanol was used for extraction. Total extraction was not affected by temperature but was affected by the composition of the solvent. Electrophoresis was conducted at pH 10.5, at which the isoflavones were weak acids and were ionized. Boric acid was added to form a negatively charged borate-isoflavone complex. A fast capillary electrophoresis method was also developed by Vanttinen and Moravcova (1999) to determine daidzein and genistein after enzyme hydrolysis in soy products. Photodiode array was used to detect the isoflavones at 254 and 268 nm, respectively. Minimum detection was 0.4 mg/L. *p*-Nitrophenol was used as an internal standard.

2.4 CONTENT IN SOYBEAN AND SOY FOODS

Many factors contribute to the variability of isoflavone contents in foods, including raw material characteristics, extraction and analysis procedures. Soybeans are affected by cultivar, location and year of cultivation. Variabilities due to analytical procedures are affected by the availability of standards, use of internal standards, instrumentation and extraction methods. Although there are 12 commonly studied isoflavones of genistein, daidzein and glycitein and their glucosides, very few studies reported all of these forms due to a lack of several commercial malonyl and acetyl conjugate standards. In addition, some early studies used only external standards for quantification, because internal standards were not available. Many extraction procedures have been used. Recoveries of standards in the food matrix have not been reported in some studies. Solvents containing HCl seem to give high recoveries, though methanol or ethanol have been implicated as effective, because they contain HCl in certain instances. To be practical, due to a lack of certain commercial standards, several studies converted the glucosides to aglycones. A database with 38 references has been established as the result of a collaboration between Dr. Patricia Murphy's lab at the Iowa State University and USDA (USDA-Iowa State University Isoflavone Database (website: www.nal.usda.gov/fnic/foodcomp/Data/isflav.html).

In the database, glucoside contents were converted to aglycones by using appropriate molecular weight adjustments. It should be noted that the physiological significance of total content may be different from the individual contents. Although glucosides of isoflavones are readily hydrolyzed by the intestinal bacteria enzymes, Setchell (1998) stated that little is known about the biological activity of individual isoflavones. The mean values of the total isoflavone content in selected soy foods as reported in the USDA-ISU database are summarized in Table 2.2. These mean values are summarized from the results of various labs using different analytical methods. For detailed information and references associated with the values, readers are recommended to consult with the database on the website.

It appears that glucoside isoflavones and their malonyl and acetyl conjugates are hydrolyzed by intestinal bacteria to aglycones prior to being absorbed. However, not everyone has the same intestinal microflora ability to hydrolyze the glucosides. A study (Xu et al., 1995) shows that two of six women studied were not able to hydrolyze the glucosides. Therefore, it is important to analyze the individual forms of isoflavones instead of total isoflavones in the foods, because foods vary widely in aglycone contents.

In general, most isoflavones in soybean are in the form of glucosides or malonyl glucoside conjugates of daidzein and genistein. Although total glycitein contributed only 5–10% of the total amount, the soy germ or hypocotyl contains a high proportion of glycitein conjugates (Kudou et al., 1991). Aglycones in nonfermented soybean and soy foods account for only a very small fraction of the total isoflavone content (Murphy et al., 1999). Aglycones in fermented soy foods are the major forms of the isoflavones (Fukutake et al., 1996). Table 2.3 shows the 12 forms of isoflavones in selected soybean and soy foods.

Reinli and Block (1996) developed a compendium of literature values with 36 references for phytoestrogens (coumestrol, daidzein, genistein, biochanin A and formononetin) in foods. They also reported the relative potency of these phytoestrogen and equol (metabolites found in urine) *in vitro* and *in vivo*. Similar to the USDA-ISU database, this database reported total daidzein and genistein, which were estimated by the hydrolysis method or by normalizing the molecular weight differences of

TABLE 2.2
Summary of the Mean Values of the Total Isoflavones in Soybean and Selected Foods Containing Soybean Products

Food	Total Isoflavones, mg/100 g Edible Portion
Meatless bacon	12.1
Soy hot dog	15.0
Meatless chicken nuggets, canned, cooked	14.6
Meatless chicken nuggets, canned, raw	12.2
Vegetable protein burger	8.2–9.3
Infant formula of various brands	2.6–26.0
Instant soy beverage, powder	109.5
Miso	42.5
Miso, dry	60.4
Natto, fermented	58.9
Soy cheese of various types	6.4–31.3
Soy fiber	44.4
Soy flour, defatted	131.2
Soy flour, textured	148.6
Soy flour, full-fat	177.9
Soy flour, full-fat, roasted	198.9
Soy meal, defatted	125.8
Soy milk	9.7
Soy film, yuba, Foo-jook, raw	193.9
Soy film, yuba, Foo-jook, cooked	50.7
Soy protein concentrate, aqueous wash	102.1
Soy protein concentrate, alcohol wash	12.5
Soy isolate	97.4
Soy sauce, shoyu from soybean and wheat	1.64
Soybean from countries other than the United States	59.8–145.0
Soybean flakes, dafatted	125.82
Soybean flakes, full-fat	129.0
Soybean immature, raw	20.4
Soybean, mature seeds, raw (U.S. food quality)	128.35
Soybean, mature seeds, raw (U.S. commodity grade)	153.4
Soybean, mature seeds, sprouted, raw	40.7
Tempeh	53.52
Tofu, fresh, soft, firm or extra firm of various brands	22.6–31.1
Tofu, frozen-dried, Kori	67.5
Tofu, fried	48.4
Tofu, pressed, semi-dry, Tau kwa	29.5

Source: Data taken from USDA-ISU Isoflavone Database (website: www.nal.usda.gov/fnic/foodcomp/Data/isflav.html).

TABLE 2.3
Isoflavone Content of Selected Soybean and Commercial Soy Products (µg/g as is Basis)

Soy Sample	Daidzin	Genistin	Glycitin	Malonyl Daidzin	Malonyl Genistin	Malonyl Glycitin	Acetyl Daidzin	Acetyl Genistin	Acetyl Glycitin	Daidzein	Genistein	Glycitein
Vinton 81 90H	690	852	56	300	743	50	1	9	nd	26	29	20
Soy flour	147	407	41	261	1023	57	trace	1	32	4	22	19
Soy isolate B	88	301	49	18	88	36	74	215	46	11	36	25
Tofu	25	84	8	159	108	nd[1]	8	1	29	46	52	12
Tempeh	2	65	14	255	104	nd	11	nd	nd	137	193	24
Hozukuri miso	72	123	18	nd	nd	22	1	11	nd	34	93	15
Fermented bean curd	nd	trace	nd	nd	nd	nd	nd	nd	nd	143	223	23
Soybean hypocotyl	3200	1180	4850	4230	1440	4450	20	1050	1050	1020	35	nd
Soybean cotyledon	450	800	nd	700	1170	nd	20	10	nd	330	480	nd

[1] nd means not detected.

Source: The soybean hypocotyl and cotyledon data were taken from Kudou et al. (1991). All other data represented the commercial products by Wang and Murphy (1994a).

glucosides and adding the normalized values to the aglycone values in the literature. The data of the compendium have been included in the USDA-ISU database.

2.5 CULTIVAR, ENVIRONMENT AND PROCESSING

2.5.1 CULTIVAR AND ENVIRONMENTAL EFFECT

Several researchers reported the effects of soybean cultivar, crop year and location (Wang and Murphy, 1994b; Mazur et al., 1998; Hoeck et al., 2000). Planting date (Aussenac et al., 1998) and climate temperature (Tsukamoto et al., 1995) also have been reported to contribute significantly to the differences in isoflavone contents. Although there are significant differences in the isoflavone profiles and total quantity among various cultivars grown in various environments, the majority of the isoflavones are in the forms of glucosides and malonyl glucosides. The aglycones and acetyl glucosides are the minor components. Several Japanese cultivars have higher ratios of malonyl isoflavones to glucosides (Wang and Murphy, 1994b) than American cultivars. Seeds harvested after cultivating in a high-temperature climate had a significant decrease in isoflavone content (Tsukamoto et al., 1995). Although hypocotyl has a high concentration, the majority of isoflavones are in the cotyledons after the weights of the hypocotyl and the cotyledons are calculated.

2.5.2 EFFECT OF ENZYMES ON ISOFLAVONE FORMS AND FLAVOR OF SOYBEAN PRODUCTS

Kudou et al. (1991) determined the bitterness taste threshold values for the 12 forms of isoflavones. Generally, the threshold values are in the order of malonyl glucosides < acetyl glucosides = aglycones < glucosides. Therefore, in the production of soy milk, measures need to be taken to reduce the malonyl forms to improve the taste of the product. Okubo et al. (1992) also reported that genistein and daidzein had higher bitterness intensities than the respective glucosides. Although saponins in soybean also contribute to bitterness and astringency, the mechanism of the undesirable taste caused by saponin was different from that caused by isoflavones.

During the production of soy milk, the natural β-glucosidase present in soybean may convert some glucosides to the aglycone forms to contribute to an increase of objectionable flavor (Matsuura et al., 1989). The enzymes should be inactivated as rapidly as possible during or after soy milk extraction to improve the quality of soy milk.

Ha et al. (1992) found that soaking in water or in 0.25% sodium bicarbonate at 50°C promoted the production of aglycones due to β-glucosidase hydrolysis and promoted the production of volatiles due to lipoxygenase-catalyzed peroxidation. Soaking in boiling water containing sodium bicarbonate inhibited the production of aglycones and the undesirable volatile flavors.

2.5.3 EFFECT OF CONVENTIONAL PROTEIN CONCENTRATION AND ISOLATION, AND TRADITIONAL FOOD PROCESSING

Eldridge (1982a) determined isoflavones in defatted soy flour, protein concentrate and isolate, and found that the total isoflavone decreased from defatted flour to concentrate and isolate. Glucosides daidzin and genistin account for 50–75% of the total isoflavones in the soy flour. Alcohol wash during the production of protein concentrate removed most of the isoflavones in the protein products. Aqueous washing had little effect on isoflavones. Approximately 50% of the isoflavones were lost during the manufacturing of soy isolate. Alcohol-treated protein concentrate had low isoflavones as reported by Coward et al. (1993) and Wang and Murphy (1994a).

Wang and Murphy (1996) reported the mass balance of isoflavones and found that soy isolate production lost 47% of the total isoflavones. Wang et al. (1998) reported that only 26% of original soy flour isoflavones were retained in the soy isolate. During three major steps (extraction at an alkaline pH, acid precipitation and aqueous washing) of soy protein isolate production, the losses of the total isoflavones were 19%, 14% and 22%, respectively. The isolate has a different isoflavone profile as compared with the soy flour. The isolate had more aglycones than the soy flour; therefore, alkaline hydrolysis of glucosides may have occurred during processing (Wang and Murphy, 1996), or the aglycones may bind to the proteins more firmly than the glucosides. Differences between these results and those of Eldridge (1982a) may in part be due to the use of different isoflavone methodologies. Wang et al. (1998) and Wang and Murphy (1996) included malonyl gluconsides in the total amount, whereas malonyl glucosides were not discovered in 1982 by the procedures used by Eldridge (1982a). Another difference is in sample preparation. Eldridge obtained all protein concentrate and isolate samples from commercial sources, and the methods for preparation were not known.

On a dry-weight basis, Coward et al. (1993) reported that Asian and American soy products, with the exception of alcohol-washed soy protein concentrate and isolates, have total isoflavone contents (aglycones + glucosides) similar to those of the whole soybean. Wang and Murphy (1996) reported the loss of isoflavones during the manufacturing of tempeh, soy milk and tofu, and protein isolate. Soaking and heating in the tempeh production caused losses of 12% and 49% of total isoflavone, respectively. Only 24% of the isoflavones were retained in the final tempeh product. From processing soybean to soy milk, there was not a significant loss of isoflavones. However, from processing soybean to tofu, only 33% of the total isofavones were retained. The coagulation step followed by whey separation during tofu-making reduced isoflavone content by 44%. Second-generation soy foods, including soy burger, hot dog, bacon, yogurt, cheese and noodles, contained only 6–20% of the isoflavones of the whole soybeans. However, total isoflavones and the individual isoflavone profiles were affected by soybean variety, processing method and other ingredient addition.

Fermentation of tempeh greatly increased the levels of the aglycones, daidzein and genistein, through β-glucosidase hydrolysis. Similar findings have also been found by others (Gyorgy et al., 1964; Murakami et al., 1984; Coward et al., 1993; Wang and Murphy, 1994a, 1996; Kaufman et al., 1997; Murphy et al., 1999).

In 1964, Gyorgy and coworkers isolated daidzein and genistein along with 6,7,4'-trihydroxyisoflavone from tempeh and found that they possessed antioxidant activities. The activity of 6,7,4'-trihydroxyisoflavone was particularly effective. An extract of the soybean without fermentation had very little activity. However, acid hydrolysis with HCl increased the soybean extract to that of tempeh. They assumed that daidzein, genistein and the 6,7,4'-trihydroxyisoflavone were produced by the metabolism of the fungi in the tempeh.

Murakami et al. (1984) compared isoflavone patterns of soymeal and tempeh, a fermented soy food obtained by fermenting cooked soybeans for 40 h with *Rhizopus oligosporus*, and found that tempeh has only aglycones and no glucosides, which were prominent in soymeal. Therefore, fungal fermentation produced β-glucosidases to hydrolyze the glucosides to aglycones. The high levels of aglycones contributed to antioxidant activity to increase the shelf life of tempeh. The production of 6,7,4'-trihydroxyisoflavone was dependent upon the type of microorganisms used for the fermentation of soybean. *Rhizopus* was reported to have no ability to produce this compound. Some microorganisms for tempeh production could (Klus et al., 1993). *Brevibacterium epidermidis* and *Micrococcus luteus* converted glycitein to this compound by *O*-demethylation. *Microbacterium aborescens* converted daidzein through hydroxylation. Klus and Barz (1995) reported that *Micrococcus* and *Arthrobacter* were able to convert glyceitin and daidzein to 6,7,4'-trihydroxylisoflavone and 7,8,4'-trihydroxyisoflavone. Klus and Barz (1998) also reported hydroxylation of biochanin A and genistein at 6 and 8 positions, respectively, by *Arthrobacter* or *Micrococcus* species from tempeh. A new antioxidant, 3-hydroxyanthranilic acid, was isolated from tempeh fermented with *Rhizopus oligosporus* (Esaki et al., 1996). This new antioxidant could prevent the auto-oxidation of soybean oil and soy powder, whereas the 6,7,4'-trihydroxyisoflavone could not.

Six patents invented by Zilliken (1980a,b, 1981, 1982a,b, 1983) reported the production of the antioxidants ergostadientriol and isoflavones. Isoflavones were extracted from dried tempeh powder or cultured fungus (*Rhizopus oryzae* or *R. oligosporus*) with 60–70% methanol. Among the isoflavones, texasin (6,7-dihydroxy-4'-methoxyisoflavone) was the most effective antioxidant. Other isoflavones included genistein, daidzein, glycitein and the 6,7,4'-trihydroxyisoflavone, which, contrary to the report of Klus and Barz (1995), also was produced by the *Rhizopus* species. The isoflavones in the methanol extract were purified by molecular sieve and silica gel chromatography.

Aside from the fermentation of tempeh, strong antioxidants, *o*-dihydroxyisoflavones (8-hydroxydaidzein and 8-dihydroxygenistein), have been reported to result from fermentation of soybean with *Aspergillus saitoi* (Esaki et al., 1999). Fermentation with koji starter molds such as *Aspergillus* and *Rhizopus* species was reported (Takebe et al., 1999) to produce various enzymes including β-glucosidases, phytase, phosphotase and protease. Because of phytase, removal of phytic acid proceeded in parallel with the hydrolysis of isoflavone glucosides. Protein hydrolysis also occurred simultaneously. Therefore, in addition to high aglycones in the final product, the content of phytic acid was reduced, and the proteins were partially hydrolyzed with improved digestibility. Production of soy cheese with *Lactobacillus casei* has also been reported to produce enzymes to convert glucosides to aglycones (Matsuda et al., 1992).

Kaufman et al. (1997) concluded that sprouting of edible beans increased aglycones levels. Legume sprouts also contain high levels of vitamin C and more soluble proteins. The sprouts are routinely used in the Asian diet and should be widely used in other diets to achieve higher nutritional benefit. The increase in daidzein, coumestrol and, to a lesser extent, genistein during a 10-day germination of soybeans was reported by Wang et al. (1990).

2.5.4 Effect of Cooking/Heating (Hot Water Extraction, Frying, Toasting and Baking) on Isoflavone Structures

Heat treatment involved in the chemical analysis of isoflavones may affect their structure. Kudou et al. (1991) separated the soybean into three fractions (seed coat, hypocotyl and the cotyledons) and extracted isoflavones using 70% aqueous ethanol at room temperature for 24 h or at 80°C for 15 h. Total isoflavone concentration in hypocotyl was 5.5 to 6 times higher than that in the cotyledon. Glycitin and its derivatives occurred only in the hypocotyl fraction. Most of the malonyl isoflavones were converted to respective glucosides after extraction at 80°C for 15 h.

Chemical structures of isoflavones can also be changed during food preparation (Coward et al., 1998a). Hot aqueous extraction as well as hot extraction in the making of soy milk and tofu converted some malonyl glucoside to β-glucoside. Toasting (dry heat) of soybean converts malonyl forms to acetyl glucosides. Production of low-fat soy milk and low-fat tofu reduced total isoflavones by 57% and 88%, respectively. Frying converted some malonyl glucosides to glucosides, acetyl glucosides and aglycones. Cookies baking for 7.5 min doubled the amount of glucosides, and 15 min of baking produced more glucosides, acetyl glucosides and aglycones. As food was burnt, total isoflavone decreased with an increase in aglycones.

2.5.5 Effect of Acid and Base Treatment

Glucosides (simple, malonyl and acetyl forms) of isoflavones can be hydrolyzed by acid treatment. Wang et al. (1990) compared hydrolysis of glucosides under three concentrations, including 1, 2 and 3 N HCl, at 98–100°C for various periods of times. Conversion of glucosides to aglycone by acid treatment was the best at 1 N HCl for 2 h. A higher concentration of acid than 1 N HCl will decrease the aglycone content beyond 2 h of hydrolysis. Daidzein was found to be more stable than genistein against boiling acid treatment. After boiling for 2 h in 1 N HCl, genistein concentration declined sharply. This may be due to structural degradation of genistein by the acid at a high temperature.

Alkaline treatments at a high temperature cause the breakdown of the malonyl or acetyl ester linkages. Several processes discussed in the following section utilize this principle to convert the malonyl conjugates to glucoside isoflavones. A summary of processing effects on chemical structures of isoflavones is illustrated in Figure 2.1 using 6″-O-malonyl genistin as an example.

2.6 INNOVATIVE PROCESSES FOR ENRICHMENT AND CONCENTRATION

Numerous patented processes have been designed recently for isoflavone enrichment and concentration and for applications in nutritional and pharmaceutical products.

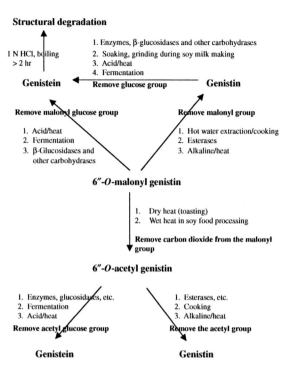

FIGURE 2.1 Summary of the chemical reactions of isoflavones that may occur during food processing as affected by acid, heat, enzymes and fermentation. Genistein and their glucosides are used for illustration.

This section describes the new methods for manufacturing the enriched or concentrated isoflavone products.

2.6.1 ISOFLAVONE OR AGLYCONE-ENRICHED PROTEIN CONCENTRATE, MOLASSES, FIBER AND PROTEIN ISOLATE

2.6.1.1 Concentrate

Protein Technologies International (PTI, St. Louis, MO, U.S.) obtained two patents (Shen and Bryan, 1995a, and 1997a; Bryan and Guevara, 1999) on the isoflavone enrichment of soy protein concentrate. The protein concentrate after removing soluble carbohydrates by pH 4.5 washing was suspended in water and reacted with β-glucosidases or esterases (pH 4–8) at 40–60°C and dried. Alternatively, defatted soy flakes could be treated with enzymes (including those naturally present in soybean or added) at pH 4.5 and followed by centrifugation to produce a aglycone-rich protein concentrate. The precipitate was dried with minimal washing to retain isoflavones. The final protein concentrate contained 1–2 mg/g genistein and 0.7–1.5 mg daidzein on a dry-weight basis. Many commercial enzymes were listed in the patent, including pectinases and lactases.

2.6.1.2 Molasses

Three patents are owned by PTI for the recovery of isoflavones from soy molasses (Waggle and Bryan, 1988, 1999a,b). In the first step, the molasses was heated at pH 6–13.5 over a range of temperature from 2 to 120°C for various periods of time to hydrolyze malonyl or acetyl glucosides to isoflavone glucosides. In the second step, the material from alkaline hydrolysis was subjected to enzymes to cleave the β-1,4 glycosidic bond to remove the glucose unit to produce aglycones. Such enzymes include saccharidase, glucoamylase, pectinase or lactase. The isoflavone-enriched, glucoside-enriched, aglycone-enriched molasses cakes were separated from the reaction slurries by adjusting to pH 4.5 followed by cooling and centrifugation or filtration.

2.6.1.3 Soy Fiber

Shen (1994a,b) treated protein extract slurry with β-glucosidase, pH 6–8, to produce aglycones, then separated the soluble protein extract with the insoluble spent flakes, which was an aglycone-enriched soy fiber. The dried fiber fraction contained 1–2 mg genistein/g and 0.7–1.7 mg soy fiber/g.

2.6.1.4 Protein Extract and Isolate

PTI obtained six patents on the production of isoflavone- or aglycone-enriched protein extract and isolate (Shen and Bryan, 1995b, 1998, 2000; Bryan and Allred, 1998; Shen et al., 2000). In general, for the production of isoflavone-rich protein isolate, the soy flakes were extracted at an alkaline pH (pH 6–10), were precipitated at pH 4.5 and were minimally washed, cooled at 40–80°F and held for 1 hr before separation. For the production of aglycone-enriched protein extract or isolate, the extract is subject to the conversion of malonyl-glucosides by heating at an alkaline pH, and the conversion of glucosides to aglycones occurs by reacting with glucosidases or esterases. Bryan et al. (1998) described a patent process involving two-step conversions of the malonyl forms to aglycones. The protein extraction at an alkaline pH could be carried out either before or after the conversion of malonyl forms to glucosides or the conversion of the glucosides to aglycones. In all of these processes, and no or minimum washing of the protein precipitate was used to retain isoflavones in the soy isolate. Based upon the total isoflavones in the raw materials, as high as 80% recovery in the isolate could be achieved. The product contained 1.5–3.5 mg daidzein and 1–3 mg genistein per g protein isolate.

A patent owned by E.I. du Pont (Crank and Kerr, 1999) described procedures for preparing isoflavone-rich soy protein products with desirable flavor and functionality. A low-stachyose soybean variety was dehulled, flaked (0.25–0.6 mm), defatted by hexane and desolventized by vacuum to yield a 90 PDI (protein dispersibility index) of soymeal. A water extraction at 80°C was carried out for 6–10 min, and it was then cooled to 40–45°C by adding more water. The extract after centrifugation was neutralized and spray-dried. The product contained 2.5 mg/g isoflavone.

2.6.1.5 Whey

PTI obtained three patents for producing aglycone-enriched protein whey and whey proteins (Shen and Bryan, 1995c, 1997b; Shen et al., 1998). Isoflavone glucosides and malonyl and acetyl conjugates were converted by enzymes (glucosidases and esterases) at pH 4–8 or by acid hydrolysis at pH 1–2, 80–90°C for 30–180 min or at pH 4.5 at 50°C for 24 hr. The whey proteins were precipitated by heating at 95°C for 1 min. The resulting whey contains 50–80% of that in the original protein materials, and the whey protein product contains 81% genistein and 69% daidzein of the defatted soy flour.

2.6.2 ISOLATION AND CONCENTRATION OF ISOFLAVONES FROM SOY MATERIALS

Iwamura (1984) designed a process for isolating saponins and flavonoids from leguminous plants, including seed, leaf, root, stem, stalk, flower or the entire plant. Alkaline extract, which was clarified by acid precipitation to get rid of proteins, followed by filtration or centrifugation, was applied to a nonpolar resin (styrene-vinyl benzene) or a slightly polar resin (such as acrylic ester resin). Water was used to elute soluble sugars, then the column was eluted with methanol to obtain a mixture of saponins and flavonoids. These two compounds were separated by acetone extraction to yield isoflavones in the soluble fraction.

Fleury et al. (1992) designed a process for obtaining genistin malonate and daidzin malonate from defatted soybean by ethanol or methanol extraction. The extract at pH 6–9 (presumably the malonyl carboxylic group is the form of R-COO⁻ and is more polar) was then extracted with a water-immiscible solvent such as butanol or ethyl acetate. The aqueous phase was adjusted to pH 2–5.4 (presumably in the ROOH form and is less polar) and then extracted with an organic solvent. The organic phase was recovered and neutralized to yield the malonate product. Further purification was accomplished with gel filtration using Sephadex LH-20, adsorption chromatography or C18 reversed-phase chromatography using 10% ethanol as eluent to separate the two malonates. Using the Rancimat method, malonyl glucosides were shown to have antioxidant activities. Malonate glucosides were decomposed by alkaline treatment to malonic acid and glucoside isoflavones.

In another patent process, acetone was used to extract genistein from soymeal (Day, 1999). The extract was evaporated and mixed with ice water to precipitate genistein and daidzein. The genistein-rich precipitate was removed by centrifugation.

Weber et al. (1996) and Hessler et al. (1997) reported a process for preparing genistein from the fermentation of soy-based substrates by *Saccharopolyspora erythraea*, which produced a mixture of genistein and erythromycin. The mixture was extracted at a high pH (9–10) using organic solvent ethyl acetate. After that, an acidified water extraction removed erythromycin. The ethyl acetate fraction contained genistein. Five kg of genistein could be produced from 2000 kg of soybean. The solubility of genistein was affected by pH. The R-OH groups may lose H⁺ at alkaline condition. Above pH 10, genistein was water soluble and insoluble in organic solvent. Acid neutralized the pHs and converted genistein to an un-ionized form that was soluble in organic solvent, such as ethyl acetate.

Differing from the hydrophobic resin separation of isoflavones, Chaihorsky (1997) claimed a process for concentrating isoflavone glucosides on a cationic column using a sulfonic polystyrene resin. The elution was done using 80% methanol. The author believed that the binding of 7-glucosyl isoflavones to the sulfonic group at an alkaline condition occurred at the 7 position of the 7-glucosyl isoflavones.

Isoflavones also have been purified using a polymethacrylate or C18 chromatography (Zheng et al., 1997). Daidzin, genistin, and glycitein were eluted sequentially with 50%, 60% and 75% methanol, respectively. The glucosides obtained could be hydrolyzed by 4 N HCl at 100°C for 5 h. Further methanol crystallization purified the aglycone forms of the isoflavones.

ADM obtained two patents in the production of isoflavone-enriched fractions from soy protein extracts and molasses (Gugger and Dueppen, 1997, 1998). The principle for the separation was based on the temperature-sensitive differential of the solubility of isoflavones. Genistin had a low solubility below 50°C, and the solubility was greatly increased at 90°C. The content of the glucosides genistin and daidzin in a hot ultrafiltration permeate was high, between 65–95°C. Ultrafiltration was carried out at temperatures above 65°C to concentrate these two glucosides. The hot permeate from ultrafiltration was subject to adsorption on a nonpolar resin column and eluted with ethanol. The effluent containing isoflavones was dried to form a powder concentrate. The product contained approximately 30% pure isoflavones. Genistein was crystallized at 0–45°C to yield a product with 70–90% or higher purity.

Kikkoman Corp. (Norda, Japan) obtained a patent process (Matsuura et al., 1998) for the production of malonyl glucosides and production of glucosides and aglycone isoflavones from malonyl glucosides. Dehulled soybeans were soaked in water at 45–65°C for 2–4 h at pH 7.5–9.0 (pH <10). Malonyl glucosides decomposed at 70°C or higher. Then, the process was followed by an acid precipitation at pH 4.3. The filtrate or supernatant after separation of the precipitated protein was applied to an adsorbent (active carbon or alumina), then the column was eluted with an aqueous alcohol or alkaline alcohol solution to yield a mixture of malonyl daidzin and malonyl genistin.

Johns et al. (2000) reported an anion exchange technology using Amberlite or Dowex or other resins to isolate phytoestrogens from plant protein materials. Carbonate or bicarbonate was used as a counterion to avoid chlorides or alkali in the final product. Protein slurry was passed through the column at pH 6–8. The bound isoflavones were released with alcohols, organic solvents or acid or base solution. The protein isolate produced was essentially free of isoflavones.

For increasing solubility and improving the taste of isoflavones, isoflavones from fermented soybean natto have been incorporated in a clathrate form with cyclodextrin (Kanaoka et al., 1998). Fifteen isoflavones, including aglycones genistein, daidzein, and glycitein and their derivatives of glucosides, acetyl glucosides, malonyl glucosides and succinyl glucosides, were clathrated with cyclodextrins. β- and γ-cyclodextrins were found to be effective with the β-form to be the best for improving solubility and reducing the bitterness of the isoflavones. Branched or methylated cyclodextrins are also effective. The clathrates were reported to be successfully applied to beverage, cookie and jelly products.

2.7 NUTRITIONAL SIGNIFICANCE

There are many excellent reviews (Messina et al., 1994; Knight and Eden, 1996; Adlercreutz and Mazur, 1997; Kurzer and Xu, 1997; Bingham et al., 1998; Humfrey, 1998; Setchell, 1998; Setchell and Cassidy, 1999) addressing the potential benefits and adverse effects of consuming diets containing isoflavones from soy foods. Although nutritional or medicinal effects are not the emphasis of this article, it is important to provide references to the current issues of health significance so that readers may obtain detailed information.

Four major potential beneficial effects of isoflavones include heart disease prevention (Anthony et al., 1998); cancer prevention, particulary with respect to breast (Peterson and Barnes, 1991; Wu et al., 1996; Messina et al., 1997; Cline and Hughes, 1998), prostate (Griffiths et al., 1998; Stephens, 1999) and colon (Messina and Bennink, 1998) cancers; bone mass density increase to prevent osteoporosis (Anderson and Carner, 1997); and reduction of postmenopausal syndromes in women (Knight et al., 1996).

The two major concerns are potential adverse effects in infants having a high intake of isoflavones from soy-based formula (Setchell et al., 1997; Whitten and Naftolin, 1998) and possible reproductive disorders in adults having high isoflavone intakes (Whitten et al., 1995). However, there is no direct evidence to show the adverse effects of isoflavones in humans.

Most of the above-mentioned potential positive or negative effects are related to the estrogenic activities of isoflavones in the body (Molteni et al., 1995). Genistein, daidzein, biochanin A and formononetin in foods and equol (Setchell et al., 1984), which is a metabolite of daidzein in animals and humans, have weak estrogenic activities, with equol having the highest activity (Reinli and Block, 1996). An earlier Australian study (Bennetts et al., 1946) showed that female sheep eating clover lost the ability to conceive, and males had reduced sperm counts. Clover has a high phytoestrogen (formononetin) content. Therefore, in humans, consuming a high content of isoflavones may cause adverse effects. However, based on the contents in the foods, it is unlikely that one would consume a high enough amount to cause any negative health concerns.

Other mechanisms, which have been hypothesized to contribute to cancer and heart disease prevention, include isoflavones serving as antioxidants (Naim et al., 1976; Fleury et al., 1992; Wei et al., 1993; Wei and Frenkel, 1993; Wei et al., 1995, 1996; Cai and Wei, 1996; Matsuo, 1997; Anderson et al., 1998; Anthony et al., 1998) and modulators in multiple biochemical pathways in the cells (Molteni et al., 1995; Messina and Bennink, 1998). Isoflavones, particularly genistein, influence the activities of protein tyrosine kinase (Akiyama et al., 1987), DNA-topoisomerases I and II (Yamashita et al., 1990), MAP kinases, S6 kinase, mitochondrial aldehyde dehydrogenase and epidermal growth factor-induced phosphatidyl-inositol turnover, angiogenesis and transforming growth factor-beta. Genistein is able to inhibit angiogenesis *in vitro* (Fotsis et al., 1993). Daidzein and genistein glucuronides *in vitro* also activate human natural killer cells at nutritionally relevant concentrations (Zhang et al., 1999).

Because of the potential benefits of isoflavones, patents have described the applications of isoflavones in the prevention of cancer (Thurn and Juang, 1999) and heart disease (Potter et al., 1999). In addition, isoflavones with the incorporation of proteins and other ingredients have been patented for preventing or treating other diseases. To give a few examples, methods and preparations containing isoflavones have been reported to inhibit or reduce macular degeneration (Jenks, 1999); to prevent hair loss and maintain hair integrity (Segelman, 2000); to decrease hyperactive, hypoactive, deficient skin conditions and manifest dermatitides (Lanzendorfer et al., 1999); to inhibit Alzheimer's disease and related dementias and for preserving cognitive function (Clarkson et al., 1999); to inhibit gram-negative bacterial cytotoxicity (Fleiszig and Evans, 1999); and to treat cystic fibrosis (Hwang et al., 1999). Isoflavones from soybean and legumes are becoming just like a modern unicorn, a magical ingredient for improving health and diseases.

2.8 CONCLUSION

There are many reports indicating the potential health benefit of soy isoflavones, however, research is still needed to establish the efficacy of various isoflavones in human health at different stages of the life cycle. More research is also needed to investigate changes of isoflavones in postharvest storage of soybeans and in the processing and storage of soy foods. Traditional soy foods are unconventional to many people in Western countries, therefore, it is critical to develop soy foods or ingredients that have isoflavone profiles for the best health effects and, in the meantime, not to negatively affect consumers' acceptability of the products containing these ingredients.

REFERENCES

Akiyama, T., Ishida, J., Nakagawa, S., Ogawara, H., Watanabe, S., Itoh, N., Shibuya, M. and Fukami, Y. 1987. Genistein: a specific inhibitor of tyrosine-specific protein kinase. *J. Biol. Chem.* 262:5592–5595.

Aldercreutz, H. 1998. Epidemiology of phytoestrogens. *Baillieres Clin. Endocrinol. Metab.* 12:605–623.

Aldercreutz, H. and Mazur, W. 1997. Phyto-oestrogens and western diseases. *Ann. Med.* 29:95–120.

Aldercreutz, H., Honjo, H., Higashi, A., Fotsis, T., Hamalainen, E., Hasegawa, T. and Okadia, H. 1991. Urinary excretion of lignans and isoflavonoid phytoestrogens in Japanese men and women consuming a traditional Japanese diet. *Am. J. Clin. Nutr.* 54:1093–1100.

Anderson, J.J.B. and Carner, S.C. 1997. The effects of phytoestrogens on bone. *Nutr. Res.* 17:1617–1632.

Anderson J.W., Diwadkar, V.A. and Bridges, S.R. 1998. Selective effects of different antioxidants on oxidation of lipoproteins from rats. *Proc. Soc. Exp. Biol. Med.* 218:376–381.

Anthony, M.S., Clarkson, T.B. and Williams, J.K. 1998. Effects of soy isoflavones on athero-sclerosis: potential mechanisms. *Am J. Clin. Nutr.* 68(suppl):1390S–1393S.

Aussenac, T., Lacombe, S., and Dayde, J. 1998. Quantification of isoflavones by capillary zone electrophoresis in soybean seeds: effects of variety and environment. *Am. J. Clin. Nutr.* 68(suppl):1480S–1485S.

Barnes, S., Kirk, M., and Coward, L. 1994. Isoflavones and their conjugates in soy foods: extraction conditions and analysis by HPLC-mass spectrometry. *J. Agric. Food Chem.* 42:2466–2474.

Barnes, S., Coward, L., Kirk, M., and Sfakianos, J. 1998. HPLC-mass spectrometry analysis of isoflavones. *Proc. Soc. Exp. Biol. Med.* 217:254–262.

Bennetts, A.W., Underwood, E.J., and Shier, F.L. 1946. A specific breeding problem of sheep on subterranean clover pastures in Western Australia. *Aust. J. Agric. Res.* 22:131–138.

Bingham, S.A., Atkinson, C., Liggins, J., Bluck, L., and Coward, A. 1998. Phytoestrogens: Where are we now? *Br. J. Nutr.* 79:393–406.

Bryan, B.A. and Allred, M.C. 1998. Aglycone isoflavone enriched vegetable protein extract and protein material and high genistein and daidzein content materials and process for producing the same. US patent 5,726,034. March 10, 1998.

Bryan, B.A. and Guevara, B.F. 1999. Isoflavone-rich protein isolate and process for producing. US patent 5,994,508. November 30, 1999.

Bryan, B.A., Allred, M.C., and Roussey, M.A. 1998. Two-step conversion of vegetable protein isoflavone conjugates to aglycones. US patent 5,827,682. October 27, 1998.

Cai, Q. and Wei, H. 1996. Effect of dietary genistein on antioxidant enzyme activities in SENCAR mice. *Nutr. Cancer.* 25:1–7.

Chaihorsky, A. 1997. Process for obtaining an isoflavone concentrate from a soybean extract. US patent 5,670,632. September 23, 1997.

Chandra, A. and Nair, M. 1996. Supercritical carbon dioxide extraction of daidzein and genistein from soybean products. *Phytochem. Anal.* 7:259–262.

Clarkson, T.B., Jr., Anthony, M.S., Pan, Y., Adams, M.R., and Waggle, D.H. 1999. Method for inhibiting the development of Alzheimer's disease and related dementias and for preserving cognitive function. US patent 5,952,374. September 14, 1999.

Cline, J.M. and Hughes, C.L. Jr., 1998. Phytochemicals for the prevention of breast and endometrial cancer. In *Biological and Hormonal Therapies of Cancer.* Eds. K.A. Foon and H.B. Muss. Boston, MA: Kluwer Academic Publishers.

Coward, L., Barnes, N.C., Setchell, K.D.R., and Barnes, S. 1993. Genistein, daidzein, and their B-glycoside conjugates: antitumor isoflavones in soybean foods from American and Asian diets. *J. Agric. Food. Chem.* 41:1961–1967.

Coward, L., Smith, M., Kirk, M., and Barnes, S. 1998a. Chemical modification of isflavones in soy foods during cooking and processing. *Am. J. Clin. Nutr.* 68(suppl):1486S–1491S.

Coward, L., Smith, M., Kirk, M., Barnes, S., Messina, M., and Erdman, J.W. 1998b. The role of soy in preventing and treating chronic disease. *Am. J. Clin. Nutr.* 68(suppl):1486–1491.

Crank, D.L. and Kerr, P.S. 1999. Isoflavone-enriched soy product and method for its manu-facture. US patent 5,858,449. January 12, 1999.

Day, C.E. 1999. Genistin-enriched fraction from soy meal. US patent 5,932,221. August 3, 1999.

Dewick, P.M. 1994. Isoflavonoids. In *The Flavonoids: Advances since 1986.* Ed. J.B. Harborne, New York, NY: Chapman-Hall, pp. 117–238.

Eldridge, A.C. 1982a. Determination of isoflavones in soybean flours, protein concentrate and isolates. *J. Agric. Food Chem.* 30:353–355.

Eldridge, A.C. 1982b. High performance liquid chromatography separation of soybean iso-flavones and their glycosides. *J. Chromatogr.* 234:494–496.

Eldridge, A.C. and Kwolek, F. 1983. Soybean isoflavones: effect of environment and variety on composition. *J. Agric. Food Chem.* 331:394–396.

Eldridge, A.C., Kalbrener, J.E., Moser, H.A., Honig, D.H., Rackis, J.J., and Wolf, W.J. 1971. Laboratory evaluation of hexane:alcohol azeotrope-extracted soybean flakes as a source for bland protein isolates. *Cereal Chem.* 48:644–646.

Esaki, H., Onozaki, H., Kawakishi, S., and Osawa, T. 1996. New antioxidant isolated from tempeh. *J. Agric. Food Chem.* 44:696–700.

Esaki, H., Watanabe, R., Onozaki, H., Kawakishi, S., and Osawa, T. 1999. Formation mechanism for potent antioxidative *o*-dihydroxyisoflavones in soybeans fermented with *Aspergillus saitoi. Biosci. Biotech. Biochem.* 63:851–858.

Farmakalidis, E. and Murphy, P. 1985. Isolation of 6″-*O*-acetylgenistin and 6″-*O*-acetyldaidzin from toasted defatted soy flakes. *J. Agric. Food Chem.* 33:385–389.

Federal Register. October 26, 1999. 64SR 57699: 21CFR Part 101. Food labeling: Health Claims; Soy Protein and Coronary Heart Disease; Final Rule.

Fenner, G.P. 1996. Low-temperature treatment of soybean (*Glycine max*) isoflavonoid aglycone extracts improves gas chromatographic resolution. *J. Agric. Food Chem.* 44:3727–3729.

Fleiszig, S.M.J. and Evans, D.J. 1999. Methods for inhibiting bacterial cytotoxicity. US patent 5,948,815. September 7, 1999.

Fleury, Y., Welti, D.H., Philippossian, G., and Magnolato, D. 1992. Soybean (malonyl) isoflavones: characterization and antioxidant properties. *ACS Symp. Ser.* 507:98–113.

Fotsis, T., Pepper, M., Adlercreutz, H., Fleischman, G., Hase, T., Montesano, R., and Schweigerer, L. 1993. Genistein, a dietary-derived inhibitor of *in vitro* angiogensis. *Proc. Natl. Acad. Sci. USA* 90:2690–2694.

Franke, A.A., Custer, L.J., Cerna, C.M., and Narala, K.K. 1994. Quantitation of phytoestrogens in legumes by HPLC. *J. Agric. Food Chem.* 42:1905–1913.

Franke, A.A., Custer, L.J., Cerna, C.M., and Narala, K.K. 1995. Rapid HPLC analysis of dietary phytoestrogens from legumes and from human urine. *Proc. Soc. Exp. Biol. Med.* 208:18–26.

Franke, A.A., Custer, L.J., Wang, W., and Shi, C.Y. 1998. HPLC analysis of isoflavonoids and other phenolic agents from foods and human fluids. *Proc. Soc. Exp. Biol. Med.* 217:263–273.

Fukutake, M., Takahashi, M., Ishida, K., Kawamura, H., Sugimura, T., and Wakabayashi, K. 1996. Quantification of genistein and genistin in soybeans and soybean products. *Food Chem. Toxicol.* 34: 457–461.

Griffiths, K., Denis, L., Turkes, A., and Morton, M.S. 1998. Phytoestrogens and diseases of the prostate gland. *Bailliere Clin. Endocrinol. Metab.* 12:625–647.

Gugger, E.T. and Dueppen, D.G. 1997. Production of isoflavone enriched fractions from soy protein extracts. US patent 5,702,752. December 30, 1997.

Gugger, E.T. and Dueppen, D.G. 1998. Production of isoflavone enriched fractions from soy protein extracts. US patent 5,792,503. August 11, 1998.

Gyorgy, P., Murata, K., and Ikehata, H. 1964. Antioxidants isolated from fermented soybeans (tempeh). *Nature.* 203:870–871.

Ha, E.Y.W., Morr, C.V., and Seo, A. 1992. Isoflavone aglycone and volatile organic compounds in soybeans: effects of soaking treatments. *J. Food Sci.* 57:414–417, 426.

Hahlbrock, K. and Grisebach, H. 1975. Biosynthesis of flavonoids. Chapter 16. In *The Flavonoids.* Eds. J.B. Harborne, T.J. Mabry, and H. Mabry, London, UK: Chapman & Hall, pp. 866–915.

Heller, W. and Forkmann, G. 1988. Biosynthesis. Chapter 11. In *The Flavonoids: Advances in Research since 1980.* Ed. J.B. Harborne, New York, NY: Chapman & Hall, pp. 399–425.

Hessler, P.E., Larsen, P.E., Constantinou, A.I., Schram, K.H., and Weber, J.M. 1997. Isolation of isoflavones from soy-based fermentations of the erythromycin-producing bacterium *Saccharopolyspora erythraea*. *Applied Microbiol. Biotech.* 47:398–404.

Hoeck, J.A., Fehr, W.R., Murphy, P.A., and Welke, G.A. 2000. Influence of genotype and environment on isoflavone contents of soybean. *Crop Sci.* 40:48–51.

Humfrey, C.D.N. 1998. Phytoestrogens and human health effects: weighing up the current evidence. *Nat. Toxins.* 6:51–59.

Hwang, T.C., Smith, A.L., Konig, P., Clarke, L.L., Price, E.M., and Cohn, L.A. 1999. Genistein for the treatment of cystic fibrosis. US patent 5,948,814. September 7, 1999.

Iwamura, J. 1984. Process for isolating saponins and flavonoids from leguminous plants. US patent 4,428,876. January 31, 1984.

Jenks, B.H. 1999. Method for inhibiting or reducing the risk of macular degeneration. US patent 6,001,368. December 14, 1999.

Johns, P.W., Suh, J.D., Daab-Krzykowski, A., Mazer, T.B., and Mei, F.I. 2000. Process for isolating phytoestrogens from plant protein. US patent 6, 202,471. February 1, 2000.

Jones, A.E., Price, K.R., and Fenwick, G.R. 1989. Development and application of a high-performance liquid chromatographic method for the analysis of phytoestrogens. *J. Sci. Food. Agric.* 46(3):357–364.

Kanaoka, S., Uesugi, T., Hirai, K., Toda, T., and Okuhira, T. 1998. Clathrate of isoflavone derivatives and edible composition comprising the same. US patent 5,847,108. December 8, 1998.

Kaufman, P.B., Duke, J.A., Brielmann, H., Boik, J., and Hoyt, J.E. 1997. A comprehensive survey of leguminous plants as sources of the isoflavones, genistein and daidzein: implications for human nutrition and health. *J. Alternate Complementary Med.* 3:7–22.

Klus, K. and Barz, W. 1995. Formation of polyhydroxylated isoflavones from the soybean seed isoflavones daizein and glycitein by bacteria isolated from tempe. *Arch. Microbiol.* 164:428–434.

Klus, K. and Barz, W. 1998. Formation of polyhydroxylated isoflavones from isoflavones genistein and biochanin A by bacteria isolated from tempe. *Phytochemistry* 47:1045–1048.

Klus, K., Boerger-Paperndorf, G., and Barz, W. 1993. Formation of 6,7,4'-trihydroxyisoflavone (factor 2) from soybean seed isoflavones by bacteria isolated from tempe. *Phytochemistry* 34:979–981.

Knight, D.C. and Eden, J.A. 1996. A review of the clinical effects of phytoestrogens. *Obstet. Gynecol.* 87:897–904.

Knight, D.C., Wall, P.L., and Eden, J.A. 1996. A review of phytoestrogens and their effects in relation to menopausal symptoms. *Aust. J. Nutr. Diet.* 53:5–11.

Kudou, S., Fleury, Y., Weiti, D., Magnotato, D., Uchida, T., Kitamura, K. and Okubo, K., 1991. Malonylisoflavone glycosides in soybean seeds. *Agric. Biol. Chem.* 55:2227–2233.

Kurzer, M.S. and Xu, X. 1997. Dietary phytoestrogens. *Annu. Rev. Nutr.* 17:353–381.

Lanzendorfer, G., Stab, F., and Untiedt, S. 1999. Agents acting against hyperactive and hypoactive, deficient skin conditions and manifest dermatitides. US patent 5,952,373. September 14, 1999.

Liggins, J., Bluck, L.J., Coward, A., and Bingham, S.A. 1998. Extraction and quantification of daidzein and genistein in food. *Anal. Biochem.* 264:1–7.

Liu, K.S. 1997. *Soybeans: Chemistry, Technology and Utilization*. New York, NY: Chapman & Hall.

Matsuda, S., Miyazaki, T., Matsumoto, Y., Ohba, R., Termoto, Y., Ohta, N., and Ueda, S. 1992. Hydrolysis of isoflavones in soybean cooked syrup by *Lactobacillus casei* subsp. Rhamnosus IFO 3425. *J. Fermentation Bioeng.* 74:301–304.

Matsuo, M. 1997. *In vivo* antioxidant activity of okara Koji, a fermented okara, by *Aspergillus oryzae*. *Biosci. Biotech. Biochem.* 61:1968–1972.

Matsuura, M., Obata, A., and Fukushima, D. 1989. Objectionable flavor of soymilk developed during the soaking of soybeans and its control. *J. Food Sci.* 54:602–605

Matsuura, M., Obata, A., Tobe, K., and Yamaji, N. 1998. Process for obtaining malonyl isoflavone glycosides and obtaining isoflavone glycosides or isoflavone aglycones from malonyl isoflavone glycosides. US patent 5,789,581. August 4, 1998.

Mazur, W.M., Duke, J.A., Wahala, K., Rasku, S., and Adlercreutz, H. 1998. Isofalvonoids and lignans in legumes: nutritional and health aspects in humans. *Nutr. Biochem.* 9:193–200.

Messina, M. 1997. Soy foods: their role in disease prevention and treatment. In *Soybeans: Chemistry, Technology and Utilization.* Ed. K.S. Liu, New York, NY: Chapman & Hall, pp. 442–477

Messina, M. and Bennink, M. 1998. Soy foods, isoflavones and risk of colonic cancer: a review of the *in vitro* and *in vivo* data. *Bailliere Clin. Endocrinol. Metab.* 12:707–728.

Messina, M.J., Persky, V., Setchell, K.D.R., and Barnes, S. 1994. Soy intake and cancer risk: a review of the *in vitro* and *in vivo* data. *Nutr. Cancer.* 21:113–131.

Messina, M., Barnes, S., and Setchell, K.D. 1997. Phytoestrogens and breast cancer. *Lancet.* 350:971–972.

Molteni, A., Brizo Molteni, L., and Persky, V. 1995. *In vitro* hormonal effects of soybean isoflavones. *J. Nutr.* 125:751S–756S.

Murakami, H., Asakawa, T., Tero, J., and Matsushita, S. 1984. Antioxidant stability of tempeh and liberation of isoflavones by fermentation. *Agric. Biol. Chem.* 48:2971–2975.

Murphy, P.A. 1981. Separation of genistin, daidzin, and their aglycones and coumestrol by gradient high-performance liquid chromatogrphy. *J. Chromatogr.* 211:166–169.

Murphy, P.A., Song, T., Buseman, G., Baru, K., Beecher, G.R., Trainer, D., and Holden, J. 1999. Isoflavones in retail and institutional soy foods. *J. Agric. Food Chem.* 47:2697–2704.

Naim, M., Gestetner, B., Zilkah, S., Birk, Y., Bondi, A. 1974. Soybean isoflavones. Characterization, determination and antifungal activity. *J. Agric. Food Chem.* 22:806–810.

Naim, M., Gestetner, B., Bondi, A., and Birk, Y. 1976. Antioxidative and antihemolytic activities of soybean isoflavones. *J. Agric. Food Chem.* 24:1174–1177.

Ohta, N., Kuwata, G., Atkhori, H., and Watanabe, T. 1979. Isoflavonoid constituents of soybeans and isolation of a new acetyldaidzin. *Agric. Biol. Chem.* 43:1415–1418.

Okubo, K., Iijima, M., Kobayashi, Y., Yoshikoshi, M., Uchida, T., and Kudou, S. 1992. Components responsible for the undesirable taste of soybean seeds. *Biosci. Biotech. Biochem.* 56:99–103.

Parthasarathi, G. and Fenner, G.P. 1999. Improved method for gas chromatographic analysis of genistein and daizein from soybean (*Glycine max*) seeds. *J. Agric. Food Chem.* 47:3455–3456.

Peterson, G. and Barnes, S. 1991. Genistein inhibition of the growth of human breast cancer cells: independence from estrogen receptors and multi-drug resistance gene. *Biochem. Biophys. Res. Commun.* 179:661–667.

Pettersson, H. and Kiessling, K.H. 1984. Liquid chromatography determination of the plant estrogens coumestrol and isoflavones in animal feed. *J. Assoc. Off. Anal. Chem.* 67:503–506.

Potter, S.M., Henley, E.C., and Waggle, D.H. 1999. Method for decreasing LDL-cholesterol concentration and increasing HDL-cholesterol concentration in the blood to reduce the risk of atherosclerosis and vascular disease. US patent 5,855,892. January 5, 1999.

Reinli, K. and Block, G. 1996. Phytoestrogen content of foods—a compendium of literature values. *Nutr. Cancer.* 26(2):123–148.

Segelman, A.B. 2000. Use of isoflavones to prevent hair loss and preserve the integrity of existing hair. US patent 6,017,893. January 25, 2000.

Setchell, K.D.R. 1998. Phytoestrogens: the biochemistry, physiology, and implications for human health of soy isoflavones. *Am. J. Clin. Nutr.* 68(suppl):1333S–1346S.

Setchell, K.D.R. and Cassidy, A. 1999. Dietary isoflavones: biological effects and relevance to human health. *J. Nutr.* 129:758S–767S.

Setchell, K.D.R. and Welsh, M.B. 1987. High-performance liquid chromatographic analysis of phytoestrogens in soy protein preparations with ultraviolet, electrochemical and thermospray mass spectrometric detection. *J. Chromatogr.* 386:315–323.

Setchell, K.D.R., Hulme, P., Kirk, D.N., and Axelson, M. 1984. Nonsteroidal estrogens of dietary origin: possible roles in hormone-dependent disease. *Am. J. Clin. Nutr.* 40:567–578.

Setchell, K.D.R., Welsh, M.B., and Lim, C.K. 1987. HPLC analysis of phytoestrogens in soy protein preparations with ultraviolet, electrochemical, and thermospray mass spectrometric detection. *J. Chromatogr.* 368: 315–323.

Setchell, K.D.R., Zimmer-Nechemisa, L., Cai, J., and Heubi, J.E. 1997. Exposure of infants to phytoestrogens from soy-based infant formula. *Lancet.* 350:23–27.

Shen, J.L. 1994a. Aglycone isoflavone-enriched vegetable protein fiber. US patent 5,350,949. June 14, 1994.

Shen, J.L. 1994b. Aglycone isoflavone-enriched vegetable protein fiber. US patent 5,352,384. October 4, 1994.

Shen, J.L. and Bryan, B.A. 1995a. An aglycone isoflavone-enriched vegetable protein concentrate and process for producing. International Patent WO 95/10529. April 20, 1995.

Shen, J.L. and Bryan, B.A. 1995b. An aglycone isoflavone enriched vegetable protein extract and isolate and process for producing. International Patent WO 95/105230. April 20, 1995.

Shen, J.L. and Bryan, B.A. 1995c. An aglycone isoflavone enriched vegetable whey, whey protein and process for producing. International Patent WO 95/10512. April 20, 1995.

Shen, J.L. and Bryan, B.A. 1997a. Aglycone isoflavone enriched vegetable protein concentrate and process for producing. US patent 5,637,562. June 10, 1997.

Shen, J.L. and Bryan, B.A. 1997b. Aglycone isoflavone enriched vegetable protein whey, and whey protein and process for producing. US patent 5,637,561. June 10, 1997.

Shen, J.L. and Bryan, B.A. 1998. Aglycone isoflavone enriched vegetable protein extract and isolate and process for producing. US patent 5,763,389. June 9, 1998.

Shen, J.L and Bryan, B.A. 2000. Aglycone isoflavone-enriched vegetable protein extract and isolate and process for producing. US patent 6,015,785. January 18, 2000.

Shen, J.L., Roussey, M.A., Bryan, B.A., and Allred, M.C. 1998. Aglycone isoflavone enriched vegetable protein whey, whey protein material aglycone isoflavone material high genistein material and high daidzein content material and process for producing the same from a vegetable protein whey. US patent 5,851,792. December 22, 1998.

Shen, J.L., Guevara, B.F., Spadafora, F.E., and Bryan, B.A. 2000. Isoflavone rich protein isolate and process for producing. US patent 6,013,771. January 11, 2000.

Song, T., Barua, K., Buseman, G., and Murphy, P.A. 1998. Soy isoflavone analysis: quality control and new internal standard. *Am. J. Clin. Nutr.* 68(suppl):1474S–1479S.

Stephens, F.O. 1999. The rising incidence of breast cancer in women and prostate cancer in men. Dietary influences: a possible preventive role for nature's sex hormone modifiers—the phytoestrogens (review). *Oncol. Reports.* 6:865–870.

Takebe, M., Ando, Y., and Kikushima, S. 1999. Process for preparing a product from a pulse crop as a starting material and a food containing the product prepared from a pulse crop as a starting material. US patent 5,885,632. March 23, 1999.

Thurn, M.J. and Juang, L.J. 1999. Methods for treating cancer with legume plant extracts. US patent 6,004,558. December 21, 1999.

Tsukamoto, C., Shimada, S., Igita, K., Kudou, S., Kokubun, M., Okubo, K., and Kitamura, K. 1995. Factors affecting isoflavone content in soybean seeds: changes in isoflavones, saponins, and composition of fatty acids at different temperatures during seed development. 43:1184–1192.

Vanttinen, K. and Moravcova, J. 1999. Phytoestrogens in soy foods: determination of daidzein and genistein by capillary electrophoresis. *Czech. J. Food Sci.* 17:61–67.

Waggle, D.H. and Bryan, B.A. 1998. Recovery of isoflavones from soy molasses. US patent 5,821,361. October 13, 1998.

Waggle, D.H. and Bryan, B.A. 1999a. Recovery of isoflavones from soy molasses. US patent 5,919,921. July 6, 1999.

Waggle, D.H. and Bryan, B.A. 1999b. Recovery of isoflavones from soy molasses. US patent 5,990,291. November 23, 1999.

Walter, E.D. 1941. Genistin (an isoflavone glucoside) and its aglycone, genistein, from soybeans. *J. Am. Chem. Soc.* 63:3273–3276.

Wang, G., Kuan, S.S., Francis, O.J., Ware, G.M., and Carman, A.S. 1990. A simplified HPLC method for the determination of phytoestrogens in soybean and its processed products. *J. Agric. Food Chem.* 38:185–190.

Wang, C., Ma, Q., Pagadala, S., Sherrard, M.S., and Krishnan, P.G. 1998. Changes of isoflavones during processing of soy protein isolates. *J. Am. Oil. Chem. Soc.* 75:337–341.

Wang, H. and Murphy, P.A. 1994a. Isoflavone content in commercial soybean foods. *J. Agric. Food. Chem.* 42:1666–1673.

Wang, H. and Murphy, P.A. 1994b. Isoflavone composition of American and Japanese soybeans in Iowa: effects of variety, crop year, and location. *J. Agric. Food Chem.* 42:1674–1677.

Wang, H. and Murphy, P.A. 1996. Mass balance study of isoflavones during soybean processing. *J. Agric. Food Chem.* 44:2377–2383.

Weber, J.M., Cinstantinou, A., and Hessler, P.E. 1996. Process for preparing genistein. US patent 5,554,519. September 10, 1996.

Wei, H. and Frenkel, K. 1993. Relationship of oxidative events and DNA oxidation in SENCAR mice to *in vivo* promoting activity of phorbol ester-type tumor promoters. *Carcinogenesis.* 14:1195–201

Wei, H., Wei, L.H., Frenkel, K., Bowen, R., and Barnes, S. 1993. Inhibition of tumor promoter induced hydrogen peroxide formation *in vitro* and *in vivo* by genistein. *Nutr Cancer.* 20:1–12.

Wei, H., Bowen, R., Cai, Q., Barnes, S., and Wang, Y. 1995. Antioxidant and antipromotional effects of the soybean isoflavone genistein. *Proc. Soc. Exp. Biol.* 208:124–130.

Wei, H., Cai, Q.Y., and Rahn, R.O. 1996. Inhibition of UV light and Fenton reaction induced oxidative DNA damage by the soybean isoflavone genistein. *Carcinogenesis.* 17(1):73–77.

Whitten, P.L. and Naftolin, F. 1998. Reproductive actions of phytoestrogens. *Baillieres Clin. Endocrinol Metab.* 12:667–690.

Whitten, P.L., Lewis, C., Russell, E., and Naftolin, F. 1995. Potential adverse effects of phytoestrogens. *J. Nutr.* 125:771S–776S.

Wu, A.H., Horn-Ross, P.L., Nomura, A.M., West, D.W., Kolonel, L.N., Rosenthal, J.F., Hoover, R.N., and Pike, M.C. 1996. Tofu and risk of breast cancer in Asian-Americans. *Cancer Epidemiol., Biomarkers Prevent.* 5:901–906.

Xu, X., Harris, K.S., Wang, H.J., Murphy, P.A., and Hendrich, S. 1995. Bioavailability of soybean isoflavones depends upon gut microflora in women, *J. Nutr.* 125:2307–2315.

Yamashita, Y., Kawada, S., and Nakano, H. 1990. Induction of mammalian topoisomerase II dependent DNA cleavage by antitumor antibiotic streptonigrin. *Cancer Res.* 50(18):5841–5844.

Zhang, Y., Song, T.T., Cunnick, J.E., and Murphy, P.A. 1999. Daidzein and genistein glucuronides *in vitro* are weakly estrogenic and activate human natural killer cells at nutritionally relevent concentrations. *J. Nutr.* 129:399–405.

Zheng, B.L., Yegge, J.A., Bailey, D.T., and Sullivan, J.L. 1997. Process for isolation and purification of isoflaovnes. US patent 5,679,806. October 21, 1997.

Zilliken, F.W. 1980a. Antioxidants, antioxidant compositions and methods of preparing and using same. US patent 4,218,489. August 19, 1980.

Zilliken, F.W. 1980b. Antioxidants, antioxidant compositions and methods of preparing and using same. US patent 4,232,122. November 4, 1980.

Zilliken, F.W. 1981. Isoflavones and related compounds, methods of preparing and using and antioxidant compositions containing the same. US patent 4,264,509. April 28, 1981.

Zilliken, F.W. 1982a. Isoflavones and related compounds, methods of preparing and using and antioxidant compositions containing same. US patent 4,366,082. December 28, 1982.

Zilliken, F.W. 1982b. Fermentation method of preparing antioxidants. US patent 4,366,2248. December 28,1982.

Zilliken, F.W. 1983. Isoflavones and related compounds, methods of preparing and using and antioxidant compositions containing same. US patent 4,390,559. June 28, 1983.

3 Flavonoids from Berries and Grapes

Grete Skrede and Ronald E. Wrolstad

CONTENTS

1-5667-6902-7/02/$0.00+$1.50
© 2002 by CRC Press LLC

3.1 INTRODUCTION

The increased interest in flavonoids and other phenolics as health-benefiting compounds has a background in ancient traditions and folk medicine. In the Nordic countries, everybody is free to walk about and pick berries in the woods, fields and seashore as long as they stay away from the farmed land. As a consequence, the use of wild fruits and berries has long traditions in home cooking as well as for therapeutic purposes for many people. Flavonoid-rich wild berries in these countries include European blueberry (*Vaccinium myrtillus* L.), bog whortleberry (*V. uliginosum* L.), lingonberry or cowberry (*V. vitis-idaea* L.), red raspberry (*Rubus idaeus*), blackberry (*R. fruticosus* L.), arctic raspberry/bramble (*R. arcticus* L.), cloudberry (*R. chamaemoru* L.), wild strawberry (*Fragaria vesca*), small cranberry (*Oxycoccus quadripetalus* also named *V. oxycoccus* L.), crowberry (*Empetrum nigrum* Coll.), elderberry (*Sambucus nigra*), red elderberry (*S. racemosa* L.), sloe plum (*Prunus spinosa* L.), bird cherry (*P. padus*) and juniper (*Juniperus communis*).

In particular, the European blueberry (*V. myrtillus* L.), (in North America and in this text called bilberry) is known as an ancient medical plant (Ulltveit, 1998).

Dried bilberries were used as treatments for diarrhea, even for small children. The tannins have been considered to be the active compounds by their contraction of the mucous membrane of the intestine. This remedy was also used for animals. Blueberries were further considered to improve night vision. Decoction of dried bilberries was used as a gargle for treating sore throat. Also, juice from lingonberry (*V. vitis-idaea*) was used for the common cold and throat complaints. The red berries made people think they would also be helpful toward blood illness. Raspberries (*R. idaeus*) were considered to strengthen the heart, sooth vomiting and protect from abortion. Wild strawberries (*F. vesca*) were used for rheumatism, gout and as a means to treat chilblains. The bog worthleberry (*V. uliginosum*) had a rather negative image and was considered to cause lepra and madness and to give ugly children. The berries were also associated with narcotics and drunkenness, most likely because they had long traditions in wine making. Small cranberries (*O. quadripetalus*) were used for urinary infections, whereas elderberries (*E. nigrum*) served as a diuretic (Ulltveit, 1998).

A variety of flavonoid-containing drugs is sold around the world (Bruneton, 1995; Kalt and Dufour, 1997). Berries named as raw material for drugs are bilberries (*V. myrtillus* L.), black currant (*Ribes nigrum* L.) and European elder (*S. nigra* L.) (Bruneton, 1995). Extracts from berry fruits rich in anthocyanins are mainly used for treatment of cutaneous capillary fragility, for symptoms of venous insufficiency and hemorrhoids. In the French "Pharmacognosie," however, it is clearly stated along with the description of the therapeutic uses of the drugs that the clinical efficacy of most flavonoids and their drugs, and in particular, those containing anthocyanins and procyanidins, has rarely been established correctly (Bruneton, 1995). Testing experiments have not always been conducted according to standard methods, and the results are more observational than actual scientific, reproducible results. The majority of the drugs available are claimed as "proposed for" the various treatments.

As will be seen in the following sections, the newly achieved scientific knowledge seems to support some of the ancient traditions by reaffirming the effects and properties described in yesterday's folklore. At the same time, a few of the qualities seem to be based on superstitions and religious beliefs that are no longer valid, and thus, most likely will never be proven by a scientific approach to the subject.

3.2 CHEMICAL AND PHYSICAL PROPERTIES OF FLAVONOIDS AND PHENOLIC ACIDS

3.2.1 FLAVONOIDS AND THEIR GLYCOSIDES

The word flavonoid comes from Latin *flavus*, which means yellow, and included at the beginning only a yellow-colored group of compounds with a flavone nucleus (Jovanovic et al., 1998). Today, the term is used in a broader context and includes colorless (flavan-3-ol) to less colored (flavanone) compounds, as well as the red and blue anthocyanins, all commonly occurring in plants. For a period, flavonoids were referred to as vitamin P because of their ability to decrease capillary fragility and permeability in humans. The term vitamin C2 has also been used, as it was known

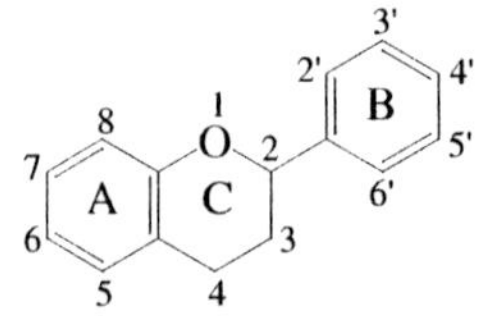

FIGURE 3.1 The C_6-C_3-C_6 (flavan) skeleton and numbering system for most flavonoids. [Source: Adapted from Spanos and Wrolstad (1992).]

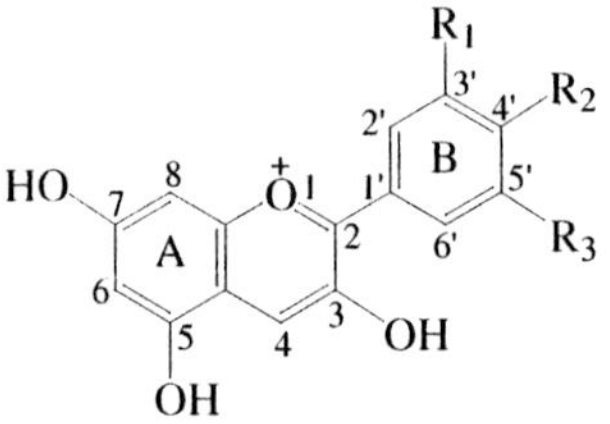

	3′	4′	5′	Color
Pelargonidin	H	OH	H	Orange
Cyanidin	OH	OH	H	Orange/red
Delphinidin	OH	OH	OH	Bluish/red
Peonidin	OMe	OH	H	Orange/red
Petunidin	OMe	OH	OH	Bluish/red
Malvidin	OMe	OMe	OMe	Bluish/red

FIGURE 3.2 Structure of anthocyanidins occurring in grapes and berry fruits. [Source: Adapted from Jovanovic et al. (1998) and Mazza and Miniati (1993).]

that some flavonoids had vitamin C-sparing activity. Today, flavonoids are not referred to as vitamins.

An important function of the flavonoids is to color plants. In addition to the intense red/bluish color from the anthocyanins, other flavonoid compounds bring white and yellow color to plants (Jovanovic et al., 1998). Additional important functions are to protect plants from deteriorating UV radiation from the sun, to protect against parasite attack, to regulate certain enzyme reactions, to function as signal substances and to provide flavor to various plant products used for human consumption (Cooper-Driver and Bhattacharya, 1998; Johnson, 1998; Jovanovic et al., 1998). The astringency associated with flavonoids originates from their ability to precipitate protein, thus forming insoluble molecular structures in the mouth.

Flavonoids constitute a group of naturally occurring plant components based on the C_6-C_3-C_6 configuration in the flavan nucleus (Figure 3.1). The flavan structure has three phenolic rings, named A, B and C. Various substituting groups are numbered according to their position on the rings. The chemical and biochemical properties of the flavonoids depend upon their chemical structure. The predominant classes of flavonoids present in fruits and berries are the anthocyanidins (Figure 3.2), the flavonols and the flavanols, flavan-3-ols (catechins) and flavan-3,4-diols (procyanidins) (Figure 3.3).

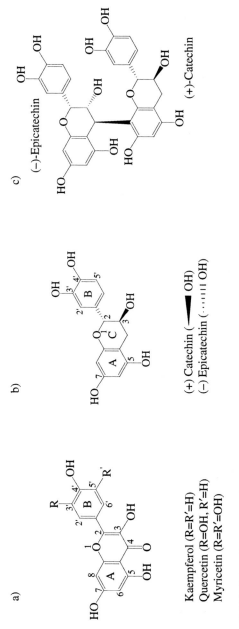

FIGURE 3.3 Structures of (a) flavonols, (b) flavan-3-ols and (c) the dimeric procyanidin B1, occurring in grapes and berry fruits. [Source: Adapted from Spanos and Wrolstad (1992).]

Most anthocyanidins and flavonols occur as glycosides in plants. The glycosides thus consist of the flavonoid unit (the aglycon) and one or more monosaccharides. The most common substitutions occur in the 3- and the 5-positions of the flavan structure for the anthocyanidins and in the 3-position for the flavonols. The sugar moiety may have one or more aliphatic or aromatic acids attached in an acyl linkage.

3.2.1.1 Anthocyanins

There are six anthocyanidins commonly occurring in grapes and berry fruits. The anthocyanidins are classified according to the number and position of their hydroxyl and methoxyl groups on the flavan nucleus and are named pelargonidin, cyanidin, delphinidin, peonidin, petunidin and malvidin (Figure 3.2).

Anthocyanins are anthocyanidins with one or more sugar moieties attached. Several combinations of type, number and position of the sugar substituents are possible, and a wide variety of anthocyanins have been identified (Jackman et al., 1987b; Mazza and Miniati, 1993; Jackman and Smith, 1996; Brouillard et al., 1997; Wrolstad, 2000). The most common sugars are glucose, galactose, xylose, rhamnose and arabinose. Di- and trisaccharides are formed by combination of the mono-saccharides into compounds such as rutinoside (glucose, rhamnose), sophoroside (glucose, glucose), sambubiose (glucose, xylose), and gentiobiose (glucose, glucose). When only one sugar substitution occurs, the sugar is always attached to the 3-position of the flavan structure (Figure 3.2). The glycosylation confers increased stability and water solubility to the anthocyanins compared with the anthocyanidins. Increasing the number of sugar residues seems to further increase stability of the anthocyanin. The variety of glycosidic substitutions accounts for much of the diversification of colors in plants and plant products and can be used for taxonomic purposes.

The sugar residues may be acylated with cinnamic (p-coumaric, caffeic, ferulic) or aliphatic (acetic, malonic, succinic) acids (Jackman and Smith, 1996; Wrolstad, 2000), which adds even more to the diversity of anthocyanins occurring in nature. Acylation, especially with cinnamic acids, greatly improves anthocyanin stability. Esters from aliphatic acids, however, are very labile to acidic hydrolyzation and are often not identified unless special precautions are taken during extraction and iso-lation of the anthocyanins from their natural sources. Generally, di-, tri-, or poly-acylated anthocyanins are more stable in neutral and slightly acidic conditions than monoacylated anthocyanins. The stabilization of acylated anthocyanins is considered to be due to intramolecular copigmentation through a "sandwich-type" stacking of the acyl groups with the flavan nucleus of the anthocyanin (Jackman and Smith, 1996). Anthocyanins also form intermolecular structures. The complex formation can occur with compounds like the flavonols, quercetin and rutin, with other antho-cyanins and also with metal ions and is known to have a protective effect on the anthocyanins. During aging of red wine, the monomeric anthocyanins are irreversibly transformed into polymeric compounds through self-association reactions. The poly-meric material is less pH-sensitive and is less susceptible to degradation by ascorbic acid and light. Skrede et al. (1992) concluded from an experiment with anthocyanin-fortified strawberry and black currant juices that the anthocyanin stability was more affected by total anthocyanin concentration than by qualitative differences among the anthocyanins of the two berry fruits.

Anthocyanins undergo reversible molecular transformations with pH change. The molecule forms an intensely colored oxonium ion at low pH, while a colorless hemiketal form exists at pH 4.5. At alkaline pH, a blue and also fairly unstable quinoidal form is generated. Acylation and position of the glycosidic groups will affect the exact pH at which these molecular changes occur. However, as a general rule, anthocyanin solutions and products should always have pH values below 4.5. This low pH occurs naturally in grapes and other berry fruits but may be exceeded in extracts, powders and other anthocyanin products.

The anthocyanins are good absorbers of visible light, thus appearing as colored substances, responsible for the characteristic orange/red/bluish colors of grapes and berries. This reflects the origin of the term anthocyanin, which is derived from Greek, and means flower and blue. The color is largely determined by the substitution pattern of the B-ring of the aglycon (Figure 3.2), compared with the pattern of glycosylation of the flavan structure, which to a smaller extent, influences color formation (Jackman and Smith, 1996).

3.2.1.2 Flavonols

The flavonols are characterized by a double bond between the 2 and 3 carbon, a hydroxyl group in the 3-position and a carbonyl group in the 4-position of the C-ring of the flavan nucleus (Figure 3.1). The most commonly occurring flavonols in fruits and berries are quercetin, myricetin and kaempferol (Hertog and Katan, 1998). Their substitution patterns are shown in Figure 3.3(a). In plants, flavonols are present as mono-, di- and triglycosides (Shahidi and Naczk, 1995). The sugars are glucose, galactose, rhamnose, arabinose, apiose and glucuronic acid. The monoglycosides are mainly 3-O-glycosides. With diglycosides, the sugar moieties may be linked to the same or to two different carbons. Among the diglycosides, 3-rutinosides with rhamnose and glucose linked together are the most common, with rutin (quercetin-3-rutinoside) being an example. Triglycosides occur less frequently. Similar to the anthocyanins, the flavonol glycosides are often acylated with phenolic acids such as p-coumaric, ferulic, caffeic, p-hydroxybenzoic and gallic acids.

The flavonols and their glycosides contribute to specific taste characteristics such as bitterness and astringency in berry fruits and their products (Shahidi and Naczk, 1995). The molecular structure of flavonols lacks the conjugated double bonds of the anthocyanins, and they are thereby colorless. They may, however, contribute to discoloration of berry fruits, as they are readily oxidized by O-phenoloxidase in the presence of catechin and chlorogenic acid. Discoloration may also occur as a consequence of complex formation with metallic ions. On the other hand, the flavonol glycoside rutin is known to form complexes with anthocyanins, thus stabilizing the color of these compounds.

3.2.1.3 Flavanols

Flavanols with the hydroxyl group in the 3-position are often referred to as flavan-3-ols or catechins (Chung et al., 1998). As these compounds have two asymmetric carbon atoms (C-2 and C-3), four isomers exist for each flavan-3-ol molecule [Figure 3.3(b)]. The flavan-3-ols most often occurring in grapes and berry fruits are

(+)-catechin and (−)-epicatechin, although gallocatechin and epigallocatechin have also been identified (Shahidi and Naczk, 1995). The flavan-3,4-diols have three asymmetric carbon atoms and, thus, eight isomers. The more rare flavan-3,4-diols are often called leucoanthocyanins, because they form anthocyanidins after heating under acidic conditions (Spanos and Wrolstad, 1992). Flavan-3-ols easily condense into oligomeric procyanidins (OPCs) (or proanthcyanidins) and polymeric (condensed tannins) compounds (Waterhouse and Walzem, 1998). The dimeric procyanidins are often referred to as the B-series [Figure 3.3(c)] and the trimeric procyanidins as the C-series procyanidins. The condensed tannins may have a degree of polymerization from 7 to 16 (Shrikhande, 2000).

3.2.2 Phenolic Acids and Their Derivatives

The most common phenolic acids in grapes and berry fruits are the cinnamic and the benzoic acids. The cinnamic acids frequently found are caffeic, p-coumaric and ferulic acids (Herrmann, 1989). The cinnamic acids seldom occur as free acids. The bound forms are esters of quinic, shikimic and tartaric acids (Shahidi and Naczk, 1995). An important compound is chlorogenic acid, which is used as a term for a group of esters between cinnamic and quinic acids (Clifford, 1999). The most commonly occurring chlorogenic acids in berry fruits are the caffeoylquinic acids [Figure 3.4(a)]. Esters between cinnamic acids and tartaric acid are common in grapes, the most common being caftaric acid (caffeoyltartaric acid) [Figure 3.4(b)] (Macheix and Fleuriet, 1998). Glucose derivatives of cinnamic acids are also found. The cinnamates may be conjugated to other molecules and occur as soluble compounds in the vacuoles of plant tissue, or they may be linked to cell-wall components and thus be insoluble. The cinnamic esters are often referred to as cinnamates or, more specifically, as hydroxycinnamates (Kroon and Williamson, 1999). The benzoic acids are also most commonly bound to organic acids or sugars. In berries, protocatechuic, p-hydroxybenzoic acids and gallic acids are the most abundant benzoic acids (Shahidi and Naczk, 1995).

3.2.3 Anthocyanins as Food Pigments

Anthocyanins give their color to grapes and most berry fruits. Thus, they are well known as the pigment of food products made from those raw materials. The anthocyanins are often present in the epidermal tissue of grapes and berry fruits, and various mechanical or enzymic treatments of skin and peels may be required to fully extract the tinctorial potential of the grapes and berries when they are processed into wine or juice. However, even with such precautions, considerable amounts of pigment are retained in the press-cake (Skrede et al., 2000a). Anthocyanin losses also occur due to the action of degrading enzymes (i.e., polyphenol oxidase) that are endogenous to the fruits (Kader et al., 1997). In products in which whole fruits or chunks are included, the pigments may be allowed to remain in the tissue and exert their coloring capacity from there.

Anthocyanins undergo reversible structural transformations with change in pH. The color is most intense at low pH, and in particular, isolated pigments change toward colorlessness when the pH passes 4.5 (Bridle and Timberlake, 1997). At

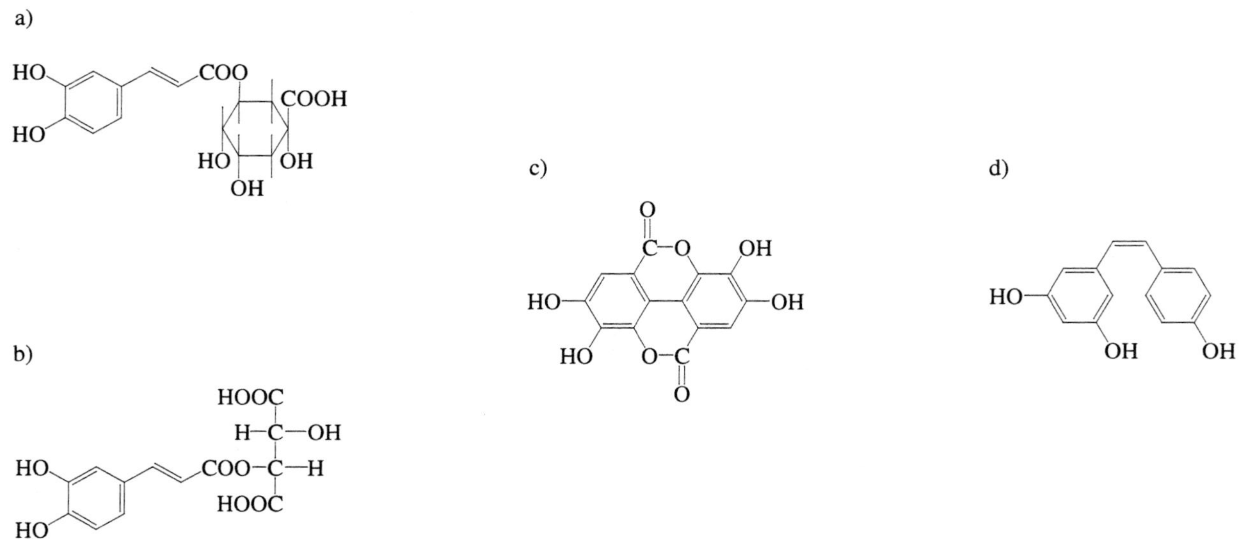

FIGURE 3.4 Structures of (a) chlorogenic acid, (b) caftaric acid, (c) ellagic acid and (d) resveratrol. [Source: Adapted from Spanos and Wrolstad (1992) and Macheix and Fleuriet (1998).]

higher pH, the color is bluish. In food products, the anthocyanins may be more colored as they copigment with other flavonoids and metal ions. The pigments are most stable at low pH. The exact pH at which structural changes occur depends upon the position of the glycosidic position and the acylation of the anthocyanins (Wrolstad, 2000). Commercial anthocyanin colorants are available based on extracts of grape juice and grape skin, elderberry, red cabbage, radish and black carrot.

3.2.4 ELLAGIC ACID DERIVATIVES

Ellagic acid is a dimeric derivative of gallic acid also known as hexahydroxydiphenic acid, which spontaneously forms a dilactone [Figure 3.4(c)]. It is the lactone form that is commonly known as ellagic acid. It exists in nature mainly in the form of ellagitannins, which are esterified with glucose. There are numerous derivatives of ellagic acid existing in plants formed by methylation and glycosylation of the phenolic groups. Ellagic acid and its derivatives have limited distribution and were first reported to be present in members of Rosaceae, specifically strawberries, raspberries and blackberries (Bate-Smith, 1959). These fruits have about 3 times as much ellagic acid as walnuts and pecans and 15 times as much as other fruits and nuts (Daniel et al., 1989; Wang et al., 1990; Maas et al., 1991a). Ellagic acids tend to be concentrated in the seeds; for example, Daniel et al. (1989) reported concentrations of 275 mg/kg in seeds and 9 mg/kg in raspberry pulp. Rommel and Wrolstad (1993) found 16 different derivatives of ellagic acid to be present in red raspberries. The mean concentration of total ellagic acid in experimental raspberries juices ($n = 45$) was 7 mg/kg, whereas the mean concentration in commercial juices ($n = 7$) was 52 mg/kg. There was considerable variation with respect to both cultivar and processing methods. Maas et al. (1991b) found large differences in ellagic acid content in strawberries with respect to both variety and maturity.

3.2.5 RESVERATROL

Resveratrol is a stilbene with a C_6-C_2-C_6 structure [Figure 3.4(d)]. It occurs in geometric *cis* and *trans* forms and as its β-glucoside derivative, piceid. Several studies report their contents in grapes and wine (Jeandet et al., 1995; Ector et al., 1996; Pezet and Cuenat, 1996). Resveratrol is a phytoalexin synthesized by grape skin cells in response to pathogenic infection (Creasey and Coffee, 1988; Jeandet et al., 1991). Resveratrol content of grapes and wine is influenced by maturity, variety, geographical origin and growing methods as well as *Botrytis* infection (Lamuela-Raventos et al., 1995; Okuda and Yokotsuka, 1996).

3.3 OCCURRENCE OF FLAVONOIDS AND PHENOLIC ACIDS IN BERRY FRUITS AND GRAPES

3.3.1 ANTHOCYANINS

Anthocyanins in grapes and berry fruits have been of interest to scientists and food technologists for a long time due to their important contributions to color and palatability of foods. A large number of books (Markakis, 1982; Ribéreau-Gayon, 1982; Timberlake and Bridle, 1982; Mazza and Miniati, 1993; Harborne, 1994; Hendry and Houghton, 1996) and literature reviews (Mazza, 1995, 1997; Bridle and Timberlake, 1997; Herrmann, 1999; Wrolstad, 2000) with a focus on qualitative and quantitative aspects have been published.

There are wide variations in the number and type of individual anthocyanins in various berry fruits. In extracts from highbush blueberries (*V. corymbosum*, cv Blue-crop), 12 different nonacylated anthocyanins, with indications of another 3 acylated anthocyanins, have been identified (Skrede et al., 2000a), whereas 25 individual anthocyanins (11 acylated) are reported for lowbush blueberries (*V. angustifolium*, cv Fundy) (Gao and Mazza, 1994, 1995). Similarly, 20 different anthocyanins, of which 11 were acylated, have been identified in grapes (*Vitis vinifera*, cv Cabernet Franc) (Mazza et al., 1999). In contrast, Evergreen blackberries (*Rubus laciniatus*) contain only one major and four minor anthocyanins (Fan-Chiang, 1999), elderberries (*Sambucus nigra*) two major and two minor (Inami et al., 1996), red raspberries two or four major depending upon cultivar (García-Viguera et al., 1998) and chokeberry (*Aronia melanocarpa*, cv Nero) one major, one intermediate and two minor anthocyanins (Strigl et al., 1995). Efforts are being made to build databases for anthocyanin composition, in which individual anthocyanins and their relative occurrence in various species and cultivars can be listed (Fan-Chiang, 1999).

With the increased interest in anthocyanins as antioxidants, studies on the relationships between anthocyanins and antioxidant activity of the fruits have appeared (Heinonen et al., 1998; Prior et al., 1998; Kalt et al., 1999a; Wang and Lin, 2000). In this context, the overall amount of anthocyanins, rather than the presence of individual anthocyanins, is the focus. The actual level of anthocyanins may be expressed on a weight basis or on a molar basis. When determining anthocyanin content, the tradition has been to select the most abundant anthocyanin and use the physical data (extinction coefficient, molecular weight) of this compound for calculation (Wrolstad, 1976). The anthocyanin content is often presented as mg per 100 g of fresh weight. In more recent studies, however, anthocyanin levels are often presented on a molar basis, likely because this more easily relates to antioxidant activity and thus can be used to compare antioxidant efficiency among various fruits. Anthocyanin levels of selected grapes and berry fruits are presented in Table 3.1.

3.3.2 FLAVONOLS, FLAVANOLS AND PHENOLIC ACIDS

Häkkinen et al. (1999) reported the phenolic profiles of 19 berries. The method used included determination of the flavonols quercetin, kaempferol and myricetin as well as the phenolic acids *p*-coumaric, caffeic, ferulic, *p*-hydroxybenzoic, gallic and ellagic after hydrolysis of glycosides and esters. Catechins and anthocyanins were not included in the analysis. The berries analyzed were cranberry (*V. oxycoccus*), lingonberry (*V. vitis-idaea*), blueberry (*V. corymbosum*), bilberry (*V. myrtillus*), green gooseberry (*R. uva-crispa*), black currant (*R. nigrum*), red currant (*Ribes x pallidum*), white currant (*Ribes x pallidum*), green currant (*R. nigrum*), chokeberry (*Aronia melanocarpa*), rowanberry (*Sorbus aucuparia*), sweet rowanberry (*Grataegosorbus mitschurinii*), strawberry (*Fragaria x ananassa*), cloudberry (*R. chamaemorus*), red

TABLE 3.1
Anthocyanin Content in Selected Grapes and Berry Fruits

Species	Genus	Anthocyanins (mg/100 g Fresh Weight)	Number of Clones/ Cultivars[1]	Source
Aronia, chokeberry	*Aronia melanocarpa* (Michx.) Elliot	307–631	1	Plocharski and Zbroszcyzk, 1992
Currants, red	*Ribes rubrum* L.	12–19	5	Mazza and Miniati, 1993
Currants, black	*Ribes nigrum* L.	110–430	12	Skrede, 1987
Bilberry	*Vaccinium myrtillus* L.	300–370	2	Prior et al., 1998; Kalt et al., 1999b
Blackberry	*Rubus* spp.	83–326	16	Mazza and Miniati, 1993
Blackberry	*Rubus fructicosus*	765	1	Heinonen et al., 1998
Blackberry, thornless	*Rubus* spp.	133–172	3	Wang and Lin, 2000
Blackberry	*Rubus eubatus*	70–201	52	Fan-Chiang, 1999
Blueberry, rabbiteye	*Vaccinium ashei* Reade	62–187	6	Prior et al., 1998
Blueberry, highbush	*Vaccinium corymbosum* L.	63–484	23	Heinonen et al., 1998; Prior et al., 1998; Kalt et al., 1999a,b; Skrede et al., 2000a
Blueberry, lowbush	*Vaccinium angustifolium* Aiton	91–255	32	Prior et al., 1998; Kalt et al., 1999a,b
Bog whortleberry	*Vaccinium uliginosum* L.	256	1	Andersen, 1987a
Cranberry, small	*Vaccinium oxycoccus* L.	78	1	Andersen, 1989
Crowberry	*Empetrum nigrum*	300–500	2	Linko et al., 1983
Grapes	*Vitis vinifera* L.	0–603	82	Mazza and Miniati, 1993; Mazza, 1995; Yi et al., 1997
Lingonberry (cowberry)	*Vaccinium vites-idaea* L.	174	1	Andersen, 1985
Raspberry, red	*Rubus idaeus* L.	41–220	6	Heinonen et al., 1998; Kalt et al., 1999a; Wang and Lin, 2000
Raspberry, black	*Rubus occidentalis* L.	197–428	4	Mazza and Miniati, 1993; Wang and Lin, 2000
Strawberry	*Fragaria x ananassa* Duch.	8–79	10	Heinonen et al., 1998; Kalt et al., 1999a; Wang and Lin, 2000
—	*Vaccinium japonicum*	113	1	Andersen, 1987b

[1] Sum of clones/cultivars in referred studies.

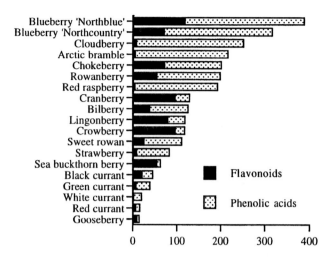

FIGURE 3.5 Total phenolic concentrations in berries. [Source: Reprinted from *Food Res. Int.*, Vol. 32, Häkkinen, S., Heinonen, M., Kärenlampi, S.O., Mykkänen, H., Ruuskanen, J., and Törrönen, R., Screening of selected flavonoids and phenolic acids in 19 berries," pp. 345–353, Copyright 1999, with permission from Elsevier Science.]

raspberry (*R. idaeus*), arctic bramble (*R. arcticus*), sea buckthorn berry (*Hippophae rhamnoides*) and crowberry (*Empetrum nigrum*). The relative levels of individual flavonols and phenolic acids varied considerably among the berries. Quercetin levels were highest in blueberry, cranberry, lingonberry, chokeberry, crowberry, sea buckthorn berry and rowanberry, with levels of about 50–100 mg/100 g seedless dry matter in decreasing order. Ellagic acid was highest in cloudberry, arctic bramble and red raspberry, with levels of 170 and 190 mg/100 g seedless dry matter. Except for strawberry with about 45 mg/100 g seedless dry matter, the remaining berries contained less than 10 mg ellagic acid per 100 g seedless dry matter. Total levels of phenols, calculated on the basis of dry matter in the berries, are shown in Figure 3.5. In a more recent study (Häkkinen and Törrönen, 2000), individual aglycon flavonols and phenolic acids in strawberry and *Vaccinium* species were reported. Strawberries from six varieties grown conventionally and organically as well as in different locations ranged from 0.2 to 9.9 mg/100 g fresh weight in kaempferol, from 0.3 to 0.5 mg/100 g fresh weight in quercetin, from 34.4 to 52.2 mg/100 g fresh weight in ellagic acid and from 0.7 to 4.8 mg/100 g fresh weight in *p*-coumaric acid. There were no effects of organic growing on total flavonols and phenolic acids. For the *Vaccinium* species, bog whortleberries (*V. uliginosum*) were particularly high in the flavonols quercetin and myricetin, with levels of 19.6 and 12.0 mg/100 g fresh weight, respectively. Blueberries and wild bilberries varied between 1.7 and 4.7 mg/100 g fresh weight for quercetin and between 0.8 and 1.8 mg/100 g fresh weight for myricetin. There were more varietal differences among the cultivated blueberries than among the strawberry cultivars. Among the flavonol aglycons, lingonberries contain mainly quercetin (169 mg/kg), black currants myricetin (104.1 mg/kg) and quercetin (52.9 mg/kg), bilberries quercetin (41.2 mg/kg) and

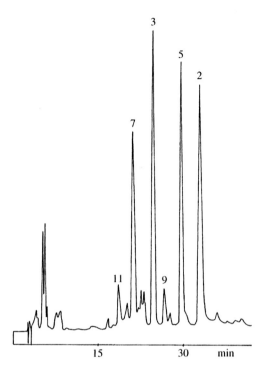

FIGURE 3.6 Flavonol-glucosides of black currant nectar separated by HPLC (diol-phase column) after treatment by polyamide and ion exchange chromatography. Peak identification: (2) myricetin-3-rutinoside, (3) myricetin-3-glucoside, (5) quercetin-3-rutinoside, (7) quercetin-3-glucoside, (9) kaempferol-3-rutinoside, and (11) kaemperol-3-glucoside. [Source: *Z. Lebensm. unters. Forsch.* "Nachweis eines Zusatzes von roten zu schwarzen Johannisbeer-Erzeugnissen über die hochdruckflüssigchromatographische Bestimmung der Flavonolglykoside," Siewek, F., Galensa, R., and Herrmann, K., Vol. 179, pp. 315–321, 1984, Copyright Springer-Verlag GmbH & Co. KG.]

strawberries kaempferol (11.8 mg/kg) and quercetin (5.2 mg/kg), whereas red raspberries contain quercetin (9.5 mg/kg) (Häkkinen et al., 2000).

Individual flavonol glycosides may be studied by high-performance liquid chromatography (HPLC) of extracts. Siewek et al. (1984) separated 14 different compounds by this technique and reported the presence of both glucosides and rutinosides of kaempferol, myricetin and quercitin in black currants (Figure 3.6). In his review, Herrmann (1989) reported levels of hydroxycinnamic and hydroxybenzoic acid compounds in various fruits. Strawberry, raspberry, blackberry, red currant, black currant, gooseberry, josta and highbush blueberry were included in the study. There were wide variations in both number and level of phenolic compounds. Except for the highbush blueberry with a chlorogenic acid level of 186–208 mg/100 g fresh weight, the levels of individual hydroxycinnamic and hydroxybenzoic acid compounds were all lower than 6 mg/100 g fresh weight, and for most fruits, lower than 1 mg/100 g fresh weight. Skrede et al. (2000a) reported 27 mg chlorogenic acid, 40 mg flavonol glucosides and 10 mg procyanidins all per 100 g fresh highbush blueberries.

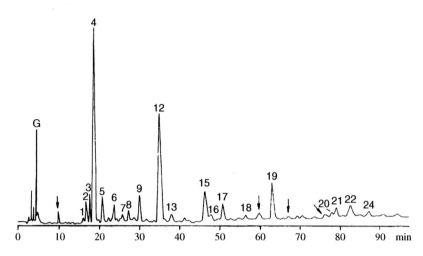

FIGURE 3.7 HPLC chromatogram recorded at 280 nm of an extract of grape seeds from *Vitis vinifera* variety Cabernet-Sauvignon. Peaks corresponding to hydrolyzable tannins are marked with an arrow. For flavan-3-ols identification (peaks numbered), see Table 3.2. [Source: Reprinted from *Food Chem.*, Vol. 53, Santos-Buelga, C., Francia-Aricha, E.M., and Escribano-Bailón, M.T. "Comparative flavan-3-ol composition of seeds from different grape varieties," pp. 197–201, Copyright 1995, with permission from Elsevier Science.]

Grapes are rich in polyphenolics, and Lu and Foo (1999) reported a total of 17 different polyphenols belonging to the groups of phenolic acids and alcohols, flavan-3-ols and flavonols from Chardonnay grape pomace. The flavonoids, including the oligomeric procyanidins, constituted about 4% of the dry grape pomace. In the grape seeds, the flavan-3-ols predominate, and a total of 24 different flavan-3-ols have been reported by Santos-Buelga et al. (1995). By investigating 17 different varieties of grapes, a total of 27 various flavan-3-ols were identified. All varieties contained flavanols esterified with gallic acid, and the authors concluded that gallic acid esters were characteristic of the grape seed flavanol composition. An HPLC chromatogram of flavan-3-ols within the B- and C-groups in seeds from *Vitis vinifera* (cv Cabernet-Sauvignon) is presented in Figure 3.7, with the identification of the various peaks given in Table 3.2.

3.3.3 Total Phenolics

With the increased focus on antioxidant activity of polyphenolics in grapes and berries, total levels of polyphenols, rather than levels of individual compounds, have been the interest of research recently. At present, there seems to be one preferred method for the analysis of total phenolics (Slinkard and Singleton, 1977). The method is based on the Folin-Ciocalteu reagent and the total amount of phenolics is calculated as any chosen phenolic substance. Most often, gallic acid equivalents (GAEs) are presented either on a weight basis or on a molar basis. The method is simple to perform but has limitations, as the various groups of phenolics respond differently to the reagent used in the assay (Kähkönen et al., 1999).

TABLE 3.2
Flavan-3-ols Isolated from Grape Seeds

Peak Number[1]	Compound
1	Catechin-(4α→8)-catechin-(4α→8)-catechin (C2)
2	Catechin-(4α→8)-catechin (B3)
3	Epicatechin-(4β→8)-catechin (B1)
4	(+)-Catechin
5	Epicatechin-(4β→8)-epicatechin-(4β→8)-catechin
6	Catechin-(4α→8)-epicatechin (B4)
7a	Catechin-(4α→8)-catechin-(4α→8)-epicatechin[2]
7b	Epicatechin-(4β→6)-epicatechin-(4β→8)-catechin
8a	Catechin-(4α→6)-catechin (B6)
8b	Epicatechin-(4β→6)-epicatechin-(4β→8)-epicatechin
9	Epicatechin-(4β→8)-epicatechin (B2)
10	Epicatechin-(4β→8)-epicatechin–3-O-gallate-(4β→8)-catechin
11	Epicatechin–3-O-gallate-(4β→8)-epicatechin (B2-3-O-gallate)
12	(−)-Epicatechin
13	Catechin-(4α→8)-epicatechin–3-O-gallate (B4-3′-O-gallate)
14	Epicatechin-(4β→8)-epicatechin-(4β→6)-catechin
15a	Epicatechin–3-O-gallate-(4β→8)-catechin (B1-3-O-gallate)
15b	Epicatechin-(4β→8)-epicatechin-3-O-gallate (B2-3′-O-gallate)
16	Epicatechin-(4B→6)-catechin (B7)
17	Epicatechin-(4β→8)-epicatechin-(4β→8)-epicatechin (C1)
18	Epicatechin-(4β→8)-epicatechin-(4β→8)-epicatechin-(4β→8)-epicatechin
19	(−)-Epicatechin-3-O-gallate
20	Epicatechin-3-O-gallate-(4β→6)-catechin (B7-3-O-gallate)[2]
21	Epicatechin-3-O-gallate-(4β→8)-epicatechin-3-O-gallate (B2-3-3′-O-digallate)
22	Epicatechin-(4β→8)-epicatechin-(4β→8)-epicatechin-3-O-gallate
23	Epicatechin-(4β→8)-epicatechin-3-O-gallate-(4β→8)-epicatechin-3-O-gallate[2]
24	Epicatechin-(4β→6)-epicatechin (B5)

[1] Numbering of peaks according to their order of elution in Figure 3.7.

[2] Flavan-3-ol identified for the first time in grape seeds.

Total phenolics in extracts from a variety of wild and cultivated berries of Finnish origin have been investigated (Kähkönen et. al., 1999). A compilation of data for total phenolics in grapes and berries is presented in Table 3.3. Due to inconsistency in methods used for analysis and calculations, direct comparisons between various studies are difficult. Results are also given either on a fresh or a dry weight basis. Only when dry matter content of the fresh material is given can recalculations be performed to compare results. When total phenolic contents of extracts are given rather than the contents of berries, calculations may be done if the exact extraction procedures are given. When testing 16 wines of various origins, Burns et al. (2000) reported total

TABLE 3.3
Total Phenolics (as Gallic Acid Equivalents, GAEs) in Selected Grapes and Berries

Species	Genus/Cultivar	Total Phenolics (GAE)	Number of Clones/Cultivars[1]	Source
		mg GAE/100 g Fresh Weight		
Bilberry	Vaccinium myrtillus L.	525	1	Prior et al., 1998
Blackberry	Rubus fructicosus, cv Chester	204–435	4	Heinonen et al., 1998; Wang and Lin, 2000
Blueberry	Vaccinium angustifolium Aiton	295–495	6	Prior et al., 1998
Blueberry	Vaccinium ashei Reade (Rabbiteye)	231–458	6	Prior et al., 1998
Blueberry	Vaccinium corymbosum	181–391	15	Prior et al., 1998; Heinonen et al., 1998
Grapes	Vitis vinifera, Vitis labrusca	44–309	14	Yi et al., 1997
Raspberry, red	Rubus idaeus	208–303	5	Heinonen et al., 1998; Wang and Lin, 2000
Raspberry, black	Rubus occidentalis L.	267	1	Wang and Lin, 2000
Strawberry	Fragaria x ananassa, Duch.	94–294	9	Heinonen et al., 1998; Wang and Lin, 2000
		μM GAE/g Fresh Weight		
Blueberry	Vaccinium angustifolium Aiton	28	20	Kalt et al., 1999a
Blueberry	Vaccinium corymbosum, cv Bluecrop	23	1	Kalt et al., 1999a
Raspberry, red	Rubus idaeus Michx., cv Nova	7	1	Kalt et al., 1999a
Strawberry	Fragaria x ananassa Duch., cv Kent	5	1	Kalt et al., 1999a
		mg GAE/g Berry Dry Weight		
Aronia	Aronia melanocarpa, cv Viking	40	1	Kähkönen et al., 1999
Black currant	Ribes nigrum, cv Öjebyn	20	1	Kähkönen et al., 1999

TABLE 3.3 (continued)
Total Phenolics (as Gallic Acid Equivalents, GAEs) in Selected Grapes and Berries

Species	Genus/Cultivar	Total Phenolics (GAE)	Number of Clones/Cultivars[1]	Source
Blackberry, thornless	*Rubus* spp.	12–15	3	Wang and Lin, 2000
Blueberry, lowbush	*Vaccinium angustifolium* Ait. cv Fandy	42[2]	1	Velioglu et al., 1998
Bilberry	*Vaccinium myrtillus*	30	1	Kähkönen et al., 1999
Cloudberry	*Rubus chamaemorus*	16	1	Kähkönen et al., 1999
Cowberry	*Vaccinium vites-idaea*	25	1	Kähkönen et al., 1999
Cranberry	*Vaccinium oxycoccus*	21	1	Kähkönen et al., 1999
Crowberry	*Empetrum nigrum*	51	1	Kähkönen et al., 1999
Gooseberry	*Ribes grossularia*	12	1	Kähkönen et al., 1999
Raspberry, red	*Rubus idaeus*, cv Ottawa	24	1	Kähkönen et al., 1999
Raspberry, red	*Rubus idaeus* L.	12–15	3	Wang and Lin, 2000
Raspberry, black	*Rubus occidentalis* L.	15	1	Wang and Lin, 2000
Red currant	*Ribes rubrum*, cv Red Dutch	13	1	Kähkönen et al., 1999
Rowanberry	*Sorbus aucuparia*	19	1	Kähkönen et al., 1999
Sea buckthorn	*Hippophae rhamnoides* L., cv Indian Summer	11[2]	1	Velioglu et al., 1998
Strawberry	*Fragaria x ananassa* Duch.	9–24	11	Kähkönen et al., 1999; Wang and Lin, 2000
Whortleberry	*Vaccinium uliginosum*	29	1	Kähkönen et al., 1999
		mg GAE/100 ml Extract[3]		
Grapes	*Vitis vinifera*	17–122	12	Meyer et al., 1997

[1] Sum of clones/cultivars in referred studies.
[2] Calculated as ferulic acid equivalents.
[3] Obtained from 2 g fresh grapes into 5 ml final extract.

phenolics of 6.5–18.6 mM gallic acid equivalents. This would correspond to 1106–3164 mg GAE/L. In a study with 16 red and 6 white Californian wines, total phenolics ranged from 1800 mg GAE/L to 4059 mg GAE/L for the red wines and from 198 mg GAE/L to 331 mg GAE/L for the white wines (Frankel et al., 1995).

3.4 PROCESSING OF BERRY FRUITS AND STABILITY OF FLAVONOIDS

3.4.1 PROCESSING

3.4.1.1 Juices and Concentrates

Commercially produced berry juice and juice concentrates can be very rich in anthocyanin pigments and polyphenolics, depending on fruit source and processing conditions. Thus, they can serve as nutraceuticals as is, or as sources for further processing into various preparations. Detailed descriptions of unit operations and processing conditions have recently been reviewed (Ashurst, 1995; Somogyi et al., 1996). Specific conditions for the processing of juices from strawberries and raspberries are reviewed by Deuel (1996), from cranberry, blueberry, currant and gooseberries by Stewart (1996) and from grapes by Morris and Striegler (1996).

In general, most berry fruits are frozen after harvest, pending juice processing (Moulton, 1995). Frozen fruits are traded around the world, the growing being dependent upon suitable climatic conditions and, in many cases, low labor costs. By using frozen fruits, the processor is free to plan year-round production, thereby utilizing the processing equipment optimally. However, with processing plants close to the growing sites, fruits may be processed fresh, chilled or after storage in a gas atmosphere. Grapes are mostly processed fresh within 4–6 h after harvest (McLellan and Race, 1995).

Prior to juice processing, the berries should be carefully thawed (Skrede, 1996) and milled. The freeze/thaw process may help in releasing pigments and other compounds from the cells (Stewart, 1996). For berry fruits, enzyme treatment is needed to release the juice from the cells and to improve pressability (Downes, 1995). Commercial enzyme preparations are available, containing pectinolytic enzymes specialized for breaking down the cell structures and dissolved pectins in the pulp, thereby releasing the juice. The maceration enzyme preparation may contain pectin methyl esterase, polygalacturonase and pectolase and also hemicellulase, cellulase and amylase (Deuel, 1996). The enzymes are added directly after milling, the pulp is heated to a temperature optimal for the enzymes and the depectinization is allowed to proceed for 1–2 h. For grapes, a hot-break process is often used, where the crushed grapes are passed through a heat exchanger and heated to 60°C (McLellan and Race, 1995; Morris and Striegler, 1996). The heating improves juice yield and yield of color, i.e., anthocyanin extraction. Pressing is performed either continuously or as a batch process. Especially with grapes, the mash is pressed while still hot. Hot pressing improves the extraction of total solids, tannins and anthocyanins, compounds that will be of main interest when producing juice for making nutraceutical products. After pressing, the juice most often needs further

TABLE 3.4
Content and Recovery of Anthocyanins, Chlorogenic Acid, Flavonol Glycosides and Procyanidins in Highbush Blueberries (*Vaccinium corymbosum* var. Bluecrop) during Juice Processing[1]

	Blueberry Fruit	Initial Pressed Juice	Press-Cake Residue	Pasteurized Juice	Concentrate	Loss[2]
Total Anthocyanin						
Sample (mg/100 g)	99.9	33.6	184	38.4	178	7.0
Blueberries (mg/100 g)	99.9a[3]	28.0d	18.2	32.0b	30.5c	7.0
Recovery (%)	100	28.0	18.2	32.0	30.3	7.0
Chlorogenic acid						
Sample (mg/100 g)	27.4	13.2	3.2	17.4	79.8	1.8
Blueberries (mg/100 g)	27.4	11.0	0.3	14.6	13.7	1.8
Recovery (%)	100	40.2	1.2	53.3	49.9	6.7
Flavonol glycosides[4]						
Sample (mg/100 g)	40.1	24.9	26.8	16.7	71.7	2.7
Blueberries (mg/100 g)	40.1	20.8	2.7	14.0	12.3	2.7
Recovery (%)	100	52.8	6.6	35.0	30.6	6.7
Procyanidins						
Sample (mg/100 g)	9.9	4.5	2.8	5.0	13.4	0.7
Blueberries (mg/100 g)	9.9	3.8	0.3	4.2	2.3	0.7
Recovery (%)	100	38.0	2.8	42.6	23.0	6.7

[1] Anthocyanin and polyphenolic contents are reported for both the actual sample (row labeled "Sample") and that derived from 100 g of berries (row labeled "Blueberries"). Juice compositional data are reported on the basis of the actual °Brix, 15.0°; to convert data to the U.S. single-strength °Brix standard of 10.0° (Anonymous, 1993), multiply by 0.666.

[2] Estimated from 7% loss in yield during processing.

[3] Numbers with different letters were significantly different ($p < 0.05$).

[4] Calculated as rutin.

Source: Adapted from Skrede et al. (2000a), with permission.

treatment to remove insoluble solids and improve clarity and stability (Downes, 1995). The treatments may include the use of enzymes and various clarifying aids prior to decanting, centrifugation or filtration. The juice is then pasteurized to kill deteriorating microorganisms and inactivate various enzymes and may then be placed in storage either as a single-strength juice or as a concentrate. Concentrates are achieved by evaporation, by which aroma components are often isolated and stored separately. By concentrating the juice, storage and transportation costs are reduced.

Extensive losses of polyphenolics can occur during processing of single-strength juice. Skrede et al. (2000a), using processing conditions typical of industry, recovered only 32% of the anthocyanin pigments in pasteurized highbush blueberry juice, whereas 18% remained in the press-cake residue after pressing of the pulp (Table 3.4). Approximately 50% of the pigments were lost. There were extensive

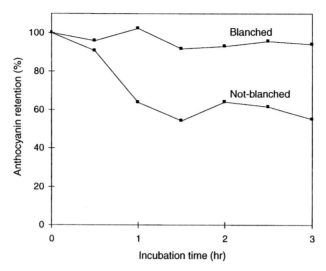

FIGURE 3.8 Anthocyanin retention in pasteurized juice from highbush blueberry (*Vaccinium corymbosum,* var. Bluecrop) incubated with blanched and not-blanched blueberry pulp for 3 h at 40°C. [Source: Skrede et al. (2000a), with permission.]

changes in the anthocyanin profile of the juice, the malvidin glycosides being the most stable and the delphinidin glycosides the least stable anthocyanins. The process involved thawing and crushing of the berries, followed by depectinization with a commercial enzyme preparation for 2 h at 43°C, pressing and pasteurization (90°C for 1 min). Similarly, only part of the initial chlorogenic acid, flavonol glucosides and procyanidins were recovered in the juice, the percent of recovery being 53%, 35% and 43%, respectively. In contrast to the anthocyanins, the press-cake residue contained only minor amounts of these phenolics, indicating that those polyphenolic compounds were more easily released from the blueberry fruit and skin tissue than the anthocyanins.

The losses in flavonoids and other phenolics during juice processing are caused by a thermolabile factor, most likely polyphenol oxidase (Kader et al., 1997). By treating pasteurized juice with fresh, unblanched blueberry homogenate, extensive reduction in anthocyanins occurred in the juice (Figure 3.8) (Skrede et al., 2000a). Treating the same juice with blanched blueberries left the anthocyanins intact. Inhibition of endogenous enzymes during processing by initial high-temperature heat treatment, for example, should be highly beneficial for the recovery of flavonoids and phenolics in blueberry juice (Stewart, 1996). The presence of enzymes able to degrade flavonoids seems to vary between types of berry fruits, as Iversen (1999) reported 91% recovery of anthocyanins in raw black currant juice after taking into account the 21% of the initial anthocyanins maintained in the pulp.

Some commercial juice processing enzyme preparations contain glycosidase side activities, which can hydrolyze the anthocyanin glycosidic linkages and liberate the unstable anthocyanidins. This will markedly accelerate anthocyanin pigment destruction. Wightman and Wrolstad (1995) tested various commercial enzyme preparations and found β-galactosidase activity to be present in some preparations,

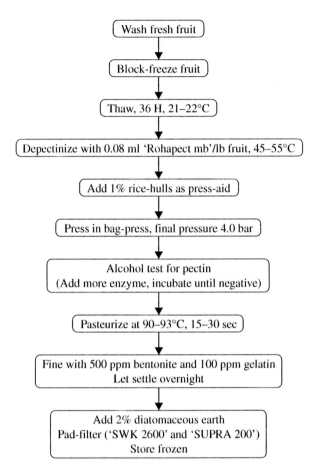

FIGURE 3.9 Flow diagram for juice processing by a standard technique. [Source: Reprinted with permission from Rommel, A. and Wrolstad, R.E. 1993. "Ellagic acid content of red raspberry juice as influenced by cultivar, processing and environmental factors," *J. Agric. Food Chem.* 41(11):1951–1960. Copyright © 1993 American Chemical Society.]

whereas β-glucosidase activity was nonexistent or present in very low levels (Wightman and Wrolstad, 1996). This is fortunate, because β-glucosides are much more widely distributed in fruits than β-galactosides. Another potential source of anthocyanin-degrading enzymes is molds, which produce extracellular enzymes with broad-spectrum glycosidase activities. Rwabahizi and Wrolstad (1988) demonstrated that mold contamination was very deleterious to the color quality of strawberry juice and concentrate.

In a study of ellagic acid in red raspberry, Rommel and Wrolstad (1993) used two different processing methods for obtaining juice. One method was standard pressing (Figure 3.9), a second centrifugation and a third diffusion extraction (Figure 3.10). It was concluded that the standard pressing method and the diffusion process produced juices with higher ellagic acid levels compared to centrifugation.

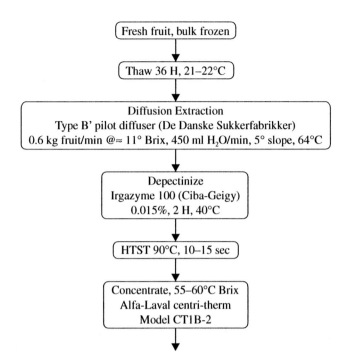

FIGURE 3.10 Flow diagram for juice processing by diffusion extraction. [Source: Rommel, A. and Wrolstad, R.E. 1993. "Ellagic acid content of red raspberry juice as influenced by cultivar, processing and environmental factors," *J. Agric. Food Chem.* 41(11):1951–1960. Copyright © 1993 American Chemical Society.]

Despite the different levels of ellagic acid in the juices produced by the three methods, the qualitative composition, i.e., the chromatographic profiles of ellagic acid and its various derivatives, was very similar.

Soluble solids (°Brix) of single-strength juices are approximately the same as that of the fruits from which they are derived. In the United States, single-strength °Brix standards exist for juices of commercial interest (Anonymous, 1993). During juice processing, flavonoids and other phenolics degrade to some extent, and a proportion of the polyphenolics will be retained in the press-cake residue. Polyphenolic levels of single-strength juice will thus always be lower than the original fruits. Juices are typically concentrated five- to sixfold with flavonoid levels being correspondingly increased. Grape juice concentrates are marketed as 55, 65 or 69 °Brix compared with the 13–15 °Brix of single-strength grape (Morris and Striegler, 1996). Skrede et al. (2000a) found that minor losses of anthocyanins, flavonol glycosides and chlorogenic acid occurred during concentration of highbush blueberry juice, whereas procyanidins were less stable with losses in the order of 20% (Table 3.4). Freeze-concentration is an alternative concentration process that should be much more protective of anthocyanins and polyphenolics. Spray-drying of juices is a common commercial practice (Abuja et al., 1998), usually with incorporation of maltodextrins as a protective agent.

3.4.1.2 Extracts and Powders

In juice processing literature, the term *extraction* is used to describe the process of obtaining juice and its water-soluble constituents from fruits. In this section, however, extraction will be extended to include processes that use various solvents for extracting the components of interest. Solvents are subsequently removed so that the final product is highly concentrated.

Mazza and Miniati (1993) and Bridle and Timberlake (1997) reviewed the procedures for extracting anthocyanins from grape pomace, the main by-product from the wine industry. Pomace consists of grape skin and is thus rich in anthocyanins and other phenolics. The traditional extraction procedure involves mild aqueous sulfur dioxide (Mazza and Miniati, 1993; Girard and Mazza, 1998). The sulfur dioxide improves the extraction of the anthocyanins and protects the pigments from oxidation and microbial spoilage. The sulfur dioxide is removed by evaporation during the final stages of the processing (Cohn and Cohn, 1996). Dilute aqueous acid, most often sulfuric acid, is also used for extracting grape pomace (Henry, 1996), as well as is acidified ethanol (Jackman and Smith, 1996). Ethanol is less efficient than methanol but is preferred due to the toxicity of methanol. To avoid degradation of anthocyanins, concentration is preferably performed under vacuum. After concentrating the extract to 20–30 °Brix, a product with 0.5–1% anthocyanins is obtained. The extracts can be dried in an oven or spray-dried, yielding a water-soluble powder. These powders may contain 4% anthocyanins. Meyer et al. (1998) reported enhanced extraction of total phenols after treating grape pomace with cell-wall degrading enzymes. Milling to reduce particle size will increase the release of phenols from the grape pomace.

Grape seeds comprise 4.4% of the weight of the grape, and they contain 5–8% phenolics by weight (Amerine and Joslyn, 1967). The seeds are particularly rich in oligomeric procyanidins, and a large number of procyanidins has been isolated (Escribano-Bailón et al., 1992; Girard and Mazza, 1998). Large quantities of grape seeds are available as by-products from the juice and wine industries. Commercial processes for extracting grape seeds are largely proprietary. However, Saito et al. (1998) briefly described two processes for preparing commercial grape seed extracts: extraction of seeds with 20% ethanol and aqueous extraction of whole seeds at 60°C for 2 h followed by hot water (90°C) extraction for an additional 2 h prior to evaporation. Labarbe et al. (1999) extracted ground grape seeds with acetone/water (60:40, v/v) followed by centrifugation and concentration under vacuum. Santos-Buelga et al. (1995) extracted ground, frozen grape seeds with methanol containing ascorbic acid as an antioxidant. Shrikhande (2000) describes the preparation of grape seed extracts, in which the seeds are extracted by aqueous acetone or ethanol, and the extracts are evaporated and fractionated by ethyl acetate to isolate the monomers and oligomeric procyanidins. The oligomers are precipitated with methylene chloride and eventually spray dried. A high-quality grape seed extract is stated to contain at least 90% total phenols as determined by the Folin-Ciocalteau method and a minimum of 10% monomeric, more than 65% oligomeric procyanidins and less than 15% polymeric or condensed tannins (Shrikhande, 2000).

3.4.2 STABILITY OF ANTHOCYANINS

The stability of anthocyanins may be viewed from the perspective of their importance as endogenous components of berry fruits and their products. In addition, a number of anthocyanin extracts are used as food colorants and may be added to various products at different stages during food processing. Several reviews of anthocyanin stability have been published over the years (Markakis, 1982; Jackman et al., 1987a; Henry, 1996; Jackman and Smith, 1996; Wrolstad, 2000).

When considering anthocyanin stability from a nutraceutical, rather than from a food, point of view, stability during processing has a direct effect on the yield of these compounds, whereas stability of the final products (powders, extracts, juices) influences product quality and, thus, sets requirements for packaging, handling, marketing and, finally, shelf life of the products. Some issues of concern to the food scientist such as consumer expectations regarding color intensity and hue may be of lesser importance or may even be irrelevant to nutraceutical manufacturers. Still, most of the information on anthocyanin stability in processed foods will be applicable and useful. In the following, a presentation of factors affecting stability of anthocyanins relevant for production, marketing and storage of nutraceuticals will be given.

3.4.2.1 Effects of Enzymes

Grapes and berry fruits contain enzymes within the group of phenolases (phenolases, polyphenolases) that can be very destructive of anthocyanins (Markakis, 1982). The activity varies considerably with different commodities, e.g., polyphenoloxidase is a major cause of anthocyanin destruction during blueberry juice processing (Skrede et al., 2000a). The anthocyanins are not favored as direct substrates for the pheno-lases; rather, there is consensus from most published works that phenolic acids, flavanols and flavonols are preferred substrates. A coupled oxidation mechanism has been proposed as an explanation for the accelerating effects of components such as chlorogenic acid, catechin and caffeoyltartaric acid on anthocyanin degradation by phenolase (Peng and Markakis, 1963; Sarni et al., 1995; Kader et al., 1997). With ascorbic acid present, the anthocyanins may be spared, as ascorbic acid rather than anthocyanins may react with the mediating phenolic component (Jackman and Smith, 1996).

Glycosidase enzymes can be very destructive of anthocyanins, but their source is unlikely to be the berries themselves. Moldy fruit is a potential source (Rwabahizi and Wrolstad, 1988) as are commercial juice-processing pectolytic enzyme prepa-rations that may have glycosidase side activities (Wrolstad et al., 1994). Native peroxidase enzymes may be a cause of anthocyanin destruction in some berry commodities. Active peroxidase may be a major contributor to anthocyanin pigment destruction in many processed strawberry products (López-Serrano and Ros Barceló, 1996; Zabetakis et al., 2000).

3.4.2.2 Effects of Molecular Structure and Copigmentation

Substitution patterns of the flavan nucleus along with glycosidation and acylation patterns influence anthocyanin stability of grapes and berry fruits to some extent.

In a study of the 3-glucosides of the six anthocyanidins occurring in grapes and berry fruits, degradation after 60 days of dark storage at pH 3.2 and 10°C ranged from 33% to 42% (Cabrita et al., 2000). Thus, for practical purposes, the effect of B-ring substitution patterns may be regarded as minor.

Inami et al. (1996) studied the effect of glycosidation and acylation on stability using anthocyanins from two elderberry species. Anthocyanins from *Sambucus canadensis* were more stable toward heat and light than anthocyanins from *S. nigra*. The differences in stability were assigned to differences in glycosidation and acylation of the anthocyanins. An additional sugar unit (3-, 5-diglycosides) stabilized the anthocyanin toward light, whereas acylation of the sugar moiety improved both heat and light stability. The effect of acylation has been explained as a folding of the acyl group over the planar pyrillium ring. Such intramolecular associations stabilize the anthocyanins (Wrolstad, 2000).

Anthocyanins form complexes with other polyphenolics (Jackman and Smith, 1996). At high concentrations, the anthocyanins may associate with themselves. Anthocyanins are more stable at higher concentrations (Wrolstad, 2000). The effect of concentration has been shown to be more important than the variation in stability caused by differences in anthocyanin structure (Skrede et al., 1992). Intermolecular copigmentation may take place between anthocyanins and catechin, amino acids, polysaccharides and metal ions. Flavonols and flavones are always found in conjunction with anthocyanins in fruits and fruit juices, and it appears that they may contribute to the stabilization of the anthocyanins (Jackman et al., 1987a). Condensation between grape tannins (condensed flavonoids) and anthocyanins has a protective effect on anthocyanins and also contributes to the color of aged red wine. Bobbio et al. (1994) found the presence of tannic acid in aqueous anthocyanin solutions to retard anthocyanin degradation during storage. In a U.S. patent, improved anthocyanin stability toward light, heat pH or a combination of these was obtained after the addition of flavonoid glucuronoids, flavanoid glucuronoids and caffeic acid derivatives to anthocyanin-based colorants (Lenoble et al., 1999). Pigment-enhancing agents are reported to originate from rosemary, sage and peppermint extracts (Bank et al., 1996).

3.4.2.3 Effects of Oxygen and Ascorbic Acid

Oxygen has a deleterious effect on anthocyanins, and it is known that anthocyanins stored under vacuum or nitrogen atmosphere are more stable than anthocyanins exposed to molecular oxygen (Jackman et al., 1987a). This implies that containers and packaging materials used for anthocyanins and their products should have high oxygen barriers. Also, headspace and packaging spare volumes should be minimized to prevent anthocyanin degradation during storage and marketing.

The oxygen may degrade anthocyanins either directly or indirectly by oxidizing compounds, which in turn, may degrade the anthocyanins (Jackman et al., 1987a). Such secondary oxidizing compounds may be metal ions and ascorbic acid. With ascorbic acid, the intermediate is thought to be hydrogen peroxide formed by oxidation of the ascorbic acid by molecular oxygen (Jackman and Smith, 1996). The deteriorating effect is most pronounced when both oxygen levels and ascorbic acid concentrations are high. The reactions are known to be accelerated by copper ions (Markakis, 1982).

In general, the effects of ascorbic acid on anthocyanins are complex and not easily predictable. In the absence of oxygen, ascorbic acid may condense with anthocyanins to form unstable products that degrade into colorless compounds (Markakis, 1982). It is assumed that condensation of anthocyanins with flavonols prevents formation of complexes between anthocyanins and ascorbic acid so that the deteriorating effect of ascorbic acid is diminished (Jackman et al., 1987a).

3.4.2.4 Effects of SO_2

Anthocyanins react with sulfur dioxide to form colorless components (Wrolstad, 2000). The reaction is reversible, and heating will release some of the sulfur dioxide from the anthocyanins, thus partially regenerating the color. Also, acidifying to a low pH regenerates the anthocyanins by liberation of the SO_2 (Jackman and Smith, 1996).

3.4.2.5 Effects of pH

Anthocyanins display their typical color at low pH values, where the molecules are in the oxonium form (Wrolstad, 2000). The oxonium form is more stable, and anthocyanins thus have higher stability in an acid environment. If pH rises to levels above 4.5, the anthocyanin instability is particularly problematic. The detailed molecular structure of the anthocyanins, i.e., acylation and the position of the glycosidic substitution, affects the exact pH at which the color changes occur. Copigmented anthocyanins are less susceptible to degradation.

A detailed study of the effects of pH on anthocyanin stability has been presented by Cabrita et al. (2000). Buffered solutions in the range of pH 1–12 of the 3-glucosides of the six common anthocyanidins were stored in the dark over a 60-day period at 10°C and 23°C. Under strong acidic conditions (pH 1–3), more than 70% of the initial concentration remained after 60 days at 10°C for all anthocyanins, while considerable losses (>90%) occurred at pH 5–6 even after 8 days. Similar stability patterns occurred at the higher temperature of 23°C, although the rates of anthocyanin degradation were higher, and only 40% of the initial anthocyanins were detectable after 60 days.

3.4.2.6 Effects of Water Activity and Sugar Content

Several studies have shown that anthocyanin stability increases with decreased water content, i.e., decreasing water activity (a_w) (Wrolstad, 2000). Dry anthocyanin powders ($a_w \leq 0.3$) are stable for several years when stored in hermetically sealed containers (Jackman and Smith, 1996). This property is advantageous when anthocyanins are produced into dry products and powders. Zajac et al. (1992) reported anthocyanin losses of 14% after 15 months of storage of a dry powder from black currant extracts at 20°C. When starch syrup or maltodextrins were included as carriers during the drying of the extracts, the anthocyanin losses were reduced to 2% and 3%, respectively.

High sugar concentrations (>20%) are protective of anthocyanins (Wrolstad et al., 1990). The effect is most likely caused by lowered water activity. Stasiak et al. (1998) reported three times longer anthocyanin half-life, i.e., the period of time required for 50% degradation of the anthocyanins, in pasteurized ascorbic-acid-enriched chokeberry extracts with 65% sucrose compared with sucrose-free extracts. At lower sugar levels, sugars and their degradation products are considered to cause accelerated anthocyanin degradation (Jackman and Smith, 1996), although slightly improved stability from lower sucrose levels (13%) has also been reported (Stasiak et al., 1998). Fructose, arabinose, lactose and sorbose are more detrimental than glucose, sucrose and maltose (Jackman and Smith, 1996). The degradation effect is associated with the degradation rate of the sugars themselves. The degradation of anthocyanins by sugar is enhanced by oxygen.

3.4.2.7   Effects of Temperature

It has been known for decades that heat is one of the most destructive factors of anthocyanins in berry fruit juices (Jackman et al., 1987a). With strawberry preserves, it was shown as early as 1953 that the half-life time was 1 h at 100°C, 240 h at 38°C and 1300 h at 20°C. In a storage experiment with concentrates and dry powder of elderberry extracts, the stability increased 6–9 times when the temperature was reduced from 20°C to 4°C (Zajac et al., 1992). Anthocyanin degradation in anthocyanin solutions increased from 30% to 60% after 60 days when storage temperatures were increased from 10°C to 23°C (Cabrita et al., 2000). High-temperature short-time processing is recommended for maximum anthocyanin retention of foods containing anthocyanins (Jackman and Smith, 1996).

3.4.2.8   Effects of Light

Visible and UV light are detrimental to anthocyanins and will even increase the rate at which anthocyanins undergo thermal degradation (Jackman and Smith, 1996). Carlsen and Stapelfeldt (1997) reported light to be a major cause of anthocyanin degradation in elderberry extracts. The effect depended upon the wavelength of irradiation, with the shorter wavelengths being the most deteriorating, and exclusion of UV light would thus greatly improve the stability of the extracts. Protection toward light may be achieved by selecting packaging material with proper light barriers in the visible and particularly in the ultraviolet range of the spectrum. Glycosidation, acylation and copigmentation of anthocyanins have been reported to improve light stability (Inami et al., 1996; Lenoble et al., 1999).

3.4.3   STABILITY OF FLAVONOLS, FLAVANOLS AND PHENOLIC ACIDS

Compared with anthocyanins, the information available on stability of flavonols, flavanols and phenolic acids is limited. As these compounds until recently had only minor interests as food components, investigations were mostly performed for the purpose of biological screening and identification, less for studying stability during processing and storage. However, the mere definition of flavonoids and phenolic acids as antioxidants indicates that the compounds react easily and should be handled as unstable compounds.

The enzyme polyphenol oxidase (PPO) is widely distributed in plants. For fruit processors, it is well known for being involved in various browning reactions. The enzyme catalyzes the formation of o-diphenols from monophenols as well as the oxidation of o-diphenols to o-quinones (Jackman and Smith, 1996; Kader et al., 1997; Jiménez and García-Carmona, 1999). The o-quinones may react with amino acids and proteins to form brown-colored polymers. In the presence of complex natural phenols such as anthocyanins, o-quinones formed by PPO from phenols such as catechin and chlorogenic acid may be regenerated through a nonenzymatic reaction. The regeneration of the initial catechin or chlorogenic acid is achieved at the expense of the anthocyanins. The degradation of flavonoids by PPO has thus been considered to take place through a coupled, indirect reaction mechanism. Jiménez and García-Carmona (1999) have demonstrated the direct oxidation of flavonols (i.e., quercetin) by PPO.

Caftaric acid has been shown to degrade faster by grape polyphenoloxidase than catechin, epicatechin, epicatechin gallate and procyanidins (Cheynier et al., 1988). In the presence of caftaric acid, the degradation rate of the flavonols increased. Coupled oxidations between caftaric acid and the flavonols, as well as formation of caftaric acid and catechin copolymers, were proposed.

Condensed tannins, the polymerized products of flavan-3-ols and flavan-3,4-diols, form complexes with proteins, starch and digestive enzymes and may cause a reduction in the nutritional value of foods (Chung et al., 1998). In fruit extracts, the tannins may cause precipitation in the presence of proteins. The high ability of the condensed tannins to form complexes with other food components reduces the extractability of those constituents from grapes and berries.

3.5 ANALYTICAL METHODS

Extensive reviews of analytical methods for anthocyanins (Francis, 1982; Jackman et al., 1987b; Strack and Wray, 1994) and other flavonoids (Williams and Harborne, 1994) as well as phenolic acids (Herrmann, 1989) have been published. In these reviews, extraction procedures, methods for fractionation of groups of polyphenols and the identification and quantification of individual components are presented. Here, a brief presentation of more recently published methods for grape and berry polyphenolic analyses is given with respect to their relationship to antioxidant activity and health benefits.

The complexity of polyphenolic composition of plant materials is such that complete identification has not been accomplished for many fruits. Often, compounds are partially characterized with respect to their compound class. Standards are often not available, but estimates of quantities can be made on the basis of reference compounds from the same class. Total polyphenolic levels can be obtained by summing the individual compounds within a class and then summing the different classes.

3.5.1 EXTRACTION AND FRACTIONATION

Anthocyanins and other polyphenolics have their highest concentrations in the epidermal tissue of grapes and many berry fruits (blueberries, chokeberries, cowberries).

Extensive disintegration and extraction of the skin and outer layers are needed for quantitative analysis of the flavonoids. In other fruits, such as raspberries and strawberries, the anthocyanins are more evenly distributed in the fruit tissue, and the flavonoids, to a much higher extent, dissolve in the juice obtained with homogenization of the fruit tissue. When quantitative assays are to be achieved, these fruits should be extracted just prior to analysis. If the primary interest is analysis of wines, juices and certain nutraceutical preparations, analysis may be performed directly on the samples after proper dilution (Wang et al., 1996; Frankel et al., 1998; Larrauri et al., 1999; Saint-Cricq de Gaulejac et al., 1999; Burns et al., 2000; Wang and Lin, 2000). Abuja et al. (1998) analyzed spray-dried elderberry extracts after dissolving the material in salt-containing phosphate buffer.

When preparing samples for analysis, care should be taken to prevent degradation of flavonoids and other phenolics prior to and during extraction. The use of antioxidants and reducing agents to prevent oxidation of phenolic compounds during extraction has been done in some investigations (Meyer et al., 1997). However, when antioxidant activities of berries are to be measured, any added antioxidants will interfere with the results and should be avoided. Freezing and subsequent grinding in liquid nitrogen is a method used to avoid oxidation during sample preparation (Labarbe et al., 1999; Skrede et al., 2000a), whereas extraction under anaerobic conditions to avoid oxidation has also been used (Meyer et al., 1997). When analyses are based on frozen material, care must be taken to avoid storage beyond several months. Several studies are based on freeze-dried grape and berry materials (Meyer et al., 1998; Häkkinen et al., 1999; Kähkönen et al., 1999; Lu and Foo, 1999; Saint-Cricq de Gaulejac et al., 1999), and the lyophilized samples are often finely ground prior to extraction. When possible, grinding of fresh grapes, grape skins and berries directly in the extraction medium is convenient (Strigl et al., 1995; Lapidot et al., 1998; Kalt et al., 1999a; Labarbe et al., 1999). Enzymic treatment of berry homogenates prior to extraction has been shown to decrease extraction yields (Heinonen et al., 1998). With many commodities, special precautions need to be taken to inactivate endogenous polyphenol oxidase (Skrede et al., 2000a).

Flavonoids and phenolic acids are hydrophilic compounds with high solubility in alcohol. The most widely used extraction media are aqueous acetone (Heinonen et al., 1998; Kähkönen et al., 1999; Labarbe et al., 1999; Karadeniz et al., 2000; Skrede et al., 2000a), aqueous methanol (Meyer et al., 1997; Heinonen et al., 1998; Lapidot et al., 1998; Kalt et al., 1999a) and aqueous ethanol (Strigl et al., 1995; Lu and Foo, 1999). Extraction is usually repeated until the pulp is colorless. After filtration or centrifugation, the extracts are combined and evaporated. Acetone extracts may be partitioned with chloroform, which removes carotenoids and lipids along with the bulk of the acetone (Karadeniz et al., 2000; Skrede et al., 2000a). Extraction of initial extracts with hexane may also be done to remove fat and fat-soluble compounds (Lu and Foo, 1999). The organic component of the extraction solvent is removed by evaporation under vacuum, taking care to keep the temperature low to prevent oxidation of the polyphenols. Samples are finally dissolved and diluted with acidified (1% v/v HCl) water or other solvents, depending upon the requirements of subsequent analysis. It is often convenient to directly extract with the solvent to be used in the analytical protocol (Prior et al., 1998; Iversen, 1999; Kalt et al., 1999a).

By using sequential extraction with different solvents, various groups of polyphenols may be separated directly during the extraction procedure (Saint-Cricq de Gaulejac et al., 1999). Bomser et al. (1996) reported an extraction scheme in which crude bilberry extracts were fractionated into purified total anthocyanins, procyanidins and various fractions of other phenolic compounds using a sequence of organic solvents (1% HCl acidified methanol, ethyl acetate, hexane and chloroform). By extracting wine with ethyl acetate, flavonols, flavanols and phenolic acids (organic phase) were separated from the anthocyanin-containing aqueous phase (Ghiselli et al., 1998). Further extraction of the organic phase with ethyl acetate at various pH values facilitated fractionation of the extract into subclasses of flavanols, flavonols and phenolic acids.

Solid-phase extraction (SPE) has been used for purifying extracts and samples prior to analysis. By using C-18 cartridges, sugar and acids are removed from crude extracts (Kähkönen et al., 1999; Skrede et al., 2000a). Furthermore, anthocyanins can be separated from other flavonoids with ethyl acetate elution (Kondo et al., 1999; Skrede et al., 2000a). This procedure is convenient if anthocyanins are interfering with the analysis of other polyphenolics. Lapidot et al. (1998) used SPE for purifying pigments from the mobile phase after chromatography.

3.5.2 HYDROLYSIS OF GLYCOSIDES AND ESTERS

Acid hydrolysis of anthocyanins (1–2 N HCl at 100°C for 30 min) will cleave the glycosidic substituents (Francis, 1982; Skrede et al., 2000a), permitting identification of anthocyanidins and individual sugars. The anthocyanidins generated are unstable, and samples should be flushed with nitrogen, stored on ice and analyzed without delay (Hong and Wrolstad, 1990). Similar conditions have been used to hydrolyze other phenolic glycosides from berry extracts using HCl and varying temperatures and times (Hertog et al., 1992; McDonald et al., 1998; Häkkinen et al., 1999). For the quantification and characterization of individual units of proanthocyanidins, acidic hydrolysis with phenyl-methanethiol (thiolysis) has been used (Labarbe et al., 1999). With this method, proanthocyanidin units are released as flavan-3-ols, and degree of polymerization and molecular weight of the proanthocyanidins can be determined.

For those anthocyanins that contain acyl substituents, the aliphatic and aromatic acids are readily released by alkaline saponification, 10% KOH for 15 min at room temperature (Rodriguez-Saona et al., 1998). The individual organic acids and the anthocyanin glycoside produced can be subsequently identified by various chromatographic procedures. Alkali has also been used to cleave conjugated caffeic acid and *p*-coumaric acid (Burns et al., 2000).

3.5.3 TOTAL PHENOLICS

The phenolics of grapes and berry fruits constitute a large group of compounds with various structural complexities. Thus, total phenolics encompass the more simple phenolic acids along with the various classes of flavonoids, some occurring as glycosides (anthocyanins, flavonols), some as esters (cinnamates, some antho-cyanins) and some in various polymeric forms (flavanols, procyanidins). For many applications, a simple measure of the total amounts of phenolics is often desired. The most common method available for this purpose is a spectroscopic method based on the Folin-Ciocalteau reagent (Slinkard and Singleton, 1977). The results are usually expressed as GAEs, however, this choice is somewhat arbitrary, and other reference standards have been used depending upon the composition of the sample and the purpose of the investigation (Ghiselli et al., 1998; Velioglu et al., 1998). It should be remembered that the method measures the number of potentially oxidizable phenolic groups. The number of phenolic groups per molecule will vary greatly both within and among different phenolic compound classes. This procedure still provides a very useful index for phenolic content, but it would not be expected to correlate with the actual weight of phenolics present. A number of investigations, however, have shown high correlation between total phenolics measured by this method and antioxidant activity as measured by various procedures.

3.5.4 Anthocyanins

The determination of monomeric anthocyanins is based on the ability of anthocya-nins to shift in color from bright red/bluish at pH 1 to nearly colorless at pH 4.5 (Wrolstad, 1976; Wrolstad et al., 1982). In contrast, the polymeric anthocyanin forms retain considerable color at pH 4.5. The absorbency at the absorbance maximum of the sample at pH 4.5 is subtracted from that at pH 1, and the total monomeric anthocyanins are calculated based on the molecular weight and extinction coefficient of the most prevalent anthocyanin. Total anthocyanin indices that also include the polymeric pigments are determined from the absorbency at pH 1, omitting measure-ment at pH 4.5 (Burns et al., 2000).

3.5.5 Total Flavanols

Total flavanols may be determined by the spectrophotometric vanillin method using (+)-catechin as the reference (Saito et al., 1998). Corrections must be made for any anthocyanins present (Broadhurst and Jones, 1978; McMurrough and Baert, 1994). Furthermore, procyanidins are transformed into anthocyanidins when treated at high temperature in acidic water-free conditions (Porter et al., 1986; Simonetti et al., 1997; Saint-Cricq de Gaulejac et al., 1999). For anthocyanin-rich samples, complete separations may be difficult to achieve, and interference between the anthocyanidins formed during the reaction and those naturally occurring in the samples provides confounding factors and limitations for this method (Skrede, 2000b).

3.5.6 HPLC Analysis

High-performance liquid chromatography (HPLC) is the method for detection, iden-tification and also quantification of flavonoids, phenolic acids and their derivatives. With this method, the sample is applied and eluted through a chromatographic column under specific conditions designed for optimum separation and resolution so that each compound or group of compound passes through the column with a different speed. With a detector at the column outlet, each compound generates a characteristic signal proportional to the amount present. By comparison with known substances either mixed with the sample (internal standard) or analyzed in a separate run (external standard), the analysis can be made quantitatively. When analyzing plant material, however, the detailed identity of each compound is seldom known, and compounds are often grouped by their typical absorbance maxima: phenolic acids and flavanols (280 nm), flavonols (260 or 365 nm) and anthocyanins (520 nm) (Meyer et al., 1997; Häkkinen et al., 1999; Skrede et al., 2000a). With a diode-array detector available, the entire UV visible spectrum of the compound may be matched with spectra of standard compounds (Larrauri et al., 1999), so that peak identification becomes more reliable.

There is a large number of chromatographic systems published for the separation of flavonoids and phenolic acids. Most methods use reversed-phase chromatography on a C-18 column equipped with a corresponding pre-column. The mobile phases are aqueous organic solvents applied most often as linear gradients with increasing proportions of the organic fractions. Elution systems are based on aqueous phosphoric acid mixed with ethanol (Frankel et al., 1998; Lapidot et al., 1998), acetonitrile (Häkkinen et al., 1999) or acetonitrile and methanol (Karadeniz et al., 2000; Skrede et al., 2000a); aqueous formic acid and methanol (Strigl et al., 1995; Cabrita and Andersen, 1999; Kalt et al., 1999b); acetonitrile (Strigl et al., 1995; Iversen, 1999); or both methanol and acetonitrile (Ghiselli et al., 1998). Also, aqueous acetic acid and acetonitrile (Larrauri et al., 1999; Lu and Foo, 1999) as well as aqueous trifluoracetic acid (TFA) and acetonitrile (Burns et al., 2000) have been used for the separation of the various groups of flavonoids and phenolic acids by HPLC.

Electrospray mass spectroscopy (ESMS) is a recent auxiliary technique shown to be very effective for more positive anthocyanin identification. The ESMS unit can be coupled directly to the HPLC outlet (Ghiselli et al., 1998) or, alternatively, used as an off-line method (Skrede et al., 2000a). A typical HPLC chromatogram and the corresponding ESMS spectra of highbush blueberries (*V. corymbosum*) anthocyanins are shown in Figure 3.11, with the explanatory peak assignments given in Table 3.5.

3.6 ANTIOXIDANT ACTIVITY OF FLAVONOIDS AND PHENOLIC ACIDS

3.6.1 ANTIOXIDANT ACTIVITY OF INDIVIDUAL COMPOUNDS

3.6.1.1 Mechanisms for Antioxidant Activity of Flavonoids

An antioxidant can be defined as a substance that significantly delays or prevents the oxidation of another substance even when present at a low concentration compared to the oxidizable substance of interest (Halliwell, 1990). For flavonoids, the antioxidant capacity has been linked to their radical scavenging ability (Bors et al., 1990). The flavonoids react readily with radicals such as hydroxyl ($\cdot$OH), azid (N3$\cdot$) and peroxyl (ROO$\cdot$), thereby forming stable flavonoid radicals. The flavonoid radicals are stabilized by extensive electron delocalization within the molecule, an ability that is crucial for the radical scavenging ability of antioxidants. The reaction

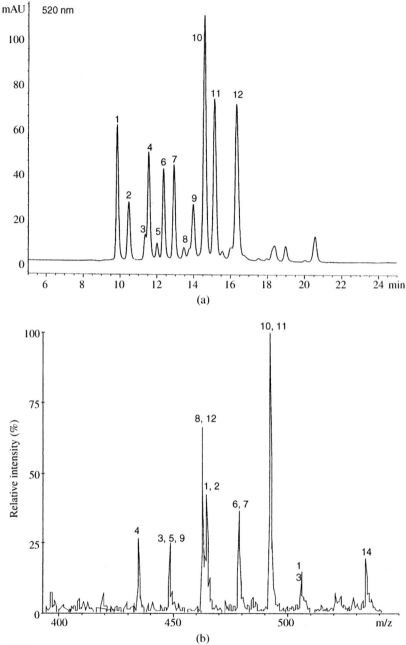

FIGURE 3.11 HPLC chromatogram (a) and ESMS spectrum (b) of anthocyanins in highbush blueberry (*Vaccinium corymbosum* var. Bluecrop) extracts. The anthocyanins were isolated by solid-phase extraction of C-18 cartridges with acidified methanol—cartridges had previously been washed with acidified water and by ethyl acetate. Peak identities for both the HPLC chromatogram and ESMS spectrum correspond to numbers in Table 3.5. [Source: Skrede et al. (2000a), with permission.]

TABLE 3.5
Peak Assignments for Highbush Blueberry (*Vaccinium corymbosum* var. Bluecrop) Anthocyanins Analyzed by HPLC and ESMS

Anthocyanin	HPLC[1] Peak #	ESMS Mass/Charge Ratio m/z
Delphinidin-3-galactoside	1	465.2
Delphinidin-3-glucoside	2	465.2
Cyanidin-3-galactoside	3	449.2
Delphinidin-3-arabinoside	4	435.0
Cyanidin-3-glucoside	5	449.2
Petunidin-3-galactoside	6	479.2
Petunidin-3-glucoside	7	479.2
Peonidin-3-galactoside	8	463.2
Petunidin-3-arabinoside	9	449.0
Malvidin-3-galactoside	10	493.2
Malvidin-3-glucoside	11	493.2
Malvidin-3-arabinoside	12	463.0
Acetylated delphinidin-galactoside/glucoside	13	507.0
Acetylated malvin-galactoside/glucoside	14	535.0

[1] Peak assignments according to Ballington et al. (1987) and Gao and Mazza (1994).

Source: Skrede et al. (2000a), with permission.

rates of flavonoids toward lipid peroxyl (LOO·) radicals and especially toward superoxide (O_2^-) are lower than for the radicals mentioned previously. In lipophilic systems, the rates of scavenger reactions may be influenced by the partition coefficients of the flavonoids between the aqueous and the lipid phases, and thereby, reduced availability of the polar flavonoids for reaction with the nonpolar LOO· radicals may result (Rice-Evans et al., 1996). The metal-chelating ability of flavonoids (Shahidi et al., 1992) and the ability to react with the α-tocopherol radical to regenerate α-tocopherol (Rice-Evans et al., 1996) are possible mechanisms for the ability of flavonoids to enhance fatty acid stability.

3.6.1.2 Methods for Assessing Antioxidant Capacity

The reactivity of flavonoids with different radicals has been utilized for assessing *in vitro* antioxidant capacity. The methods are based on the reactivity of the polyphenols with a stable radical (Yoshida et al., 1989). In order for a radical to serve as a suitable reagent in measuring the general antioxidant capacity of flavonoids, it should react with most types of flavonoids without discrimination. If the transformation of the radical into the reduced form or of the flavonoid into the radical form is accompanied by any changes in measurable properties, suitable instrumentation can be used to monitor the extent of the reaction.

At present, the most preferred method developed to study radical scavenging effects of flavonoids seems to be the oxygen radical absorbing capacity (ORAC) method that utilizes a peroxyl radical (ROO·) generator and the water-soluble fluorescent indicator phycoerythrin (Cao et al., 1993). Recently, 6-carboxyfluorescein has been proposed as a more stable dye for a fluorometric assay for measuring the oxygen radical scavenging activity of water-soluble antioxidants (Naguib, 2000). In addition, total antioxidant activity (TAA) and Trolox equivalent antioxidant capacity (TEAC) (Rice-Evans et al., 1996) assays, as well as the DPPH (1,1-diphenyl-2-picrylhydrazyl radical or 2,2-diphenyl-1-picrylhydrazyl) (Yoshida et al., 1989; Brand-Williams et al., 1995; Fukumoto and Mazza, 2000) assay are widely used. Both ORAC and TEAC report their results relative to Trolox, a water-soluble tocopherol analogue, whereas the results from the DPPH method are reported relative to the amount of DPPH required for a complete reaction (Fukumoto and Mazza, 2000) or for a 50% reduction in DPPH (Brand-Williams et al., 1995). Fukumoto and Mazza (2000) modified the DPPH method by using microtitration plates so that a large number of samples can be screened simultaneously. The radical scavenger ability of flavonoids has further been monitored by electron spin resonance (ESR) spectroscopy (Gardner et al., 1999), by chemiluminescence (Kondo et al., 1999) and by methods based on the aggregation of human blood platelets (Ghiselli et al., 1998) and the oxidation of rat liver microsomes and rabbit erythrocyte membrane systems (Tsuda et al., 1994).

The ability of flavonoids to enhance the resistance to oxidation and to terminate free-radical chain reactions in lipophilic systems can be monitored using low-density lipoproteins (LDL) as a model (Rice-Evans et al., 1996). The LDL oxidation is initiated either by copper or by a peroxyl radical [2,2′azobis(2-amidinopropane hydrochloride) (AAPH)] (Abuja et al., 1998). Hexanal liberated from the decomposition of oxidized n-6 polysaturated fatty acids in LDL may be determined by static headspace gas chromatography (Frankel and Meyer, 1998). Also, bleaching of β-carotene (Velioglu et al., 1998; Fukumoto and Mazza, 2000) and the tracing by HPLC (Fukumoto and Mazza, 2000) of malonaldehyde formed in lipid emulsion systems in the presence of iron (Tsuda et al., 1994) have been used to measure antioxidants in lipophilic systems.

The antioxidant activity of a compound depends upon which free radical or oxidant is used in the assay (Halliwell and Gutteridge, 1995), and a different order of antioxidant activity is therefore to be expected when analyses are performed using different methods. This has been demonstrated by Tsuda et al. (1994) in their study of antioxidative activity of an anthocyanin (cyanidin-3-O-β-D-glucoside) and an anthocyanidin (cyanidin) in four different lipophilic assay systems. Both compounds had antioxidative activity in all four systems, but the relative activity between them and their activity, compared with Trolox, varied with the method used. Fukumoto and Mazza (2000) reported that antioxidant activity of compounds with similar structures gave the same trends, although not always the same results, when measured by β-carotene bleaching, DPPH and HPLC detection of malonaldehyde formation in linoleic acid emulsion.

3.6.1.3 Structural Requirements for Antioxidant Activity of Individual Flavonoids

Extensive research has been carried out to study the relationship between molecular structure and radical scavenger properties of flavonoids. In short, the availability of a phenolic hydrogen from a hydrogen-donating radical scavenger such as the flavonoids, is predictive of the antioxidant activity of an individual flavonoid (Rice-Evans et al., 1996). Three criteria have been identified as important for facilitating hydrogen donation, and thereby for effective radical scavenging and antioxidant potential in flavonoids (Bors et al., 1990). These are the *o*-dihydroxy structure in the B ring; the 2,3-double bond in conjugation with a 4-oxo-function in ring C; and the additional presence of the 3- and 5-OH groups in rings A and C (Figure 3.12).

The individual differences in antioxidant efficiency observed within each group of flavonoids result from the variation in number and arrangement of the hydroxyl groups on the flavonoid structure and from the type and extent of alkylation and glycosylation of the hydroxyl groups. A hierarchy can thus be made for the anti-oxidant potential of flavonoids against radicals generated in the aqueous phase relative to that of Trolox (Cooper-Driver and Bhattacharya, 1998). In accordance with this, quercetin (2,3-double bond in conjugation with the 4-oxo-function in ring C) is a more effective antioxidant than catechin (a flavanol lacking these functionalities). With the TEAC method for antioxidant activity, Rice-Evans et al. (1996) reported a TEAC value for quercetin of 4.7 mM versus 2.4 mM for catechin. Similarly, quercetin (flavonol, flavone-3-ol), catechin (flavanol, 3-hydroxyflavan) and cyanidin (anthocyanidin) all have the same hydroxyl arrangements and differ only in the saturation of the heterocyclic C ring. When measured by the TEAC method, quercetin and cyanidin have similar values (4.7 mM and 4.4 mM, respectively), whereas the antioxidant activity of catechin is lower (2.4 mM). Glycosylation of a flavonoid reduces radical scavenger activity compared with the aglycon, as it reduces the ability of the flavonoid radical to delocalize electrons. In accordance with this, Fukumoto and Mazza (2000) reported increased antioxidant activity with an increase in the hydroxyl groups and decreased antioxidant activity with glycosylation of anthocyanidins. A full understanding of the correlation between antioxidant activity and chemical structure is not yet clear.

In the lipophilic phase, the 3′,4′-*o*-dihydroxy configuration in the B ring and the 4-carbonyl group in the C ring are particularly important, while the 2,3-double bond in the C ring appears to be less important for antioxidant activity (Rice-Evans et al., 1996). Tamura and Yamagami (1994) reported enhanced antioxidant activity of malvidin-3- and -3,5-glucosides through acylation with *p*-coumaric acid, compared to the same malvidin glucosides without acylation. Thus, molecular structure also influences the antioxidant ability of flavonoids in lipophilic systems. In systems that measure polyphenols directly as radical scavengers and by resistance to oxidation of LDL, all the major polyphenolic constituents in food show greater efficacy as antioxidants on a mole to mole basis than the antioxidant nutrients vitamin C, vitamin E and β-carotene (Rice-Evans et al., 1996).

3.6.2 ANTIOXIDANT ACTIVITY OF GRAPES AND BERRIES AND THEIR PRODUCTS

Due to the different methods used for evaluating antioxidant capacity of flavonoid from grapes and berries, direct comparisons between various species and products often cannot be made. However, a wide range of berry extracts, wines and juices

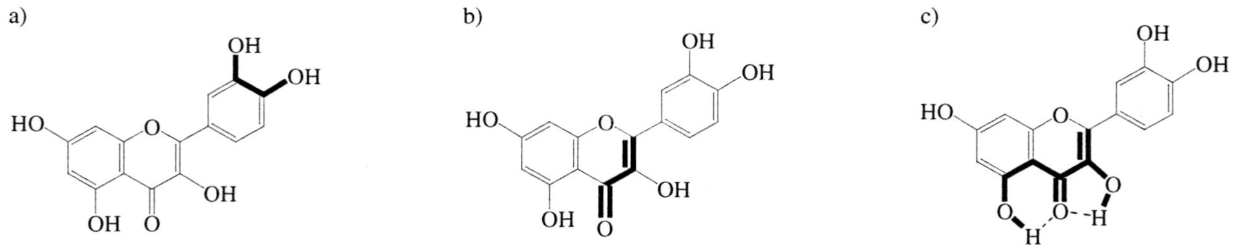

FIGURE 3.12 Flavonoid structures favoring free-radical formation in flavonoids. [Source: Bors et al. (1990), with permission.]

have recently been studied, and significant antioxidant activities have been assigned to most of them by one or another analytical method.

In nature, flavonoids usually occur as mixtures of several compounds rather than as pure solutions of one single polyphenol. To simulate this, attempts have been made to deduce the antioxidant capacity in fruits and berries and their products from the antioxidant capacity of their individual polyphenols. Frankel et al. (1995) calculated the antioxidant capacity of 14 red and 6 white wines from the antioxidant activity of the individual polyphenols determined in the wines. The calculated antioxidant activity accounted for only 25% of the measured value. The difference was partly ascribed to unidentified polyphenols and polyphenolic acids and their polymers. Thus, there apparently is still no complete understanding of the factors contributing to the antioxidative capacity of a product. The best strategy at present seems to be to measure the products of interest directly.

3.6.2.1 Grape and Berry Extracts

The high phenolic content of grapes makes these fruits interesting as sources of antioxidants. The results from an extensive study of the antioxidant potential of grape extracts from different grape varieties have been reported (Meyer et al., 1997; Yi et al., 1997). The selection of varieties included both red and white wine grapes as well as table grapes. The grapes were harvested at commercial maturity on the basis of the sugar content. Antioxidant potential was tested in extracts obtained after extraction of the grapes with 60% methanol in water, removal of the methanol by evaporation and final dilution with water. Total phenolic content of extracts varied from 168 to 1236 mg GAE/L in the two studies. For the red grapes, anthocyanins accounted for the main proportion of phenolic compounds, whereas flavonols dominated in the white grape extracts. Antioxidant activities were evaluated by a copper-initiated lecithin liposome oxidation assay using the inhibition of conjugated diene hydroperoxides and hexanal formation as the criteria (Yi et al., 1997) as well as by a copper-initiated human LDL oxidation assay similarly based on hexanal (Meyer et al., 1997). The results revealed significant differences in antioxidant activity among grape varieties with all methods. However, the relative order of antioxidant activity among the varieties depended upon the method used for evaluation. When compared at the same phenolic concentration (20 µM GAE), the grape extracts inhibited the formation of conjugated diene hydroperoxides by 25.1–67.9% and hexanal formation by 49.3–97.8% in the lecithin liposome system and by 62–91% in the human LDL system. The relative percentage inhibition of all methods correlated with total phenols of the extracts.

Wang and Lin (2000) measured antioxidant activity (ORAC) and total phenolic and anthocyanin contents of thornless blackberry, strawberry and red and black raspberry fruits and found linear relationships between both ORAC and total phenolics and between ORAC and anthocyanins of ripe fruits. Spray-dried elderberry juice with high amounts of anthocyanin glucosides caused prolongation of the lag-phase for Cu-induced oxidation of human LDL, while the maximum oxidation rate remained unchanged (Abuja et al., 1998). For peroxyl-radical-driven LDL oxidation, however, both prolongation of lag time and reduction of maximum oxidation rate occurred.

Heinonen et al. (1998) studied the antioxidant activity of extracts from black-berries, highbush blueberries, red raspberries, strawberries and sweet cherries. The phenolic compounds were extracted by using either aqueous methanol (60%) or acetone (70%), evaporated and finally diluted with water. The antioxidant activity was determined by a human LDL oxidation assay or by a lecithin liposome oxidation assay. With the LDL system, the antioxidant activity was monitored as the inhibition of hexanal formation, while in the liposome system, both the inhibitions of hexanal and hydroperoxide formations were detected. When tested at concentrations of 20 µM GAE/L, all extracts except strawberries and blueberries completely inhibited LDL oxidation. At 10 µM GAE/L, the inhibition ranged from 53.9 to 83.9% and was in the decreasing order of blackberries > red raspberries > sweet cherries > blueberries > strawberries. In the liposome system, the orders of antioxidant activity among the berries changed, even between the conjugated dienes and the hexanal inhibition. Inhibitions by 27.4–68.8% with the dienes and by 60.5–92.6% with the hexanal were seen at extract concentrations of 10 µM GAE/L. With conjugated diene inhibition, the antioxidant activity decreased in the order of sweet cherries > red raspberries > blackberries > blueberries > strawberries, whereas the order was sweet cherries > blueberries > red raspberries > blackberries > strawberries with hexanal inhibition. In this study, all berry extracts inhibited both LDL and liposome oxida-tion. However, the relative antioxidant activity was different when determined by the different methods, and the differences could not easily be explained by the phenolic content of the berry extracts. While the antioxidant activity in the LDL system was directly associated with anthocyanins and indirectly by flavonols of the berries, the activity was correlated with the hydroxycinnamates for the liposome method. It was concluded that berries are good sources of phenolic antioxidants, but that more research is needed to understand the antioxidant potential of phenolic compounds as they occur in nature.

Among 28 Canadian plant products, antioxidant activities of methanolic extracts of blueberries, sea buckthorn fruit and sweet cherries were determined (Velioglu et al., 1998). A method utilizing bleaching of β-carotene was applied. Both blueberries and buckthorn showed high antioxidant activities, whereas sweet cherries showed lower activity. Among all plant products tested, the anthocyanin-containing products pos-sessed high antioxidant activities. In a recent study, phenolic content and antioxidant activity of blackberries, black currants, blueberries and Saskatoon berries were com-pared (Fukumoto and Mazza, 2000). Methanol extracts (80%) of the berries were comparable in total phenolics (347–405 mg chlorogenic acid/100 g), while there were larger discrepancies in anthocyanins (149–233 mg malvidin 3-glucoside/100 g). The authors concluded that the four extracts had similar antioxidative activity when mea-sured by three different methods (β-carotene bleaching, DPPH and malonaldehyde determination by HPLC).

In a study of 92 Finnish edible and nonedible plant materials, a total of 15 berry samples were screened for antioxidant activity (Kähkönen et al., 1999). The activity was compared with the total phenolic content of the samples. Inhibition of methyl linoleate oxidation was used as a measure of antioxidant activity of berry acetone extracts and total phenolics determined by the Folin-Ciocalteau method as GAE/g dry matter of extracts. Berry extracts were high in antioxidant activity; among the species tested, crowberry, cloudberry, whortleberry, cowberry, aronia, rowanberry and cranberry had the highest activities. The lowest antioxidant activities were seen for red currant, black currant, strawberry and raspberry, whereas bilberry and goose-berry had intermediate antioxidant activities. No significant correlation was obtained between total phenolic content and antioxidant activity in this study. The authors conclude that the antioxidant activity of an extract cannot be predicted from the content of total phenols. They ascribe this partly to varying response of the phenolic subgroups in the Folin-Ciocalteau method and to the fact that the antioxidant activity of phenols depends upon their chemical structure.

A commercial *V. myrtillus* anthocyanin extract inhibited lipid peroxidation (IC_{50} = 50.28 µg/ml) and scavenged superoxide anions (IC_{50} < 25 µg/ml) and hydroxyl radicals (Martín-Aragón et al., 1998). Pigments isolated by high-speed counter-current chromatography from commercial black currant (*Ribes niger*) extract exerted antioxidant activities comparable to those of pure anthocyanins on a molar basis (Degenhardt et al., 2000). The analyses were performed using the TEAC method at pH 7.4, a pH value reflecting the physiological condition.

With the ORAC method, comparisons between various studies can be performed (Table 3.6). At present, the availability of data is still rather limited (Wang et al., 1996; Prior et al., 1998; Kalt et al., 1999a). When ranked on a fresh weight basis, lowbush blueberries have the highest ORAC values, followed by highbush blueberry, bilberry, rabbiteye blueberries, red raspberry, strawberry and red and white grapes. Judged on a dry matter basis, the ranking is bilberry, lowbush blueberry, highbush blueberry, rabbiteye blueberry, strawberry and red and white grapes. It is interesting to note that for blueberries, germ plasm is available with higher activity than the principal varieties in commercial usage today.

In a study with acetone/acetic acid extracted berries of different cultivars within four different *Vaccinium* species, ORAC values comparable to those of strawberry pulp were obtained (Prior et al., 1998). Berries from six commercial cultivars of *V. corymbosum* L. (Northern Highland) averaged 107.2 µmol TE/g on a dry weight basis—the cultivars ranged from 63.2 to 182.8 µmol TE/g. Berries from three commercial cultivars of *V. ashei* Reade (rabbiteye) had ORAC values ranging from 85.0 to 206.5 µmol TE/g, the average being 136.2 µmol TE/g. Among cultivars not commercially available, even higher ORAC values were obtained. The average for five cultivars of *V. corymbosum* L. was 173.4 µmol TE/g, and for five cultivars of *V. angustifolium* (lowbush), it was 229.8 µmol TE/g. These ORAC values were comparable to the ORAC value of 282.3 µmol TE/g obtained for wild bilberries, *V. myrtillus* L., in the same study.

Prior et al. (1998) reported that ORAC values of blueberries increased with maturity of the fruits, while various growing places throughout the United States did not affect the ORAC values. A linear relationship between ORAC values, i.e., antioxidant capacity as determined by the ORAC method, was obtained with the anthocyanin (r_{xy} = 0.77) and total phenolics (r_{xy} = 0.92) content.

During storage of fresh strawberries, raspberries and highbush and lowbush blueberries at 0, 10, 20 and 30°C for up to 8 days, the antioxidant capacity was found to be stable or even to increase (Kalt et al., 1999a). The antioxidant capacity was strongly correlated with the content of total phenolics (r = 0.83) and anthocyanins

TABLE 3.6
ORAC Values (μmol Trolox Equivalents/g Fresh Fruit) of Selected Grapes and Berry Fruits

Species	Genus	ORAC (μmol TE/g) Fresh Weight	Dry Weight	Number of Clones/ Cultivars[1]	Source
Fruits					
Bilberry	*Vaccinium myrtillus* L.	45	282	1	Prior et al., 1998[2]
Blueberry, lowbush	*Vaccinium angustifolium* Aiton	26–64	83–277	26	Prior et al., 1998[2]; Kalt et al., 1999a[3]
Blueberry, highbush	*Vaccinium corymbosum* L.	17–60	63–231	15	Prior et al., 1998[2]; Kalt et al., 1999a[3]
Blueberry, rabbiteye	*Vaccinium ashei* Reade	14–38	85–207	6	Prior et al., 1998[2]
Blueberry, highbush	*Vaccinium corymbosum* L.	17–37	63–183	8	Prior et al., 1998[2]
Strawberry	*Fragaria x ananassa* Duch.	15–21	154	4	Wang et al., 1996; Kalt et al., 1999a[3]
Blueberry, lowbush	*Vaccinium angustifolium* Aiton	26	83	1	Prior et al., 1998[2]
Raspberry, red	*Rubus idaeus* Michx.	21	—	1	Kalt et al., 1999a[3]
Grape, red	*Vitis*	7.4	36	3	Wang et al., 1996
Grape, white	*Vitis*	4.5	26	3	Wang et al., 1996
Juices					
Strawberry	*Fragaria x ananassa* Duch.	12–17	121–172	9	Wang et al. 1996[3]; Wang and Lin, 2000
Blackberry, thornless	*Rubus* spp.	20–22	121–146	3	Wang and Lin, 2000
Raspberry, black	*Rubus occidentalis* L.	28	136	1	Wang and Lin, 2000
Raspberry, red	*Rubus idaeus* L.	16–20	91–115	3	Wang and Lin, 2000

[1] Sum of clones/cultivars in referred studies.
[2] ORAC as μmol TE/g dry weight calculated from fresh weight and dry matter.
[3] ORAC as μmol TE/g fresh weight only.

($r = 0.90$) in the berries. The ORAC values obtained for strawberries and blueberries deviated considerably from values reported previously using the same method (Wang et al., 1996; Prior et al., 1998). The authors explain this as being due to genetic and environmental factors influencing both total phenolics and anthocyanin contents as well as antioxidant capacities of the fruits.

3.6.2.2   Red Wine

Red wines have been shown to have higher antioxidant capacity than rosé and white wines when analyzed by a linoleic acid peroxidation assay based on measuring chemiluminescence (Kondo et al., 1999). The higher antioxidant activity was related to the higher concentrations of polyphenols such as anthocyanins and flavanols in the red wines.

In products with complex combinations of various types of flavonoids and other phenolic compounds, effort has been made to ascribe the antioxidant activity of the product to different classes of polyphenols. Frankel et al. (1995) studied 20 selected California wines and related the antioxidant activity to the polyphenolic components of the wines, rather than to resveratrol. Antioxidant activity was measured by the ability of the wines to inhibit copper-catalyzed oxidation of human LDL. The correlation coefficient between antioxidant activity and total phenolic components of the wines was $r = 0.94$. Individual phenolic compounds (gallic acid, catechin, myricetin, quercetin, caffeic acid, rutin, epicatechin, cyanidin, malvidin-3-glucoside) contributed to the antioxidant activity. The correlation coefficient for the compounds ranged from $r = 0.92$ to $r = 0.38$ in descending order.

In red wine, Ghiselli et al. (1998) reported the anthocyanin fraction to be the most effective fraction against both hydroxyl and peroxyl radicals, *in vitro* inhibition of LDL oxidation and platelet aggregation, compared with the phenolic acids and quercetin-3-glucuronide fraction and with the catechin and quercetin-3-glucoside fraction of the wines. The high efficiency of the anthocyanin fraction was explained both by its high concentration and by its antioxidant molar efficiency calculated as mM Trolox equivalents/mM phenolic concentration. The molar antioxidant efficiency was 0.6 for the anthocyanin fraction, 0.22 for the phenolic acid fraction and 0.12 for the catechin fraction. Gardner et al. (1999), also studying red wine, similarly concluded that it is unlikely that quercetin and myricetin are the main cause of the health benefits assumed to arise from moderate consumption of red wine. The flavonols accounted for only 2% of total phenolic content and for less than 1.5% of the total antioxidant capacity in the red wine. Also, Saint-Cricq de Gaulejac et al. (1999) concluded that the anthocyanin fraction showed high free radical activity compared with the other tannic fractions when using an enzymatic method (hypoxanthin–xanthin oxidase) for determination of free radical scavenger activities.

Total antioxidant activities in 13 typical Italian wines were determined with an average of 12.3 μM Trolox equivalents for red wines and 1.6 μM for white wines (Simonetti et al., 1997). The values correlated well with total phenols ($r = 0.9902$) and flavanols ($r = 0.9270$) and clearly demonstrated that red wine polyphenols are significant *in vitro* antioxidants. The antioxidant capacity of 16 red wines from several countries was tested using electron spin resonance (ESR) spectroscopy (Burns et al., 2000). The antioxidant activity ranged from 4.13×10^{21} to 9.29×10^{21} number of Fremy's radicals reduced by 1 L of wine. The antioxidant capacity was associated with the phenolic content of the wines, determined either by Folin-Ciocalteau or by summation of the individual phenolics determined by HPLC or by spectroscopy. Total phenolics as measured by the Folin-Ciocalteau method ranged from 6.47 to 18.6 mM GAE per liter. The anthocyanin levels ranged from 101.5 to 325.7 μM. There were substantial differences in the proportion of polymeric pigments in the wines.

The development of free radical scavenging capacity as determined by the DPPH method was studied during aging of selected red Spanish wines (Larrauri et al.,

1999). The antiradical efficiency increased during aging up to a certain time, then leveled off. The main polyphenolics identified were tannic acid, oenin and gallic acid. The changes were explained as a gradual transition from monomeric anthocyanins to oligomers to polymeric pigments. The polyphenols were characterized as slow reacting compared with other antioxidants analyzed by this method.

3.6.2.3 Juices

Wang and Lin (2000) measured antioxidant activity (ORAC) and total phenolic and anthocyanin contents of juice from thornless blackberry, strawberry and red and black raspberry fruits and found linear relationships between ORAC and total phenolics and between ORAC and anthocyanins of ripe fruits. The ORAC values are presented in Table 3.6. Calculated on a wet weight basis, black raspberries and blackberries were the best sources for antioxidants. However, when calculated on a dry weight basis, strawberries had the highest ORAC activities, followed by blackberries, black raspberries and red raspberries. Where comparisons can be made, the ORAC values of the juices were comparable to those obtained by extraction with organic solvents (Wang et al., 1996; Prior et al., 1998; Kalt et al., 1999a). Wang et al. (1996) reported increases in ORAC values of 12.4 to 15.4 for strawberry when the pulp remaining after juice production was extracted and added. The corresponding increases were 4.0 to 7.4 ORAC values for red grapes and 2.9 to 4.5 for white grapes. The antioxidant activity increased most in the red grapes, where the polyphenols are strongly incorporated into the skin.

ORAC values during ripening of fruits have also been reported (Wang and Lin, 2000). Blackberry, black raspberry and strawberry revealed the highest ORAC values during their green stages, while the ripe fruits had the highest ORAC values in red raspberries. With all species, the pink stage of ripening had the lowest ORAC values.

The antioxidant activities of red and white commercial grape juices have been studied using *in vitro* inhibition of the copper-catalyzed oxidation of human LDL (Frankel et al., 1998). The correlation between total phenols, expressed as GAE, and relative percent of inhibition of LDL oxidation was $r = 0.99$. In red Concord grape juices, the antioxidant activity was related to the anthocyanin levels. In the white grape juice, the antioxidant activity was associated with the levels of hydroxycinnamates (caffeic acid) and flavanols (catechin). When compared at the same level of total phenolics (10 μM GAE), the antioxidant activities of the grape juices were comparable to the antioxidant activities of red wine (Frankel et al., 1995). Laplaud et al. (1997) found protective action of copper-mediated LDL oxidation in aqueous *V. myrtillus* extracts. On a molar base, the extracts were more efficient than ascorbic acid and BHT in inhibiting LDL oxidation.

Spray-dried elderberry juice with high amounts of anthocyanin glucosides caused prolongation of the lag phase of copper induced oxidation of human LDL, while the maximum oxidation rate remained unchanged (Abuja et al., 1998). For peroxyl-radical-driven LDL oxidation, however, both prolongation of lag time and reduction of maximum oxidation rate occurred. When the extract was added after LDL oxidation was initiated, the elderberry extract demonstrated a pro-oxidative action. The pro-oxidative effect was more pronounced the longer the LDL oxidation was allowed to proceed before the extract was added.

Miller and Rice-Evans (1997) studied total antioxidant activity (TAA) and reduction in vitamin C during storage of black currant drinks containing 16.7% and 25% black currant extract. The anthocyanin fraction of the black currant drinks contributed only partly to the TAA values, and the authors concluded that there may be additional antioxidants present in the black currant drinks. The black currant drinks demonstrated vitamin C sparing ability.

3.6.2.4 Grape Seeds

Grape seeds are rich in polyphenols, especially flavanols (catechin, epicatechin) and gallic acid (Meyer et al., 1997). When grape extracts are prepared after crushing of the seeds, increased levels of phenols are obtained compared with extracts prepared with intact grape seeds. In particular, flavan-3-ol levels increased significantly. The antioxidant activity, as measured by the ability to inhibit LDL oxidation, also increased when the seeds were crushed prior to extraction.

Bagchi et al. (1997) studied the oxygen free-radical scavenger ability of grape seed proanthocyanidins using a chemiluminescence assay and cytochrome c reduction. The results revealed that on a weight basis (100 mg/L), the grape seed proanthocyanidin extract was a better inhibitor of both superoxide anion and hydroxyl radical than vitamin C and vitamin E succinate.

3.6.2.5 Other Materials

Leaves of blackberry, raspberry and strawberry have been shown to exert high antioxidant activity when measured by the ORAC method (Wang and Lin, 2000). The activities were higher than those of the corresponding fruits. ORAC values of fresh leaves ranged from 205.0 to 728.8 μmol TE/g dry matter, while the values for fruits ranged from 35.0 to 162.1 μmol TE/g dry matter. The ORAC values differed significantly among species and leaf age. ORAC values and total phenolic content decreased as the leaves grew older.

3.7 HEALTH EFFECTS OF ANTHOCYANINS AND PHENOLIC ACIDS

3.7.1 INTESTINAL TRACT ABSORPTION

The information about absorption, metabolism and excretion of individual flavonoids in humans is scarce (Cook and Samman, 1996). The flavonoids are absorbed from the intestinal tracts of humans and animals and are excreted either unchanged or as flavonoid metabolites in the urine or feces. Most studies have dealt with the absorption of flavonoids after oral administration of pharmacological doses of individual flavonoids rather than dietary levels of the flavonoids. Furthermore, most experiments have been conducted using flavonoids without glycosidic substitution rather than the glycosylated flavonoids, which predominate in plants.

Das (1971) reported that within 6 hours after the administration of about 92.3 mg/kg body weight (=4.2 g) catechin to healthy men, the compound was detectable in the plasma. The catechin was excreted in the urine in the free form and as conjugated and derived forms. Approximately 19% was recovered in the feces. In a similar study, however, Gugler et al. (1975) were unable to detect quercetin, also an aglycon, in plasma or urine of persons given an oral dose of 4 g quercetin. About 53% of the dose was recovered unchanged in feces, and it was concluded that only about 1% (40 mg) of quercetin had been absorbed. More recently, Okushio et al. (1995, 1996) confirmed the presence of (–)-epicatechin and (–)-epigallocatechin and their respective gallates in portal blood after oral administration of these tea catechins to rats. The findings clearly indicate at least partial absorption of the administered catechins into the rat portal vein. There are indications that larger tea flavonoids may be absorbed in humans, but it is not yet clear whether these compounds are absorbed as such or whether they undergo degradation prior to absorption (Hollman et al., 1997). Donovan et al. (1999) found sulfate and sulfate glucuronide catechin conjugates and little free catechin in plasma of human subjects after the intake of alcoholized and dealcoholized red wine. The conjugates disappeared from the blood with a half-time of about 4 h. Thus, the catechins appeared to be well absorbed but rapidly transformed into various metabolites. There are suggestions as to whether the parent compounds or the metabolites are important for the biological activity of catechins.

It was long considered that only free flavonoids without one or more sugar molecules were able to pass the gut wall. As no enzymes that can split the β-glucosidic bonds of the flavonoid-glycosides have been reported to be secreted into the gut or are present in the intestinal wall, microbial hydrolysis can occur only in the colon (Hollman and Katan, 1998). The bacteria in the colon hydrolyze the flavonoid glycosides, and the aglycons are supposed to be absorbed. However, the bacteria are also known to substantially degrade the flavonoid moiety by cleavage of the heterocyclic ring, and it has been postulated that various phenolic acids or lactones are formed. The phenolic acids are absorbed and excreted in the urine. The significance of these reactions in man is not clear. The absorption of quercetin aglycon has been confirmed by Hollman et al. (1995) in their study with ileostomy subjects, where no colon microbiological activity occurs. After oral administration of 100 mg pure quercetin aglycon, 24% quercetin aglycon was absorbed.

From a study with ileostomy subjects given flavonoid glucosides from onions, Hollman et al. (1995) concluded that flavonoid glucosides can be absorbed in man without previous hydrolysis to the aglycon. Absorption rates of flavonol glucosides vary from moderate to rapid in man (Hollman et al., 1997). While quercetin glucosides from onions were rapidly absorbed, the corresponding rutinoside was only slowly absorbed. Similarly, quercetin glucosides from apples were absorbed at a lower rate than those from onions.

More recently, studies on the absorption of flavonoid glycosides typical of berries and grapes are reported. Lapidot et al. (1998) traced anthocyanins in human urine after the intake of 300 ml red wine, corresponding to 218 mg of anthocyanins. Totals of 1.5–5.1% of the anthocyanins were recovered in the urine within 12 h after wine consumption, two compounds were unchanged, whereas other compounds seemed to have undergone molecular modifications. The anthocyanin levels of the urine reached a peak within 6 h of consumption. Serafini et al. (1998) tested the effects of the intake of the nonalcoholic fraction of red or white wine on plasma antioxidant capacity, measured as total radical-trapping parameter (TRAP), and on the presence of plasma phenolics in humans. The red nonalcoholic wine caused significantly increased TRAP values and plasma phenolic levels, in contrast to the white nonalcoholic wine in which no increases were observed. Also, Cao et al. (1998) reported increased serum antioxidant capacity, measured as ORAC, TEAC and TRAP, after the consumption of strawberries or red wine.

Miyazawa et al. (1999) studied the incorporation of the red fruit anthocyanins, cyanidin-3-glucoside and cyanidin-3, 5-diglucoside, into human and rat plasma. The anthocyanins were obtained from spray-dried extracts of elderberry (*Sambucus nigra*) and black currant (*Ribes nigrum*) juice concentrates. The human volunteers were given 2.7 mg/kg body weight of cyanidin-3-glucoside and 0.25 mg/kg body weight of cyanidin-3,5-diglucoside. After 30 min, cyanidin-3-glucoside (11 μg/L) and a trace amount of cyanidin-3,5-diglucoside were detected in plasma. The rats were given higher doses (320 mg/kg body weight of cyanidin-3-glucoside and 40 mg/kg body weight of cyanidin-3,5-diglucoside), and the plasma anthocyanin recovery was higher (1563 μg/L for cyanidin-3-glucoside and 195 μg/L for cyanidin-3,5-diglucoside) than for humans. The results thus indicated that anthocyanins are incorporated as intact glucosides from the digestive tract into the blood circulation system of mammals.

Tsuda et al. (1998) studied oxidation resistance during serum formation in rats after feeding the animals a diet containing 2 g cyanidin-3-glucoside per kg body weight for 14 days. The dietary anthocyanins caused a significant decrease in the generation of 2-thiobarbituric acid-reactive substances (TBARS) during serum formation. When tested with the oxidizing reagents AAPH and Cu^{2+}, a lower susceptibility to further lipid peroxidation was demonstrated compared with the control group.

Grapes, wine and other berry products are acidic, and the anthocyanins are in the flavylium form. A hypothesis that the anthocyanins may be pH-transformed into their carbinol pseudo-base and quinoidal base, or the chalcone, in the intestine and blood system during digestion has been proposed by Lapidot et al. (1999). These compounds have been shown to have antioxidant activity and are also most likely absorbed from the gut into the blood system. The pseudo-base and the quinoidal base of malvidin and malvidin-3-glucoside remained as very effective antioxidants, both when tested by a linoleate-oxidizing assay and by a microsomal lipid peroxidation assay.

3.7.2 Effects on Metabolism

Flavonoids are regarded as being antiallergenic, anti-inflammatory, antiviral, antiproliferant and anticarcinogenic and to act in certain metabolic pathways in mammals (Middleton and Kandaswami, 1994). A mixture of two flavonoid compounds, called citrin, was previously considered to have a vitamin-like activity. The compounds were called vitamin P and were known to decrease capillary permeability and fragility and to help under vitamin C-deficit conditions. Later, the vitamin classification of citrin was abandoned. However, with the current research on flavonoids and their participation in biological systems being generated, it is likely that the importance of flavonoids in human health will be redefined in the near future.

Flavonoids are very active toward a great number of enzyme systems in mammals. Middleton and Kandaswami (1994) have given an extensive review of the present knowledge of various enzyme reactions and biological mechanisms related to flavonoids. The effects are believed to depend upon the ability of the flavonoids to scavenge free radicals, chelate metal ions and exert antioxidant activity under a variety of conditions. Most studies of flavonoid actions in biological studies seem to have been performed with a small number of individual flavonoids and some of their glycosides. Among the most tested are flavonols (quercetin) and flavanols (catechin). Often, the biological effects depend upon the exact molecular structure of the compound tested, and general conclusions cannot necessarily be drawn from one flavonoid compound to the whole group of flavonoids in general. Also, Dillard and German (2000) reviewed phenolic acids, condensed tannins and flavonoids as compounds with beneficial effects on human health and listed the compounds' antioxidant properties and their modulating effects on a variety of metabolic and signaling enzymes as mechanisms for their action. Reference is made to biological activities of flavonoids against inflammation, allergies, platelet aggregation, microbes, ulcers, viruses, tumors and hepatotoxins. Similarly, Mazza (2000) refers to several recent studies that show anticarcinogenic, anti-inflammatory, antihepatotoxic, antibacterial, antiviral, antiallergenic, antithrombotic and antioxidant effects of flavonoids and other phenolics present in fruits and vegetable products.

In the following, a few theories that have recently received considerable attention as being explanatory to some of the putative health effects of flavonoids are presented.

3.7.2.1 Antioxidants in Human Health

Free radicals and other reactive oxygen species (ROS) are continuously formed in the human body, either deliberately to serve important biological functions or accidentally (Aruoma, 1998; Halliwell, 1994). Targets for their actions are DNA, proteins and lipids (Diplock, 1991). Living organisms have developed complex antioxidant systems to control production and reduce damage from free radicals (Papas, 1996). The antioxidant system may be endogenous or dietary. Endogenous factors include specific enzymes, metal-binding proteins and some low-molecular weight substances such as glutathione, uric acid, tocopherol (vitamin E) and ascorbate (vitamin C), while the dietary antioxidants include tocopherols (vitamin E), ascorbate (vitamin C), carotenoids, Se and other metals essential for the function of the antioxidant systems (Aruoma, 1998).

Oxidative stress refers to an imbalance between the generation of ROS and the activity of the antioxidant defenses in the body (Aruoma, 1998). Severe oxidative stress can cause cell damage and death. Oxidative stress may occur as a consequence of pollution (cigarette smoke, ozone and nitrogen oxides). Free radical formation may also be the side effect of certain drugs or disease treatments (radiation therapy) (Halliwell, 1997). An excess of free radicals may induce changes that ultimately lead to the development of various diseases.

Recent emphasis has focused on the role of antioxidants in the general aging of humans. As pointed out by Stähelin (1999), key factors in the development of the aging processes are associated with a species capacity of repairing DNA and of the antioxidant defense of the body. Various organs of the body may respond to the oxidative stress at different rates and with different defense mechanisms.

TABLE 3.7
Diseases Associated with Free Radicals and
Advanced Glycation End Products

Atherosclerosis	Reperfusion injury
Cancer	Senile macular degeneration
Complications of diabetes	Rheumatoid arthritis
Cataract	Respiratory distress syndrome
Alzheimer's disease	Sepsis
Parkinson's disease	Ulcerative colitis
Amyotrophic lateral sclerosis	Crohn's disease

Source: Reprinted from Stähelin (1999), with permission.

Atherosclerosis, cancer and degenerative brain diseases may result from specific processes in an organ or a cell system and at the same time may be the result of the universal aging process. Damage to DNA by radicals may be a significant contributor to the age-dependent development of cancer (Halliwell, 1994). Diseases associated with oxidative stress, free radicals and metabolites generating free radicals in the body are listed in Table 3.7 (Stähelin, 1999). It is not yet clear, however, whether the oxidative stress is the primary cause of the diseases or whether formation of radicals is a secondary effect of tissue damage caused by the disease.

There is increasing evidence that in addition to the traditional antioxidative vitamins, other compounds may also have important antioxidative functions in the human body. The antioxidant vitamins C and E, together with other antioxidants from products such as tea, wine and other foods of vegetable origin may thus play a critical role in the reduction or in the delay of the onset of degenerative diseases, including coronary heart diseases (CHD) (Biesalski, 1999). It is not yet well established as to what extent antioxidants have protective properties beyond the levels obtained through a balanced diet (Halliwell, 1997; Stähelin, 1999).

3.7.2.2 Coronary Heart Diseases (CHD)

Several cohort studies have been performed in which the relationships between flavonoid intake and the risk of coronary heart disease have been investigated. The studies have shown that the mortality from coronary heart diseases (CHD) is inversely correlated with the intake of flavonoids in the diet. Hollman and Katan (1998) summarize that in three out of five cohort studies, in addition to one cross-cultural study, flavonoids from the flavonol and flavone subgroups demonstrated a protective role toward cardiovascular disease. The protective effect of the flavonoids is partly explained by the inhibition of LDL oxidation and by reduced platelet aggregability. As reviewed by Cook and Samman (1996), there are several possible routes as to how LDL is oxidized by free radicals generated in the cells and how the oxidized LDL initiates and promotes atherosclerosis in the human body.

The occurrence of plaques is accepted as an intermediate end point in the development of CHD (Biesalski, 1999). The blood platelets aggregate and adhere to blood vessels, thus favoring the development of thrombosis and atherosclerosis (Cook and Samman, 1996). Special flavonoids have been shown to inhibit the development of blood platelet aggregation and blood vessel adhesion, thereby reducing the risk of these thrombotic reactions. The mechanisms for the flavonoid actions appear to be very complex. The flavonoids may bind to the platelet membranes and scavenge free radicals from this position. The inhibitory effect of flavonoids on platelet function is linked to structural features of the flavonoid molecule, and flavonoids may thus vary in their efficiency in preventing conditions favorable for development of CHD.

In addition to the antiaggregatory effects, flavonoids have been shown to increase vasodilation by inducing vascular smooth muscle relaxation. For quercetin, it has been suggested that the flavonoid affects the dilation of the smooth muscle tissue by altering the Ca^{2+} availability or by acting on tissue enzymes (Middleton and Kandaswami, 1994). Recently, a study of the relationships between vasodilation capacity, antioxidant activity and phenolic content of 16 red wines was presented (Burns et al., 2000). The total phenol content, measured by HPLC and colorimetry, correlated well with vasodilation capacity and antioxidant activity of the wines. Only anthocyanins were correlated with vasodilation capacity, whereas the antioxidant capacity was associated with gallic acid, total resveratrol and total catechin. Thus, different phenolic profiles of wines can produce varying vasodilation capacities and antioxidant activities. The *ex vivo* method used for assessing vasodilation capacity was based on the ability of the wines to relax precontracted rabbit thoracic aorta. From the close relationship between the chemical methods and the *ex vivo* biological vasodilation assay, the authors suggest that the chemical method has biological relevance. Other recent studies support that the vasodilation activity is connected to skin-derived compounds in the grapes. Andriambeloson et al. (1998) found that only the anthocyanin and oligomeric condensed tannin-containing fractions of red wine showed a vasorelaxant activity comparable to the original polyphenol fraction of the red wine. The phenolic acid derivatives, hydroxycinnamic acids and flavanol classes tested failed to induce this type of response.

3.7.2.3 Cancer

The antioxidant properties and the inhibitory role in various stages of tumor development in animal studies make flavonoids interesting as cancer-preventing agents (Hollman and Katan, 1998). Several reviews have been published with the conclusions that fruits and vegetables have preventive effects on various types of cancer, although the actual chemical components remain to be identified (Block et al., 1992; Steinmetz and Potter, 1996; WCRF/AICR, 1997). The most well-known epidemiological study of the effect of flavonoids (flavonols and flavones) on cancer is the Zutphen Elderly Study (Hertog et al., 1993a). In this study, the diets of 805 elderly men aged 65–84 years were followed for 5 years, after which their morbidity and mortality data were collected from their health records. The authors concluded that no associations could be detected between the flavonol and flavone intake and total cancer mortality. The results further showed that special types of cancer, such as lung cancer, were not associated with their intake of flavonols and flavones. The flavonols and flavones were considered the main sources of flavonoids in the diets, and the average intake per person was 23 mg/day (Hertog et al., 1993b). Hollman and Katan (1998), in their review, further refer to a larger Dutch cohort study consisting of 120,850 men and women aged 55–69 years, in which no association between flavonol and flavone intake and cancer in the stomach, colon or lung were revealed during a 4.3-year period. Similarly, Dragsted et al. (1997) summarized in a review that flavonoid-rich foods had no uniform protective effects against cancer. It therefore appears that the flavonoids are not the major cause of protective effects on cancer development from high intakes of fruits and vegetables.

Quercetin has been considered to have mutagenic effects in bacterial systems (Hollman and Katan, 1998). *In vivo* mutagenicity of flavonoids in mammals has not been shown, and tests for carcinogenicity have, in general, proved negative except for one case (Middleton and Kandaswami, 1994). On the other hand, quercetin is known to inhibit the mutagenic action of a benzo-[*a*]-pyrene and many reactions associated with tumor promotion. Also, anthocyanins from *Aronia melanocarpa* have been shown to inhibit the mutagenic activity of benzo-[*a*]-pyrene as well as 2-amino fluorene in the Ames test (Gąsiorowski et al., 1997). The anthocyanins inhibited the generation and release of superoxide radicals by human granulocytes, and the anthocyanins are suggested to exert their antimutagenic effect by their free radical scavenging ability. Inhibition of enzymes activating promutagens and converting mutagens to DNA-reacting derivatives has been proposed as a possible mechanism. Flavonoids also show antiproliferation effects toward tumor cells. Ellagic acid has been reported to exert anticarcinogenic properties in mice and rats and is also of interest in cancer-preventive treatment of humans (Chung et al., 1998). The protective effects of ellagic acid on cancer are postulated to include radical scavenger activity and protection of DNA from damage.

3.7.2.4 Visual Acuity

There are anecdotal reports that British RAF pilots were administered a diet enriched with bilberry (*Vaccinium myrtillus*) products for improved night vision. A recent study in Japan investigated the effect of blueberry extract on eyesight (Kajimoto, 1999). Subjects were administered 125 mg of blueberry extract (31.2 mg anthocyanin pigments) twice a day for 28 days, and a control group was given a placebo. After 28 days, the materials given the groups were reversed, with the experiment continuing for an additional 28 days. Conclusions were that symptoms associated with weak eyesight (fatigue, pain, etc.) improved with the group consuming blueberry extract. While beneficial for "weak eyesight," they concluded that the extract had little effect on cataracts. Aldose reductase is an enzyme involved in the development of cataracts by reducing glucose and galactose to their polyols that may accumulate in large quantities in the lens, thereby increasing lens opacity (Middleton and Kandaswami, 1994). Certain flavonoids have been shown to inhibit aldose reductase.

Hodge et al. (1999) studied the effect of various flavonoids and phenolic acids on the intraocular pressure (IOP)-lowering activity in rabbits after intravenous injections. Among the compounds tested, only special structural configurations demonstrated lowering of IOP. The authors suggest the requirement of specific structures of the flavonoids, rather than general antioxidant activity, for interaction with the ocular system. The protein- or proteoglycan-binding abilities of the active compounds were also considered as being important in the IOP-lowering activity.

3.7.2.5 Microbial Infections

Middleton and Kandaswami (1994) reported that among other flavonoids, several of those present in berries and grapes (such as quercetin, rutin, leucocyanidin, pelargonidin chloride and catechin) have been shown to possess antiviral activity for several types of viruses. The antiviral activity was related to the flavonoid aglycon, and hydroxylation in the 3-position seemed to be necessary for the antiviral activity. Several condensed tannins have also been shown to have inhibitory effects on the human immunodifiency virus (HIV) (Chung et al., 1998). In their review of antimicrobial effects of tannins, Chung et al. (1998) infer that proanthocyanidins from strawberry inhibit growth of *Botrytis cineriaa*, a deteriorating microorganism infecting strawberries.

3.7.2.6 Diabetes, Inflammation, Ulcers and Immune Activity

Orally administrated anthocyanins from *V. myrtillus* have been shown to exert a significant preventive and curative antiulcer activity (Cristoni and Magistretti, 1987). The effect was partly attributed to increases in gastric mucus. Keracyanin (cyanidin-3-rhamnoglucoside) may contribute to reduce the detrimental effect of UV radiation on cells (Kano and Miyakoshi, 1976). Diabetes affects several compounds of connective tissues, including capillary basal laminae and arterial walls (Boniface et al., 1985). Anthocyanins from *V. myrtillus* have been shown to reduce the biosynthesis of polymeric collagen and of structural glycoproteins that are responsible for the thickening of capillaries in diabetes. Long-term treatment of patients with diabetes with anthocyanins (600 mg/day for 6 months) reduced the frequency of capillaries with lesions.

Flavonoids affect the immune system and the body's inflammatory cells. This function is partly accomplished through the effects on enzymes involved in the immune response and the generation of inflammatory cells. Balentine et al. (1999) recently reported that only the aglycon cyanidin, not the anthocyanin glycoside, showed inhibitory effects on enzymes involved in inflammatory activity. The anthocyanins can be metabolized into their glycoside-free constituents either before or after absorption in the gut and may thereby act as anti-inflammatory agents in the body. Middleton and Kandaswami (1994) stress the fact, however, that as most of these studies have been carried out as *in vitro* experiments, care should be taken not to draw too extensive conclusions for the *in vivo* actions of flavonoids. Fundamental biochemical research is needed before the mechanisms behind the diverse actions of flavonoids are fully understood.

3.7.3 INTERACTIONS WITH OTHER NUTRIENTS

Condensed tannins, flavan-3-ols (catechins) and flavan-3,4-diols (leucoanthocyanins) can form complexes with important nutrients as proteins, starch and digestive enzymes (Hollman et al., 1997; Chung et al., 1998). The ability to precipitate proteins depends upon the molecular weight of the condensed tannins ($n > 2$). When condensed tannins react with the proteineous taste receptors in the mouth, a sensation of astringency is produced, and these compounds thus have a bitter taste (Cheeke, 1999). Tannins form insoluble complexes with divalent ions, resulting in less availability of the minerals during digestion. The present hypothesis, however, is that the major adverse effects of high intake of condensed tannins involve inhibition of metabolic pathways within the body, rather than formation of less-digestible complexes with proteins or digestive enzymes (Chung et al., 1998).

3.8 CONCLUSIONS

Grapes and berry fruits are rich sources of flavonoids and polyphenolics. Labor-intensive investigations over the years, initially from botanists and later from chemists and food scientists, provide a vast amount of knowledge about chemical structure, occurrence and stability of these compounds. As a consequence, food products, extracts, tinctures and dry powders are available with documented levels of flavonoids and other phenolics.

The renewed interest in flavonoids as health-benefiting compounds creates the necessity to search for their exact function in the human body. From a health point of view, the main interest at present is concentrated on the ability of the flavonoids to act as free radical scavengers. Methodologies for the assessment of these and other functions are currently being developed. These *in vitro* methods clearly demonstrate that flavonoids are effective in scavenging various types of free radicals, in inhibiting lipid oxidation and the activity of certain enzymes as well as in possessing antiviral effects.

Simultaneous with the *in vitro* research, reports are appearing from the nutritional and human medical areas focusing on bioavailability of flavonoids and the importance of dietary antioxidants in preventing diseases such as coronary heart disease, cancer, vision deterioration and aging *in vivo*. When these two approaches meet in the hopefully not-too-distant future, the results should reveal scientifically sound conclusions regarding the health benefits of flavonoids. Although there is still much to be learned, there are indications that the scientific approach may reaffirm the basis for many of the remedies known from traditional therapeutic use of grape and berry products in folk medicine.

REFERENCES

Abuja, P.M., Murkovic, M. and Pfannhauser, W. 1998. Antioxidant and prooxidant activities of elderberry (*Sambucus nigra*) extract in low-density lipoprotein oxidation, *J. Agric. Food Chem.* 46(10):4091–4096.

Amerine, A. and Joslyn, M.A. 1967. Composition of grapes, in *Table Wines, the Technology of their Production*, 2nd ed., Berkeley, CA: University of California Press, pp. 234–238.

Andersen, Ø.M. 1985. Chromatographic separation of anthocyanins in cowberry (lingonberry) *Vaccinium vites-idaea* L., *J. Food Sci.* 50(5):1230–1232.

Andersen, Ø.M. 1987a. Anthocyanins in fruits of *Vaccinium uliginosum* L. (bog whortleberry). *J. Food Sci.* 52(3):665–666, 680.

Andersen, Ø.M. 1987b. Anthocyanins in fruits of *Vaccinium japonicum. Phytochemistry* 26 (4):1220–1221.

Andersen, Ø.M. 1989. Anthocyanins in fruits of *Vaccinium oxycoccus* L. (small cranberry), *J. Food Chem.* 54(2):383–384, 387.

Andriambeloson, E., Magnier, C., Haan-Archipoff, G., Lobstein, A., Anton, R., Beretz, A., Stoclet, J.C., and Andriantsitohaina, R. 1998. Natural dietary polyphenolic compounds cause endothelium-dependent vasorelaxation in rat thoracic aorta. *J. Nutr.* 128:2324–2333.

Anonymous. 1993. Food Labeling: Declaration of Ingredients. *Federal Register* 21CFR 101.30, 58, pp. 2925–2926.

Aruoma, O.I. 1998. Free radicals, oxidative stress, and antioxidants in human health and disease. *J. Am. Oil. Chem. Soc.* 75(2):199–212.

Ashurst, P.R. 1995. *Production and Packaging of Non-Carbonated Fruit Juices and Fruit Beverages.* 2nd ed., Glasgow, UK: Blackie Academic & Professional, pp. 1–429.

Bagchi, D., Garg, A., Krohn, R.L., Bagchi, M., Tran, M.X., and Stohs, S.J. 1997. Oxygen free radical scavenging abilities of vitamins C and E, and a grape seed proanthocyanidin extract *in vitro. Res. Commun. Mol. Pathol. Pharmacol.* 95(2):179–189.

Balentine, D.A., Albano, M.C., and Nair, M.G. 1999. Role of medicinal plants, herbs, and spices in protecting human health, *Nutr. Rev.* 57(9):(II)S41–S45.

Ballington, J.R., Ballinger, W.E., and Maness, E.P. 1987. Interspecific differences in the percentage of anthocyanins, aglycones, and aglycone-sugars in the fruit of seven species of blueberries, *J. Am. Soc. Hort. Sci.* 112(5):859–864.

Bank, V.R., Bailey, D.T., Lenoble, R., and Richheimer, S.L. 1996. Enhanced stability and improved color of anthocyanin-based natural colorants by the use of natural extracts. *IFT Annual Meeting: Book of Abstracts,* p. 69.

Bate-Smith, E.C. 1959. Plant Phenolics in Foods, in: *The Pharmacology of Plant Phenolics,* ed., J.W. Fairbairn, New York: Academic Press, pp. 133–147.

Biesalski, H.K. 1999. The role of antioxidative vitamins in primary and secondary prevention of coronary heart disease. *Int. J. Vitam. Nutr. Res.* 69(3):179–186.

Block, G., Patterson, B., and Subar, A. 1992. Fruit, vegetables, and cancer prevention: A review of the epidemiological evidence. *Nutr. Cancer.* 18(1):1–29.

Bobbio, F.O., do Nascimento Varella, M.T., and Bobbio, P.A. 1994. Effect of light and tannic acid on the stability of anthocyanin in DMSO and in water. *Food Chem.* 51:183–185.

Bomser, J., Madhavi, D.L., Singletary, K., and Smith, M.A.L. 1996. *In vitro* anticancer activity of fruit extracts from *Vaccinium* species, *Planta Medica.* 62:212–216.

Boniface, R., Miskulin, M., Robert, L., and Robert, A.M. 1985. Pharmacological properties of *myrtillus* anthocyanosides: correlation with results of treatment of diabetic microangiopathy, in *Flavonoids and Bioflavonoids,* eds., L. Farkas, M. Gábor, and F. Kállay, Amsterdam: Elsevier, pp. 293–301.

Bors, W., Heller, W., Michel, C., and Saran, M. 1990. Flavonoids as antioxidants: determination of radical-scavenging efficiencies. *Methods Enzymol.* 186:343–355.

Brand-Williams, W., Cuvelier, M.E., and Berset, C. 1995. Use of a free radical method to evaluate antioxidant activity. *Lebensm.-Wiss. u.-Technol.* 28(1):25–30.

Bridle, P. and Timberlake, C.F. 1997. Anthocyanins as natural food colours—selected aspects. *Food Chem.* 58(1–2):103–109.

Broadhurst, R.B. and Jones, W.T. 1978. Analysis of condensed tannins using acidified vanillin. *J. Sci Food Agric.* 29(9):788–794.

Brouillard, R., George, F., and Fougerousse, A. 1997. Polyphenols produced during red wine ageing. *BioFactors.* 6:403–410.

Bruneton, J. 1995. *Pharmacognosy, Phytochemistry, Medicinal Plants.* Andover, UK: Intercept Ltd., pp. 266–288.

Burns, J., Gardner, P.T., O'Neil, J., Crawford, S., Morecroft, I., McPhail, D.B., Lister, C., Matthews, D., MacLean, M.R., Lean, M.E.J., Duthie, G.G., and Crozier, A. 2000. Relationship among antioxidant activity, vasodilation capacity, and phenolic content of red wines. *J. Agric. Food Chem.* 48(2):220–230.

Cabrita, L. and Andersen, Ø.M. 1999. Anthocyanins in blue berries of *Vaccinium padifolium*. *Phytochemistry* 52(8):1693–1696.

Cabrita, L., Fossen, T., and Andersen, Ø.M. 2000. Colour and stability of the six common anthocyanidin 3-glucosides in aqueous solutions. *Food Chem.* 68:101–107.

Cao, G., Alessio, H.M., and Cutler, R.G. 1993. Oxygen-radical absorbance capacity assay for antioxidants. *Free Radic. Biol. Med.* 14:303–311.

Cao, G., Russell, R.M., Lischner, N., and Prior, R.L. 1998. Serum antioxidant capacity is increased by consumption of strawberries, spinach, red wine or vitamin C in elderly women. *J. Nutr.* 128:2383–2390.

Carlsen, C. and Stapelfeldt, H. 1997. Light sensitivity of elderberry extract. Quantum yields for photodegradation on aqueous solution. *Food Chem.* 60(3): 383–387.

Cheeke, P.R. 1999. *Applied Animal Nutrition. Feeds and Feeding.* 2nd ed., Upper Saddle River, NJ: Prentice Hall, p. 38.

Cheynier, V., Osse, C., and Rigaud, J. 1988. Oxidation of grape juice phenolic compounds in model solutions. *J. Food Sci.* 53(6):1729–1732, 1760.

Chung, K.-T., Wong, T.Y., Wei, C.-I., Huang, Y.-W., and Lin, Y. 1998. Tannins and human health: A review. *Crit. Rev. Food Sci. Nutr.* 38(6):421–464.

Clifford, M.N. 1999. Chlorogenic acids and other cinnamates—nature, occurrence and dietary burden. *J. Sci. Food Agric.* 79(2):362–372.

Cohn, R. and Cohn, A.L. 1996. The by-products of fruit processing, in *Fruit Processing*, eds., D. Arthey and P.R. Ashurst. Glasgow, UK: Blackie Academic & Professional, pp. 198–220.

Cook, N.C. and Samman, S. 1996. Flavonoids—Chemistry, metabolism, cardioprotective effects, and dietary sources. *J. Nutr. Biochem.* 7(2):66–76.

Cooper-Driver, G.A. and Bhattacharya, M. 1998. Role of phenolics in plant evolution. *Phytochemistry* 49(5):1165–1174.

Creasey, L.L. and Coffee, M. 1988. Phytoalexin production potential of grape berries. *J. Am. Soc. Hort. Sci.* 113:230–234.

Cristoni, A. and Magistretti, M.J. 1987. Antiulcer and healing activity of *Vaccinium myrtillus* anthocyanosides. *Il Farmaco, Edizione practica.* 42(2):29–43.

Daniel, E.M., Krupnick, A.S., Heur, Y.-H., Blinzler, J.A., Nims, R.W., and Stoner, G.D. 1989. Extraction, stability and quantitation of ellagic acid in various fruits and nuts. *J. Food Comp. Anal.* 2(4):338–349.

Das, N.P. 1971. Studies on flavonoid metabolism. Absorption and metabolism of (+)-catechin in man. *Biochem. Pharmacol.* 20:3435–3445.

Degenhardt, A., Knapp, H., and Winterhalter, P. 2000. Separation and purification of anthocyanins by high-speed countercurrent chromatography and screening for antioxidant activity. *J. Agric. Food Chem.* 48(2):338–343.

Deuel, C.L. 1996. Strawberries and raspberries, in *Processing Fruits: Science and Technology, Vol. 2: Major Processed Products*, eds. L.P. Somogyi, D.M. Barrett, and Y.H. Hui. Lancaster, PA: Technomic Publishing Co., Inc., pp. 117–157.

Dillard, C.J. and German, J.B. 2000. Phytochemicals: nutraceuticals and human health. *J. Sci. Food Agric.* 80:1744–1756.

Diplock, A.T. 1991. Antioxidant nutrients and disease prevention: An overview. *Am. J. Clin. Nutr.* 53:189S–193S.

Donovan, J.L., Bell, J.R., Kasim Karakas, S., German, J.B., Walzem, J.B., Hansen, R.J., and Waterhouse, A.L. 1999. Catechin is present as metabolites in human plasma after consumption of red wine. *J. Nutr.* 129(9):1662–1668.

Downes, J.W. 1995. Equipment for extraction and processing of soft and pome fruit juices, in *Production and Packaging of Non-Carbonated Fruit Juices and Fruit Beverages*, 2nd ed., ed., P.R. Ashurst, Glasgow, UK: Blackie Academic & Professional, pp. 197–220.

Dragsted, L.O., Strube, M., and Leth, T. 1997. Dietary levels of plant phenols and other non-nutritive components: Could they prevent cancer? *Eur. J. Cancer Prev.* 6:522–528.

Ector, B.J., Magee, J.B., Hegwood, C.P., and Coign, M.J. 1996. Resveratrol concentration in Muscadine berries, juice, pomace, purees, seeds and wines. *Am. J. Enol. Vitic.* 47:57–62.

Escribano-Bailón, T., Gutiérrez-Fernández, Y., Rivas-Gonzalo, J., and Santos-Buelga, C. 1992. Characterization of procyanidins of *Vitis vinifera* variety Tinta del País grape seeds. *J. Agric. Food Chem.* 40(10):1794–1799.

Fan-Chiang, H.-J. 1999. Anthocyanin pigment, nonvolatile acid and sugar composition of blackberries. M.S. thesis, Oregon State University, Corvallis, OR.

Francis, F.J. 1982. Analysis of anthocyanins, in *Anthocyanins as Food Colors*, ed., P. Markakis, New York: Academic Press, Inc., pp. 181–207.

Frankel, E.N. and Meyer, A.S. 1998. Antioxidants in grapes and grape juices and their potential health effects. *Pharmaceut. Biol.* 36:14–20.

Frankel, E.N., Waterhouse, A.L., and Teissedre, P.L. 1995. Principal phenolic phytochemicals in selected California wines and their antioxidant activity in inhibiting oxidation of human low-density lipoproteins. *J. Agric. Food Chem.* 43(4):890–894.

Frankel, E.N., Bosanek, C.A., Meyer, A.S., Silliman, K., and Kirk, L.L. 1998. Commercial grape juices inhibit the *in vitro* oxidation of human low-density lipoproteins. *J. Agric. Food Chem.* 46(3):834–838.

Fukumoto, L.R. and Mazza, G. 2000. Assessing antioxidant and prooxidant activities of phenolic compounds. *J. Agric. Food Chem.* 48(8):3597–3604.

Gao, L. and Mazza, G. 1994. Quantitation and distribution of simple and acylated anthocyanins and other phenolics in blueberries. *J. Food Sci.* 59(5):1057–1059.

Gao, L. and Mazza, G. 1995. Characterization of acylated anthocyanins in lowbush blueberries. *J. Chromatogr.* 18(2):245–259.

García-Viguera, C., Zafrilla, P., Artés, F., Romero, F., Abellán, P., and Tomás-Barberán, F.A. 1998. Colour and anthocyanin stability of red raspberry jam. *J. Sci. Food Agric.* 78:565–573.

Gardner, Pr.T., McPhail, D.B., Crozier, A., and Duthie, G.G. 1999. Electron spin resonance (ESR) spectroscopic assessment of the contribution of quercetin and other flavonols to the antioxidant capacity of red wines. *J. Sci. Food Agric.* 79:1011–1014.

Gąsiorowski, K., Szyba, K., Brokos, B., Kołaczyńska, B., Jankowiak-Włodarczyk, M., and Oszmiański, J. 1997. Antimutagenic activity of anthocyanins isolated from *Aronia melanocarpa* fruits. *Cancer Lett.* 119:37–46.

Ghiselli, A., Nardini, M., Baldi, A., and Scaccini, C. 1998. Antioxidant activity of different phenolic fractions separated from an Italian red wine. *J. Agric. Food Chem.* 46(2):361–366.

Girard, B. and Mazza, G. 1998. Functional grape and citrus products, in *Functional Foods. Biochemical and Processing Aspects*, ed., G. Mazza, Lancaster, PA: Technomic Publishing Co., Inc., pp. 139–191.

Gugler, R., Leschik, M., and Dengler, H.J. 1975. Disposition of quercetin in man after single oral and intravenous doses. *Eur. J. Clin. Pharmacol.* 9:229–234.

Häkkinen, S., Heinonen, M., Kärenlampi, S., Mykkänen, H., Ruuskanen, J., and Törrönen, R. 1999. Screening of selected flavonoids and phenolic acids in 19 berries. *Food Res. Int.* 32:345–353.

Häkkinen, S. and Törrönen, A.R. 2000. Content of flavanols and selected phenolic acids in strawberries and *Vaccinium* species: Influence of cultivar, cultivation site and technique. *Food Res. Intern.* 33:517–524.

Häkkinen, S., Kärenlampi, S.O., Mykkänen, H.M., and Törrönen, A.R. 2000. Influence of domestic processing and storage on flavonol contents in berries. *J. Agric. Food Chem.* 48(7):2960–2965.

Halliwell, B. 1990. How to characterize a biological antioxidant. *Free Radic. Res. Comm.* 9(1):1–32.

Halliwell, B. 1994. Free radicals and antioxidants: A personal view. *Nutr. Rev.* 52(8):253–265.

Halliwell, B. 1997. Antioxidants and human disease: A general introduction. *Nutr. Rev.* 55(1):S44–S52.

Halliwell, B. and Gutteridge, J.M.C. 1995. The definition and measurement of antioxidants in biological systems. *Free Radic. Biol. Med.* 18(1):125–126.

Harborne, J.B. 1994. *The Flavonoids. Advances in Research since 1986.* London: Chapman & Hall, pp. 1–676.

Heinonen, I.M., Meyer, A.S., and Frankel, E.N. 1998. Antioxidant activity of berry phenolics on human low-density lipoprotein and liposome oxidation. *J. Agric. Food Chem.* 46(10):4107–4112.

Hendry, G.A.F. and Houghton, J.D. 1996. *Natural Food Colorants.* 2nd ed., Glasgow, UK: Blackie Academic & Professional, pp. 1–348.

Henry, B.S. 1996. Natural food colours, in *Natural Food Colorants.* 2nd ed., eds., G.A.F. Hendry and J.D. Houghton, Glasgow, UK: Blackie Academic & Professional, pp. 40–79.

Herrmann, K. 1989. Occurrence and content of hydroxycinnamic and hydroxybenzoic acid compounds in foods. *Crit. Rev. Food Sci. Nutr.* 28(4):315–347.

Herrmann, M.-E. 1999. [Sekundäre Pflanzenunhaltsstoffe in Obst: Carotenoids und Flavonoids als Antioxidantien], Phytochemicals in fruits: Carotenoids and flavonoids as antioxidants. *Erwerbsobstbau.* 41:213–217.

Hertog, M.G.L., Hollman, P.C.H., and Katan, M.B. 1992. Content of potentially anticarcinogenic flavonoids of 28 vegetables and 9 fruits commonly consumed in The Netherlands. *J. Agric. Food Chem.* 40(12):2379–2383.

Hertog, M.G.L., Feskens, E.J.M., Hollman, P.C.H., Katan, M.B., and Kromhout, D. 1993a. Dietary antioxidant flavonoids and risk of coronary heart disease: The Zutphen Elderly Study. *Lancet.* 342:1007–1011.

Hertog, M.G.L., Hollman, P.C.H., Katan, M.B., and Kromhout, D. 1993b. Intake of potentially anticarcinogenic flavonoids and their determinants in adults in The Netherlands. *Nutr. Cancer.* 20(1):21–29.

Hertog, M.G.L. and Katan, M.B. 1998. Quercetin in foods, cardiovascular disease, and cancer, in *Flavonoids in Health and Disease*, eds., C.A. Rice-Evans and L. Packer, New York: Marcel Dekker, Inc., pp. 447–467.

Hodge, L.C., Kearse, E.C., and Green, K. 1999. Intraocular pressure-lowering activity of phenolic antioxidants in normotensive rabbits. *Curr. Eye Res.* 19(3):234–240.

Hollman, P.C.H., de Vries, J.H.M., van Leeuwen, S.D., Mengelers, M.J.B., and Katan, M.B. 1995. Absorption of dietary quercetin glycosides and quercetin in healthy ileostomy volunteers. *Am. J. Clin. Nutr.* 62:1276–1282.

Hollman, P.C.H., Tijburg, L.B.M, and Yang, C.S. 1997. Bioavailability of flavonoids from tea. *Crit. Rev. Food Sci. Nutr.* 37(8):719–738.

Hollman, P.C.H. and Katan, M.B. 1998. Bioavailability and health effects of dietary flavonols in man. *Arch. Toxicol. Suppl.* 20:237–247.

Hong, V. and Wrolstad, R.E. 1990. Use of HPLC separation/photodiode array detection for characterization of anthocyanins. *J. Agric. Food Chem.* 38(3):708–715.

Inami, O., Tamura, I., Kikuzaki, H., and Nakatani, N. 1996. Stability of anthocyanins of *Sambucus canadensis* and *Sambucus nigra. J. Agric. Food Chem.* 44(10):3090–3096.

Iversen, C.K. 1999. Black currant nectar: Effect of processing and storage on anthocyanin and ascorbic acid content. *J. Food Sci.* 64(1):37–41.

Jackman, R.L. and Smith, J.L. 1996. Anthocyanins and betalains, in *Natural Food Colorants.* 2nd ed., eds., G.A.F. Hendry and J.D. Houghton, Glasgow, UK: Blackie Academic & Professional, pp. 244–309.

Jackman, R.L., Yada, R.Y., Tung, M.A., and Speers, R.A. 1987a. Anthocyanins as food colorants—a review. *J. Food Biochem.* 11:201–247.

Jackman, R.L., Yada, R.Y., and Tung, M.A. 1987b. A review: Separation and chemical properties of anthocyanins used for their qualitative and quantitative analysis. *J. Food Biochem.* 11:279–308.

Jeandet, P., Bessis, R., and Gaugheron, B. 1991. The production of resveratrol (3,5,4′-trihydroxystilbene) by grape berries in different developmental stages. *Am. J. Enol. Vitic.* 42(1):41–46.

Jeandet, P., Bessis, R., Maume, B.F., Meunier, P., Peyron, D., and Trollat, P., 1995. Effect of oenological practices on the resveratrol isomer content of wine. *J. Agric. Food Chem.* 43(2):316–319.

Jiménez, M. and García-Carmona, F. 1999. Oxidation of the flavonol quercetin by polyphenol oxidase. *J. Agric. Food Chem.* 47(1):56–60.

Johnson, I. 1998. Use of food ingredients to reduce degenerative diseases, in *Nutritional Aspects of Food Processing and Ingredients*, eds., C.J.K. Henry and N.J. Heppell, Gaithersburg, MD: Aspen Publishers, Inc., pp. 136–165.

Jovanovic, S.V., Steenken, S., Simic, M.G., and Hara, Y. 1998. Antioxidant properties of flavonoids: Reduction potentials and electron transfer reactions of flavonoid radicals, in *Flavonoids in Health and Disease*, eds., C.A. Rice-Evans and L. Packer, New York: Marcel Dekker, Inc., pp. 137–161.

Kader, F., Rovel, B., Girardin, M., and Metche, M. 1997. Mechanism of browning in fresh highbush blueberry fruit (*Vaccinium corymbosum* L.). Role of blueberry polyphenol oxidase, chlorogenic acid and anthocyanins. *J. Sci. Food Agric.* 74:31–34.

Kähkönen, M.P., Hopia, A.I., Vuorela, H.J., Rauha, J.-P., Pihlaja, K., Kujala, T.S., and Heinonen, M. 1999. Antioxidant activity of plant extracts containing phenolic compounds. *J. Agric. Food Chem.* 47(10):3954–3962.

Kajimoto, O. 1999. Eye Care. *Food Style* (English translation). 3: 30–35.

Kalt, W. and Dufour, D. 1997. Health functionality of blueberries. *HortTechnol.* 7(3):216–221.

Kalt, W., Forney, C.F., Martin, A., and Prior, R.L. 1999a. Antioxidant capacity, vitamin C, phenolics, and anthocyanins after fresh storage of small fruits. *J. Agric. Food Chem.* 47(11):4638–4643.

Kalt, W., McDonald, J.E., Ricker, R.D., and Lu, X. 1999b. Anthocyanin content and profile within and among blueberry species. *Can. J. Plant Sci.* 79(4):617–623.

Kano, E. and Miyakoshi, J. 1976. UV protection effect of keracyanine, an antocyanine derivative, on cultured L-cells. *J. Radic. Res.* 17:55, abstract 3-D-8.

Karadeniz, F., Durst, R.W., and Wrolstad, R.E. 2000. Polyphenolic composition of raisins. *J. Agric. Food Chem.* 48(11):5343–5350.

Kondo, Y., Ohnishi, M., and Kawaguchi, M. 1999. Detection of lipid peroxidation catalyzed by chelating iron and measurement of antioxidant activity in wine by a chemiluminescence analyzer. *J. Agric. Food Chem.* 47(5):1781–1785.

Kroon, P.A. and Williamson, G. 1999. Hydroxycinnamates in plants and food: Current and future perspectives. *J. Sci. Food Agric.* 79:355–361.

Labarbe, B., Cheynier V., Brossaud, F., Souquet, J.-M., and Moutounet, M. 1999. Quantitative fractionation of grape proanthocyanidins according to their degree of polymerization. *J. Agric. Food Chem.* 47(7):2719–2723.

Lamuela-Raventos, R.M., Romero-Perez, A.I., Waterhouse, A.L., and Carmen de la Torre-Boronat, M. 1995. Direct HPLC analysis of *cis-* and *trans*-resveratrol and piceid isomers in Spanish red *Vitis vinifera* wines. *J. Agric. Food Chem.* 43(2):281–283.

Lapidot, T., Harel, S., Granit, R., and Kanner, J. 1998. Bioavailability of red wine anthocyanins as detected in human urine. *J. Agric. Food Chem.* 46(10):4297–4302.

Lapidot, T., Harel, S., Akiri, B., Granit, R., and Kanner, J. 1999. pH-Dependent forms of red wine anthocyanins as antioxidants. *J. Agric. Food Chem.* 47(1):67–70.

Laplaud, P.M., Lelubre, A., and Chapman, M.J. 1997. Antioxidant action of *Vaccinium myrtillus* extract on human low density lipoproteins *in vitro*: Initial observations. *Fundam. Clin. Pharmacol.* 11:35–40.

Larrauri, J.A., Sánchez-Moreno, C., Rupérez, P., and Saura-Calixto, F. 1999. Free radical scavenging capacity in the aging of selected red Spanish wines. *J. Agric. Food Chem.* 47(4):1603–1606.

Lenoble, R., Richheimer, S.L., Bank, V.R., and Bailey, D.T. 1999. Pigment composition containing anthocyanins stabilized by plant extracts. US 5,908,650. (FSTA 1999-11-A1880).

Linko, R., Kärppä, J., Kallio, H., and Ahtonen, S. 1983. Anthocyanin content of crowberry and crowberry juice. *Lebensm.-Wiss. U.-Technol.* 16(6):343–345.

López-Serrano, M. and Ros Barceló, A. 1996. Purification and characterization of a basic peroxidase isoenzyme from strawberries. *Food Chem.* 55(2):133–137.

Lu, Y. and Foo, L.Y. 1999. The polyphenol constituents of grape pomace. *Food Chem.* 65(1):1–8.

Maas, J.L., Galletta, G. J., and Stoner, G.D. 1991a. Ellagic acid, an anticarcinogen in fruits, especially in strawberries: A review. *Hort. Sci.* 26(1):10–14.

Maas, J.L., Wang, S.Y., and Galletta, G.J. 1991b. Evaluation of strawberry cultivars for ellagic acid content. *Hort. Sci.* 26(1):66–68.

Macheix, J.-J. and Fleuriet, A. 1998. Phenolic acids in fruits, in *Flavonoids in Health and Disease*, eds., C.A. Rice-Evans and L. Packer, New York: Marcel Dekker, Inc., pp. 35–59.

Markakis, P. 1982. *Anthocyanins as Food Colors*. New York: Academic Press, Inc., pp. 1–263.

Martín-Aragón, S., Basabe, B., Benedí, J.M., and Villar, A.M. 1998. Antioxidant action of *Vaccinium myrtillus* L. *Phytother. Res.* 12:S104–S106.

Mazza, G. 1995. Anthocyanins in grapes and grape products. *Crit. Rev. Food Sci. Nutr.* 35(4):341–371.

Mazza, G. 1997. Anthocyanins in edible plant: A qualitative and quanitiative assessment, in *Antioxidant Methodology*, eds, O.I. Aruoma and S.L. Cuppett, Champaign, IL: AOCS Press, pp. 119–140.

Mazza, G. 2000. Health aspects of natural colors, in *Natural Food Colorants. Science and Technology*, eds., G.J. Lauro and F.J. Francis, New York: Marcel Dekker, Inc., pp. 289–314.

Mazza, G. and Miniati, E. 1993. *Anthocyanins in Fruits, Vegetables, and Grains*. Boca Raton, FL: CRC Press, Inc., pp. 1–362.

Mazza, G., Fukumoto, L., Delaquis, P., Girard, B., and Ewert, B. 1999. Anthocyanins, phenolics and color of Cabernet Franc, Merlot, and Pinot Noir wines from British Colombia. *J. Agric. Food Chem.* 47(10):4009–4017.

McDonald, M.S., Hughes, M., Burns, J., Lean, M.E.J., Matthews, D., and Crozier, A. 1998. Survey of the free and conjugated myricetin and quercetin content of red wines of different geographical origins. *J. Agric. Food Chem.* 46(2):368–375.

McLellan, M.R. and Race, E.J. 1995. Grape juice processing, in *Production and Packaging of Non-Carbonated Fruit Juices and Fruit Beverages*, 2nd ed., ed., P.R. Ashurst, Glasgow, UK: Blackie Academic & Professional, pp. 88–105.

McMurrough, I. and Baert, T. 1994. Identification of proanthocyanidins in beer and their direct measurement with a dual electrode electrochemical detector. *J. Inst. Brew.* 100:409–416.

Meyer, A.S., Yi, O-S., Perason, D.A., Waterhouse, A.L., and Frankel, E.N. 1997. Inhibition of human low-density lipoprotein oxidation in relation to composition of phenolic antioxidants in grapes (*Vitis vinifera*). *J. Agric. Food Chem.* 45(5):1638–1643.

Meyer, A.S., Jepsen, S.M., and Sørensen, N.S. 1998. Enzymatic release of antioxidants for human low-density lipoprotein from grape pomace. *J. Agric. Food Chem.* 46(7):2439–2446.

Middleton, E. and Kandaswami, C. 1994. The impact of plant flavonoids on mammalian biology: Implications for immunity, inflammation and cancer, in *The Flavanoids: Advances in Research since 1986*, ed., J.B. Harborne, London: Chapman & Hall, pp. 619–652.

Miller, N.J. and Rice-Evans, C. 1997. The relative contributions of ascorbic acid and phenolic antioxidants to the total antioxidant activity of orange and apple fruit juices and blackcurrant drink. *Food Chem.* 60(3):331–337.

Miyazawa, T., Nakagawa, K., Kudo, M., Muraishi, K., and Someya, K. 1999. Direct intestinal absorption of red fruit anthocyanins, cyanindin-3-glucoside and cyanidin-3,5-diglucoside, into rat and humans. *J. Agric. Food Chem.* 47(3):1083–1091.

Morris, J.R. and Striegler, K. 1996. Grape juice: Factors that influence quality, processing technology, and economics, in *Processing Fruits: Science and Technology, Vol. 2: Major Processed Products*, eds. L.P. Somogyi, D.M. Barrett, and Y.H. Hui. Lancaster, PA: Technomic Publishing Co., Inc., pp. 197–234.

Moulton, M.F. 1995. Growing and marketing soft fruit for juice and beverages, in *Production and Packaging of Non-Carbonated Fruit Juices and Fruit Beverages*, 2nd ed., ed. P.R. Ashurst, Glasgow, UK: Blackie Academic & Professional, pp. 129–152.

Naguib, Y.M.A. 2000. A fluorometric method for measurement of oxygen radical-scavenging activity of water-soluble antioxidants. *Anal. Biochem.* 284:93–98.

Okuda, T. and Yokotsuka, K. 1996. Trans-resveratrol concentration in berry skins and wines from grapes grown in Japan. *Am. J. Enol. Vitic.* 47(1):93–99.

Okushio, K., Matsumoto, N., Suzuki, M., Nanjo, F., and Hara, Y. 1995. Absorption of (–)-epigallocatechin gallate into rat portal vein. *Biol. Pharm. Bull.* 18(1):190–191.

Okushio, K., Matsumoto, N., Kohri, T., Suzuki, M., Nanjo, F., and Hara, Y. 1996. Absorption of tea catechins into rat portal vein. *Biol. Pharm. Bull.* 19(2):326–329.

Papas, A.M. 1996. Determinants of antioxidant status in humans. *Lipids.* 31:S77–S82.

Peng, C.Y. and Markakis, P. 1963. Effect of phenolase on anthocyanins. *Nature.* 199:597–598.

Pezet, R. and Cuenat, P. 1996. Resveratrol in wine: Extraction from skin during fermentation and post-fermentation standing of must from Gamay grapes. *Am. J. Enol. Vitic.* 47(3):287–290.

Plocharski, W. and Zbroszcyzk, J. 1992. Aronia fruit (*Aronia melanocarpa*, Elliot) as a natural source for anthocyanin colorants. *Flüssiges Obst.* 59(6):85–89.

Porter L.J., Hrstich, L.N., and Chan, B.G. 1986. The conversion of procyanidins and prodel-phinidins to cyanidin and delphinidin. *Phytochemistry* 25(1):223–230.

Prior, R.L., Cao, G., Martin, A., Sofic, E., McEwen, J., O'Brien, C., Lischner, N., Ehlenfeldt, M., Kalt, W., Krewer, G., and Mainland, C.M. 1998. Antioxidant capacity as influenced by total phenolic and anthocyanin content, maturity, and variety of *Vaccinium* species. *J. Agric. Food Chem.* 46(7):2686–2693.

Ribéreau-Gayon, P. 1982. The anthocyanins of grapes and wines, in *Anthocyanins as Food Colors*, ed. P. Markakis. New York: Academic Press, Inc., pp. 209–244.

Rice-Evans, C.A., Miller, N.J., and Paganga, G. 1996. Structure-antioxidant activity relationships of flavonoids and phenolic acids. *Free Radic. Biol. Med.* 29(7):933–956.

Rodriguez-Saona, L.E., Giusti, M.M., and Wrolstad, R.E. 1998. Anthocyanin pigment composition of red-fleshed potatoes. *J. Food Sci.* 63(3):458–465.

Rommel, A. and Wrolstad, R.E. 1993. Ellagic acid content of red raspberry juice as influenced by cultivar, processing and environmental factors. *J. Agric. Food Chem.* 41(11):1951–1960.

Rwabahizi, S. and Wrolstad, R.E. 1988. Effects of mold contamination and ultrafiltration on the color stability of strawberry juice and concentrate. *J. Food Sci.* 53(3):857–861, 872.

Saint-Cricq de Gaulejac, N., Glories, Y., and Vivas, N. 1999. Free radical scavenging effect of anthocyanins in red wines. *Food Res. Int.* 32:327–333.

Saito, M., Hosoyama, H., Ariga, T., Kataoka, S., and Yamaji, N. 1998. Antiulcer activity of grape seed extracts and procyanidins. *J. Agric. Food Chem.* 46(4):1460–1464.

Santos-Buelga, C., Francia-Aricha, E.M., and Escribano-Bailón, M.T. 1995. Comparative flavan-3-ol composition of seeds from different grape varieties. *Food Chem.* 53:197–201.

Sarni, P., Fulcrand, H., Souillol, V., Souquet, J.-M., and Cheynier, V. 1995. Mechanisms of anthocyanin degradation in grape must-like model solutions. *J. Sci. Food Agric.* 69:385–391.

Serafini, M., Maiani, G., and Ferro-Luzzi, A. 1998. Alcohol-free red wine enhances plasma antioxidant capacity in humans. *J. Nutr.* 128:1003–1007.

Shahidi, F. and Naczk, M. 1995. *Food Phenolics: Sources, Chemistry, Effects, Application.* Lancaster, PA: Technomic Publishing Co. Inc., pp. 75–107.

Shahidi, F., Janitha, P.K., and Wanasundara, P.D. 1992. Phenolic antioxidants. *Crit. Rev. Food Sci. Nutr.* 32(1):67–103.

Shrikhande, A.J. 2000. Wine by-products with health benefits. *Food Res. Intern.* 33:469–474.

Siewek, F., Galensa, R., and Herrmann, K. 1984. [Detection of adulteration of black currant products by products from red currants by means of HPLC of flavonol glycosides,] Nachweis eines Zusatzes von roten zu schwarzen Johannisbeer-Erzeugnissen über die hochdruckflüssigchromatographische Bestimmung der Flavonolglykoside. *Z. Lebensm. Unters. Forsch.* 179:315–321.

Simonetti, P., Pietta, P., and Testolin, G. 1997. Polyphenol content and total antioxidant potential of selected Italian wines. *J. Agric. Food Chem.* 45(4):1152–1155.

Skrede, G. 1987. Evaluation of colour quality in blackcurrant fruits grown for industrial juice and syrup production. *Norw. J. Agric. Sci.* 1:67–74.

Skrede, G. 1996. Fruits, in *Freezing Effects on Food Quality.* ed., L.E. Jeremiah, New York: Marcel Dekker, Inc., pp. 183–245.

Skrede, G., Wrolstad, R.E., Lea, P., and Enersen, G. 1992. Color stability of strawberry and blackcurrant syrups. *J. Food Sci.* 57(1):172–177.

Skrede, G., Wrolstad, R.E., and Durst, R.W. 2000a. Changes in anthocyanins and poly-phenolics during juice processing of highbush blueberries (*Vaccinium corymbosum* L.). *J. Food Sci.* 65(2):357–364.

Skrede, G. 2000b. Unpublished.

Slinkard, K. and Singleton, V.L. 1977. Total phenol analysis: Automation and comparison with manual methods. *Am. J. Enol. Vitic.* 28(1):49–55.

Somogyi, L.P., Ramaswamy, H.S., and Hui, Y.H. (eds.). 1996. *Processing Fruits: Science and Technology. Vol. 1: Biology, Principles, and Applications.* Lancaster, PA: Technomic Publishing Co., Inc., pp. 1–510.

Spanos, G.A. and Wrolstad, R.E. 1992. Phenolics of apple, pear, and white grape juices and their changes with processing and storage—A reveiw. *J. Agric. Food Chem.* 40(9):1478–1487.

Stähelin, H.B. 1999. The impact of antioxidants on chronic disease in ageing and in old age. *J. Vitam. Nutr. Res.* 69(3):146–149.

Stasiak, A., Pawlak, M., Sosnowska, D., and Wilska-Jeszka, J. 1998. Rate of degradation of anthocyanin pigments and ascorbic acid in aqueous solutions containing various additions of sucrose. *Przemysl Fermenacyjny i Owocowo Warzywny.* 42(12):26–34 (FSTA 1999-12-T0845).

Steinmetz, K.A. and Potter, J.D. 1996. Vegetables, fruit, and cancer prevention: A review. *J. Am. Diet. Assoc.* 96(10):1027–1039.

Stewart, K. 1996. Processing of cranberry, blueberry, currant, and gooseberry, in *Processing Fruits: Science and Technology, Vol. 2: Major Processed Products.* eds. L.P. Somogyi, D.M. Barrett, and Y.H. Hui. Lancaster, PA: Technomic Publishing Co., Inc., pp. 159–195.

Strack, D. and Wray, V. 1994. The anthocyanins, in *The Flavonoids. Advances in Research since 1986.* ed., J.B. Harborne, London: Chapman & Hall, pp. 1–22.

Strigl, A.W., Leitner, E., and Pfannhauser, W. 1995. [Qualitative and quantitative analyses of the anthocyans in black chokeberry (*Aronia melanocarpa* Michx. Ell.) by TLC, HPLC and UV/VIS-spectrometry,] Qualitative und Quantitative Analyse der Anthocyane in Schwarzen Apfelbeeren (*Aronia melanocarpa* Michx. Ell.) mittels TLC, HPLC und UV/VIS-Spektrometrie. *Z. Lebensm. Unters. Forsch.* 201:266–268.

Tamura, H. and Yamagami, A. 1994. Antioxidative activity of monoacylated anthocyanins isolated from Muscat Baily A grape. *J. Agric. Food Chem.* 42(8):1612–1615.

Timberlake, C.F. and Bridle, P. 1982. Distribution of anthocyanins in food plants, in *Anthocyanins as Food Colors.* ed., P. Markakis, New York: Academic Press, Inc., pp. 126–162.

Tsuda, T., Watanabe, M., Ohshima, K., Norinobu, S., Choi, S.-W., Kawakishi, S., and Osawa, T. 1994. Antioxidative activity of the anthocyanin pigments cyanidin 3-O-β-D-glucoside and cyanidin. *J. Agric. Food Chem.* 42(11):2407–2410.

Tsuda, T., Horio, F., and Osawa, T. 1998. Dietary cyanidin 3-O-β-D-glucoside increases *ex vivo* oxidation resistance of serum in rats. *Lipids.* 33(6):583–588.

Ulltveit, G. 1998. [*Wild berries.*] *Ville bær.* 2nd ed., Oslo, Norway: Teknologisk forlag, N.W. Damm & Søn A.S., pp. 1–166.

Velioglu, Y.S., Mazza, G., Gao, L., and Oomah, B.D. 1998. Antioxidant activity and total phenolics in selected fruits, vegetables, and grain products. *J. Agric. Food Chem.* 46(10):4113–4117.

Wang, S.Y. and Lin, H.-S. 2000. Antioxidant activity in fruit and leaves of blackberry, raspberry, and strawberry varies with cultivar and development stage. *J. Agric. Food Chem.* 48(2):140–146.

Wang, S.Y., Maas, J.L., Daniel, E.M., and Galletta, G. J. 1990. Improved HPLC resolution and quantification of ellagic acid from strawberry, blackberry and cranberry. *Hort. Sci.* 25:1078.

Wang, H., Cao, G., and Prior, R.L. 1996. Total antioxidant capacity of fruits. *J. Agric. Food Chem.* 44(3):701–705.

Waterhouse, L.A. and Walzem, R.L. 1998. Nutrition of grape phenolics, in *Flavonoids in Health and Disease*. eds., C.A. Rice-Evans and L. Packer. New York: Marcel Dekker, Inc., pp. 359–385.

WCRF/AICR. 1997. *Food, Nutrition and the Prevention of Cancer: A Global Perspective*, World Cancer Research Fund and American Institute for Cancer Research, Washington DC: Am. Inst. Cancer Res., pp. 1–670.

Wightman, J.D. and Wrolstad, R.E. 1995. Anthocyanin analysis as a measure of glycosidase activity in enzymes for juice processing. *J. Food Sci.* 60(4):862–867.

Wightman, J.D. and Wrolstad, R.E. 1996. β-glucosidase activity in juice-processing enzymes based on anthocyanin analysis. *J. Food Sci.* 61(3):544–547, 552.

Williams, C.A. and Harborne, J.B. 1994. Flavone and flavonol glycosides, in *The Flavonoids. Advances in Research since 1986*, ed., J.B. Harborne, London: Chapman & Hall, pp. 337–385.

Wrolstad, R.E. 1976. Color and pigment analyses in fruit products. Agric. Exp. Station Bulletin 624. Corvallis, OR. OSU Agr. Exp. Station, Oregon State University.

Wrolstad, R.E. 2000. Anthocyanins, in *Natural Food Colorants. Science and Technology*. eds., G.J. Lauro and F.J. Francis. New York: Marcel Dekker, Inc., pp. 237–252.

Wrolstad, R.E., Culbertson, J.D., Cornwell, C.J., and Mattick, L.R. 1982. Detection of adulteration in blackberry juice concentrates and wines by sugar, sorbitol, nonvolatile acid and pigment analyses. *J. Assoc. Off. Anal. Chem.* 65:1417–1423.

Wrolstad, R.E., Skrede, G., Lea, P., and Enersen, G. 1990. Influence of sugar on anthocyanin pigment stability in frozen strawberries. *J. Food Sci.* 55(4):1064–1065, 1072.

Wrolstad, R.E., Wightman, J.D., and Durst, R.W. 1994. Glycosidase activity of enzyme preparations used in fruit juice processing. *Food Technol.* 48(11):90, 92–94, 96, 98.

Yi, O.-S., Meyer, A.S., and Frankel, E.N. 1997. Antioxidant activity of grape extracts in a lecithin liposome system. *J. Am. Oil Chem. Soc.* 74(10):1301–1307.

Yoshida, T., Mori, K., Hatano, T., Okumura, T., Uehara, I., Komogoe, K., Fujita, Y., and Okuda, T. 1989. Studies on inhibition mechanism of autoxidation by tannins and flavonoids. V. Radical-scavenging effects of tannins and related polyphenols on 1,1-diphenyl-2-picrylhydrazyl radical. *Chem. Pharmacol. Bull.* 37(7):1919–1921.

Zabetakis, I., Leclerc, D., and Kajda, P. 2000. The effect of high hydrostatic pressure on the strawberry anthocyanins. *J. Agric. Food Chem.* 48(7):2749–2754.

Zajac, K.B., Wilska-Jeszka, J., Mamak, S., and Juszczak, M. 1992. Anthocyanins in foods. III. Potential for commercial manufacture of anthocyanins from black currants. *Przemysl Fermenacyjny i Owocowo Warzywny.* 36(5):21–24 (FSTA 1994-03-J0059).

4 Lycopene from Tomatoes

John Shi, Marc Le Maguer and Mike Bryan

CONTENTS

4.1 INTRODUCTION

Lycopene, a phytochemical, has attracted considerable attention over the past few years. This has fueled research on the biological properties of lycopene and on the numerous factors that control these properties. Possible applications of lycopene include not only those in traditional foods and ingredients, but also in novel functional food preparations specifically formulated to promote human health, as well as applications in pharmaceutical products. A multidisciplinary approach to this research has proven quite beneficial, building on the expertise not only of chemists and biochemists, but also of molecular biologists, agronomists, food technologists and engineers, and health professionals. Current information about prominent research groups and activities in the area is available on the Internet (http://www.lycopene.org; http://www.tomato-news.com; http://www.leffingwell.com/lycopene.html).

This chapter provides an overview of lycopene bioavailability and the potential benefits of dietary lycopene on human health. Lycopene chemistry and its effect on functionality will be discussed, followed by the effects of processing on lycopene obtained from tomatoes. Finally, methods of analysis and of extraction from tomatoes are discussed.

4.2 STRUCTURE–FUNCTION RELATIONSHIP IN LYCOPENE

4.2.1 LYCOPENE AND RELATED CAROTENOID PIGMENTS

Lycopene is a lipophilic (oil-soluble) carotenoid pigment. Carotenoids are widely distributed in fruits and vegetables. While not exclusive to higher plants, they are generally referred to as "plant pigments" because of their biosynthetic origin and basic functions in photosynthesis. Carotenoids also occur frequently in algae, bacteria, fungi, and in animals, especially birds, fish and invertebrates, which rely exclusively on plant sources for carotenoids. From the plant products commonly consumed by humans, more than 600 different carotenoids (the number includes *cis-trans* isomeric forms) have been isolated to date. More than 20 carotenoids have been characterized in tomato fruit alone.

Carotenoid pigments consist of two major classes; namely, the hydrocarbon carotenes (including lycopene) and the oxygenated xanthophylls. All chemical structures of the carotenoids may be formally derived from the acyclic $C_{40}H_{56}$ structure of lycopene as shown in Figure 4.1. A polyisoprenoid backbone and a series of centrally located conjugated double bonds characterize the carotenoids. Lycopene and the other carotenes (e.g., α-, β-, γ-, and ζ-carotene) do not contain oxygen. These polyene hydrocarbons are nonpolar and are orange to red in color. Xanthophylls

FIGURE 4.1 Molecular structure of lycopene.

TABLE 4.1
Physical Properties of Lycopene

Molecular formula	$C_{40}H_{56}$
Molecular weight	536.85 Da
Melting point	172–175°C
Crystal form	Long red needles separate from a mixture of carbon disulfide and ethanol
Powder form	Dark reddish-brown
Solubility	Soluble in chloroform, hexane, benzene, carbon disulfide, acetone, petroleum ether and oil
	Insoluble in water, ethanol and methanol
Stability	Sensitive to light, oxygen, high temperatures, acids, catalysts and metal ions

such as β-cryptoxanthin, lutein, and zeaxanthin are more polar than the carotenes, because they are oxygenated. Xanthophylls impart the yellow and brown colors to autumn leaves.

4.2.2 PHYSICOCHEMICAL CHARACTERISTICS OF LYCOPENE

Lycopene is the major carotenoid pigment found in tomatoes, along with lesser amounts of α-, β-, γ-, and ζ-carotene, phytoene, phytofluene, neurosporene, and lutein (Trombly and Porter, 1953; Kargl et al., 1960). The basic physicochemical information on lycopene is fairly well established and is outlined in Table 4.1. Lycopene is dispersible in edible oils and soluble in apolar organic solvents. In aqueous systems, lycopene tends to aggregate and to precipitate as crystals; this behavior is suspected to inhibit the bioavailability of lycopene in humans (Zumbrunn et al., 1985). In fresh tomatoes, the crystalline form of lycopene is responsible for the typical bright red of the ripe fruits.

Structurally, lycopene is a highly unsaturated aliphatic hydrocarbon. Its chain contains 13 carbon-carbon double bonds, 11 of which are conjugated and arranged in a linear array. The two central methyl groups of lycopene are in a 1,6-position relative to each other. All other methyl groups are in the 1,5-position. The extensively conjugated polyene system (i.e., the extended series of alternated double bonds) is the key to lycopene biological activity, which includes its susceptibility to oxidative degradation. The ruby color of lycopene, its solubility in oils, and its antioxidant activities stem from its chemical structure. Unlike many other carotenes, lycopene has no vitamin A activity, because it lacks a β-ionone ring structure. Cyclization of lycopene end groups and formation of distinct carotene products can result from the action of specific enzymes in the chloroplasts (Hill and Rogers, 1969).

With very few exceptions, the natural configuration of lycopene in plants is all-*trans*. Thermodynamically, this corresponds to the most stable configuration (Emenhiser et al., 1995; Wilberg and Rodriguez-Amaya, 1995). Upon exposure to high temperatures, light, catalysts, or active surfaces, seven of the double bonds of lycopene can isomerize to the less stable mono- or poly-*cis*- conformations. Stereoisomeric forms of lycopene have also been described with special emphasis on the light absorption properties of different molecular structures (Zechmeister, 1962).

Stability is a critical aspect of lycopene functionality. As a highly conjugated polyene, lycopene can undergo two types of change—isomerization and oxidation. Physical and chemical factors known to contribute to the degradation of other carotenoids in general also affect lycopene. These include elevated temperatures, exposure to light, oxygen, metallic ions (e.g., Cu^{2+} and Fe^{3+}), extremes in pH and active surfaces (Davies, 1976; Moss and Weedon, 1976; Crouzet and Kanasawud, 1992; Scita, 1992; Henry et al., 1998). This means that important modifications of lycopene may occur during the processing and preparation of tomato-based foods.

4.2.3 ISOMERIC FORMS OF LYCOPENE AND FACTORS INFLUENCING THEIR PRESENCE AND DISTRIBUTION

Lycopene is known to exist in a variety of geometric isomers, including all-*trans*, mono-*cis*, and poly-*cis* forms. The basis of our understanding of lycopene susceptibility to isomerization stems from early research by Zechmeister and his collaborators (1962).

The structures of some prominent *cis*-isomers of lycopene are shown in Figure 4.2. *Cis*-isomers have distinct physicochemical characteristics (and hence, bioactivity and bioavailability) compared to their all-*trans* counterparts. In general, the *cis*-isomers are more soluble in oil and hydrocarbon solvents than their all-*trans* counterparts. They are less prone to crystallization because of their kinked structures. They also are less intense in color, which may influence the consumers' perception of food quality. The appearance of a distinct absorption maximum in the UV region ("*cis*-peak") is useful for distinguishing between the different isomers.

The all-*trans* configuration predominates in fresh tomatoes and gradually isomerizes to *cis*- configurations upon processing and storage of tomato products. Isomerization of lycopene has been shown to take place both in foods and in model systems with "pure" lycopene. *Trans*- to *cis*- isomerization typically occurs during processing. The storage of processed foods favors reversion from *cis*- to *trans*- because the *cis*-isomers are in a relatively unstable state compared to the *trans*-isomer, which is in the stable ground state.

The 5-*cis*, 9-*cis* and 15-*cis*-isomers of lycopene have been assayed in various foods and in human tissues using nuclear magnetic resonance (NMR) spectroscopy (Zumbrunn et al., 1985). In the various tomato-based foods surveyed by Schierle et al. (1996), the all-*trans* isomer represented between 35% to 96% of the total lycopene. In their survey, the proportion of 5-*cis*-isomer in tomato products ranged from 4% to 27%, with considerably lower amounts of the other *cis*-isomers. In human serum and tissues, the *cis*-isomers of lycopene were found to contribute more than 50% of total lycopene (Krinsky et al., 1990).

4.2.4 BIOLOGICAL ACTIVITY OF LYCOPENE, ANTIOXIDANT PROPERTIES

With its acyclic structure, large array of conjugated double bonds and important hydrophobicity, lycopene exhibits a range of unique and distinct biological properties. Of these properties, its antioxidant properties continue to arouse substantial interest. The system of conjugated double bonds allows lycopene molecules to

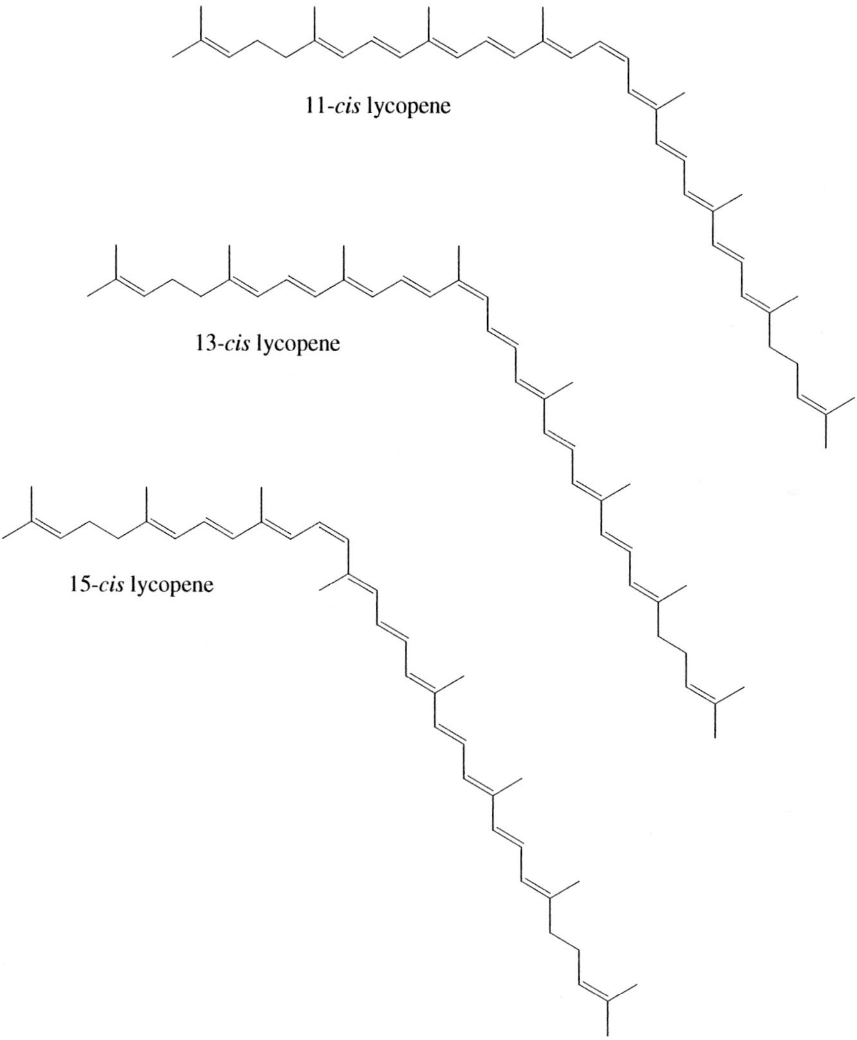

11-*cis* lycopene

13-*cis* lycopene

15-*cis* lycopene

FIGURE 4.2a Molecular structures of some *cis*-isomers (Nguyen and Schwartz, 1999).

efficiently quench the energy of notably deleterious forms of oxygen (singlet oxygen) and to scavenge a large spectrum of free radicals. Of all naturally occurring carotenoids, lycopene is the most efficient quenchers of singlet oxygen (Di Mascio et al., 1989; Conn et al., 1991).

The antioxidant activities of lycopene and of other carotenoids are related to their ability to quench singlet oxygen ($O_2^{\sqrt{}}$) and to trap peroxyl radicals (ROO·) (Foote and Denny, 1968; Burton and Ingold, 1984). The quenching activity of the different carotenoids depends essentially on the number of conjugated double bonds. It is modulated by end groups or the nature of the substituents in the carotenoids that contain cyclic end groups (Foote and Denny, 1968; Stahl et al., 1993). This explains the important differences in the quenching rate constants (K_q) of the different

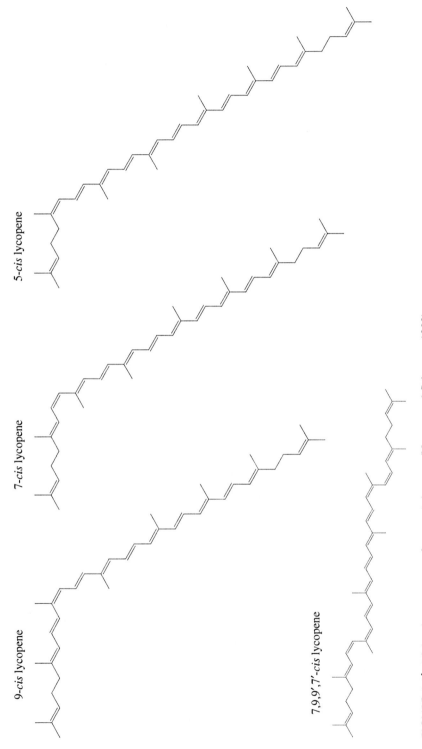

FIGURE 4.2b Molecular structures of some *cis*-isomers (Nguyen and Schwartz, 1999).

TABLE 4.2
Comparison of the Antioxidant Activities
(Quenching of Singlet Oxygen) of
Different Carotenoids

Carotenoid	Rate Constant for Quenching of Singlet Oxygen $K_q \times 10^9$ (mol^{-1}s^{-1})
Lycopene	31
γ-Carotene	25
α-Carotene	19
β-Carotene	14
Lutein	8
Astaxanthin	24
Bixin	14
Canthaxanthin	21
Zeaxanthin	10

Source: Data from Di Mascio et al. (1989, 1991),
Conn et al. (1991, 1992) and Miller et al. (1996).

carotenoids (Table 4.2). Lycopene's superior ability to quench singlet oxygen vs. that of γ- and β-carotene is related to the opening of the β-ionone ring.

Interactions between lycopene and oxygen radicals can be considered second-order rate reactions. Lycopene is less efficient, and electron transfer is observed in both directions (Conn et al., 1992). The potential reduction of the antioxidant property of lycopene is related to the formation of the superoxide radical anion, $O_2^{\surd}$ (Palozza, 1998).

Lycopene also allows for the formation of peroxyl radicals capable of acting as pro-oxidants and of undergoing auto-oxidation themselves. The proposed pathway for lycopene oxidative degradation is shown in Figure 4.3. Oxygen is introduced into lycopene in at least two ways: oxidation of a methyl or methylene group or addition to a carbon-carbon double bond. Oxidative degradation can occur at either end of the C_{40}-carbon skeleton. On the basis of widely accepted nomenclature rules, a product of degradation that does not retain the C_{20} and attached $C_{20'}$-methyl group of the original C_{40} structure is no longer deemed a carotenoid. Ultimately, the final noncarotenoid fragments of lycopene oxidative degradation form as the result of direct oxidative cleavage of carbon-carbon double bonds. The scheme accepted for the stepwise desaturation of phytoene to lycopene is shown in Figure 4.4.

4.3 OCCURRENCE OF LYCOPENE IN NATURE AND IN FOODS

4.3.1 NATURAL OCCURRENCE OF LYCOPENE

Lycopene is widely spread in nature. Because carotenoids are photosynthesized by plants and microorganisms, they are the source of all carotenoids found in the

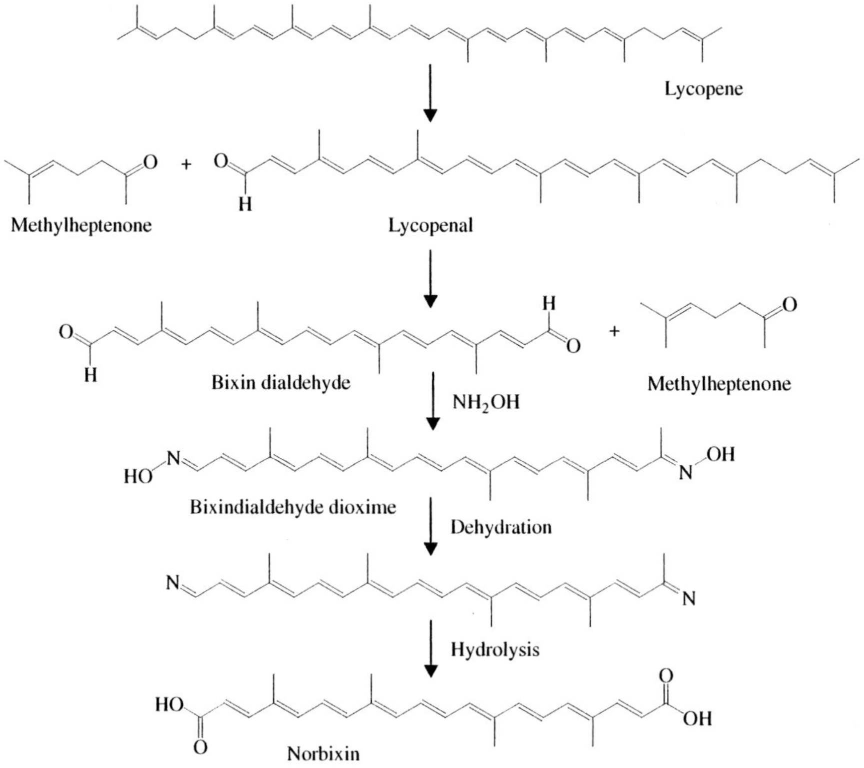

FIGURE 4.3 Schematic of lycopene degradation pathway. [Modified from Karrer and Jucker (1950).]

animal kingdom. One of the functions of lycopene and related pigments is to absorb light during photosynthesis, thereby protecting plants against photosensitization. The bright color of lycopene can be masked by green chlorophyllic pigments (e.g., in green vegetables and leaves). In a number of cases, the chlorophyll content decreases as plants mature, leaving lycopene and other carotenoids responsible for the bright colors of most fruits (e.g., pineapple, orange, lemon, grapefruit, strawberry, tomato, paprika, rose hip) and many flowers (e.g., eschscholtzia, narcissus). Lycopene and other carotenoids also contribute to the attractive colors of some birds (e.g., flamingo and canary feathers), insects, and marine animals (e.g., shrimp, lobster and salmon).

4.3.2 COMMON BOTANICAL AND FOOD SOURCES OF LYCOPENE

The characteristic deep-red color of ripe tomato fruit and related products is mainly due to lycopene. This color serves as an important indicator of their quality. Tomatoes and tomato foods are an important source of carotenoids for humans and are a major source of lycopene in the Western diet. Other important sources of dietary lycopene include watermelon, guava, rose hip, papaya and pink grapefruit (Gross, 1987, 1991; Beerh and Siddappa, 1959; Mangels et al., 1993) (Table 4.3).

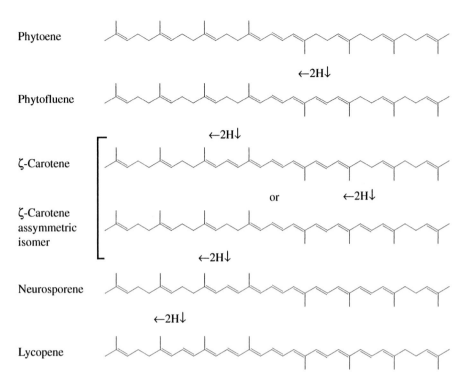

FIGURE 4.4 Scheme for the stepwise desaturation of phytoene to lycopene in carotenoid biosynthesis. (Goodwin, 1980. With permission.)

TABLE 4.3
Lycopene Content of Some Commonly Consumed Fruits and Vegetables

Material	Lycopene Content mg/100 g, Wet Basis
Tomato	0.72–20
Watermelon	2.3–7.2
Guava (pink)	5.23–5.50
Grapefruit (pink)	0.35–3.36
Papaya	0.11–5.30
Rose hip	0.68–0.71
Carrot	0.65–0.78
Pumpkin	0.38–0.46
Sweet potato	0.02–0.11
Apple pulp	0.11–0.18
Apricot	0.01–0.05

Source: Data from Beerh and Siddappa (1959), Gross (1987, 1991) and Mangels et al. (1993).

4.3.3 LYCOPENE IN FRESH TOMATOES

4.3.3.1 Content and Distribution

Lycopene is the most abundant carotenoid in ripe tomatoes, making up approximately 80–90% of the total carotenoids in the common variety of tomatoes, *Lycopersicon esculentum*. Other carotenoids such as α-carotene, β-carotene, lutein and β-cryptoxanthin are present in negligible amounts (Curl, 1961). Lycopene is found predominantly in the chromoplasts of plant tissues. In ripening tomatoes, it is the last carotenoid to form. Its biosynthesis increases dramatically after the breaker stage (i.e., after the color of the berries starts changing from green to pink) as chloroplasts undergo transformation to chromoplasts (Kirk and Tilney-Bassett, 1978). Chromoplasts rich in lycopene predominate in the outer part of the pericarp. They contain voluminous sheets in which lycopene is integrated into so-called LHCs (light-heaping complexes). These complexes consist of sequences of hydrophobic membrane-linked proteins that interact with several pigment molecules coagulated in the form of elongated needle-shaped crystals. At the cellular level, lycopene is localized in the chloroplasts of tomato fruits and can be found among the thylakoid membranes in the photosynthetic pigment-protein complex. The final stage of chromoplast development is the formation of lycopene crystals, which occupy a large portion of chromoplast and appear as voluminous red sheets in the chromoplasts. The largest concentrations of lycopene are found in the pericarp. There, lycopene exists as small globules, suspended in the tomato pulp throughout the fruit. After tomatoes reach the mature stage, elongated crystals or crystalloids of lycopene form in association with extended thylakoids and appear in the chromoplasts. Lycopene may exist in the form of lipid dispersions and may be bound to structural elements. Lycopene is found predominantly in the chromoplasts of tomato fruit tissues. Lycopene biosynthesis increases during the ripening process as chloroplasts undergo transformation to chromoplasts (Kirk and Tilney-Basset, 1978). The undulating structures associated with elongated thylakoids are probably crystalloids of lycopene. Thus, crystalloids are associated with globules and thylakoids. The globules in the plastids appear light orange under microscopy and probably contain most of the lycopene and other carotenoids. Lycopene is concentrated in the globules because of its lipophilic nature.

The skin and the pericarp of the tomato fruit are particularly rich in lycopene and other carotenoids (Al-Wandawi et al., 1985; D'Souza et al., 1992; Sharma and Le Maguer, 1996). They also are the tissues that are usually discarded upon industrial processing of tomatoes. According to Al-Wandawi et al. (1985), tomato skin contains 12 mg of lycopene per 100 g (wet basis), whereas whole mature tomato contains only 3.4 mg lycopene per 100 g (wet basis).

4.3.3.2 Effects of Agricultural Practices on Lycopene Content

The amount of lycopene in fresh tomato fruits is influenced by variety, agricultural practices, maturity and the environmental conditions under which the fruits matured. Tomatoes of the common variety *Lycopersicon esculentum* typically contain about 3 to 10 mg of lycopene per 100 g of ripe fresh fruit (Hart and Scott, 1995; Tonucci et al., 1995).

Higher amounts can be found in certain varieties. In *L. pimpinellifolum,* levels as high as 40 mg per 100 g of fresh tissue have been reported, accounting for more than 95% of the total carotenoid content of these tomatoes (Porter and Lincoln, 1950). Nitrogen fertilizers and calcium sprays (0.2%) were reported to increase the lycopene content of tomatoes. Potassium deficiency may lower lycopene concentration; however, very high potassium applications (>800 kg/ha) would cause a marked decrease in color uniformity disorders (Grolier, 2000).

Liu and Luh (1977) reported that the maturity at harvest affected the lycopene content of tomato paste. Lycopene levels increase as tomatoes mature. Heinonen et al. (1989) reported that the lycopene concentration in tomatoes was highest in the summer (from June to August) and lowest in the winter (from October to March). For a given variety, lycopene level increases in tomatoes as they are transferred from being grown under glass, then under plastic, and finally out to the field. Tomato fruits grown in greenhouses either in summer or winter usually contain less lycopene than the fruits grown outdoors during the summer. Also, the fruits picked green and ripened in storage are substantially lower in lycopene than vine-ripened fruits (Gould, 1994). Lurie et al. (1996) reported that the formation of lycopene is inhibited at temperatures above 32°C. Soil temperatures between 12 and 32°C are required for optimal biosynthesis of lycopene. Outside of this range, the synthesis of lycopene precursors is inhibited, which prevents the production of lycopene. The accumulation of lycopene is hindered by excessive sunlight. The best conditions for production include adequately high temperature together with a relatively dense foliage to protect the growing fruits from direct exposure to the sun. Following harvesting, a temperature of *ca* 37°C stimulates the formation of tomato lycopene.

Lycopene synthesis in the *rin* mutant was enhanced by high O_2 in the presence of 10 ppm ethylene (Frenkel and Garrison, 1976). On the other hand, ethanol inhibits ripening and the synthesis of tomato lycopene (Saltveit and Mencarelli, 1988). In addition, Sheehy et al. (1988) found that a reduction in polygalacturonase did not affect biosynthesis of lycopene. Lampe and Watada (1971) and Mohr (1979) pointed out that the lycopene content in tomato fruits may be increased by improving the practices at times of selection of varieties, fertilization and harvest.

In summary, preprocessing factors that affect the level of carotenoids in tomatoes include variety, water (irrigation) and fertilization practices, temperature and light, degree of maturity at harvest, harvest time and postharvest storage.

4.4 METHODS FOR ASSAYING LYCOPENE

Determination of the lycopene content in tomatoes and some tomato products can be carried out by physical and chemical methods. Physical methods depend on the relation of color parameters to lycopene concentration in the samples. In chemical analysis, lycopene is extracted from the tomato fruits and products and quantified.

Traditionally, lycopene concentrations were determined spectrophotometrically at 460 to 470 nm (Lovric et al., 1970; Edwards and Lee, 1986; Mencarelli and Saltveit, 1988; Tan and Soderstrom, 1988). However, this method measured only total lycopene.

Chromatographic separation and subsequent detection of lycopene offers the best choice for analysis. Identification of *trans-* and *cis-* stereoisomers of lycopene may be possible. These methods include column chromatography, thin-layer chromatography, gas chromatography, and high-performance liquid chromatography (HPLC). The widely used AOAC chromatographic method (AOAC, 1995) for the determination of lycopene and other carotenoids fails to separate the *cis*-isomers from the all-*trans* isomer. Both normal-phase and reversed-phase HPLC methods have been used to separate and quantitate provitamin A carotenoids in fruits and vegetables (Chandler and Schwartz, 1987, 1988; Quackenbush, 1987; Saleh and Tan, 1988; Godoy and Rodriguez-Amaya, 1989). Reversed-phase HPLC methods utilizing C_{18} stationary phases allow for the partial separation and detection of *cis*- and *trans*-isomers of provitamin A carotenoids. Some rapid and highly efficient HPLC methods with minimum oxidation and isomerization have been developed and studied extensively to quantitate lycopene and its isomers from tomatoes and tomato-based foods (Schwartz and Patroni-Killam, 1985; Bureau and Bushway, 1986; Daoud et al., 1987; Zonta et al., 1987; Craft et al., 1990; Sadler et al., 1990; Stahl and Sies, 1992; Emenhiser et al., 1995; Clinton et al., 1996; Schierle et al., 1996; Gartner et al., 1997; Nguyen and Schwartz, 1998; Shi and Le Maguer, 1999a,b). A polymeric C_{30} stationary phase has been developed that can efficiently separate the geometric isomers (Sander et al., 1994; Emenhiser et al., 1995, 1996; Lessin et al., 1997).

4.5 BIOAVAILABILITY OF LYCOPENE FOR HUMANS, A KEY TO LYCOPENE EFFECTIVENESS

Beyond our knowledge about lycopene content in foods, the knowledge about lycopene availability to the human organism, i.e., its absorbability and efficient use by the body, is very important. The U.S. FDA's definition of bioavailability of a drug is "the rate and extent to which the active substance or therapeutic moiety is absorbed from a drug product and becomes available at the site for action" (Benet and Shiner, 1985). More generally, others (Macrae et al., 1993; Jackson, 1997) have defined bioavailability as the fraction of an ingested constituent that is available to the body through absorption for utilization in normal physiological functions and for metabolic processes. The concept of bioavailability of a constituent or microconstituent is thus closely related to the concept as it applies to pharmaceutical compounds.

The absorption of dietary lycopene can be affected by a number of factors, including culinary practices and food properties. Such factors include the amount of lycopene consumed in a meal, the nature of the food matrix in which lycopene is incorporated (Deshmukh and Ganguly, 1964; Kemmerer et al., 1974; Jayaarahan et al., 1980), the physical form or location of lycopene within the matrix (lycopene-protein complexes of cell chloroplasts vs. crystalline form in chromoplasts) (De Pee et al., 1995), the particle size of the material (Clinton et al., 1996; Nguyen and Schwartz, 1998), molecular linkage, interactions with other carotenoids and nutrients, xanthophyll and chlorophyll contents (Prince et al., 1991; Van Vliet et al., 1996), co-ingestion of fat as a delivery medium (Stahl and Sies, 1992; De Pee et al., 1995), co-ingestion of high amounts of dietary fiber, content of dietary proteins (Erdman et al., 1986; Rock and Swendseid, 1992), factors that interfere with proper micelle formation in the gastrointestinal tract, bioconversion (isomerization) and idiosyncratic factors (e.g., modifications of absorption functions and genetic factors) (Bowen et al., 1993; Olson, 1994; Parker, 1996; Castenmiller and West, 1998; Dimitrov et al., 1988).

In human plasma, lycopene occurs as a 50/50 mixture of *cis*- and *trans*-isomers. This is the case in human and animal tissues, because this mixture corresponds to an equilibrium between the *trans*- and *cis*-isomers (Boileau et al., 1999). Among the different geometrical isomers of lycopene, the *cis*-isomers (5-*cis*, 9-*cis*, 13-*cis* and 15-*cis*) are better absorbed by the human body than the naturally occurring all-*trans* form (Stahl and Sies, 1992; Boileau et al., 1999). The *cis*-isomers of lycopene are better absorbed than the all-*trans* isomer (Sakamoto et al., 1994; Britton, 1995; Stahl and Sies, 1996; Boileau et al., 1999). This may be due to the greater solubility of *cis*-isomers in mixed micelles, possibly to the preferential incorporation into chylomicrons, and a lower tendency of *cis*-isomers to aggregate. *Cis*-isomers are less likely to crystallize, are more efficiently solubilized in lipophilic solutions and are more readily transported within cells or tissue matrix.

The composition and structure of food have an important bearing on lycopene bioavailability. Food characteristics may affect the release of lycopene from the tomato tissue matrix. Practices such as cooking or fine grinding can increase bio-availability by physically disrupting or softening plant cell walls and by disrupting lycopene-protein complexes (Hussein and El-Tohamy, 1990).

Lycopene in fresh tomatoes is less bioavailable than lycopene in heat-treated tomato products (Figure 4.5). Lycopene serum concentrations are indeed greater following the consumption of heat-processed tomato-based foods than after the consumption of unprocessed fresh tomatoes (Giovannucci et al., 1995; Gartner et al., 1997). Stahl and Sies (1992) found that 20–30% of total lycopene consisted of *cis*-isomers when tomatoes were heated at 100°C for 1 h.

The above observations could be attributed to a lower availability of lycopene from the raw material, where it is probably bound in the matrix. Thermal processing

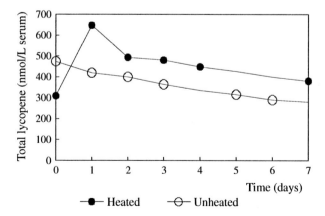

FIGURE 4.5 Serum concentrations of total lycopene after the consumption of heated (●) or unheated (○) tomato juice. The lycopene dose was 2.5 μmol/kg body weight (Modified from Stahl and Sies, 1992).

(cooking included) and mechanical texture disruption (e.g., homogenization) are convenient ways to enhance bioavailability by breaking down sturdy cell wall structures, by disrupting chromoplast membranes and cellular integrity, thereby improving the accessibility of lycopene. The food/fruit matrix (i.e., the lipids and other constituents of chromoplasts, as well as the fiber contained in tomato fruits) may contribute greatly to the stability of the all-*trans* form of lycopene in fresh fruits.

The bioavailability of lycopene in tomato-based foods vs. that in fresh tomatoes increases even further when lycopene is consumed with oil. In studies by Stahl and Sies (1992, 1996), ingestion of tomato juice cooked in an oil medium resulted in a two- to threefold increase in lycopene serum concentrations 1 day after ingestion. An equivalent consumption of unprocessed tomato juice caused no rise in lycopene plasma concentration. This indicates that thermal pretreatment and an oil medium were beneficial for extracting lycopene into the lipophilic phase. Solubilization of lycopene in a lipophilic matrix is expected to considerably enhance its availability and its bioactivity. This is likely to boost its effectiveness as an antioxidant. However, this higher reactivity also renders lycopene more vulnerable to the detrimental effects of factors such as air, temperature and interactions with other components of the food.

Various types of dietary fiber were found to reduce carotenoid bioavailability in foods (Erdman et al., 1986). Rock and Swendseid (1992) tested the inhibitory effect of pectin, a typical dietary fiber. Pectin was shown to affect the absorption of dietary carotenoids in humans. High-methoxyl pectin is strongly associated with the hypo-cholesterolemic effect of dietary fibers and with relatively low absorption of dietary lycopene (Castenmiller et al., 1998; Van het Hof et al., 2000). Absorption of lycopene seems to be most efficient at relatively low concentrations, and it seems to be boosted by the co-ingestion of β-carotene (Johnson, 1997). Lycopene can undergo an *in vivo* isomerization. Van Vliet (1996) carried out the dioxygenase assay with β-carotene as substrate, both alone and together with increasing amounts of lutein or lycopene. He found that lutein lowered retinal formation from β-carotene, whereas lycopene had no effect. Prince et al. (1991) reported a marked decrease in serum lycopene levels following supplementation with a high dose of β-carotene; another study of the effects of high-dose supplementation found a decrease in low-density lipoprotein (LDL) lycopene content (Gaziano et al., 1995). Evidence for carotenoid interactions during absorption has been reported in ferrets, in which supplementation with either canthaxanthin or lycopene limited the appearance of β-carotene in the plasma after 24 h compared with administration of β-carotene alone (White et al., 1993; Kostic et al., 1995).

The induction of *trans*-to-*cis* isomerization (reversion) during processing includes the release of lycopene by thermal processing that induces disruption of the tomato tissue structure and cell walls, change of the bonding forces between lycopene and tissue matrix, dissociation or weakening of protein-lycopene complexes, dissolution or dispersion of crystalline lycopene complexes and heat-improved extraction of lycopene into the oily phase of the mixture when oil is used as a vehicle. Gartner et al. (1997) pointed out that heat treatment can improve the bioavailability of lycopene without significantly changing the *cis*-isomer content of the heat-treated foods. The lycopene bioavailability from tomato-based foods may be enhanced in two ways: extraction of lycopene from the food matrix into the

TABLE 4.4
Lycopene Retention in Tomato Puree after Different Cooking Times

	Lycopene Content mg/100 g Fresh Puree	Relative Loss (%)
Fresh tomato puree	6.62	—
5 min cooking	6.61	1.61
10 min cooking	6.57	2.41
30 min cooking	6.41	5.11
60 min cooking	6.09	8.76

Source: Shi and Le Maguer (1999b).

lipophilic phase (Brown et al., 1989; Zhou et al., 1996), and thermal processing and mechanical disruption of tomato tissue cells. Effect of cooking time on lycopene content is shown in Table 4.4. Heat treatment leads to an increased bioavailability of lycopene after cooking, which means that food processing may improve the availability of lycopene in tomato-based foods. Heat treatment promotes *trans-cis* isomerization of lycopene. The loosely bound lycopene in tomatoes is released by the heat during processing, which makes it more easily absorbable by the body. The degree of isomerization is directly correlated with the intensity and duration of heat processing (Schierle et al., 1996; Shi and Le Maguer, 1999a,b).

It is now generally accepted that lycopene in a lipid medium is more bioavailable than in fresh tomatoes. Cooking and reduction of particle size by grinding or homogenization can also reduce the unfavorable effects of the food matrix. However, because thorough destruction of the matrix (for example, by extensive cooking) can also destroy lycopene, optimum processing technology parameters ought to be determined to maximize destruction of the matrix and minimize destruction of lycopene.

4.6 HEALTH-PROMOTING EFFECTS OF DIETARY LYCOPENE

There has been a growing interest in exploring the role of lycopene in the prevention of a variety of ailments in humans, including some cancers and cardiovascular diseases. Recent suggestions that the consumption of lycopene-rich foods may reduce the risk of such diseases has prompted in-depth studies of the levels of lycopene in foods and of the nature of the correlation between dietary lycopene and the onset of certain diseases (Clinton, 1998). Although it has no provitamin A activity, lycopene is able to function as an antioxidant, and it efficiently quenches singlet oxygen *in vitro* (Boileau et al., 1999). Under appropriate conditions, lycopene can function as an effective antioxidant at the relatively low concentrations found *in vivo*, not only against O_2^- but also against lipid peroxidation and the highly destructive hydroxyl radical HO·, which is implicated in many diseases. The antioxidant properties of lycopene most likely contribute to limiting the risk for some diseases (Sies et al., 1992). Clinical evidence is mounting in support of the role of lycopene as an important microconstituent. It may provide some degree of protection against cancers of the prostate, breast, digestive tract, bladder, cervix, lung, and other epithelial cancers (Olson, 1986; Micozzi et al., 1990; Levy et al., 1995).

Three major types of research have emerged (Levy et al., 1995): epidemiological studies involving patients with various malignancies, studies on the direct effect of lycopene on the proliferation of various tumors in cell lines and in animal models, and studies on the biochemical and immunological mechanisms of lycopene action.

4.6.1 PROTECTION FROM SOME CANCERS AND CARDIOVASCULAR DISEASES

4.6.1.1 Various Types of Cancers

The incidence of cancer (Helzlsouer et al., 1989; Van Eenwyk et al., 1991) has been inversely correlated with the serum level of lycopene and the dietary intake of tomatoes. A study conducted in Italy with 2706 cases of cancer of the oral cavity and pharynx, esophagus, stomach, colon and rectum matched with 2879 controls showed that protection for all sites of digestive tract cancers was associated with an increased intake in tomato-based food (Franceschi et al., 1994). The correlation between consumption of tomatoes and the diminished cancer risk was related to an increased supply of dietary lycopene. A reduced risk of cancers at other sites, such as the digestive tract, pancreas and bladder (Gerster, 1997) has been found to be associated with the dietary intake of lycopene. A study released by the University of Milan suggested that people who ate at least one serving of a tomato-based product per day had 50% less chance of developing digestive tract cancer than those who did not eat tomatoes (Franceschi et al., 1994). A Harvard University study found that older Americans who regularly ate tomatoes were less likely to die from any form of cancer (Colditz et al., 1985). A Harvard School of Public Health report suggested that men who ate 10 or more servings per week of tomato products, including tomatoes, tomato sauce and pizza sauce, were up to 34% less likely to develop prostate cancer (Giovannucci et al., 1995). This study monitored dietary habits and the rate of incidence of prostate cancer in 48,000 men for 4 years and assessed more than 46 different fruits and vegetables and related products on the basis of their consumption frequency. The researchers found that a lower risk of prostate cancer was most strongly associated with tomato sauce consumption. The protective effects were even stronger when the analysis focused upon the risk of more advanced or aggressive prostate cancers.

Lycopene has been reported to increase the survival rate of mice exposed to x-ray radiation (Forssberg et al., 1959). Ribaya-Mercado et al. (1995) reported on the protective effects of lycopene toward oxidative stress-mediated damage of the human skin upon irradiation with UV light. Peng et al. (1998) examined the levels of different carotenoids, including lycopene, and vitamins A and E in plasma and cervical tissues obtained from 87 women subjects (27 cancerous, 33 precancerous and 27 noncancerous). Women with cancer had lower plasma levels of lycopene, other carotenoids, vitamin A and E compared to pre- and noncancerous women. The results indicate that women who have high levels of lycopene are less likely to suffer from cervical cancer than those having lower lycopene levels in their body. In a study by Goodman et al. (1998) involving 147 patients with confirmed cervical cancer and 191 noncancerous subjects, only lycopene was found to be significantly lower in patients with cancer. Kanetsky et al. (1998) studied 32 non-Hispanic, black women with cervical cancer and 113 noncancerous women. The authors measured lycopene levels in the blood. It was found that women with the highest levels of blood lycopene had consumed higher levels of lycopene and vitamin A, and they had one-third less chance of developing cervical cancer.

4.6.1.2 Atherosclerosis and Coronary Heart Disease

Riso and Porrini (1997) concluded that the consumption of tomato-based foods may reduce the susceptibility of lymphocyte DNA to oxidative damage. Lycopene had a preventive effect on atherosclerosis by protecting plasma lipids from oxidation. Lower blood lycopene levels were also associated with increased risk of coronary heart disease according to studies of Lithuanian and Swedish subjects (Kristenson et al., 1997). Kohlmeir et al. (1997) measured the relationship between antioxidant levels and acute heart disease in a case study of people from 10 different European countries. They found that the consumption of lycopene in fruits and vegetables may reduce the likelihood of developing heart disease. Lycopene is believed to prevent oxidation of LDL cholesterol and to reduce the risk of developing atherosclerosis and coronary heart disease. In Agarwal and Rao's study (1998), the daily consumption of tomato products providing at least 40 mg of lycopene was enough to substantially reduce LDL oxidation.

Inhibition of cancer cell growth by lycopene has been extensively demonstrated in tissue culture experiments. Wang et al. (1989) studied an *in vivo* model of glioma cells transplanted in rats and showed that lycopene was as effective as β-carotene in inhibiting the growth of glioma cells. Breast Cancer Serum Bank (Columbia, MO, U.S.) tissue samples were analyzed to evaluate the relationship between carotenoid levels (including lycopene level) and selenium and retinol and breast cancer (Dorgan et al., 1998). Only lycopene was found to reduce the risk for developing breast cancer. Other carotenoids did not reduce the risk of breast cancer. In studies of cell cultures, the activities of lycopene in inhibiting breast cancer tumors were compared with those of α- and β-carotene using several human cancer cells (Levy et al., 1995). It was found that the cell cultures that were enriched with lycopene had reduced growth of MCF-7 breast cancer cells, and that α- and β-carotene were far less effective inhibitors of growth than was lycopene.

4.6.2 Mechanisms of Protection by Lycopene

4.6.2.1 Antioxidant Effects and Interactions with Free Radicals

Lycopene has been recognized as the most efficient singlet oxygen quencher among the carotenoids of biological importance (Di Mascio et al., 1989, 1991). Its antioxidant properties are associated with *in vitro* lowering of oxidative damage to the DNA, slowing down of the transformation to malignant state, and lowering biological oxidative damage of proteins, lipids and other cells. Lycopene has also been found to increase gap-junctional communication between cells and to induce the synthesis of connexin-43 (Zhang et al., 1992). The loss of gap-junctional communication may be important in the process of malignant transformation; its restoration may reverse the process.

4.6.2.2 Modulation of the Immune System

Evidence is now being obtained that suggests that lycopene may have direct stimulatory effects on the response of the immune system. This may involve an antioxidant action, and it could form the basis of a protective action against cancer and also against human immunodifiency virus (HIV) and acquired immune deficiency syndrome (AIDS). The free radical quenching constant of lycopene was found to be more than double that of β-carotene and 10 times more than that of α-tocopherol, which makes lycopene's presence in the diet of considerable interest (Di Mascio et al., 1989, 1991; Conn et al., 1991; Devasagayam et al., 1992; Ribaya-Mercado et al., 1995).

4.7 PROCESSING EFFECTS ON LYCOPENE IN TOMATOES

The main causes of lycopene degradation during food processing are oxidation and isomerization. It is widely believed that lycopene generally undergoes isomerization upon thermal treatment. Changes in lycopene content and in the distribution of *trans*- and *cis*-isomers result in modifications of its biological properties (Zechmeister, 1962). Determination of the extent of lycopene isomerization would provide better insights into the potential health effects of processed food products. In processed foods, oxidation is a complex process that depends upon many factors, such as processing conditions, moisture, temperature, and the presence of pro- or anti-oxidants and of lipids. For example, use of fine screens in juice extraction enhances the oxidation of lycopene due to the large surface exposed to air and metal. The amount of sugar, acids and amino acids also affects lycopene degradation in processed food products by leading to the formation of brown pigments (Gould, 1994). The deterioration in color that occurs during the processing of various tomato products results from exposure to air at high temperatures during processing, causing the naturally occurring all-*trans* lycopene to be isomerized and oxidized. Coupled with exposure to oxygen and light, heat treatments that disintegrate tomato tissue can result in substantial destruction of lycopene.

4.7.1 HEAT

The effects of heating on total lycopene and *cis*-isomer content in tomato puree are shown in Figures 4.6 and 4.7 (Shi and Le Maguer, 1999a). Increasing the temperature from 90 to 150°C caused a greater loss of total lycopene. Total lycopene concentration decreased over treatment time, but *cis*-isomers mostly appeared within the first hour of heating. After 2 h of heating, the rate of *cis*-isomer accumulation decreased.

Temperature increase from 90°C to 150°C caused a 35% decrease in total lycopene content. The greater percentage loss of lycopene as compared to the less

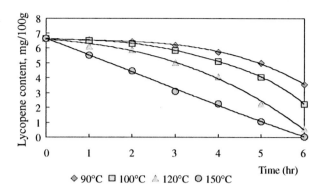

FIGURE 4.6 The effect of heat treatment on total lycopene degradation (Shi and Le Maguer, 1999b).

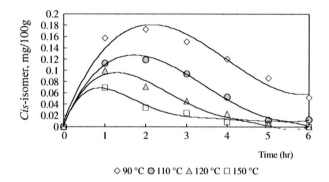

FIGURE 4.7 The effect of heat treatment on *cis*-isomer degradation (Shi and Le Maguer, 1999b).

gain in *cis*-isomer suggests that oxidation of lycopene was the main mechanism for the lycopene loss when heated at temperatures greater than 100°C. Fewer *cis*-isomers were present in the low temperature treatment material. The *cis*-isomers formed during the first 1 to 2 h of heating in processed tomato samples had degraded quickly. Thus, food processing can foster *cis*-isomerization in tomato-based foods if an optimum heating temperature is used.

The results of lycopene retention in tomato puree cooked at 100°C are shown in Table 4.4. The length of cooking time had little or no effect on the degradation of lycopene if the heating temperature was lower than 100°C. These results suggest that the duration of heating, rather than cooking temperature, is the critical factor affecting degradation of lycopene. High temperatures such as more than 120°C, will result in more lycopene loss.

When tomato-based products are subjected to thermal processing, the changes in lycopene content and the conversion of *trans*- to *cis*-isomers will change its bioactivity and bioavailability (Khachik et al., 1992b; Emenhiser et al., 1995; Wilberg and Rodriguez-Amaya, 1995; Stahl and Sies, 1996). High temperature will break down lycopene molecules into small fractions (Figure 4.8).

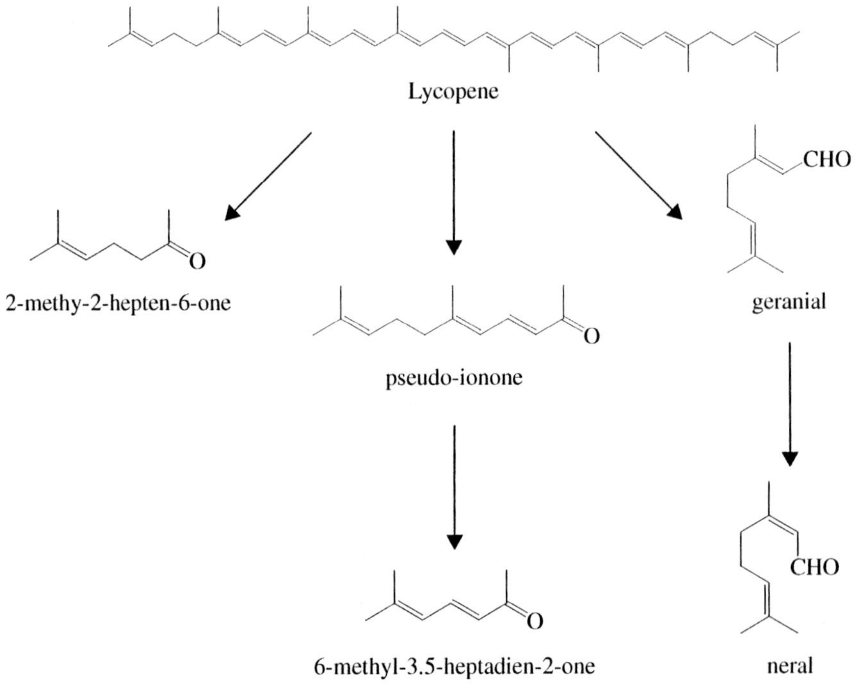

FIGURE 4.8 Reaction sequence for the formation of volatile compounds during heat treatment of lycopene. [Kanasawud, P. and Crouzet, J.C. 1990, *J. Agric. Food Chem.*, 38: 1238–1242. Copyright 1990 American Chemical Society. With permission.]

4.7.2 LIGHT

The effects of light irradiation on the content of total lycopene and *cis*-isomer in tomato puree are shown in Figures 4.9 and 4.10. The increase in *cis*-isomers began to occur at the onset of exposure to light. Total lycopene loss as well as *cis*-isomer loss increased with light irradiation time.

The losses of *cis*-isomers were further increased by exposure to light. This suggests that the light irradiation causes losses of total lycopene content. The amount of *cis*-isomer formed during light irradiation appeared to decrease quickly as the intensity of light increased, suggesting that *cis*-isomer oxidation was the main reaction pathway. A possible explanation for this phenomenon is that the *trans*-isomer first isomerizes into *cis*-isomers, which then preferentially follow the oxidative pathway. The rate of *cis*-isomer oxidation was much greater under heat and light treatment than the rate of *cis*-isomer formation.

4.7.3 OXYGEN

Monselise and Berk (1954) first reported on the oxidative destruction of lycopene during the processing of tomato purée. Cole and Kapur (1957b) showed that more than 30% of lycopene was degraded upon heat treatment at 100°C for 3 h in the

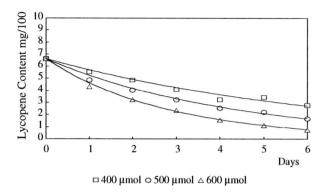

FIGURE 4.9 The effect of light irradiation on total lycopene degradation (Shi and Le Maguer, 1999b).

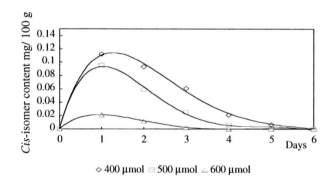

FIGURE 4.10 The effect of light irradiation on *cis*-isomer degradation (Shi and Le Maguer, 1999b).

presence of oxygen, whereas only 5% was lost in the presence of CO_2 after 3 h of treatment (Table 4.5).

4.7.4 ISOMERIZATION DURING PROCESSING

The contents of various lycopene isomers in some commercial tomato products are shown in Table 4.6. Detailed studies of the degree of isomerization resulting from food processing are limited. Color changes, which are usually used as a quality index, are only partly due to lycopene isomerization to *cis*-isomers. It is generally accepted that the all-*trans* form of lycopene has higher stability, and the *cis*-isomers have lower stability. Biological activity depends on the extent of isomerization and oxidation as well as the stability (Zechmeister, 1962; Khachik et al., 1992a; Stahl and Sies, 1992; Emenhiser et al., 1995; Wilberg and Rodriguez-Amaya, 1995). *Trans-cis* isomerization leads to faster degradation of lycopene. Although the processing of tomatoes by cooking, freezing, or canning does not usually cause a serious loss in total lycopene content, it is widely assumed that lycopene undergoes important isomerization during processing. Heat, light, acids and other factors have been

TABLE 4.5
Effect of Oxygen on Rate of Loss of
Lycopene on Heating Tomato Pulp at 100°C

Condition	Heating Time, Hours	Loss (%)
Dark and CO_2	0	0
	0.5	4.6
	1	4.9
	2	5.0
	3	5.1
Dark and O_2	0	0
	1	14.0
	2	23.7
	3	30.1
Daylight and CO_2	0	0
	1	5.4
	2	8.6
	3	11.3
Daylight and O_2	0	0
	1	15.1
	2	25.9
	3	33.1

Source: Data from Cole, E.R. and Kapur, N.S. 1957a. "The stability of lycopene. I. Degradation by oxygen." *J. Sci. Food Agric.*, 8: 360–365. Copyright Society of Chemical Industry. Reproduced with permission. Permission is granted by John Wiley & Sons Ltd. on behalf of the SCI.

reported to cause isomerization of lycopene (Zechmeister, 1962; Wong and Bohart, 1957; Lovric et al., 1970; Boskovic, 1979; Schierle et al., 1996; Nguyen and Schwartz, 1998; Shi and Le Maguer, 1999a,b). A true assessment of the relationship between nutritional quality and health benefits of dietary lycopene depends not only on the total lycopene content, but also on the distribution of lycopene isomers. Better characterization and quantification of lycopene isomers would provide better insight into the potential nutritional quality and health benefits. The control of lycopene isomerization during production and storage can be of benefit in improving the retention of product color, overall quality and biological activity.

Lycopene isomerization increases as a function of processing time using heat. The results in Table 4.7 show that food processing can enhance *cis*-isomerization in tomato-based foods. Heating tomato-based foods in oil caused increased lycopene isomerization vs. heating in water. This would indicate that not only the duration and temperature of heat treatment, but also the food matrix components (such as oil or fat) further influence the lycopene isomerization. An outline of the lycopene degradation pathway has been proposed by Boskovic (1979) (Figure 4.11).

TABLE 4.6
Lycopene Isomers in Various Commercial Tomato Products

Sample	Total Lycopene Wet Basis (mg/100 g)	All-trans (%)	5-cis (%)	9-cis (%)	13-cis (%)	Other cis (%)
Tomato paste ("*Tomatenmark*," Panocchia, Italy)	52	96	4	<1	<1	<1
("*Miracoli*," Kraft, Germany)	3.7	91	5	1	2	<1
Tomato ketchup (*"Hot Ketchup,"* Del Monte, Italy)	9.5	88	7	2	3	1
(*"Hot Ketchup,"* Heinz, U.S.)	3.0	77	11	5	7	1
Instant meal (*"Eier-Ravioli,"* Hero, Switzerland)	0.6	76	8	5	6	5
Sauce (*"Hamburger Relish,"* Heinz, The Netherlands)	3.0	93	5	<1	3	<1
(*"Sauce Bolognaise,"* Barilla, Italy)	9.2	67	14	6	5	8
Canned tomatoes (*"Chris,"* Roger Sud, Italy)	7.1	84	5	3	5	3

Data from Schierle et al. (1996).

TABLE 4.7
Effect of Heating Treatment on Lycopene *Trans-Cis* Isomerization in Aqueous and Oily Dispersions of Tomato Paste (70°C)

Heating time (min)	All-trans (%)	5-cis (%)	9-cis (%)	15-cis (%)	Other cis (%)
In Water					
0	92.6	4.5	0.9	1.6	0.5
15	92.3	4.4	0.9	1.6	0.5
30	88.1	5.1	2.1	2.3	2.5
60	87.1	5.2	2.2	2.7	3.0
120	86.2	5.5	2.7	2.6	3.1
180	83.4	6.1	3.6	3.2	3.8
In Olive Oil					
0	87.4	4.8	4.3	3.0	0.5
30	85.2	5.8	5.5	2.9	0.5
90	83.5	6.2	5.9	3.3	1.2
120	80.3	7.0	6.9	3.2	2.6
180	76.7	8.1	8.8	3.1	3.3

Source: Data from Schierle et al. (1996).

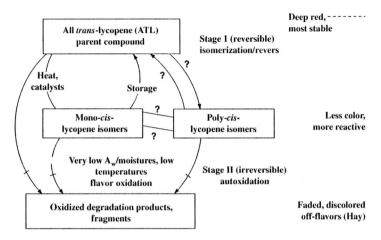

FIGURE 4.11 Fate of lycepene *in situ* during production and storage of tomato powder. Propose outline of reaction pathway. (Boskovic, 1979. With permission.)

TABLE 4.8
Retention of Total Lycopene in Tomato Powder Stored under Different Atmospheres for Different Periods of Time

Storage Period (days)	Storage Conditions	Retention of Total Lycopene (%)
0	Fresh Tomato Powder	100
30	N_2, 20°C	90.0
	Air, 20°C	46.3
80	N_2, 20°C	78.5
	Air, 20°C	28.7
160	N_2, 20°C	76.5
	Air, 20°C	25.5
210	N_2, 20°C	69.8
	Air, 20°C	23.0
385	N_2, 20°C	65.8
	Air, 20°C	21.8

Source: Data from Lovric, R., Sablek, Z. and Boskovic, M. 1970. "*Cis-trans* isomerization of lycopene and color stability of foam-mat dried tomato powder during storage." *J. Sci. Food Agric.*, 21: 641–647. Copyright Society of Chemical Industry. Reproduced with permission. Permission is granted by John Wiley & Sons Ltd. on behalf of the SCI.

4.7.5 STABILITY DURING STORAGE

The fate of lycopene in processed tomato products is notably influenced by storage factors. Lycopene content of tomato powder decreased in the presence of oxygen during storage (Table 4.8). A study on vacuum-dried tomato powder showed that in-package desiccation and packaging in an inert atmosphere (e.g., nitrogen) favored the retention of color, while the presence of air caused a loss of lycopene and color fading due to oxidation (Kaufman et al., 1957). Analyses of lycopene content in stored tomato powder samples by Wong and Bohart (1957) showed that air-packed samples retained the lowest levels of lycopene; all air-packed samples showed a progressive decrease in lycopene content during the storage period. The most important factor contributing to lycopene degradation was oxygen availability during storage. With careful selection of storage conditions to protect the products from air (e.g., by storing in an inert atmosphere or under vacuum), it is possible to protect initial color during storage. The mechanism of lycopene loss depends upon many parameters during food processing and storage. The main cause of damage to lycopene during food processing and storage is oxidation. Application of suitable antioxidants (e.g., ethoxyquin, ascorbic acid, sodium acid pyrophosphate) at appropriate levels can be beneficial (Granado et al., 1992; Clinton et al., 1996; Porrini et al., 1998). Low storage temperature, low oxygen contents and avoidance of light exposure in storage will also limit the extent of the oxidation of lycopene.

4.8   EXTRACTION OF LYCOPENE FROM TOMATOES

Industrial production of lycopene from tomatoes appears to be in high demand by pharmaceutical and food companies for the development of functional foods. At present, large quantities of tomato skin and outer pericarp tissue are discarded as a waste product from the peeling operation. To be used as a nutraceutical and functional food product, lycopene must satisfy regulatory requirements and have optimum stability. An environmentally friendly extraction and purification procedure on an industrial scale with minimal loss of bioactivity is highly desirable for food and pharmaceutical applications. High-purity lycopene products that meet food safety regulations would offer potential benefits to the food industry. New technologies are currently under development to separate lycopene from tomatoes. High-purity lycopene can be obtained from tomato fruits by various purification and separation processes including solvent extraction, supercritical fluid extraction, distillation, membrane separation, chromatography, and crystallization. Extraction with organic solvents is a well-established method in the food industry. One of the most important considerations in developing new processes and technology for extraction is the safety issue of final products when used as food. A critical and demanding step is the selection of proper safe solvents. The supercritical CO_2 fluid extraction (SFE-CO_2) process has gained increasing importance in the food industry in recent years (King, 1994; Eckert et al., 1996). This process can be used to solve some of the problems associated with more conventional separation techniques using organic solvents. It is particularly relevant to food and pharmaceutical applications involving processing and handling of complex, thermally-sensitive, oxygen-sensitive, bioactive components. Lycopene extraction by SFE-CO_2 technology is still at the laboratory scale.

4.9   SUMMARY

Lycopene is presently of particular interest for the food and pharmaceutical industries, both as a natural colorant and as a nutraceutical. People knowledge and curiosity of the health benefits of lycopene have skyrocketed. Nutraceuticals such as lycopene in the human diet have significant protective and beneficial physiological effects. Lycopene can play an important role in improving human health and can provide protection against a broad range of epithelial cancers.

Lycopene degradation in food processing and storage should be carefully minimized. Lycopene-containing products need to be protected from excessive heat, extreme pH conditions and exposure to oxygen and light. Processing technology should be optimized to prevent lycopene oxidation. The availability of lycopene in foods is influenced not only by its isomeric form, but also by the food matrix, the presence of sufficient bulk lipid for solubilization of released lycopene, and the presence of interfering factors in the lumen, such as pectin and other dietary fibers. These factors should be evaluated in an attempt to maximize lycopene absorption from the diet.

Tomato lycopene and other carotenoids have been used as natural food constituents for a long time without toxicological considerations, in the same manner as other vegetable and fruit products. Lack of sound information regarding toxicology and bioavailability data limits the development of food safety regulations. Further studies are necessary to provide toxicological evaluation of lycopene and information on how lycopene interacts with other constituents.

Consumer demand for healthful food products provides an opportunity to develop a market for food-grade and pharmaceutical-grade lycopene products. Industrial production of lycopene from tomatoes is of great interest to pharmaceutical and food companies, as a nutraceutical and for functional food development. Very little information on commercial production of lycopene, either isolated or synthetic, is available. Some technologies such as membrane separation, supercritical fluid CO_2 extraction, and solvent extraction are being applied to lycopene production. Environmentally friendly extraction and purification industrial procedures with minimal loss of bioactivity are highly desirable for the food, feed and pharmaceutical industries. High-quality lycopene or lycopene-rich products that meet food safety regulations will offer significant benefits to the people. A successful commercialization of high-value lycopene production may improve the competitiveness of tomato-based products and lycopene products in the global market.

4.10 ACKNOWLEDGMENTS

Help and suggestions from Marijan Boskovic (Kraft Foods, U.S.), Humayoun Akhtar, John Kramer, Greg Poushinsky, and Chris Young (Agriculture and Agri-Food Canada), AAFC/FRP Series No. S098, Yukio Kakuda and Carole Tranchant (University of Guelph, Canada) are appreciated.

REFERENCES

Agarwal, S. and Rao, A.V. 1998. Tomato lycopene and low density lipoprotein oxidation: a human dietary intervention study. *Lipids*. 33:981–984.

Al-Wandawi, H., Abdul-Rahman, M. and Al-Shaikhly, K. 1985. Tomato processing waste as essential raw materials source. *J. Agric. Food Chem.* 33:804–807.

AOAC. 1995. *Official Methods of Analysis*. Association of Official Analytical Chemists, 16th ed., Arlington, VA.

Beerh, O.P. and Siddappa, G.S. 1959. A rapid spectrophotometric method for the detection and estimation of adulterants in tomato ketchup. *Food Technol.* 7:414–418.

Benet, L.Z. and Shiner, L.B. 1985. Pharmacodynamics: the dynamics of drug absorption, distribution and elimination. In *The Pharmacological Basis of Therapeutics*, 7th Ed., A. Goodman-Gilman, L.S. Goodman, T.W. Rall, and F. Murad (eds.). New York: MacMillan Publishers.

Boileau, A.C., Merchen, N.R., Wasson, K., Atkinson, C.A. and Erdman, J.W. 1999. *Cis*-lycopene is more bioavailable than *trans*-lycopene *in vitro* and *in vivo* in lymph-cannulated ferrets. *J. Nutr.* 129:1176–1181.

Boskovic, M.A. 1979. Fate of lycopene in dehydrated tomato products: carotenoid isomerization in food system. *J. Food Sci.* 44:84–86.

Bowen, P.E., Mobarhan, S. and Smith, J.C. 1993. Carotenoid absorption in humans. *Methods Enzymol.* 214:3–17.

Britton, G. 1995. Structure and properties of carotenoids in relation to function. *FASEB J.* 9:1551–1558.

Brown, E.D., Micozzi, M.S. and Craft, N.E. 1989. Plasma carotenoids in normal men after a single ingestion of vegetables or purified β-carotene. *Am. J. Clin. Nutr.* 49:1258–1265.

Bureau, J.L. and Bushway, R.J. 1986. HPLC determination of carotenoids in fruits and vegetables in the United States. *J. Food Sci.* 51:128–130.

Burton, G.W. and Ingold, K.U. 1984. β-carotene: an unusual type of lipid antioxidant. *Science.* 224:569–573.

Castenmiller, J.J.M. and West, C.E. 1998. Bioavailability and bioconversion of carotenoids. *Ann. Rev. Nutr.* 18:19–38.

Chandler, L.A. and Schwartz, S.J. 1987. HPLC separation of *cis-trans* carotene isomers in fresh and processed fruits and vegetables. *J. Food Sci.* 52:669–672.

Chandler, L.A. and Schwartz, S.J. 1988. Isomerization and losses of *trans*-β-carotene in sweet potatoes as affected by processing treatments. *J. Agric. Food Chem.* 36:129–133.

Clinton, S.K. 1998. Lycopene: chemistry, biology, and implications for human health and disease. *Nutr. Rev.* 56(2):35–51.

Clinton, S.K., Emenhiser, C., Schwartz, S.J., Bostwick, D.G., Williams, A.W., Moore, B.J. and Erdman, J.W. 1996. *Cis-trans* lycopene isomers, carotenoids, and retinol in the human prostate. *Cancer Epidemiol., Biomarkers Prevent.*, 5:823–833.

Colditz, G.A., Branch, L.G., and Lipnic, R.J. 1985. Increased green and yellow vegetable intake and lowered cancer death in an elderly population. *Am. J. Clin. Nutr.* 41:32–36.

Cole, E.R. and Kapur, N.S. 1957a. The stability of lycopene. I. Degradation by oxygen. *J. Sci. Food Agric.* 8:360–365.

Cole, E.R. and Kapur, N.S. 1957b. The stability of lycopene. II. Oxidation during heating of tomato pulp. *J. Sci. Food Agric.* 8:366–368.

Conn, P.F., Schalch, W. and Truscott, T.G. 1991. The singlet oxygen and carotenoid interaction. *J. Photochem. Photobiol. B. Biol.* 11:41–47.

Conn, P.F., Lambert, C., Land, E.J., Schalch, W. and Truscott, T.G. 1992. Free Radicals. *Res. Commun.*, 16:401–408.

Craft, N.E., Sander, L.C. and Pierson, H.F. 1990. Separation and relative distribution of all-*trans*-β-carotene and its *cis*-isomers in β-carotene preparation. *J. Micronutrient Anal.*, 8:209–221.

Crouzet, J. and Kanasawud, P. 1992. Formation of volatile compounds by thermal degradation of carotenoids. *Methods Enzymol.* 213:54–62.

Curl, A.L. 1961. The xanthophylls of tomatoes. *J. Food Sci.* 26:106–111.

Daoud, H.G., Biacs, P.A., Hoschke, A., Vinkler, M.H. and Hajdu, F. 1987. Separation and identification of tomato fruit pigments by TLC and HPLC. *Acta Alimentaria.* 16(4):339–350.

Davies, B.H. 1976. Carotenoids, in *Chemistry and Biochemistry of Plant Pigments*, Vol. 2, 2nd ed., T.W. Goodwin (ed.). New York: Academic Press, pp. 38–165.

De Pee, S., West, C.E., Muhilal, Karyadi, D. and Hautvast, J.G.A.J. 1995. Lack of improvement in vitamin A status with increased consumption of dark-green leafy vegetables. *Lancet.* 346:75–81.

Deshmukh, D.S. and Ganguly, J. 1964. Effect of dietary protein contents on the intestinal conversion of β-carotene to vitamin A in rats. *Indian J. Biochem.* 1:204–207.

Devasagayam, T.P.A., Werner, T., Ippendorf, H., Martin, H.D. and Sies, H. 1992. Synthetic carotenoids, novel polyene polyketones and new capsorubin isomers as efficient quenchers of singlet molecular oxygen. *Photochem. Photobiol.* 55:511–514.

Di Mascio, P., Kaiser, S. and Sies, H. 1989. Lycopene as the most efficient biological carotenoid singlet oxygen quencher. *Arch. Biochem. Biophys.* 274:532–538.

Di Mascio, P., Murphy, M.C., and Sies, H. 1991. Antioxidant defense systems, the role of carotenoids, tocopherols and thiols. *Am. J. Clin. Nutr. Suppl.* 53:194–200.

Dimitrov, N.V., Meyer, C. and Ullrey, D.E. 1988. Bioavailability of β-carotene in humans. *Am. J. Clin. Nutr.* 48:298–304.

Dorgan, J.F., Sowell, A., Swanson, C.A., Potischman, N., Miller, R., Schussler, N., Stephenson, Jr., H.E. 1998. Relationships of serum carotenoids, retinol, α-tocopherol and selenium with breast cancer risk: results from a prospective study in Columbia, Missouri (United States). *Cancer Causes Control.* 9:89–97.

D'Souza, M.C., Singha, S. and Ingle, M. 1992. Lycopene concentration of tomato fruit can be estimated from chromaticity values. *Hort. Sci.* 27:465–466.

Eckert, C.A., Knutson, B.L. and Debenedetti, P.G. 1996. Supercritical fluid as solvents for chemical and materials processing. *Nature.* 383(26):313–318.

Edwards, C.G. and Lee, C.Y. 1986. Measurement of provitamin A carotenoids in fresh and canned carrots and green peas. *J. Food Sci.* 51(2):534–535.

Emenhiser, C., Sander, L.C. and Schwartz, S.J. 1995. Capability of a polymeric C_{30} stationary phase to resolve *cis-trans* carotenoids in reversed phase liquid chromatograph. *J. Chromatogr. A.* 707:205–216.

Emenhiser, C., Simunovic, N., Sander, L.C. and Schwartz, S.J. 1996. Separation of geometrical carotenoid isomers in biological extracts using a polymeric C_{30} column in reversed-phase liquid chromatography. *J. Agric. Food Chem.* 44:3887–3893.

Erdman, J.W., Fahey, G.C. and White, C.B. 1986. Effect of purified dietary fiber sources on β-carotene utilization by the chick. *J. Nutr.* 116:2415–2423.

Foote, C.S. and Denny, R.W. 1968. Chemistry of singlet oxygen. VII. Quenching by β-carotene. *J. Am. Chem. Soc.* 90:6233–6235.

Forssberg, A., Lingen, C., Ernster, L., and Lindberg, O. 1959. Modification of the X-irradiation syndrome by lycopene. *Exp. Cell Res.* 16:7–14.

Franceschi, S., Bidoli, E., La Vecchia, C. Talamini, R., D'Avanzo, B. and Negri, E. 1994. Tomatoes and risk of digestive-tract cancers. *Int. J. Cancer.* 59:181–184.

Frenkel, C. and Garrison, S.A. 1976. Initiation of lycopene synthesis in the tomato mutant rin influenced by oxygen and ethylene interaction. *Hort. Sci.* 11(1):20–21.

Gartner, C., Stahl, W. and Sies, H. 1997. Lycopene is more bioavailable from tomato paste than from fresh tomatoes. *Am. J. Clin. Nutr.* 66:116–122.

Gaziano, J.M., Johnson, E.J. and Russell, R.M. 1995. Discrimination in absorption or transport of β-carotene isomers after oral supplementation with either all-*trans* or 9-*cis*-β-carotene. *Am. J. Clin. Nutr.* 61:1247–1252.

Gerster, H. 1997. The potential role of lycopene for human health. *J. Am. Coll. Nutr.* 16:109–126.

Giovannucci, E., Ascherio, A., Rimm, E.B., Stampfer, M.J., Colditz, G.A., and Willett, W.C. 1995. Intake of carotenoids and retinol in relation to risk of prostate cancer. *J. Natl. Cancer Inst.* 87:1767–1776.

Godoy, H.T and Rodriguez-Amaya, D.B. 1989. Carotenoid composition of commercial mangoes from Brazil. *Lebensmittel-Wissenschaft und-Technologie.* 22:100–103.

Goodman, M.T., Kiviat, N., McDuffie, K., Hankin, J.H., Hernandez, B., Wilkens, L.R., Franke, A., Kuypers, J., Kolonel, L.N., Nakamura, J., Ing, G., Branch, B., Bertram, C.C., Kamemoto, L., Sharma, S. and Killeen, J. 1998. The association of plasma micronutrients with the risk of cervical dysplasia in Hawaii. *Cancer Epidemiol. Biomark. Prev.* 7:537–544.

Goodwin, T.W. 1980. *The Biochemistry of the Carotenoids.* 2nd ed., New York: Chapman and Hall.

Gould, W.V. 1994. *Tomato Production, Processing and Technology.* Baltimore, MD: CTI Publ.

Granado, F., Olmedilla, B., Blanco, I., and Rojas-Hidalgo, E. 1992. Carotenoid composition in raw and cooked Spanish vegetables. *J. Agric. Food Chem.* 40:2135–2140.

Grolier, P. 2000. Tomato antioxidants and biosynthesis. In *Summary of the White Book: The Antioxidants in Tomatoes and Tomato Products and their Health Benefits,* Amiton (ed.), Avignon Cedex, France, p. 6.

Gross, J. 1987. *Pigments in Fruits.* London: Academic Press.

Gross, J. 1991. *Pigments in Vegetables.* New York: Van Nostrand Reinhold.

Hart, D.J. and Scott, K.J. 1995. Development and evaluation of an HPLC method for the analysis of carotenoids in foods, and the measurement of the carotenoid content of vegetables and fruits commonly consumed in the UK. *Food Chem.* 54:101–111.

Heinonen, M.I., Ollilainen, V., Linkola, E.K., Varo, P.T. and Koivistoinen, P.E. 1989. Carotenoids in Finnish foods, vegetables, fruits, and berries. *J. Agric. Food Chem.* 37:655–659.

Helzlsouer, K.J., Comstock, G.W. and Morris, J.S. 1989. Selenium, lycopene, α-tocopherol, β-carotene, retinol, and subsequent bladder cancer. *Cancer Res.* 49:6144–6148.

Henry, L.K., Catignani, G.L. and Schwartz, S.J. 1998. Oxidative degradation kinetics of lycopene, lutein, 9-*cis* and all-*trans* β-carotene. *J. Am. Oil Chem. Soc.* 75:823–829.

Hill, H.M. and Rogers, L.J. 1969. Conversion of lycopene into β-carotene by chloroplasts of higher plants. *Biochem. J.* 113:31–32.

Hussein, L. and El-Tohamy, M. 1990. Vitamin A potency of carrot and spinach carotenes in human metabolic studies. *Int. J. Vit. Nutr. Res.* 60:229–235.

Jackson, M.J. 1997. The assessment of bioavailability of micronutrients: introduction. *Eur. J. Clin. Nutr.* 51:S1–S2.

Jayaarahan, P., Reddy, V. and Mahanran, M. 1980. Effect of dietary fat on absorption of β-carotene from green leafy vegetables in children. *Indian J. Med.* 71:53–56.

Johnson, E.J., Qin, J., Krinsky, N.I. and Russell, R.M. 1997. Ingestion by men of a combined dose of beta-carotene and lycopene does not affect the absorption of beta-carotene but improves that of lycopene. *J. Nutr.* 127:1833–1837.

Kanasawud, P. and Crouzet, J.C. 1990. Mechanism of formation of volatile compounds by thermal degradation of carotenoids in aqueous medium. 2. Lycopene degradation. *J. Agric. Food Chem.* 38:1238–1242.

Kanetsky, P.A., Gammon, M.D., Mandelblatt, J., Zhang, Z.F., Ramsey, E., Dnistrian, A., Norkus, E.P., Wright, Jr., T.C. 1998. Dietary intake and blood levels of lycopene: association with cervical dysplasia among non-Hispanic, black women. *Nutr. Cancer.* 31:31–40.

Kargl, T.E., Quackenbush, F.W. and Tomes, M.L. 1960. The carotene polyene system in a strain of tomatoes high in delta-carotene and its comparison with eight other tomato strains. *Proc. Am. Soc. Hortic. Sci.* 75:574–578.

Karrer, P. and Jucker, E. 1948. Carotenoids (English translation by E.A. Braude, 1950) New York: Elsevier.

Kaufman, V.F., Wong, F.F., Taylor, D.H. and Talburt, W.F. 1957. Problems in the production of tomato juice powder by vacuum. *Food Technol.* 9:120–123.

Kemmerer, A.R., Fraps, G.S. and de Mother, J. 1974. The effect of xanthophylls, chlorophylls, sulfasuxidine and α-tocopherol on the utilization of carotene by rats. *Arch. Biochem.* 12:135–138.

Khachik, F., Beecher, G.R., Goli, N.B., Luby, W.R. and Smith, J.C. 1992a. Separation and identification of carotenoids and their oxidation products in the extracts of human plasma. *Anal. Chem.* 64:2111–2122.

Khachik, F., Goli, N.B., Beecher, G.R., Holden, J., Luby, W.R., Tenorio, M.D., and Barrera, M.R. 1992b. Effect of food preparation on qualitative and quantitative distribution of major carotenoid constituents of tomatoes and several green vegetables. *J. Agric. Food Chem.* 40:390–398.

King, J.W. 1994. *Application of Preparative Scale SFE/SFC to Food and Natural Products.* Proceedings of the 3rd International Symposium on Supercritical Fluids, 421–428.

Kirk, J.T.O. and Tilney-Basset, R.A.E. 1978. *The Plastids. Their Chemistry, Structure, Growth and Inheritance*, 2nd ed. Amsterdam: Elsevier.

Kohlmeir, L., Kark, J.D., Gomez-Garcia, E., Martin, B.C., Steck, S.E., Kardinal, A.F.M., Ringstad, J., Thamm, M., Masaev, V., Riemersma, R., Martin-Moreno, J.M., Huttunen, J.K. and Kok, F.J. 1997. Lycopene and myocardial infarction risk in the EURAMIC study. *Am J. Epidemiol.* 146:618–626.

Kostic, D., White, W.S. and Olson, J.A. 1995. Intestinal absorption, serum clearance, and interactions between lutein and β-carotene when administered to human adults in separate or combined oral does. *Am. J. Clin. Nutr.* 62:604–610.

Krinsky, N.I., Russett, M.D., Handeman, G.J. and Snodderly, D.M. 1990. Structural and geometrical isomers of carotenoids in human plasma. *J. Nutr.* 120:1654–1662.

Kristenson, M., Zieden, B., Kucinaskiene, Z., Elinder, L.S., Bergdahl, B., Elwing, B., Abaravicius, A., Razinkoviene, L., Calkauskas, H. and Olsson, A. 1997. Antioxidant state and mortality from coronary heart disease in Lithuanian and Swedish men: concomitant cross-sectional study of men aged 50. *BMJ.* 314:629–633.

Lampe, C. and Watada, A.E. 1971. Post-harvest quality of high pigment and crimson tomato fruit. *J. Am. Soc. Hort. Sci.* 96(4):534–535.

Lessin, W., Catigani, G. and Schwartz, S.J. 1997. Quantification of *cis-trans* isomers of provitamin A carotenoids in fresh and processed fruits and vegetables. *J. Agric. Food Chem.* 45:3728–3732.

Levy, J., Bisin, E., Feldman, B., Giat, Y., Minster, A., Danilenko, M. and Sharoni, Y. 1995. Lycopene is a more potent inhibitor of human cancer cell proliferation then either α-carotene or β-carotene. *Nutr. Cancer.* 24:257–266.

Liu, Y.K. and Luh, B.S. 1977. Effect of harvest maturity on carotenoids in pastes made from VF-145-7879 tomato. *J. Food Sci.* 42:216–220.

Lovric, T., Sablek, Z. and Boskovic, M. 1970. *Cis-trans* isomerization of lycopene and color stability of foam-mat dried tomato powder during storage. *J. Sci. Food Agric.* 21:641–647.

Lurie, S., Handros, A., Fallik, E. and Shapira, M. 1996. Reversible inhibition of tomato fruit gene expression at high temperature. *Plant Physiol.* 110:1207–1214.

Macrae, R., Robinson, R.K. and Sadler, M.J. 1993. *Encyclopedia of Food Science, Food Technology and Nutrition.* London: Academic Press.

Mangels, A.R., Holden, J.M., Beecher, G.R., Forman, M.R. and Lanza, E. 1993. Carotenoids in fruits and vegetables: an evaluation of analytic data. *J. Am. Diet. Assoc.* 93:284–296.

Mencarelli, F. and Saltveit, M.E. 1988. Ripening of mature-green tomato fruit slices. *J. Am. Soc. Hort. Sci.* 113(5):742–745.

Micozzi, M.S., Beecher, G.R., Taylor, P.R. and Khachik, F. 1990. Carotenoid analyses of selected raw and cooked foods associated with a lower risk for cancer. *J. Natl. Cancer Inst.* 82:282–288.

Miller, N.J., Sampson, J., Candeias, L.P., Bramley, P.M. and Rice-Evans, C.A. 1996. Antioxidant activities of carotenes and xanthophylls. *FEBS Lett.* 384:240–242.

Mohr, W.P. 1979. Pigment bodies in fruits of crimson and high pigment lines of tomatoes. *Ann. Bot.* 44:427–434.

Monselise, J.J. and Berk, Z. 1954. Some observations on the oxidative destruction of lycopene during the manufacture of tomato puree. *Bull. Res. Counc. Israel.* 4:188–191.

Moss, G.P. and Weedon, B.C.L. 1976. Chemistry of the carotenoids. In *Chemistry and Biochemistry of Plant Pigments*, T.W. Goodwin, (ed.), Vol. 1, 2nd ed., New York: Academic Press. pp. 149–224.

Nguyen, M. and Schwartz, S. 1998. Lycopene stability during food processing. *Proc. Soc. Exp. Biol. Med.* 218:101–105.

Nguyen, M. and Schwartz, S. 1999. Lycopene: chemical and biological properties. *Food Technol.* 53(2):38–45.

Olson, J. 1986. Carotenoid, vitamin A and cancer. *J. Nutr.* 116:1127–1130.

Olson, J. 1994. Absorption, transport, and metabolism of carotenoids in humans. *Pure Appl. Chem.*,66:1011–1016.

Palozza, P. 1998. Prooxidant actions of carotenoids in biologic systems. *Nutr. Rev.* 56(9):257–265.

Parker, R.S. 1996. Absorption, metabolism and transport of carotenoids. *FASEB J.* 10:542–551.

Peng, Y.M., Peng, Y.S., Childers, J.M., Hatch, K.D., Roe, D.J., Lin, Y. and Lin, P. 1998. Concentrations of carotenoids, tocopherols and retinol in paired plasma and cervical tissue of patients with cervical cancer, precancer and noncancerous diseases. *Cancer Epidemiol. Biomark. Prev.* 7:347–350.

Porrini, M., Riso, P., and Testolin, G. 1998. Absorption of lycopene from single or daily portions of raw and processed tomato. *Br. J. Nutr.* 80:353–361.

Porter, J.W. and Lincoln, R.E. 1950. Lycopersicon selections containing a high content of carotenes and colorless polyenes. The mechanism of carotene biosynthesis. *Arch. Biochem.* 27:390.

Prince, M.R., Frisoli, J.K. and Goetschkes, M.M. 1991. Rapid serum carotene loading with high-dose β-carotene: clinic implications. *J. Cardiovasc. Pharmacol.* 17:343–347.

Quackenbush, F.W. 1987. Reverse phase HPLC separation of *cis-* and *trans-* carotenoids and its application to β-carotenes in food materials. *J. Liq. Chromatogr.* 10:643–653.

Ribaya-Mercado, J.D., Garmyn, M., Gilchrest, B.A. and Russell, R.M. 1995. Skin lycopene is destroyed preferentially over β-carotene during ultraviolet irradiation in humans. *J. Nutr.* 125:1854–1859.

Riso, P. and Porrini, M. 1997. Determination of carotenoids in vegetable foods and plasma. *Intl. J. Vit. Nutr. Res.* 67:47–54.

Rock, C.L. and Swendseid, M.F. 1992. Plasma β-carotene response in humans after meals, supplemented with dietary pectin. *Am. J. Clin. Nutr.* 55:96–99.

Sadler, G., Davis, J. and Dezman, D. 1990. Rapid extraction of lycopene and β-carotene from reconstituted tomato paste and pink grapefruit homogenates. *J. Food Sci.* 55:1460–1461.

Sakamoto, H., Mori, H., Ojima, F., Ishiguro, Y., Arimoto, S., Imae, Y., Nanba, T., Ogawa, M. and Fukuba, H. 1994. Elevation of serum carotenoids after continual ingestion of tomato juice. *J. Jpn Soc. Nutr. Food Sci.* 47:93–99.

Saleh, M.H. and Tan, B. 1988. HPLC-diode array detection technique for the identification and quantification of carotene formulations. *Anal. Biochem.* 33:247–252.

Saltveit, M.E. and Mencarelli, F. 1988. Inhibition of ethylene synthesis and action in ripening tomato fruit by ethanol vapors. *J. Am. Soc. Hort. Sci.* 113(4):572–576.

Sander, L.C., Sharpless, K.E., Craft, N.E. and Wise, S.A. 1994. Development of Engineered Stationary Phases for the Separation of Carotenoids Isomers, *Anal. Chem.* 66:1667–1674.

Schierle, J., Bretzel, W., Buhler, I., Faccin, N., Hess, D., Steiner, K. and Schuep, W. 1996. Content and isomeric ratio of lycopene in food and human blood plasma. *Food Chem.* 59(3):459–465.

Schwartz, S.J. and Patroni-Killam, M. 1985. Detection of *cis-trans* carotene isomers by two-dimensional thin-layer and high-performance liquid chromatography. *J. Agric. Food Chem.* 33:1160–1163.

Scita, G. 1992. Stability of beta-carotene under different laboratory conditions. *Methods Enzymol.* 213:175–185.

Sharma, S.K. and Le Maguer, M. 1996. Lycopene in tomatoes and tomato pulp fractions. *Intl. J. Food Sci.* 2:107–113.

Sheehy, R.E., Kramer, M. and Hiatt, W. 1988. Reduction of polygalacturonase activity in tomato fruit by antisense RNA. *Proc. Natl. Acad. Sci. USA.* 85:8805–8809.

Shi, J. and Le Maguer, M. 1999a. Lycopene as quality index in tomato processing. *Industrial Application of Osmotic Treatment Seminar Proceeding*, Milano, Italy.

Shi, J. and Le Maguer, M. 1999b. Stability of lycopene bioactivity in tomato dehydration. *Proceeding of Industrial Application*, Valencia, Spain.

Sies, H., Stahl, W. and Sundquist, A.R. 1992. Antioxidant function of vitamins. Vitamins E and C, β-carotene, and other carotenoids. *Ann. NY Acad. Sci.* 669:7–20.

Stahl, W. and Sies, H. 1992. Uptake of lycopene and its geometrical isomers is greater from heat-processed than from unprocessed tomato juice in humans. *J. Nutr.* 122:2161–2166.

Stahl, W. and Sies, H. 1996. Perspectives in biochemistry and biophysics. *Arch. Biochem. Biophys.* 336(1):1–9.

Stahl, W., Sundquist, A.R., Hamusch, M., Schwarz, W. and Sies, H. 1993. Separation of β-carotene geometrical isomers in biological samples. *Clin. Chem.* 39(5):810–814.

Tan, B. and Soderstrom, D.N. 1988. Qualitative aspects of UV-VIS spectrophotometry of β-carotene and lycopene. *J. Chem. Ed.* 22:21–31.

Tonucci, L.H., Holden, J.M., Beecher, G.R., Khachik, F., Davis, C. and Mulokozi, G. 1995. Carotenoid content of thermally processed tomato-based food products. *J. Agric. Food Chem.* 43:579–586.

Trombly, H.H. and Porter, J.W. 1953. Additional carotenes and a colorless polyene of lycopersicon species and strains. *Arch. Biochem. Biophys.* 43:443–457.

Van Eenwyk, J., Davis, F.G., and Bowen, P.E. 1991. Dietary and serum carotenoids and cervical intraepithelial neoplasia. *Int. J. Cancer.* 48:34–38.

Van het Hof, K.H., West, C.E., Weststrate, J.A. and Hautvast, J.G.A.J. 2000. Dietary factors that affect the bioavailability of carotenoids. *J. Nutr.* 130:503–506.

Van Vliet, T. 1996. Absorption of β-carotene and other carotenoids in humans and animal models. *Eur. J. Clin. Nutr.* 50(3):S32–S37.

Wang, C.J., Chou, M.Y. and Lin, J.K. 1989. Inhibition of growth and development of the transplantable C-6 glioma cells inoculated in rats by retinoids and carotenoids. *Cancer Lett.* 48:135–142.

White, W.S., Peck, K.M., Bierer, T.L., Gugger, E.T. and Erdman, Jr., J.W. 1993. Interactions of oral β-carotene and canthaxanthin in ferrets. *J. Nutr.* 123:1405–1413.

Wilberg, V.C. and Rodriguez-Amaya, B.D. 1995. HPLC quantitation of major carotenoids of fresh and processed guava, mango and papaya. *Lebensmittel-Wissenschaft und Technol.* 28:474–480.

Wong, F.F. and Bohart, G.S. 1957. Observations on the color of vacuum-dried tomato juice powder during storage. *Food Technol.* 5:293–296.

Zechmeister, L. 1962. *Cis-trans Isomeric Carotenoids, Vitamin A and Arylpolyenes.* New York: Academic Press.

Zhang, L.X., Cooney, R.V. and Bertram, J.S. 1992. Carotenoids up-regulate connexin-43 gene expression independent of their provitamin A or antioxidant properties. *Cancer Res.* 52:5707–5712.

Zhou, J.R., Gugger, E.T. and Erdman, Jr., J.W. 1996. The crystalline form of carotenes and the food matrix in carrot root decrease the relative bioavailability of β- and α-carotene in the ferret model. *J. Am. Coll. Nutr.* 15:84–91.

Zonta, F., Stancher, B. and Marletta, G. 1987. Simultaneous high-performance liquid chromatographic analysis of free carotenoids and carotenoid esters. *J. Chromatogr.* 403:207–209.

Zumbrunn, A., Uebelhart, P. and Eugster, C.H. 1985. HPLC of carotenes with y-end groups and (Z)-configuration at terminal conjugated double bonds, isolation of (5Z)-lycopene from tomatoes. *Helv. Chim. Acta.* 68:1540–1542.

5 Limonene from Citrus

*Amparo Chiralt, Javier Martínez-Monzó,
Teresa Cháfer and Pedro Fito*

CONTENTS

5.1 INTRODUCTION

Citrus fruits (particularly oranges, mandarins, lemons and limes) contain significant quantities of limonene in the peel and smaller quantities in the pulp. Citrus is one of the most heavily produced crops in the world (103,639,732 metric tonnes in 2000). As shown in Table 5.1, orange production represents 63% of the citrus world production, and more than a quarter is produced in Brazil and the United States. Lemons and limes are produced mainly in Argentina, India and Mexico (FAO, 2000). Recently, the industrialized countries have seen a marked increase in the level of processed citrus fruits, which has generated a large number of by-products derived from peel. These include pectin, concentrated hesperidin and naringin, dried peel, molasses, alcohol, carotenoids, seed oil and citric acid and essential oils, which contain a great proportion of limonene (Cohn and Cohn, 1996; Braddock, 1995). In Table 5.2, the oil and limonene contents of some citrus fruits are summarized.

Traditionally, essential oils were used as beverage flavorings (soft drinks), in the perfume industry and for other chemical uses. In this sense, other applications such as solvents and pesticides have been documented (Braddock, 1999). Furthermore, useful chemicals have been obtained from chemical reactions of limonene (Thomas and Bessière, 1989).

TABLE 5.1
World Citrus Production: Oranges and Lemons and Limes (2000)

Country	Production of Citrus		Orange		Lemons and Limes	
	Metric Tonnes	%[1]	Metric Tonnes	%	Metric Tonnes	%
Argentina	2,380,000	2.3	780,000	1.2	1,050,000	11
Brazil	23,987,200	23	22,692,200	34.6	470,000	4.8
China	11,781,703	11.4	3,407,000	5,2	287,191	3
Egypt	2,276,100	2.2	1,550,000	2.4	270,000	2.7
India	3,192,000	3	2,000,000	3	1,000,000	10.2
Italy	3,092,944	3	1,993,600	3	563,844	5.8
Mexico	5,108,349	5	3,390,371	5.2	1,296,978	13.2
Spain	5,310,000	5	2,500,000	3.8	880,000	9
United States	15,698,260	15.2	11,896,000	18.1	695,320	7
Total World Production	103,639,732		65,673,011	63.4	9,802,407	9.5

[1] Percentage of the total production.

Adapted from FAO (2000).

TABLE 5.2
Distribution of Citrus Production and Level of Processed, Oil Content and Approximate *d*-Limonene in Different Citrus Oils

Citrus Fruit	Distribution of Citrus Production[1] %	Percentage of Citrus Processed[2]	Oil Content[2] Lb/Mt	Limonene Content[3,4] % v/v
Oranges	71	52–93	8–14	>95
Mandarins	13	5	—	90–95
Lemons	10[5]	98	15	75
Limes		53	8	50
Grapefruit	6	38–97	5.6–6.5	90–95

[1] Kale and Adsule (1995)
[2] Sinclair (1984)
[3] Primo (1997)
[4] *d*-limonene concentration in the essential oil
[5] Lemons and limes

The most widespread use of limonene has been as a raw material for the manufacture of adhesives, such as the glue on labels and envelopes. Terpene monomers used for resin production are pinene, dipentene from turpentine and *d*-limonene. Waterless hand cleaners produced from *d*-limonene were among the first to replace solvents such as mineral spirits. Although *d*-limonene is more expensive than mineral spirits and kerosene, the former is used because of the pleasant citrus aroma and its claimed biodegradability (Coleman, 1975; Kutty et al., 1994). Many flavor chemicals

TABLE 5.3
d-Limonene Content of Various Citrus Oils

Citrus Oils	Variety/ Origin	d-Limonene (% in oil)	References
Lemon	California	72.2	(4)
	Arizona	79.8	(4)
	Meyer, Texas	82.9	(4)
	Lisbon	64.92	(1)
Orange	California	98.0	(4)
	Florida	98.5	(4)
Grapefruit	Texas	97.3	(4)
	Osbeck	93.83	(2)
Lime	Key	49.352	(3)
	Iran	56.553	(3)

References: (1) Njoroge et al., 1994; (2) Mondello et al., 1996; (3) Dugo et al., 1997; and (4) Sinclair, 1984.

are also synthesized from limonene (Thomas and Bessière, 1989). Pesticide products containing d-limonene are used for flea and tick control on pets, as insecticide sprays, outdoor dog and cat repellents, fly repellent tablecloths and mosquito larvicide insect repellent for use on humans (Braddock, 1999).

All of these applications are usually characterized as being barely profitable and, therefore, do not justify the additional extraction costs during processing. In the last few years, however, numerous research studies have shown a relationship between d-limonene and the prevention of some forms of cancer (Girard and Mazza, 1998; Braddock, 1999). This gives these essential oils functional characteristics and attracts great interest to the incorporation of these oils in food products.

In this chapter, properties and stability of limonene, sources and recovery methods as well as the reported toxicity and health properties are described.

5.2 OCCURRENCE OF LIMONENE IN CITRUS

Oil yield, properties and d-limonene content are affected by many aspects such as seasonal factors, ripening, soil types, climatic conditions, cultural practices, cultivar and varietal factors, scion and rootstock, etc. (Sinclair, 1984). Variation of oil content was also partially correlated with fruit size or peel thickness (Sinclair, 1984). Nitrogen fertilization increases the oil yield, while potassium fertilization had no significant effect on peel oil content (Sinclair, 1984). Limonene content of different kinds of citrus from different origins can be observed in Table 5.3.

Properties of essential oils depend on citrus fruit and variety, in an extraction oil-dependent manner. The differences in some of their properties can be observed in Table 5.4. The aldehyde content in lemon and lime cold-pressed oils is considerably higher than in orange oil but is close to the content of grapefruit distilled oil (Food Chemical Codex, 1996). The variations in aldehyde content are related to the

TABLE 5.4
Some Properties of Citrus Peel Distilled Oil

	Orange	Grapefruit	Lemon		Lime	
References[3]	(a)	(a)	(a)	(b)	(b)	(c)
Method of Extraction	D.O.[1]	D.O.[1]	C.P.O.[2]	D.O.[1]	C.P.O.[2]	C.P.O.[2]
Property						
Aldehydes (%)[4]	1.15–2.48	2.30–4.06	2.2–5.5	1.5–2.4	5.0–6.5	—
Specific gravity (20°C)	0.840–0.846	0.842–0.854	0.849–0.855	0.857–0.868	0.874–0.882	—
Refractive index (20°C)	1.473–1.473	1.471–1.475	1.471–1.475	1.475–1.477	1.483	1.482–1.485
Optical rotation (20°C)	+95.9, +98.6	+91.5, +96.5	+91.5, +96.5	+46.0, +48.9	+37.5, +43.8	+37.0, +47.9

[1] D.O.: Distilled Oil/C.P.O.
[2] Cold-Pressed Oil.
[3] References: (a) Sinclair, 1984; (b) Braddock, 1999; and (c) Dugo et al., 1997.
[4] Expressed as decanal/100 g of essential oil.

TABLE 5.5
**Physicochemical Properties of Cold-Pressed Orange Oil (var. Comuna)
at Different Harvesting Times**

Date	Evaporation Residue (%)	Ester Index[1]	Aldehydes and Ketones[2]	Acidity[1]	Carotenoids[3]
December	3.54	2.40	1.73	1.42	3.90
January	5.09	6.40	1.43	1.87	5.99
February	6.52	7.58	1.51	2.63	12.59
March	8.79	—	1.60	—	15.88
April	13.87	9.38	1.41	3.49	30.54

[1] Expressed as mg KOH/100 g of essential oil.

[2] Expressed as decanal/100 g of essential oil.

[3] Expressed as mg of β-carotene/100 g of essential oil.

Adapted from López (1995).

concentration of coumarins, particularly bergapten, in the peel (Braddock, 1999). Some of these compounds may cause photosensitive skin reactions in people.

Harvesting time affects some chemical properties of oil, as is shown in Table 5.5. Furthermore, smaller differences have been observed in other properties such as specific gravity, refractive index and optical rotation in line with harvesting time (López, 1995).

5.3 MOLECULAR STRUCTURE AND CHEMICAL PROPERTIES

Limonene is a monocyclic monoterpene [1-methyl-4-(1-methylethenyl) cyclohex-1-ene] that appears to have been formed by the union of two isoprene molecules (Figure 5.1). Because carbon-4 in limonene is asymmetric, this terpene naturally occurs in an optically active form as the *d*- or *l*-limonene. Each stereoisomer (enantiomer) has the ability to rotate a plane of polarized light to the right (+) (dextrorotary, historically referred to as *d*), or left (–) [levorotatory (*l*)], by a certain number of degrees (α). A racemic mixture of *d*- and *l*-limonene is known as dipentine (Sinclair, 1984). Limonene's molecular structure may be altered chemically, by oxidation yielding alcohols, aldehydes, ketones or acids, or reduction and isomerization, resulting in the formation of a great variety of simple terpenes or metabolites of the essential oils as is shown in Figure 5.1.

Various chemical processes of limonene, which lead to the obtainment of useful chemicals and some analytical methods, are based on these reactions. Many flavor chemicals are synthesized from limonene by reaction with water, sulfur and halogens, or hydrolysis, hydrogenation, boration, oxidation and epoxide formation (Thomas and Bessière, 1989). Hydroperoxides have also been studied and isolated because of their effect on off-flavor development in products containing citrus oil flavoring agents (Clark et al., 1981; Schieberle et al., 1987). Hydration of *d*-limonene produces alpha-terpineol, a compound that gives off an undesirable aroma in citrus-flavored products. It is also possible to produce alpha-terpineol and other useful value-added compounds

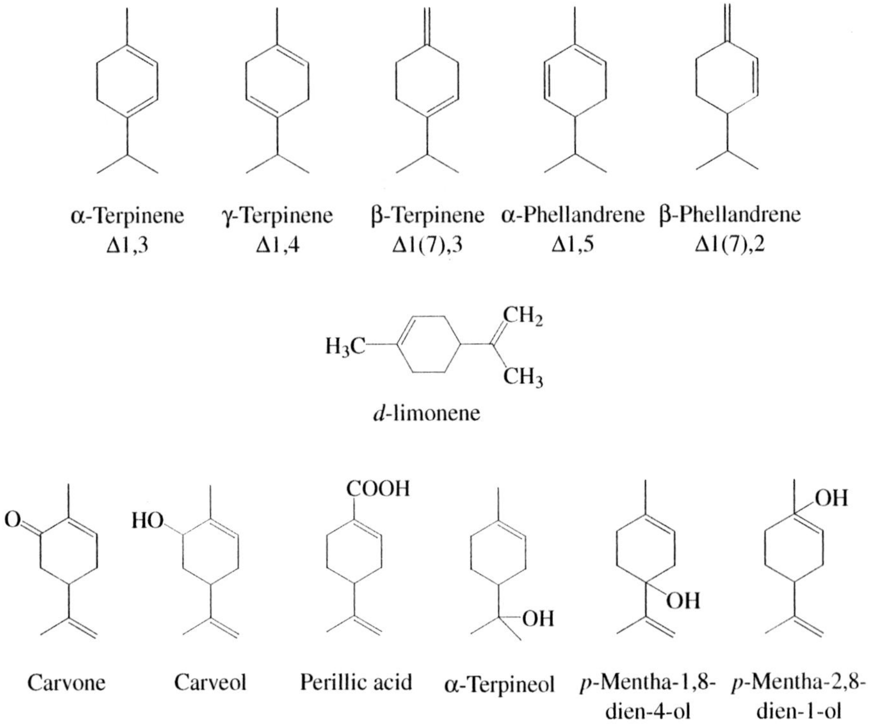

α-Terpinene　　γ-Terpinene　　β-Terpinene　α-Phellandrene　β-Phellandrene
Δ1,3　　　　　Δ1,4　　　　　Δ1(7),3　　　Δ1,5　　　　Δ1(7),2

d-limonene

Carvone　　　Carveol　　　Perillic acid　　α-Terpineol　　p-Mentha-1,8-　p-Mentha-2,8-
　　　　　　　　　　　　　　　　　　　　　　　　　　　　dien-4-ol　　　dien-1-ol

FIGURE 5.1 Structure of d-limonene, oxidated forms and derived methadienes.

through microbial bioconversion using d-limonene as a raw substrate (Rama-Devi and Battacharyya, 1977; Braddock and Cadwallader, 1995). Maintaining the purity of the d-limonene during resin synthesis is very important.

Physical and chemical properties of limonene are summarized in Table 5.6. The chemical reactivity of d-limonene implies that this compound should degrade in the environment, but the specifics of such degradation in a natural environment are not well defined in the case of a large spill of d-limonene or disposal of excessive amounts into the sewage or waste treatment system. Inherent biodegradation of organic substances such as d-limonene can be tested (Braddock, 1999). The degradation time period may be several months for some compounds. For d-limonene, some of the properties described in Table 5.6, such as water solubility, vapor pressure and inhibition of certain degradative microorganisms have to be considered. d-Limonene has limited water solubility, is volatile and can inhibit certain bacteria. For these reasons, conditions of d-limonene biodegradability in the environment need to be established.

5.4　EXTRACTION AND ANALYSIS

Citrus peel is a natural source of limonene. However, it is also present in very small quantities in other essential plant oils. Citrus peel contains essential oils in oblate-shaped oil glands in the flavedo (0.5–1 ml essential oil/100 cm^2 flavedo).

TABLE 5.6
Chemical and Physical Properties of *d*-Limonene

Property	Values
Commercial purity	94 (crude)–99.6 (high purity)
Color	colorless
Odor	citrus, odorless at high purity
Molecular wt. (g/mol)	136.23
Boiling point (°C, 763 mm Hg)	178 (high purity)
Freezing point (°C)	–96.9
Vapor pressure (mm Hg at 0–25°C)	0.41–2.1
Specific heat (cal/g° C)	0.438
Flash point (°C, open cup)	46 (crude)–49 (high purity)
Dielectric constant (10 MHz at 20°C)	2.37
Peroxide number	2–5
Optical rotation	+96 (crude)
Solubility (mg/L at 25°C)	
In pure water	13.8
In 20% sucrose	13.5
Viscosity (cP)	
At –50°C	3.5
At 0°C	1.5
At 25°C	0.9
At 50°C	0.7
At 178°C	0.25

Source: CRC (1986), Perry (1973), Lange (1985), International Critical Tables (1993), Braddock et al. (1986), and Braddock (1999).

During industrial citrus processing, a part of the essential oil passes to the juice, depending on the extraction method, and contributes to the aroma. In the juice, essential oils are present in the range of 0.0005–0.004%, and these levels can increase to 0.01–0.1 during extraction due to contamination from the flavedo (Primo, 1997). Modern juice extraction methods tend to minimize oil contamination.

Juice extraction represents 55% (w/w) of the initial weight of oranges, whereas the essential oils and essences extracted constitute only 0.2–0.22% (Sinclair, 1984). The citrus juice industry can recover essential oils from different intermediate process steps to obtain limonene or crude essences. In Figure 5.2, the different possibilities for obtaining essential oils with different properties are summarized (Girard and Mazza, 1998; Braddock, 1999).

5.4.1 METHODS FOR ESSENTIAL OIL EXTRACTION

The current citrus industry uses oil extraction machinery associated with the specific type of juice extractor in use. In general, peel oil extraction is carried out with Food

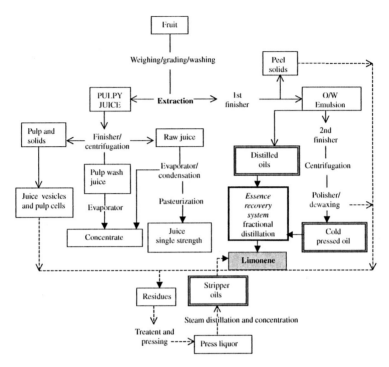

FIGURE 5.2 Some operations of citrus fruit processing in relationship with recovery of essential oils and essences. Adapted from Girard and Mazza (1998) and Braddock (1999).

Machinery Corporation (FMC), Brown or Italian juice extractor systems. Water is the solvent or carrier for the peel oil as it is extracted from the oil glands and enters the recovery system. Because the oil components are mostly water insoluble, recovery involves handling oil-water emulsions and suspensions; therefore, the process of water recycling is very important (Matthews and Braddock, 1987). In Figure 5.2, some citrus processing operations related to the different essential oils and essence recovery are shown (Girard and Mazza, 1998; Braddock, 1999). Essential oils are extracted at different stages during citrus processing, obtaining three main types of oils:

- Cold-pressed oils obtained from oil-water emulsion centrifugation
- Distilled oils obtained by evaporating and distilling the juice volatile fraction
- Stripper oils obtained by steam distillation of citrus press liquor

The quality of these essential oils is related to the extraction type. Table 5.7 summarizes the range of some properties (aldehyde content, specific gravity, refractive index and optical rotation) of the essential oils that contribute to their final quality for orange oil. The aldehyde content is quite different for the different kinds of oils—distilled oils are richer in aldehydes. The range of the other properties shown in Table 5.7 overlaps for the different kinds of oils. The recovery level of *d*-limonene is intimately tied to the type of oil obtained in citrus peel processing, as shown in Table 5.8 for calabrian bergamote. In distilled oils, the process carried out at atmospheric pressure yields a

TABLE 5.7
Some Properties of the Different Kinds of Orange Essential Oils

Property	Cold-Pressed[2]	Distilled[3]	Stripper Oil[3]
Aldehydes (%)[1]	1.12	1.15–2.48	0.47–1.50
Specific gravity (20°C)	0.8468	0.8400–0.8464	0.8398–0.8433
Refractive index (20°C)	1.4730	1.4725–1.4732	1.4713–1.4721
Optical rotation (20°C)	+94.54	+95.92–+98.56	+95.55–+98.90

[1] Expressed as citral.
[2] López (1995).
[3] Sinclair (1984).

TABLE 5.8
Percentage of *d*-Limonene (Including β-Phellandrene) in Calabrian Bergamote Cold-Pressed and Distilled Oils

Essential Oil		*d*-Limonene Content (%)[1]
Distilled oil	at reduced pressure	23.14
	at atmospheric pressure	35.036
Cold-pressed oil		25.4–45.4

[1] Percentage of *d*-limonene in volatile fraction analyzed by gas chromatography.

Adapted from Verzera et al. (1998).

greater recovery degree than that performed under vacuum. Cold-pressed oils yield greater recovery levels of limonene than distilled oils (Verzera et al., 1998).

5.4.1.1 Cold-Pressed Oils

The mechanical cold-pressing method extracts oil from citrus peels by rupturing the oil glands using pressure or abrasion so that the oil is ejected and washed away with a spray of water (Sinclair, 1984). Examples of these procedures are found in commonly used oil-extracting equipment: Pipkin roll, screw press, Fraser-Brace excoriator, AMC scarifier, Brown peel shaver and, the most popular, the in-line extractor from FMC (Food Machinery Corporation, Chicago, Illinois). This machine recovers juice and oil in one operation at high speed and is used in Florida to extract approximately 60% of the citrus essential oils. The screw press and the Brown peel shaver are used for the extraction of the remaining 40% of oils (Sinclair, 1984). Some differences in oil properties are obtained by these two most commonly used equipments (Brown and FMC) and are summarized in Table 5.9. Oils obtained from FMC in-line equipment show greater values in aldehyde content and evaporation residue, while optical rotation and refractive index may be lower in this kind of oil.

The distillation of cold-pressed oils under reduced pressure is carried out to recover the low boiling point components, and the vacuum is increased to recover

TABLE 5.9
Properties of Cold-Pressed Orange Oil Produced by Various Methods (FMC In-line and Brown Shaver)

Property	FMC In-Line	Brown Shaver
Specific gravity (20°C)	0.8424–0.8438	0.8427–0.8435
Refractive index (20°C)	1.4725–1.4731	1.4730
Optical rotation (20°C)	+95.32–+97.08	+97.18–+97.32
Aldehydes (%)[1]	1.54–1.96	0.86–1.66
Evaporation residue (%)	2.45–3.08	2.17–2.56

[1] Expressed as citral.

Adapted from Sinclair (1984).

d-limonene. Careful attention is required to separate the d-limonene from close boiling impurities such as myrcene and octanal. If additional purity is required, distillation over dilute NaOH or carbonyl-adduct agent can remove aldehydes, producing commercial d-limonene (99.5%). Activated silica may be used as an adsorbent to remove carbonyls and oxygenated impurities. After the purification, d-limonene will rapidly react with oxygen in air and, if not protected, will revert to crude d-limonene of 95–96% purity (Braddock, 1999).

Generally, the amount of limonene recovered is dependent on the efficiency of the cold-pressed oil recovery process. From reported data, actual recovery of d-limonene is about 0.05–0.07 kg/box (40.9 kg oranges processed/box) (Braddock, 1995).

5.4.1.2 Distilled Oils

Essential oils or d-limonene are recovered from oil-water emulsions by steam distillation at a reduced temperature. This is not a typical fractional distillation, as practiced when distilling cold-pressed oil, but rather, it is a bulk separation of the volatile compounds in the condensed vapor. When this distillation is applied to pure d-limonene, the mass ratio of water:d-limonene at atmospheric pressure is in the range of 8–10:1. In the actual oil emulsions, considerably more water must be distilled to recover a given quantity of d-limonene, because this component is adsorbed by the pulp particles in the emulsion. If the distillation is performed in a continuous manner under pressure, the ratio of water removed per amount of d-limonene recovered can be reduced to 3–4:1 (Braddock, 1999). However, when the steam temperature rises above 120°C, recovery of d-limonene decreases due to formation of some water-soluble alcohols and epoxides, which are soluble in the aqueous phase and are not recovered in the d-limonene phase above the condensate.

5.4.1.3 Stripper Oils

The peel sent to the feed mill contains the remainder of the oils that were not extracted in the cold-pressed oil recovery process. After liming and pressing, the press liquor is the source of the d-limonene recovered during molasses concentration

TABLE 5.10
Standard Methods in Essential Oil Characterization

Analyzed Property	Method of Determination	Values in Orange Essential Oils
Specific gravity (20°C)	AOAC 19.078 (1975)	0.840–0.855
Refractive index (20°C)	AOAC 19.079 (1975)	1.4710–1.4770
Optical rotation (20°C)	AOAC 19.080 (1975)	+85 and 99°
Aldehydes (%)	Hydroxilamine chlorhydrate (AOAC, 1975)	1.1%[1]
	Colorimetric method of N-hydroxy-benzene-	(Spanish oils)
	sulfonamide (Petrus et al., 1970)	1.1–1.9%[1]
	Capillary GLC (Wilson and Shaw, 1984)	(oils of Florida)
Carotenoids (%)	Absorbance at 450 nm	3.90–30.54[2]
	Color intensity between 370–550 nm	
UV absorption (CD value)	Palermo method (Cultrera et al., 1952)	U.S.P./F.C.C.[3]
		0.130 (CA)[4]–0.240 (FL)[5]
Ester index	Saponification (Guenther, 1949)	2.40–9.38[6]

[1] Expressed as decanal/100 g of essential oil.
[2] Expressed as mg of β-carotene/100 g of essential oil.
[3] U.S.P.: United States Pharmacopeia/F.C.C.: Food Chemical Codex.
[4] UV absorption (260–400 nm).
[5] Major value of UV absorption at 330 ± 3 nm.
[6] Expressed as mg KOH/100 g of essential oil.

Adapted from López (1995). CA: California, FL: Florida.

in the waste heat evaporator. The degree of *d*-limonene concentration in the press liquor is influenced by many process variables, but it is usually in the range of 0.1–0.5% (v/v). Press liquor water vapor from the first effect tubes is enriched with *d*-limonene, condensed and then passed to a tank at the bottom of the evaporator (Odio, 1993). This tank contains baffles and material for increasing the surface area to break the emulsion of *d*-limonene and condensate. This is a continuous process allowing the *d*-limonene to float and be recovered by decanting, where it is sent to a storage tank. The product oil is primarily composed of the chemical *d*-limonene, which is the name under which it is traded. The decant tank underflow is usually sent to waste (Braddock, 1999).

5.4.2 ANALYTICAL METHODS

Characterization of essential oils must include three kinds of analysis: sensory analysis; determination of physicochemical properties such as specific gravity (20°C), refractive index, optical rotation, aldehyde and carotenoid contents and solubility; spectroscopic properties (UV and IR); and chromatographic analysis. In Table 5.10, the main analytical determinations that can be carried out in the quality control of essential oils are summarized. The ranges of values for each analytical parameter of essential oils from oranges are also shown.

Although *d*-limonene is commonly analyzed by gas chromatography/mass spectroscopy (GC/MS), wet chemical analysis is still used for certain applications, particularly by the industries using citrus *d*-limonene as a chemical starting material (e.g., for synthesis of adhesive resins). Values of some of the useful properties of crude *d*-limonene are listed in Table 5.6.

Among the different analytical variables listed in Table 5.10, specific gravity, refractive index and optical rotation are widely used as quality indexes. The values of these properties allow us to estimate the *d*-limonene content, general composition and adulterations in essential oils. Aldehyde content is also a widely used index in the characterization of essential oils, because it is a measure of the aromatic fraction; values increase in mature fruit and decrease during fruit storage (López, 1995). The UV absorption index indicates general quality and adulterations, and shows differences between the different types of essential oils (López, 1995).

The most common quantitative wet chemical analysis for *d*-limonene, referred to as the Scott test, is based on an edible oil bromination reaction to determine the number of fatty acid double bonds. The test is commonly used to measure the peel oil content of citrus juices by adding a small amount of isopropanol to the juice, co-distilling the alcohol and *d*-limonene from the juice, adding acid and indicator (methyl orange), followed by titration with the KBr-$KBrO_3$ reagent to the end point (color disappearance) (López, 1995; Braddock, 1999). This estimation of the oil content, based on the *d*-limonene reaction, is accurate in orange, grapefruit and tangerine juice, because the oil from these fruits contains more than 95% *d*-limonene. Minor variations may be obtained for lemon and lime juices, the oils of which have 5% alpha-pinene and 4% citral (with only one double bond) (Scott and Veldhuis, 1966). Precision of this method for juices has been determined to be close to 10 ppm *d*-limonene (Braddock, 1999).

5.5 IMPORTANCE OF LIMONENE IN THE HUMAN DIET —HEALTH EFFECTS

Human consumption of *d*-limonene occurs mainly through citrus fruits and juices and flavored beverages. The daily per capita consumption from both its natural occurrence in food and as a flavoring agent is estimated to be 0.27 mg/kg body weight/day (Stofberg and Grundschober, 1987). Table 5.11 shows the human metabolic pathway of limonene. Oral consumption implies a complete absorption, and its half-life in humans, when doses were 100 mg/kg, was 24 h (Crowell et al., 1994; Crowell and Gould, 1994). After ingestion, five metabolites were identified in the blood (Crowell et al., 1994). Excretion of normal doses occurs mainly by renal function, although in people with liver diseases, high levels of limonene in the expired air were detected (Friedman et al., 1994). Certain amounts of limonene may also accumulate in fatty tissues (Crowell et al., 1994).

When inhaled, limonene is readily metabolized, and a long half-life in blood was observed in a slow elimination phase, which indicates accumulation in adipose tissues. About 1% of the total uptake was removed unaltered in the air exhaled after the exposure, while approximately 0.003% was eliminated in the urine (Falk-Filipsson, 1993).

TABLE 5.11
Human Metabolic Pathway of Limonene

Consumption	Estimated to be 0.27 mg/kg body weight/day.[1]
Absorption	Oral limonene is completely absorbed.[2,3]
	Half-life is approximately 24 h (humans, 100 mg/kg).[2,3]
Distribution	Some preference shown for fatty tissues.[2]
Metabolism	Five metabolites identified in blood after ingestion.[2]
	Metabolites have greater pharmacological potency than limonene in some systems.[2]
Excretion	Eliminated slowly in subjects exposed to d-limonene by inhalation.[4]
	Higher levels in expired air of persons with liver disease.[5]

[1] Stofberg and Grundschober (1987).
[2] Crowell et al. (1994).
[3] Crowell and Gould (1994).
[4] Falk-Filipsson (1993).
[5] Friedman et al. (1994).

5.5.1 PHYSIOLOGICAL EFFECTS

Numerous *in vivo* and *in vitro* studies have been reported on the physiological effects of *d*-limonene (Crowell and Gould, 1994). Most of the *in vivo* studies have been carried out with rats, because both humans and rats metabolize limonene in a similar manner (Crowell et al., 1994). The high therapeutic ratio of limonene in the chemotherapy of rodent cancers suggests that limonene may be an efficacious chemotherapeutic agent for human malignancies (Crowell et al., 1994; Elson and Yu, 1994). Table 5.12 summarizes some of the reported physiological actions deduced from *in vivo* and *in vitro* studies.

Although few studies have been carried out into the physiological actions of limonene as a solvent, it has been demonstrated that a solvent *d*-limonene preparation, injected directly into the biliary system, can dissolve or disintegrate the retained cholesterol gallstones in humans (Igimi et al., 1976; Schenk et al., 1980). Also, when used as a solvent, it enhances the percutaneous absorption of drugs such as anti-inflammatory agents (Okabe et al., 1994).

Nevertheless, the main physiological action reported for limonene concerns cancer prevention and therapy. In fact, there is some evidence to suggest that *d*-limonene may afford protection from chemical tumor formation. Limonene has been shown to block and suppress carcinogenic events, because certain biochemical pathways in tumor tissues may be sensitive to the inhibitory action of *d*-limonene and certain terpene constituents of essential oils (Elson and Yu, 1994).

Endogenous antioxidant systems such as glutathione *S*-transferase (GST) are involved in the detoxification pathways of carcinogens and therefore in the prevention of carcinogenesis. The addition of limonene and limonene-rich citrus oils (orange, lemon, grapefruit or tangerine) to a semipurified diet resulted in induction of GST activity in the liver and small bowel mucosa (Hocman, 1989; Wattenberg et al., 1986). Likewise, studies carried out on rats have demonstrated that these oils

TABLE 5.12
Physiological and Pathological Actions of Limonene

Cholesterol Gallstones

Injected into the biliary system, they can dissolve cholesterol gallstones[1,2]

Absorption of Drugs

Enhanced percutaneous absorption of drugs[3]

Cancer Chemoprevention

Effect	Mechanism
Prevents the carcinogenesis process at the initiation and promotion-progression stages[4]	Induction of GST activity[5,6]
Prevents mammary, liver and lung cancers[5,6]	Inhibition of carcinogenic metabolites (NNK, NDEA, DMBA, NMU)[7,8,9]
Long-term administration causes carcinogenesis in male rats but not in female rats or male and female mice[11]	Alteration of gene expression through the inhibition of protein isoprenylation[4,10]
	Reduced AOM-induced aberrant crypt foci[12]

GST: glutathione S-transferase; NNK: 4-(methylnitrosamino)-1-(3-pyridyl)-1-butanone; NDEA: N-nitrosodiethylamine; DMBA: N-methyl-N-nitrosurea; NMU: nitrosomethylurea; AOM: azoxymethane

[1] Schenk et al. (1980).
[2] Igimi et al. (1994).
[3] Okabe et al. (1994).
[4] Gould (1997).
[5] Hocman (1989).
[6] Wattenberg et al. (1986).
[7] Wattenberg and Coccia (1991).
[8] Maltzman et al. (1989).
[9] Haag et al. (1992).
[10] Crowell et al. (1991).
[11] Flamm and Lehman-McKeeman (1991).
[12] Kawamori et al. (1996).

inhibited the effects of carcinogenic metabolites such as NNK [4-(methyl-nitrosamino)-1-(3-pyridyl)-1-butanone], NDEA (N-nitrosodiethylamine), DMBA (N-methyl-N-nitrosurea) and NMU (nitrosomethylurea) (Maltzman et al., 1989; Wattenberg and Coccia, 1991; Haag et al., 1992).

Limonene inhibits the isoprenylation of a class of cellular proteins of 21–26 kDa, including p21ras and possibly other small GTP-binding proteins, in a dose-dependent manner in both cell lines (Crowell et al., 1991). Such inhibitors could alter signal transduction and result in altered gene expression (Gould, 1997). The two major rat serum metabolites of limonene, perillic acid and dihydroperillic acid, were more potent than limonene in the inhibition of isoprenylation (Crowell et al., 1991). These actions could explain the chemotherapeutic acitivity of limonene oils.

Previous studies provided data demonstrating that monoterpenoids such as limonene represented a novel class of anticancer drugs with the potential to cause

TABLE 5.13
Toxicity of *d*-Limonene

Humans	Rats
In GRAS[6] list:	
Does not pose carcinogenic or nephrotoxic risk to humans[1,2]	Kidney tumors in male rats[2]
Decreased vital capacity when exposed to 450 mg/m³ by inhalation[3]	Renal tubular tumors[5]
Dermal contact allergenic properties of oxidized forms[4]	

[1] Flamm and Lehman-McKeeman (1991).
[2] Hard and Whysner (1994).
[3] Falk-Filipsson et al. (1993).
[4] Falk et al. (1991).
[5] Kimura et al. (1996).
[6] GRAS: generally recognized as safe by the Food and Drug Adminstration.

tumor regressions with limited toxicity. Consumption of these oils further resulted in the inhibition of forestomach, lung and mammary tumors (Wattenberg et al., 1986; Hocman, 1989). *d*-Limonene was effective in preclinical models of breast cancer, causing more than 80% of carcinomas to regress with little host toxicity (Crowell et al., 1994). Likewise, azoxymethane-induced colonic aberrant crypt foci has been demonstrated to be significantly reduced in rats when 5% limonene was included in the drinking water (Kawamori et al., 1996).

5.5.2 Toxicity

The monoterpene *d*-limonene, widely used as a flavor and fragrant agent, is in the list of substances generally recognized as safe (GRAS) in food by the Food and Drug Administration (21 CFR 182.60 in the Code of Federal Regulations). However, recently, *d*-limonene has been shown to cause some toxicity effects (Table 5.13).

A decrease in vital capacity was observed after exposure to *d*-limonene at a high exposure level (450 mg/m³). Nevertheless, the subjects experienced neither irritative symptoms nor symptoms related to the central nervous system (Falk-Filipsson et al., 1993).

d-Limonene has been found to cause tumors at high doses in the kidney of the male rat in association with the development of hyaline droplet nephropathy (Flamm and Lehman-McKeeman, 1991). The same authors also found *d*-limonene to cause a significant incidence of renal tubular tumors in male rats upon chronic exposure. In contrast, neither kidney tumors nor the associated nephropathy have been found in female rats or mice at much higher doses (Hard and Whysner, 1994).

Although *d*-limonene is not carcinogenic in female rats or male and female mice given much higher dosages, the male rat-specific nephrocarcinogenicity of *d*-limonene may raise some concern regarding the safety of *d*-limonene for human consumption. The mechanism of *d*-limonene in tumor development appears unlikely in humans based on its mode of action in male rats (Flamm and Lehman-McKeeman, 1991; Hard and Whysner, 1994). Three major lines of evidence appear to support the safety of *d*-limonene for humans: male rats are specifically sensitive to nephrotoxicity and carcinogenicity; alpha 2u-globulin plays a unique role in the toxicity of male rats, while in other species, there is a complete lack of toxicity despite the presence of structurally similar proteins; and the lack of genotoxicity of both *d*-limonene and *d*-limonene-1,2-oxide in male rats supports the concept of a nongenotoxic mechanism that is sustained by renal cell proliferation (Flamm and Lehman-McKeeman, 1991).

5.5.3 ALLERGENICITY OF *d*-LIMONENE—EFFECT OF *d*-LIMONENE ON SKIN

Most reported cases of citrus peel allergy are due to the essential oils (Cardullo et al., 1989; Karlberg and Dooms-Goossens, 1997). *d*-Limonene of high purity gave no significant allergic reactions, whereas *d*-limonene exposed to air for 2 months sensitized the animals (Karlberg et al., 1991). Several of the products formed by oxidation upon prolonged exposure to air of *d*-limonene exhibit strong dermal contact allergenic properties (e.g., the hydroperoxides) such as purpuric rash (Falk et al., 1991). Contact allergy to citrus peel oil should be considered in patients with hand dermatitis who are occupationally exposed to citrus fruits (Cardullo et al., 1989). In these cases, it is important to limit the product storage time and to optimize handling conditions. Cold and dark storage of *d*-limonene in closed vessels prevents auto-oxidation for 1 year without the addition of an antioxidant (Karlberg et al., 1994).

Cats are susceptible to poisoning by ingestion of insecticidal products containing *d*-limonene, linalool and crude citrus oil extracts. Signs of toxicosis include hypersalivation, muscle tremors, ataxia, depression and hypothermia (Hooser, 1990).

REFERENCES

AOAC. 1975. Association of Official Analytical Chemist Official Methods of Analysis. Washington, D.C.

Braddock, R.J. 1995. By-products of citrus juice. *Food Technol.* 49(9):74, 76–77.

Braddock, R.J., ed. 1999. *Handbook of Citrus By-Products and Processing Technology*. New York: John Wiley and Sons, Inc.

Braddock, R.J. and Cadwallader, K.R. 1995. Bioconversion of citrus *d*-limonene, in *Fruit Flavors: Biogenesis, Characterisation and Authentication*, ACS Symp. No. 596. R.L. Rouseff and M.M. Leahy, eds. Washington, DC: American Chemical Society, pp. 142–148.

Braddock, R.J., Temmelli, F., and Cadwallader, K.R. 1986. Citrus essential oils—a dossier for material safety data sheets. *Food Technol.* 40(11):114–116.

Cardullo, A.C., Ruszkowski, A.M. and DeLeo, V.A. 1989. Allergic contact dermatitis resulting from sensitivity to citrus peel, geraniol and citral. *J. Am. Acad. Dermatol.* 21(2Pt2):395–397.

Clark, Jr., B.C., Jones, B.B. and Iacobucci, G.A. 1981. Characterisation of the hydroperoxides derived from singlet oxygen oxidation of (+)-limonene. *Tetraeldron.* 37(suppl 1):405–409.

Cohn, R. and Cohn, A.R. 1996. Subproductos del procesado de cítricos, in *Procesado de Frutas*. D. Arthey and P.R. Ashurst, eds. Zaragoza: Acribia, S.A., pp. 149–190.

Coleman, R.L. 1975. *d*-Limonene as a decreasing agent. *Citrus Ind. Mag.* 56(11):23–25.

CRC. 1986. *CRC Handbook of Chemistry and Physics*. Boca Raton, FL: CRC Press.

Crowell, P.L. and Gould, M.N. 1994. Chemoprevention and therapy of cancer by *d*-limonene. *Crit. Rev. Oncogen.* 5:1–22.

Crowell, P.L., Chang, R.R., Ren, Z., Elson, C.E. and Gould, M.N. 1991. Selective inhibition of isoprenylation of 21–26 kDa proteins by the anticarcinogen *d*-limonene and its metabolites. *J. Biol. Chem.* 266:17679–17685.

Crowell, P.L., Elson, C.E., Bailey, H.H., Elegbede, A., Haag, D.J. and Gould, M.N. 1994. Human metabolism of experimental cancer therapeutic agent *d*-limonene. *Cancer Chemother. Pharmacol.* 35:31–37.

Cultrera, R., Bufa, A. and Trifiro, E. 1952. Palermo method. *Conserve Der Agrum.* 1(2):18.

Dugo, P., Mondello, L., Lamonica, G. and Dugo, G. 1997. Characterization of cold-pressed Key and Persian lime oils by gas chromatography, gas chromatography/mass spectroscopy, high performance liquid chromatography and physicochemical indices. *J. Agric. Food Chem.* 45(9):3608–3616.

Elson, C.E. and Yu, S.G. 1994. The chemoprevention of cancer by mevalonate-derived constituents of fruit and vegetables. *J. Nutr.* 124:607–614.

Falk, A., Fisher, T. and Hagberg, M. 1991. Purpuric rash caused by dermal exposure to *d*-limonene. *Contact Dermatitis.* 25(3):198–199.

Falk-Filipsson, A., Lof, A., Hagberg, M., Hjelm, E.W. and Wang, Z. 1993. *d*-Limonene exposure to humans by inhalation: uptake, distribution, elimination and effects on the pulmonary function. *J. Toxicol. Environ. Health.* 38(1):77–88.

FAO. 2000. Statistical Databases. http://apps.fao.org/inicio.htm.

Flamm, W.G. and Lehman-McKeeman, L.D. 1991. The human relevance of the renal tumor-inducing potential of *d*-limonene in male rats: implications for risk assessment. *Reol. Toxicol. Pharmacol.* 13:70–86.

FMC. 1998. By-product capabilities. http://www.fmcfoodtech.com.

Food Chemical Codex. 1996. *Monographs.* Washington, DC: National Academy Press.

Friedman, M.I., Preti, G., Deems, R.O., Friedman, L.S., Munoz, S.J. and Maddrey, W.C. 1994. Limonene in expired lung air of patients with liver disease. *Dig. Dis. Sci.* 39:1672–1676.

Girard, B. and Mazza, G. 1998. Functional grape and citrus products, in *Functional Foods: Biochemical and Processing Aspects.* G. Mazza, ed. Lancaster, PA: Technomic Publishing Co., Inc., pp. 139–191.

Gould, M.N. 1997. Cancer Chemoprevention and therapy by monoterpenes. *Environ. Health Perspect.* 104(4):977–979.

Guenther, E. 1949. *The Essential Oils.* I, II and III. New York: Van Nostrand.

Haag, J.D., Lindstrom, M.J. and Gould, M.N. 1992. Limonene-induced regression of mammary carcinomas. *Cancer Res.* 52:4021–4026.

Hard, G.C. and Whysner, J. 1994.Risk assessment of *d*-limonene: an example of male rat-specific renal tumorigens. *Crit. Rev. Toxicol.* 24:231–254.

Hocman, G. 1989. Prevention of cancer: vegetables and plants. *Comp. Biochem. Physiol.*, 93:201–212.

Hooser, S.B. 1990. Toxicology of selected pesticides, drugs, and chemicals. *d*-Limonene, linalool, and crude citrus oil extracts. *Vet. Clin. North Am. Small Anim. Pract.* 20(2):383–385.

Igimi, H., Hisatsugu, T. and Nishimura, M. 1976. The use of *d*-limonene preparation as a dissolving agent of gallstones. *Am. J. Dig. Dis.* 21(11):926–939.

International Critical Tables. 1993. *International Critical Tables of Numerical Data, Physics, Chemistry and Technology.* New York: McGraw-Hill Inc., pp. 460 (Vol. 4) and 163 (Vol. 5).

Kale, P.N. and Adsule, P.G. 1995. Citrus, in *Handbook of Fruit Science and Technology. Production, Composition, Storage and Processing.* D.K. Salunkhe and S.S. Kadam, eds. New York: Marcel Dekker, Inc., pp. 39–67.

Karlberg, A.T. and Dooms-Goossens, A. 1997. Contact allergy to oxidized *d*-limonene among dermatitis patients. *Contact Dermatitis.* 36(4):201–206.

Karlberg, A.T., Boman, A. and Melin, B. 1991. Animal experiments on the allergenicity of d-limonene—the citrus solvent. *Ann. Occup. Hyg.* 35(4):419–426.

Karlberg, A.T., Magnusson, K. and Nilsson, U. 1994. Influence of an antioxidant on the formation of allergenic compounds during auto-oxidation of d-limonene. *Ann. Occup. Hyg.* 38(2):199–207.

Kawamori, T., Tanaka, T., Hirose, Y., Ohnishi, M. and Mori, H. 1996. Inhibitory effects of d-limonene on the development of colonic aberrant crypt foci induced by azoxymethane in F344 rats. *Carcinogenesis.* 17(2):369–372.

Kimura, J., Takahashi, S., Ogiso, T., Yoshida, Y. and Akagi, K. 1996. Lack of chemoprevention effects of the monoterpene d-limonene in a rat multi-organ carcinogenesis model. *Jpn. J. Canc.* 87:589–594.

Kutty, V., Braddock, R.J. and Sadler, G.D. 1994. Oxidation of d-limonene in presence of low density polyethylene. *J. Food Sci.* 59(2):402–405.

Lange, N.A. 1985. *Lange's Handbook of Chemistry.* J.A. Dean, ed. New York: McGraw-Hill, Inc.

López, J. 1995. *La naranja, composición y cualidades de sus zumos y esencias.* Conselleria de Agricultura, Pesca y Alimentación, ed. Valencia: Servicio de Publicaciones de la Generalitat Valenciana.

Maltzman, T.H., Hurt, L.M., Elson, C.E., Tanner, M.A. and Gould, M.N. 1989. The prevention of nitrosomethylurea-induced mammary tumors by d-limonene and orange oil. *Carcinogenesis.* 4(10):781–783.

Matthews, R.F. and Braddock, R.J. 1987. Recovery and applications of essential oils from oranges. *Food Techn.* 41(1):57–61.

Mondello, L., Dugo, P. and Cavazza, A. 1996. Characterization of essential oil of pummelo (cv. Chandler) by GS/MS, HPLC and physicochemical indices. *J. Essential Oil Res.* 8:311–314.

Njoroge, H., Ukeda, H., Kusunose, H. and Sawamura, M. 1994. Volatile components of Japanese Yuzu and lemon oils. *Flavor Fragrance J.* 9:159–166.

Odio, C.E. 1993. Tank reaction system for citrus peel. *J. Food Sci.* 59(2):402–405.

Okabe, H., Suzuki, E., Sayito, T., Takayama, K. and Nagai, T. 1994. Development of novel transdermal system containing d-limonene and ethanol as absorption enhancers. *J. Control Rel.*, 32:243–247.

Perry, R.H. 1973. *Chemical Engineers' Handbook.* New York: McGraw-Hill, Inc.

Petrus, D.R., Dougherty, M.H. and Wolford, R.W. 1970. A quantitative total aldehydes test useful in evaluating and blending essences and concentrated citrus products. *J. Agric. Food Chem.* 18:908–910.

Primo, E. 1997. Cítricos y derivados, in *Química de los Alimentos.* Madrid: Síntesis, D.L., pp. 230–276.

Rama-Devi, J. and Battacharyya, P.K. 1977. Microbial transformation of terpenes: Part XXIV. Pathways of degradation of geraniol, nerol and limonene by *Pseudomonas incognita* (linalool strain). *Indian J. Biochem. Biophys.* 14:359–363.

Schenk, J., Dobronte, Z., Koch, H. and Stolte, M. 1980. Studies on tissue compatibility of d-limonene as a dissolving agent of cholesterol gallstones. *Z. Gastroenterol.* 18(7):389–394.

Schieberle, P., Maier, W., Firi, J. and Grosch, W. 1987. HRGC separation of hydroperoxides formed during the photosensitized oxidation of d-(+)-limonene. *J. High Resolution Chromatog. Commun.* 10(11):588–593.

Scott, W.C. and Veldhius, M.K. 1966. Rapid estimation of recoverable oil in citrus juices by bromate titration. *J. Am. Oil Chem. Soc.* 49(3):628–633.

Sinclair, W.B. 1984. *The Biochemistry and Physiology of the Lemon.* Sacramento: California. Division of Agriculture and Natural Resources.

Stofberg, J. and Grundschober, F. 1987. Consumption ratio and food predominance of flavoring materials. *Perfumer Flavorist.* 12:27.

Thomas, A.F. and Bessière, Y. 1989. Limonene. *Natural Products Rep.* 6(3):291–309.

Verzera, A., Trozzi, A., Stagno D'Alcontres, I., Mondello, L., Dugo, G. and Sebastiani, E. 1998. The composition of the volatile fraction of calabrian Bergamot essential oil. *Rivista Italiana Eppos.* 20:17–38.

Wattenberg, I.W., Hanley, A.B., Barany, G., Sparnins, V.L., Lam, L.K.T. and Fenwick, G.R. 1986. Inhibition of carcinogenesis by some minor dietary constituents, in *Diet, Nutrition and Cancer*, Y. Hayashi ed. Tokyo, Sci. Soc. Press, pp. 193–203.

Wattenberg, L.W. and Coccia, J.B. 1991. Inhibition of 4-(methylnitrosamino)-1-(3-pyridyl)-1-butanone carcinogenesis in mice by *d*-limonene and citrus fruit oils. *Carcinogenesis.* 1(12):115–117.

Wilson, C.W. and Shaw, P.E. 1984. Quantitation of individual and total aldehydes in citrus cold-pressed oils by fussed silica capillary gas chromatography. *J. Agric. Food Chem.* 32:399–401.

6 Phenolic Diterpenes from Rosemary and Sage

Karin Schwarz

CONTENTS

6.1 INTRODUCTION

Diterpenes have diverse functional roles in plants, acting as hormones (gibberellins), photosynthetic pigments (phytol) and regulators of wound-induced responses (abietic acid) (Hanson, 1995; McGarvey and Croteau, 1995). The presence of phenolic abietane diterpenes, also known as "phenolic diterpenes" has been reported so far only for *Rosmarinus officinalis* and some species of the genus *Salvia*. In addition, the presence of carnosol in *Lepechinia hastata* was reported by Dimayuga et al. (1991).

Several species of *Salvia* and subspecies *R. officinalis* belong since antiquity to the most popular medical plants. Many indications for the therapeutical use were given in the literature of the old Greeks and Romans and in the literature of the Middle Ages (e.g., Hildegard von Bingen). The Latin name of sage stems from the words *salvare* = cure and *officinalis* = pharmacy/chemist's. Extracts of *R. officinalis* have been used in folk medicine as a diuretic, an emenagogue, and an antispasmodic. Its aqueous extracts do not present toxicity to humans, but they present abortive

1-5667-6902-7/02/$0.00+$1.50
© 2002 by CRC Press LLC

effects. Experiments with Wistar rats suggest that rosemary extracts may present an anti-implantation effect, but the effect compared to the control group was not significant. Rosemary extract did not interfere with the normal development of the embryo after implantation (Lemonica et al., 1996).

Around 800 species have been found to belong to *Salvia*, whereas until today, *R. officinalis* is the only well-known species in the genus *Rosmarinus*. However, some authors divide *Rosmarinus* into two or three species (Hickey and King, 1997; Kew database of the Royal Botanic Garden, London). The part of the plants used is the dried leaf, which supplies the spice. Rosemary has thick, aromatic, linear leaves and can reach a height of up to 2 m. Sage leaves are characterized by compound or simple leaves with one to two pairs of lateral segments and a large terminal segment. Rosemary and sage are native to the Mediterranean region and South America, and today, they are cultivated worldwide.

The leaves of rosemary and sage contain valuable essential oils rich in mono- and sesquiterpenes, such as borneol, camphor, caryophyllene, cineol, humulene, linalool, α- and β-pinene, and thujone "salviol." The essential oils are located in special secretory glands on the leaves (Venkatachalam et al., 1984). Diterpenes are in the leaves, and the cuticula is rich in triterpenes (Hegnauer, 1996). The essential oil is prepared by steam distillation of the fresh flowering top from rosemary and sage. Major oil-producing countries include Spain, France, and Tunisia (Leung and Foster, 1996). After removal of the essential oil, the remaining plant material represents a rich source of phenolic diterpenes.

6.2 CHEMISTRY AND BIOSYNTHESIS

Diterpenoids from *Salvia* and *Rosmarinus* belong to two main groups: abietane and clerodane. In general, the diterpenoids with abietane skeletons usually occur in Asiatic and European species, whereas the most common diterpenoids in South American species are of the clerodane type (Al-Hazimi and Miana, 1994).

As well as other terpenoids, abietane diterpenes are presumably synthesized in plastids via the pyruvate/glyceraldehyde 3-phosphate pathway with geranylgeranylpyrophosphate as the mother compound (Eisenreich et al., 1996; Lichtenthaler, 1999; Schneider and Hiller, 1999). It has been discussed that interactions between organelles occur in the transfer of abietic acid metabolites from plastids to sites of secondary transformation, such as endoplasmatic reticulum-bound cytochrome P-450 oxygenases and various cytosolic redox enzymes (Kleinig, 1989; McGarvey and Croteau, 1995). Figure 6.1 shows phenolic diterpenes that have been reported for *R.* and *S. officinalis* (Table 6.1) and, in addition, phenolic diterpenes that have been found in other species of the genus *Salvia*, according to Al-Hazimi et al. (1994) (Table 6.2). The main phenolic diterpene of *R.* and *S. officinalis* is carnosic acid. Richheimer et al. (1996) reported concentrations of 1.7–5.5% carnosic acid in rosemary biomass from different geographical regions. The carnosic acid concentration of dried *Salvia officinalis* L. plant material from different countries varied from 1.2% to 3.9%, and in most cases, the concentration was around 2% (data not published).

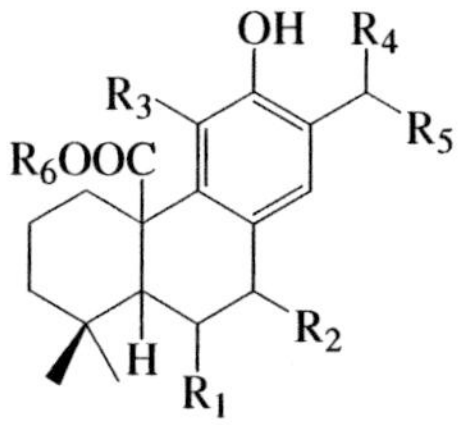

(1a) Carnosic acid (Salvin): R_1=H, R_2=H, R_3=OH, R_4=CH$_3$, R_5=CH$_3$, R_6=H
(1b) 16-Hudroxycarnosic acid: R_1=H, R_2=H, R_3=OH, R_4=CH$_3$, R_5=CH$_2$OH, R_6=H
(1c) 11-Acetoxycarnosic acid: R_1=H, R_2=H, R_3=OAc, R_4=CH$_3$, R_5=CH$_3$, R_6=H
(1d) Methylcanosoate: R_1=H, R_2=H, R_3=OH, R_4=CH$_3$, R_5=CH$_3$, R_6=Me
(1e) 7-oxocarnosic acid: R_1=H, R_2=O, R_3=OH, R_4=CH$_3$, R_5=CH$_3$, R_6=H
(1f) 6-oxo-7-beta-hydroxy carnosic acid: R_1=O, R_2=OH, R_3=OH, R_4=CH$_3$, R_5=CH$_3$, R_6=H

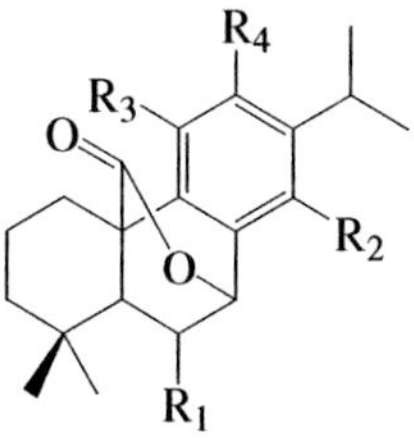

(2a) Carnosol (Picrosalvin): R_1=H, R_2=H
(2b) Isorosmanol: R_1=OH, R_2=H
(2c) 12-Hydroxyisocarnosol: R_1=H, R_2=OH
(2d) 11,12-Methyoxy-isorosmanol: R_1=OH, R_2=H, R_3, R_4=H

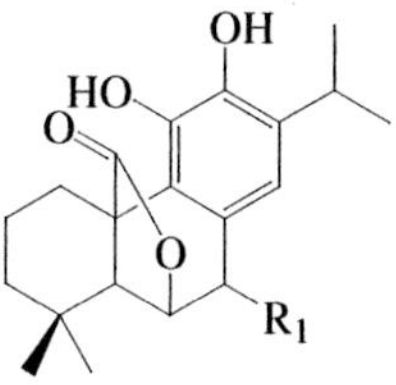

(3a) Rosmanol: R_1=αOH
(3b) Epirosmanol: R_1=βOH
(3c) 7-Monomethylepirosmanol: R_1=βOMe

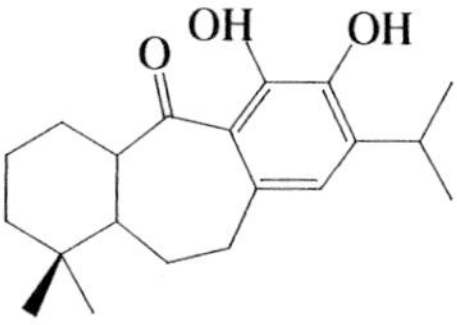

Rosmaridiphenol

FIGURE 6.1 Phenolic diterpenes in rosemary and sage.

5. 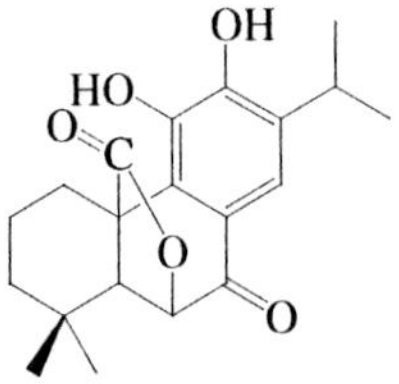 Galdosol

6. 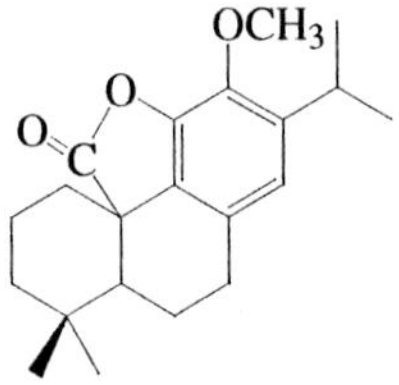 Carnosic acid 12-methylether-delta-lactone

7. 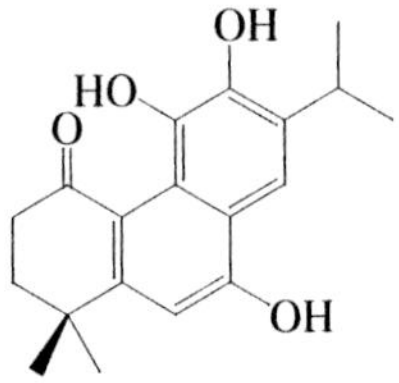 Arucatriol

8. 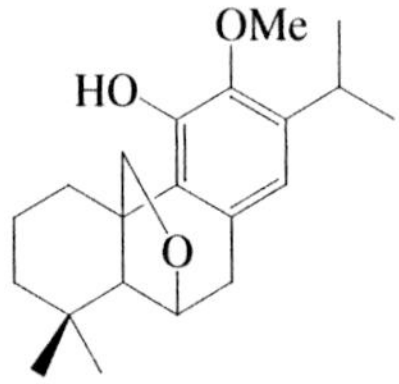 Deoxocarnosol 12-methylether

9. 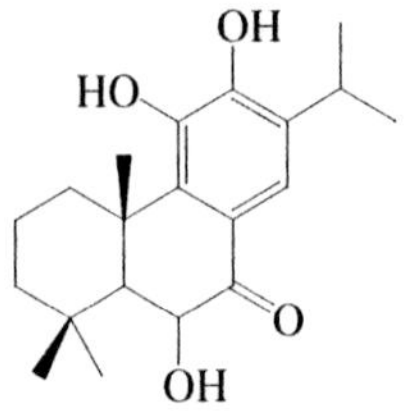 6-alpha-Hydroxydemethylcryptojaponol

10. 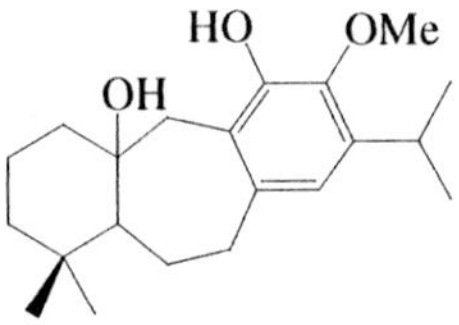 Salvicanol

FIGURE 6.1 (continued)

11. 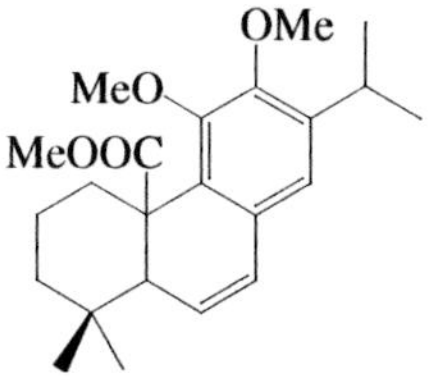 11,12-Dimethoxy-6,8,11,13-tetraen-20-oic acid methylester

12. 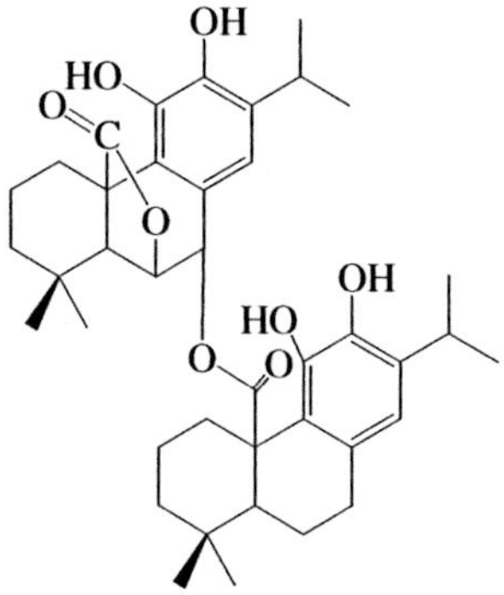 Rosmanol-carnosoate

13. 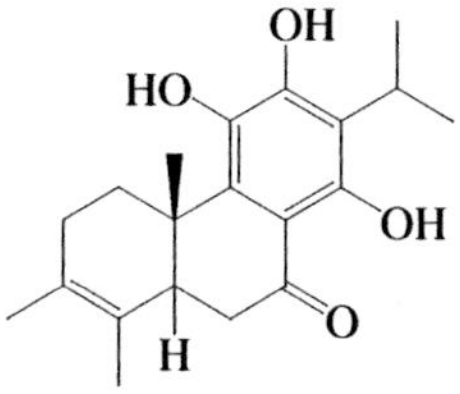 Candesalvone A

14. 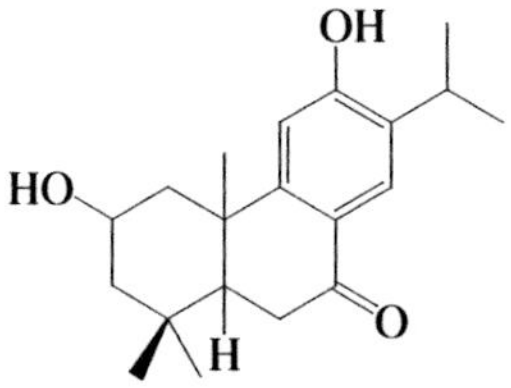 2-alpha-Hydroxysugiol

15. 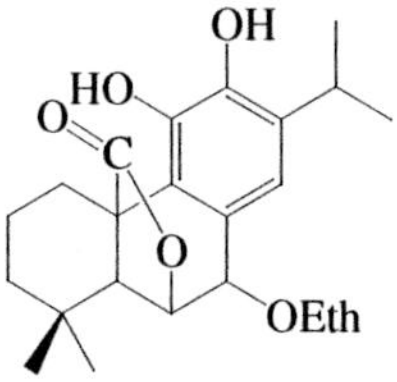 7-Ethoxyrosmanol

16. 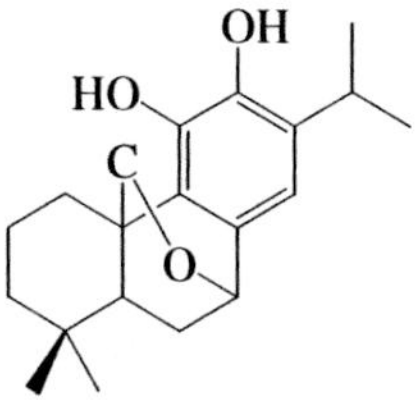 20-deoxocarnosol

FIGURE 6.1 (continued)

17.

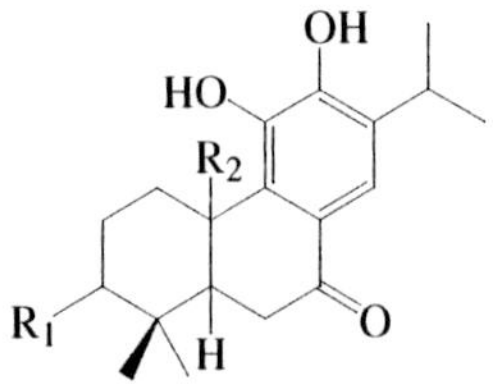

Cryptanol: R$_1$=H, R$_2$=CH$_3$

18. 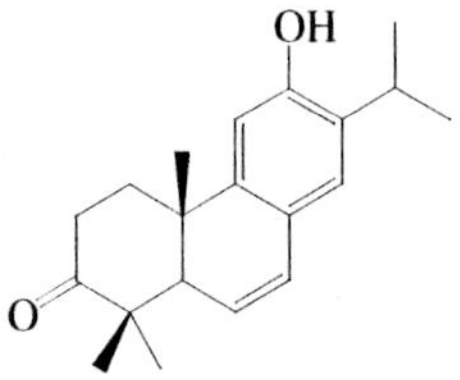   Hypargenin D

19. 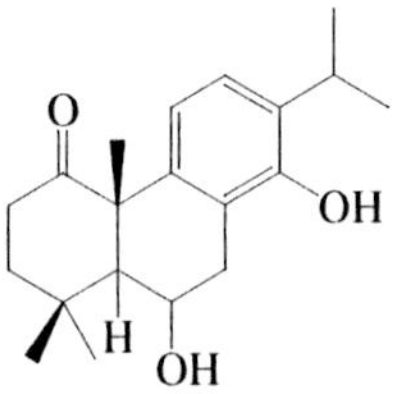   Hypargenin E

20. 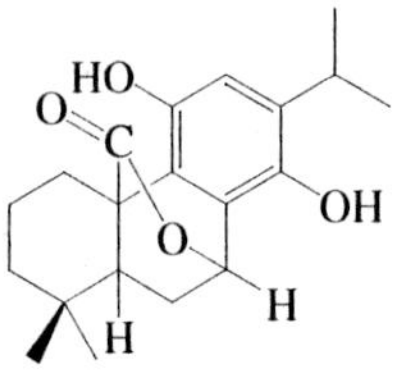   Isocarnosol

21. 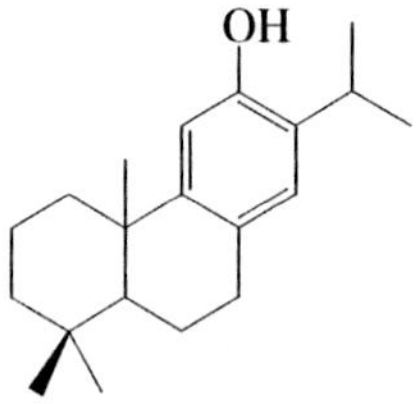   Ferruginol

22. 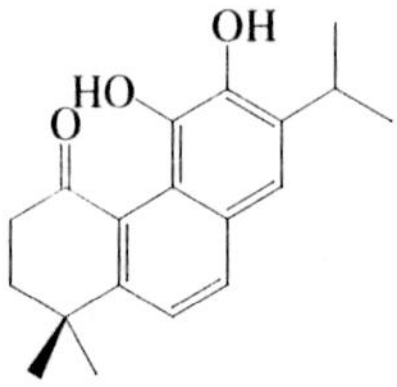   Miltidiol

FIGURE 6.1 (continued)

23. 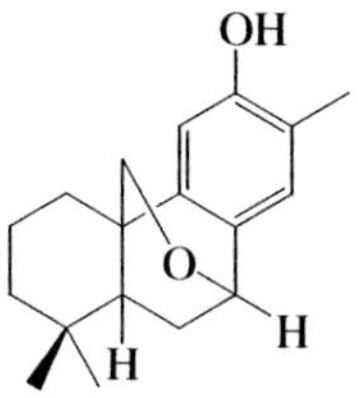 Norsalvioxide

24.

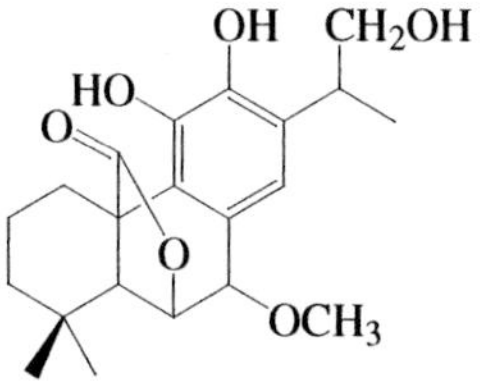

16-Hydroxy-7-methoxy-rosmanol

25.

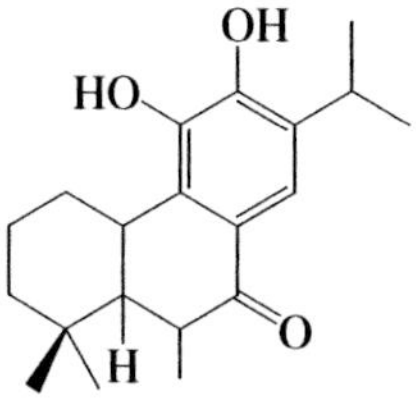

Demethylcrptojapano

26. 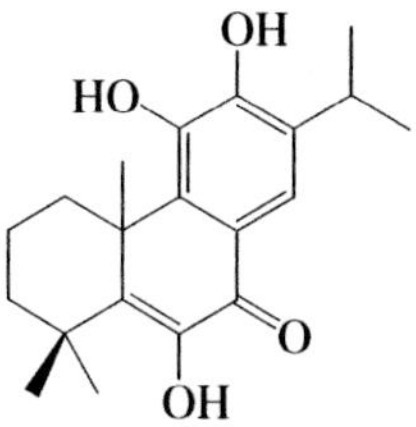 14-Desoxy-Coleon U

27.

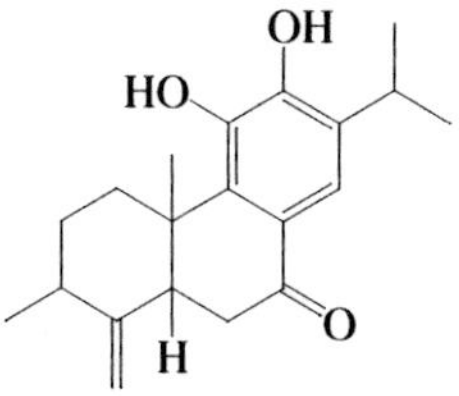

19(4,3)abeo-O-demethylcryptojaponol

FIGURE 6.1 (continued)

28.

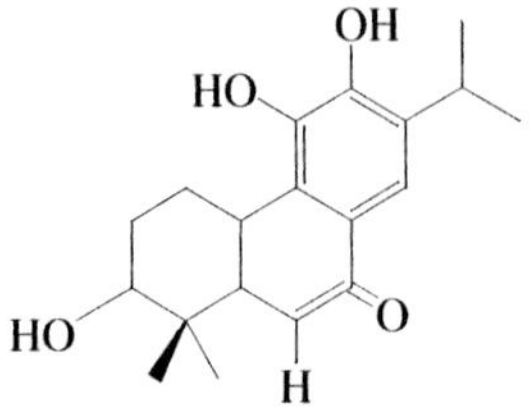

3-beta-Hydroxy-demethylcryptojapan

29. 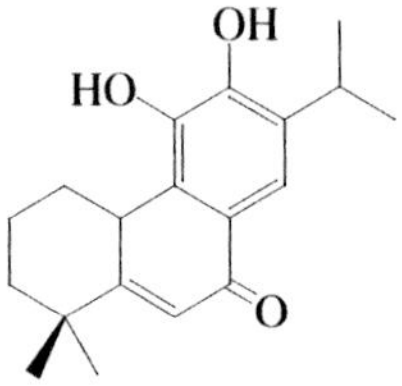   Salvinolone

30. 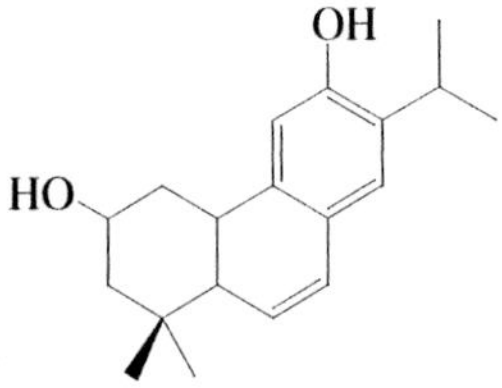   6,7-Dehydrosalviol

FIGURE 6.1 (continued)

6.3 SEASONAL VARIATION IN THE PLANT

As carnosic acid is the major phenolic diterpene in rosemary, its concentration in plant material is of interest for further processing applications. Observing the formation of phenolic antioxidants in plant material during the season and evaluating the influence of climatic conditions should indicate growing conditions resulting in high phenolic diterpene concentration.

The concentration of carnosic acid in rosemary plants investigated under Mediterranean field conditions in Barcelona, Spain ranged from 2.8 in August to 4.8 mg g^{-1} in December (Figure 6.2). In contrast, α-tocopherol present in the plants showed a reversed trend, suggesting the lowest concentration in Mediterranean winter and the highest concentration during summer (Munné-Bosch et al., 1999). The relationship between antioxidant formation vs. relative water content of the leaves due to precipitation, solar radiation and temperature is shown in Figure 6.2. Carnosic acid contrasts with the behavior of most phenolic compounds that obey a positive relationship between the solar radiation and the quantity formed. The humidity is apparently a crucial parameter for the formation of carnosic acid, enhancing its concentration in the leaf material. During the Mediterranean summer, the low humidity can cause low relative water content (RWC) in the plant that may then result in a water stress situation for the plant.

TABLE 6.1
Phenolic Diterpenes in *Rosmarinus* and *Salvia officinalis*

R. officinalis:	Carnosic acid, Salvin (1a)
	Methylcarnosoate (1d)
	Carnosol (Picrosalvin) (2a)
	Isorosmanol (2b)
	11,12-Dimethoxy-isorosmanol (2d)
	Rosmanol (3a)
	Epirosmanol (3b)
	7-Monomethylepirosmanol (3c)
	Rosmaridiphenol (4)
	Galdosol (5)
S. officinalis:	Carnosic acid, Salvin (6)
	Carnosol (Picrosalvin) (2a)
	Rosmanol (3a)
	Carnosic acid 12-methylether-delta-lactone (6)

Source: Brieskorn et al. (1964), Wenkert et al. (1965), Al-Hazimi et al. (1984), Houlihan et al. (1984), Nakatani et al. (1984), Fraga et al. (1985), Schwarz and Ternes (1992b), Cuvelier et al. (1994), Richheimer et al. (1996), and Munné-Bosch et al. (2000).

Numbers in parentheses refer to the structure formula in Figure 6.1.

6.4 EXTRACTION TECHNOLOGY

Selection of a suitable extraction procedure can increase the concentration of phenolic diterpenes relative to the plant material. In addition, undesirable components can be removed prior to adding extracts to foods. For example, chlorophylls present in plant material may reduce the light stability of food during storage. Several extraction techniques have been patented (Nakatani et al., 1984; Aeschbach and Philippossian, 1989) using solvents with different polarity, such as petrol ether, hexane, toluene, acetone, methanol and ethanol. To obtain tailored extracts, the successive use of different solvents has been applied.

One important trend in the food industry is the increased demand for natural food ingredients free of chemicals. Therefore, special attention has been paid to alternative processes directed toward extraction solvents and techniques with both GRAS and GMP labels (Ibánez et al., 1999). Supercritical CO_2-extraction (SFC CO_2) has been used (Weinreich, 1989; Nguyen et al., 1991; Nguyen et al., 1994; Ibánez et al., 1999). Tena et al. (1997) noted that extracts from rosemary obtained by SFC CO_2 (35 bar at 100°C) were the cleanest extracts and provided the highest recovery of carnosic acid compared to solvent extracts (acetone, hexane, dichlormethane and methanol) after bleaching with active carbon. Bicchi et al. (2000) reported a fractionated SFC CO_2 method to selectively isolate carnosol and carnosic acid at 250 atm and 60°C in the second fraction. The authors used 5% methanol to modify the dissolution power of SFC CO_2.

TABLE 6.2
Phenolic Diterpenes in Other Species of the Genus *Salvia*

S. apiana Jeps.:	16-Hydroxycarnosic acid (1b)
S. canariensis:	7-oxocarnosic acid (1e)
	6-oxo-7-beta-hydroxy carnosic acid (1f)
	Arucatriol (7)
	Deoxocarnosol 12-methyl ether (8)
	6-alpha-Hydroxydemethylcryptojaponol (9)
	Salvicanol (10)
	11,12-Dimethoxy-6,8,11,13-tetraen-20oic acid methylester (11)
	Rosmanol-carnosoate (12)
S. candelabrum Boiss:	Candesalvone A (13)
S. cardiophylla Benth:	2-alpha-Hydroxysugiol (14)
S. columbariae:	7-ethoxyrosmanol (15)
	20-deoxocarnosol (16)
	11-Acetoxycarnosic acid (1c)
	Miltidiol (22)
S. cryptantha:	Cryptanol (17)
S. hypargeia:	Hypargenin D (18)
	Hypargenin E (19)
S. lanigera:	12-Hydroxyisocarnosol (2c)
	Isocarnosol (20)
S. miltiorrhiza Bunge:	Ferruginol (21)
	Miltidiol (22)
	Norsalvioxide (23)
S. munzii:	16-hydroxy-7-methoxy-rosmanol (24)
S. phlomoides:	Demethylcryptojaponol (25)
	14-Desoxy-Coleon U (26)
S. pubescens:	19(4,3)abeo-*O*-demethylcryptojaponol (27)
	3-beta-Hydroxy-demethylcryptojaponol (28)
S. prionitis:	Salvinolone (29)
S. texana:	6,7-Dehydrosalviol (30)

According to (Al-Hazimi and Miana, 1986; Luis and Grillo, 1993; Al-Hazimi et al., 1994; Luis et al., 1994)

Numbers in parentheses refer to the structure formula in Figure 6.1.

The comparison of six commercial extracts (Figure 6.3) showed that CO_2 extraction results in the highest proportion of carnosic acid among the phenolic diterpenes (Schwarz and Ternes, 1992b). Extraction solvents used for the extracts shown in Figure 6.3 were ethanol (extract 1); hexane and acetone (extract 2); ethanol (extract 3); methanol (extract 4); hexane, ethanol and methanol (extract 5); and super critical CO_2 (extract 6).

Schwarz et al. (1992), Cuvelier et al. (1994), and Richheimer et al. (1996) showed that carnosic acid degrades in solvents at ambient conditions, forming carnosol, 7-epirosmanol, rosmanol and methyl carnosic acid. Bracco et al. (1981) and Aeschbach et al. (1994) have explored mechanical extraction using medium-chain triglycerides

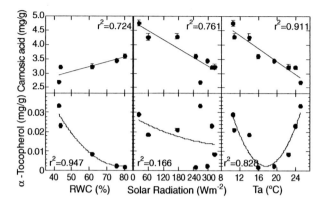

FIGURE 6.2 Formation of carnosic acid and α-tocopherol in rosemary during the Mediterranean summer (RWC = relative water content; Ta= air temperature). Data taken from Munné-Bosch et al. (2000).

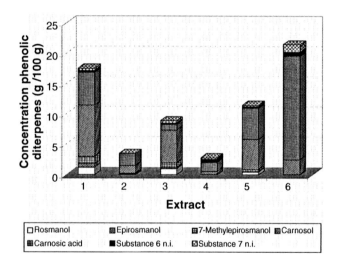

FIGURE 6.3 The phenolic diterpene composition of commercial extracts from rosemary and sage (for extraction solvents used, see text). Data taken from Schwarz et al. (1992).

as carriers to avoid solvent residues. The antioxidants dissolved in the lipid phase were collected by two-stage falling film molecular distillation to separate the lipid phase (to be recycled) from the active, low-molecular-weight fraction (Bracco et al., 1981).

Dapkevicius et al. (1998) compared yields and antioxidant activities of four different extracts from rosemary and sage leaves: an acetone, a water extract (both from deodorized plant material), and an acetone and SFC CO_2 extract (both from nondeodorized plant material). The yields (g per kg dry matter) ranged from 50.2 for the SFC CO_2 to 90.8 for the water extract from deodorized plant material. High antioxidant activity was found for the SFC CO_2 and the acetone extracts, but low activity was determined for all water extracts. This emphasizes the importance of carnosol and carnosic acid that are extracted from leaves with water-ethanol solvent containing more than 42% ethanol; this is the reason why teas from sage rarely taste bitter (Brieskorn et al., 1958).

6.5 COMMERCIAL PREPARATIONS

The accessibility of natural phenolic compounds has opened up the intriguing possibility that one could apply these compounds in foods, cosmetics, and other lipid-containing systems. Rosemary and sage are well known for the highest antioxidant activity among herbs, and their extracts are used in foods because of the strong antioxidant properties.

The regulatory climate and growing consumer preference for natural products caused the U.S. and European food industries to search for natural sources of antioxidants rather than invest in synthetic antioxidants.

Krishnakumar and Gordon (1996) estimated that the Western European market for feed/food antioxidant in 1995/1996 was split into approximately 50% synthetic antioxidants and 50% natural antioxidants, including vitamin C types, vitamin E types, carotenoids and spice extracts. The current legal status of herb extracts and other plant extracts is that of ingredients, but it can be assumed that extracts used as antioxidants are likely to be legislated as additives in the future.

Since the ancient times, spices have been added to different types of food to improve their flavor and to enhance their storage stability. The intake of herbs and spices is regulated by themselves by means of the flavor intensity of the essential oil. However, antioxidant extracts with high contents of phenolic diterpenes do not necessarily contain essential oils. Particularly, plant material from essential oil production for cosmetic or pharmaceutical products is an interesting side product to be used for the preparation of antioxidative extracts. However, data published by Richheimer et al. (1996) indicated that the deoiled biomass contains markedly less carnosic acid than the dried, nondeoiled plant material.

A range of commercial products containing extracts of rosemary and sage is available now; some of the products are water dispersible, others are oil soluble, and, in order to exploit the synergistic effect, some of them are combined with tocopherols. Rosemary and sage extracts have been used in many food products.

Of particular relevance for the application of antioxidant extracts from rosemary and sage, are meat products.

Madsen and Bertelsen (1995) reviewed the antioxidant activities of rosemary and sage extracts in meat products (Table 6.3) measured by the inhibition of thiobarbituric acid reactive substances (TBARS). In lard and meat products, rosemary and sage display the highest activity among other herbs. However, in oil-in-water emulsion, other spices such as clove, mace, turmeric and cinnamon are more effective.

6.6 BIOACTIVITIES AND BIOAVAILABILITY

In the following section, bioactivities of abietane diterpenes and related activities of rosemary and sage are discussed.

In view of the very few reports on allergenicity, their common use as a medical drug and herb can be considered as a plant with low allergen potential. In 1997, a

TABLE 6.3
Effect of Rosemary Extracts on the Oxidative Stability of Meat Products Measured by the Inhibition of Thiobarbituric Acid Reactive Substances (TBARS)

Oleoresin/ Extract	Foodstuff	Storage Conditions	Reduction of TBARS %	References[1]
Oleoresin	Raw restructured beef steaks	–20°C, 6 months	39	Stoick et al., 1991
Oleoresin	Cooked restructured beef steaks	4°C, 6 days	14	Stoick et al., 1991
Oleoresin	Restructured chicken nuggets	4°C, 6 days	11	Lai et al., 1991
Oleoresin	Restructured chicken nuggets	–20°C, 6 months	1	Lai et al., 1991
Extract	Frankfurters	4°C, 18–35 days	0–72	Resurreccion et al., 1990
Extract	Cooked beef patties	4°C, 2 days	56	St. Angelo et al., 1990
Extract	Cooked beef patties	4°C, 2 days	71	St. Angelo et al., 1990

[1] References cited in Madsen and Bertelsen (1995).

first case of contact dermatitis from a rosemary extract was reported. Carnosol was the identified compound responsible for the allergic reaction (Fernandez et al., 1997; Hjorther et al., 1997).

A study by Krause and Ternes (2000) indicates the oral absorption of carnosic acid with the feed. The bioavailability of carnosic acid in the egg yolk of Leghorn hens amounted to 0.0024% and 0.0025% at doses of 50 mg and 100 mg carnosic acid per 100 g feed, respectively.

6.6.1 ANTIOXIDANT ACTIVITY

The strong antioxidant activity of plant material from rosemary and sage leaves compared to other herbs was already recognized by Chipault et al. (1952). The antioxidant properties of rosemary and sage are extensively documented and well related to the phenolic diterpenes.

Brieskorn and Dömling (1969) suggested that the phenolic diterpenes carnosic acid and carnosol are mainly responsible for the antioxidant activity. Rosmarinic acid, a derivative of caffeic acid, which was earlier believed to make a major contribution to the activity of rosemary and sage (Herrmann, 1961) is, however, widely present in the Labiatae family. In the 1980s, the research group of Inatani-Nakatani published several papers on phenolic diterpenes with γ-lactone structures that possess high antioxidant activity (Inatani et al., 1983; Nakatani and Inatani, 1981, 1983; Nakatani et al., 1984). In 1992, Schwarz et al. demonstrated the relationship between the content of phenolic diterpenes in rosemary and sage extracts with the antioxidant activity. Cuvelier et al. (1994) reported strongest activity for carnosic acid (0.24) followed by methyl carnosoate (0.14), rosmadial (0.13), carnosol (0.13), and rosmanol (0.10) compared to BHT (1.0) when inhibiting the linoleic acid degradation at 100°C. Also, Richheimer et al. (1999) reported strongest activity for carnosic acid (1.0) followed by 7-methoxyrosmanol (0.52) and carnosol (0.4) in the rancimat test at 100°C, whereas the activity of 12-O-methyl carnosic acid was near zero. However, the same author indicated that the activity of carnosic acid was approximately seven times higher than that of BHT and BHA. This is in accordance with the findings of Schwarz et al. (1996) comparing BHA and carnosic acid in the Schaal oven test at 60°C. In corn oil emulsion, carnosol and methyl carnosoate were more effective than carnosic acid, but both antioxidants were less effective than α-tocopherol (Hopia et al., 1996; Huang et al., 1996). In mixtures, carnosol decreased and carnosic acid increased the oxidative stability of α-tocopherol in corn oil. During oxidation, carnosic acid and carnosol converted into unidentified compounds that exhibited antioxidant activity (Hopia et al., 1996).

Heterocyclic amines are consumed with fried meat products prepared at temperatures above 150°C. It has been concluded from epidemiological studies that heterocyclic amine formation may be a risk factor of colon and other types of cancer (Murkovic et al., 1998). The formation of individual heterocyclic amines was reduced by the addition of rosemary and sage between 38% and 75%. As radicals play an important role in the formation of heterocyclic amines, the effect is attributable to the antioxidant properties of the spice (Murkovic et al., 1998). Strong reduction of heterocyclic amines was found by Balogh et al. (1995) when adding oleoresin rosemary. The effect was, however, stronger using vitamin E.

6.6.2 ANTIMICROBIAL ACTIVITY

Liu and Nakano (1996) compared the antibacterial activity of alcoholic extracts from 17 spices and 5 herbs. *Bacillus stearothermophilus*, which produces heat-resistant spores, was highly sensitive to most of the spices tested. At 0.05%, sage perfectly inhibited the growth. Against *B. coagulans*, sage and rosemary inhibited the growth, and the minimum inhibitory concentrations (MICs) were 0.05% and 0.2%, respectively. Both *Escherichia coli* and *Salmonella* were not inhibited by 0.5% concentration. Sage extract also showed inhibitory effects against *B. subtilis*, *B. cereus*, and *Staphylococcus aureus* at 0.2%, whereas rosemary extract was a weak inhibitor against *S. aureus* at 0.5%.

Antimicrobial effects on the growth of *Yersinia enterocolitica* were found for ground sage and sage extracts in a dose-dependent manner (Bara and Vanetti, 1995).

For powdered sage, antilisterial effects were at a concentration of 0.7–1%, resulting in a strong decrease of the population within 1 to 14 days at 4°C in broth (Hefnawy et al., 1993; Ting and Deibel, 1992). The highest antilisterial activities were found for rosemary among 18 spices. Antioxidant extracts of rosemary (0.3–0.5%) were effective, whereas 1% oil was ineffective (Pandit and Shelef, 1994).

Extracts from sage (free of essential oil) inhibited the growth of *E. coli* and *Shigella dysenteriae* almost completely (Petkov, 1986). Shelef et al. (1984) found decreasing resistence in the following order: *Salmonella typhimurium* > *Pseudomonas* spp. > *S. aureus* > *B. cereus*. Although the antimicrobial activity of sage increased with an increase in the volatile oil fraction, essential oils alone had limited inhibitory effects. Salt and water levels increased bacterial sensitivity, whereas fat and protein decreased it. Liu and Nakano (1996) demonstrated that the MIC of sage was reduced from 0.05% to 0.005% when pH was decreased from 7.3 to 6.0 and the NaCl concentration was increased from 0.5% to 1.5%. The highest antimicrobial activity (MIC = 0.1 – 1%) was found in broth and decreased in rice (MIC 0.4 to >2.5%) and chicken and noodles (MIC of 1.0 to >2.5%) (Shelef et al., 1984).

For carnosol bacteriostatic and fungistatic effects were found (Shelef et al., 1984). The following MICs where found for carnosol: 1:10,000 (*S. aureus*) and 1:1000 (*E. coli*).

Dimayuga et al. (1991) reported activity of carnosol against *B. subtilis* and *Candida albicans*, and supported results on *S. aureus* and *E. coli*. The authors used 4 mg per disc. Significant activity was also found for carnosic acid and its 12-*O*-methyl ester to inhibit *S. aureus*. The inhibitive action is due to the inhibition of the nucleic acid biosynthesis pathway, as carnosic acid prevents the incorporation of thymidine and uridine into the DNA and RNA of *S. aureus* (Wagner, 1993).

Azzouz and Bullerman (1982) compared 16 spices and found only minor effects for 2% addition of rosemary on the growth of *Aspergillus* and *Penicillium* strains and no effect for sage. Also, fungicidal activity was found against *Aspergillus* and *Penicillium* and seven other test fungi by Schmitz et al. (1993). Kunz (1994) even found a promoting effect of sage on mold growth in wheat bread.

Also, low antimicrobial activity was reported against *A. flavus*, *Lactobacillus plantarum*, and *Pediococcus acidlacitic* (Llewellyn et al., 1981; Zaika et al., 1983). The MIC of an ethanolic rosemary leaf extract against *Clostridium botulinum* growth was 500 mg/kg related to the leaf material (Huhtanen, 1980).

6.6.3 ANTIVIRAL ACTIVITY

Antiviral activity of a sage extract was observed by measuring an inhibition plaque toward *Herpes hominis* HVP 75 (type 2), *Influenza* A2 *mannheim* 57 and *Vaccine virus*.

The inhibitory activity of carnosol, carnosic acid, rosmanol, 1-*O*-methylrosmanol, 7-*O*-ethylrosmanol, 11,12-*o,o*-dimethylrosmanol against HIV-1 protease was tested (Pukl et al., 1992; Paris et al., 1993). HIV-1 protease is a virus-specific aspartic protease that is essential for polyprotein processing and viral infectivity of the acquired immune deficiency syndrome (AIDS). Only carnosic acid showed a strong inhibitory effect (IC_{90} = 0.08 µg/ml). Furthermore, carnosic acid showed antiviral activity at IC_{90} = 0.32 µg/ml, which is, however, very close to the cytotoxic dose of $EC_{90} \neq 0.36$ µg/ml (on H9 lymphocytes in culture). In fact, a wide range of anti-oxidants has been claimed to inhibit HIV expression in cell lines (Schreck et al., 1992).

A rosemary extract containing 0.6% carnosol and 4.4% carnosic acid and a spice cocktail from rosemary, sage, thyme and oregano (0.28% carnosol, 1.2% carnosic acid, 0.25% carvacrol and 0.25% thymol) were tested for their ability to inhibit HIV-1 replication in C8166 cells and also chronically infected H9 cells. Both extracts exhibited slight anti-HIV activity in C8166 cells at EC_{50} of 0.01% and 0.04%. However, the cytotoxic concentration (TC_{50}) was in the same range. EC_{50} and TC_{50} represent the concentration that reduces the viral antigen gp120 by 50% in infected cultures and cell growth by 50% (Aruoma et al., 1996). Of the pure compounds tested, only carnosol showed weak but definite anti-HIV activity at EC_{50} 2.64 µg/ml. The cytotoxic dose was five times higher.

The formation of plaques by *Herpes simplex* virus Type 2 in embryonic fibroblasts was dose-dependently inhibited by a rosemary extract (2–100 µg/ml). The antiviral activity was further tested in four fractions of the ethanol extract and was found in the hexane fraction (Romero et al., 1989).

6.6.4 ANTIMUTAGENIC AND ANTICANCEROGENIC ACTIVITY

The inhibitory effect was tested in a photomutagenesis assay (UV radiation at 356 nm) in Ames tester strain *Salmonella typhimurium* (TA 102 and TA 98) with 8-methoxypsoralen (8-MOP) or benzo(a)pyrene as initiator. The formation of revertants in the presence of 8-MOP was 600 (rosemary extract at 100 µg/ml), 250 (β-carotene 100mg/ml) and 800 for the control. The formation of revertants in the presence of benzo(a)pyrene was 270 (rosemary extract at 100 µg/ml), 230 (β-carotin 100 mg/ml) and 370 for the control (Santamaria et al., 1987). Aqueous extracts from sage were found to decrease the mutagenicity of sodium azide from 26% to 49% in *S. typhimurium* TA 100 (Karakaya and Kavas, 1999). Further studies with *S. typhimurium* TA 98 showed the dependence of the extraction technique on the antimutagenic effect in the presence of Trp-P-2, a carcinogen formed in meat during cooking (Samejima et al., 1995) (Table 6.4). Flavonoids such as luteolin, galagin and quercetin were attributed to the effect. The activity of inhibiting the formation of revertants is considered to correspond with the antioxidant activity (Tateo et al., 1988; Samejima et al., 1995). The highest activity was found for the polar fraction of a sage extract obtained by 50% ethanol (Tateo et al., 1988). When mutagenicity in the strain TA102 was induced by the generation of oxygen-reactive species (using tert-butyl-hydroperoxide or hydrogen peroxide), carnosic acid was identified as the antimutagenic agent responsible for the activity of rosemary extract (Minnunni et al., 1992).

TABLE 6.4
Antimutagenicity of Sage Extracts
against 20 ng of Trp-P-2

Solvent	Antimutagenicity (%)	Yield (g)
Hexane	76	1.9
Methylene chloride	91	2.2
Ethyl acetate	93	0.9
Acetone	85	1.0
Methanol	7	20.5
Water	nonactive	13

Data taken from Samejima et al. (1995).

6.6.4.1 Tumorgenesis

Application of rosemary to mouse skin inhibited covalent binding of benzo(a)pyrene to epidermal DNA and inhibited tumor initiation by benzo(a)pyrene [B(a)P] and 7,12-dimethylbenz(a)anthracene (DMBA). Topical application of 1, 3 or 10 μmol carnosol together with 5 nmol 12-O-tetradecanoylphorbol-13-acetate (TPA) twice weekly for 20 weeks to the backs of mice previously initiated with DMBA inhibited the number of skin tumors per mouse by 38, 63 and 78%, respectively.

Studies with female CD-1 mice with 2% rosemary methanol extract diet increased liver microsomal oxidation and glucuronidation of estradiol and estrone and inhibited uterotropic action (Zhu et al., 1998). It is suggested that stimulating certain pathways of estrogen metabolism may be beneficial for the prevention of estrogen-dependent tumors in the target organs of humans and animals.

Rosemary extract at 1% in the diet decreased the DMBA-induced mammary tumors incidence in female Sprague-Dawley rats by 47% (Singletary and Nelshoppen, 1991).

Supplementation of diets for 2 weeks with rosemary extract (0.5%) but not carnosol (1.0%) or ursolic acid (0.5%) resulted in a significant decrease in the *in vivo* formation of rat mammary DMBA-DNA adducts compared to the control. When injected intraperitoneally for 5 days at 200 mg/kg body wt., rosemary and carnosol, but not ursolic acid, significantly inhibited mammary adduct formation and were associated with a significant decrease of 74 and 65%, in the number of DMBA-induced mammary adenocarcinomas per rat (Singletary et al., 1996).

Amagase et al. (1996) demonstrated that the benefits of rosemary in the diet of rats exposed to DMBA are dependent on the source and concentration of dietary lipids. 1% rosemary but not 0.5% rosemary powder reduced the DMBA-induced DNA adduct in a diet containing 5% corn oil, whereas 0.5% rosemary powder was effective in a 20% corn oil diet. Further, the effect of rosemary was lower when the dietary lipid consisted of 5% corn oil and 15% coconut oil.

6.6.5 CHEMOPREVENTIVE ACTIVITIES

In light of the potential impact of rosemary and sage on carcinogenesis, the cellular mechanisms of carnosol and carnosic acid were studied. Human epithelial cell lines, derived from target tissues in carcinogenesis such as lung, liver and colon, were exposed to procarcinogens such as nitrosamines, mycotoxins and heterocyclic amines. Carnosol and carnosic acid inhibited DNA adduct formation. At least two mechanisms are involved in the anticarcinogenic action: first, inhibition of the metabolic activation of procarcinogens catalyzed by the phase I cytochrome P450 enzymes, and second, induction of the detoxification pathway catalyzed by the phase II enzymes such as glutathione S-transferase (GST) (Offord et al., 1995, 1997; Mace et al., 1998). Tawfiq et al. (1994) screened more than 145 extracts of fruits, herbs, spices and beverages that are consumed regularly in the European diet. The results indicated that some vegetables and herbs such as rosemary, basil and tarragon induced NAD(P)H-quinone reductase (QR) activity in murine cells.

Similar effects were found in animal models. Female rats fed rosemary extracts at concentrations from 0.25% to 1.0% showed a 3- to 4.5-fold increase in liver GST and QR activities. Carnosol in the diet did not enhance GST activity. Intraperitoneal administration of carnosol (100–400 mg/kg) was associated with 1.6- to 1.9-fold increase in the GST activity and a 3.1- to 4.8-fold increase in the QR activity. Rosemary extract (200 mg/kg) administered i.p. increased GST and QR activities 1.5-fold and 3.2-fold, respectively (Singletary, 1996). Whereas GST and QR in the liver were significantly induced, only GST activity was increased in the stomach, and no effect was found for lung GST and QR when rosemary was fed to female mice at concentrations of 0.3 and 0.6% (Singletary and Rokusek, 1997).

6.6.6 IMMUNOENHANCING ACTIVITY

No generalized immunoenhancing effect was found for rosemary tested in mononuclear cell proliferation derived from Weanling male Sprague-Dawley rats. Rosemary probably has an immunoenhancing effect in some stressed conditions, such as protein or antioxidant deficiency (Babu et al., 1999).

6.7 ANALYSIS OF PHENOLIC DITERPENES

The quantification of phenolic diterpenes has been mostly carried out by high-performance liquid chromatography (HPLC) based on the method reported by Schwarz and Ternes (1992a) using a gradient system based on methanol, water, citric acid and tetraethylammoniumhydroxide. Methods based on acetonitrile provide sufficient separation by isocratic elution (Richheimer et al., 1996). The detection of phenolic diterpenes was carried out by UV at 230 and 280 nm or by electrochemical detection using +800 mV (Schwarz and Ternes, 1992a; Richheimer et al., 1996). Bicchi et al. (2000) summarized different detection methods for phenolic diterpenes including UV, evaporated light scattering detection, and a particle beam interface linked to a mass spectroscopy device (MS).

Figure 6.4 shows a chromatogram for a methanol extract from rosemary extract using UV detection (230 nm; spectrophotometer HP 1100, Hewlett-Packard, Haldbronn, Germany). The specifications for the column were Hypersil ODS-5 µm Säule (250 × 4 mm, Knauer, Berlin, Germany). The eluents consisted of water/acetonitril

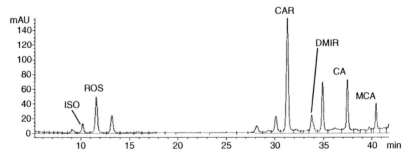

FIGURE 6.4 HPLC chromatogram of phenolic diterpenes in commercial rosemary extracts (ISO: isorosmanol; ROS: rosmanol; CAR: carnosic acid; CA: carnosol; DMIR: dimethyl-isorosmanol; and MCA: 12-*O*-methylcarnosic acid).

(70/30, v:v) containing 0.5% *o*-phosphoric acid and acetonitrile containing 0.25% *o*-phosphoric acid. The gradient program for the eluents was 78% A: 0–20 min, 78% → 41% A: 20–34 min, 41% → 5% A: 34–40 min; the flow rate was 0.6 ml min^{-1} (Müller et al., 2001). This method has been established in the author's laboratory for routine analysis.

REFERENCES

Aeschbach, R., Bächler, R., Rossi, P., Sandoz, L. and Wille, H.J. 1994. Mechanical extraction of plant antioxidants by means of oils. *Fat Sci. Technol.* 96:441–443.

Aeschbach, R. and Philippossian, G. 1989. Preparation of a spice extract having antioxidant action. European Patent Application. EP 0 307 626 A 1.

Al-Hazimi, M.G. and Miana, G.A. 1986. The diterpenoids of *Salvia* species. *J. Chem. Soc. Pak.* 8:549–569.

Al-Hazimi, H.M.G. and Miana, G.A. 1994. The diterpenoids of *Salvia* species (Part II). *J. Chem. Soc. Pakistan.* 16:46–63.

Amagase, H., Sakamoto, K., Segal, E.R. and Milner, J.A. 1996. Dietary rosemary suppresses 7,12-dimethylbenz(a)anthracene binding to rat mammary cell DNA. *J. Nutr.* 126:1475–80.

Aruoma, O.I., Spencer, J.P.E., Rossi, R., Aeschbach, R., Khan, A., Mahmood, N., Munoz, A., Murcia, A., Butler, J. and Halliwell, B. 1996. An evaluation of the antioxidant and antiviral action of extracts of rosemary and provencal herbs. *Food Chem. Toxicol.* 34:449–456.

Azzouz, M.A. and Bullerman, L.B. 1982. Comparative antimycotic effects of selected herbs, spices, plant components and commercial antifungal agents. *J. Food Protection.* 45:1298–1301.

Babu, U.S., Wiesenfeld, P.L. and Jenkins, M.Y. 1999. Effect of dietary rosemary extract on cell-mediated immunity of young rats. *Plant Foods Hum. Nutr.* 53:169–74.

Balogh, Z., McAnlis, G.T. and Gray, J.I. 1995. The Effects of Selected Antioxidants Including Vitamin E and Oleoresin Rosemary on Heterocyclic Amine Formation. In *IFT Annual Meeting.*

Bara, M.T.F. and Vanetti, M.C.D. 1995. Antimicrobial effect of spices on the growth of *Yersinia enterocolitica. J. Herbs, Spices Medic. Plants.* 3:51–58.

Bicchi, C., Binello, A. and Rubiolo, P. 2000. Determination of phenolic diterpene antioxidants in Rosemary (*Rosmarinus officinalis* L.) with different methods of extraction and analysis. *Phytochem. Anal.* 11:236–242

Bracco, U., Löliger, J. and Viret, J.-L. 1981. Production and use of natural antioxidants. *J. Am. Oil Chem. Soc.* June:686–690.

Brieskorn, C.H. and Dömling, H.-J. 1969. Carnosolsäure, der wichtige antioxydativ wirksame Inhaltsstoff des Rosmarin- und Salbeiblattes. *Z. Lebensm.-unters. Forsch.* 141:10–16.

Brieskorn, C.H., Fuchs, A., Bredenberg, B., McChesney, J.D. and Wenkert, E. 1964. The structure of carnosol. *J. Org. Chem.* 29:2293–2298

Brieskorn, C.H., Leiner, U. and Thiele, K. 1958. 10. Mitteilung über die Inhaltsstoffe con Salvia officinalis L. Isolierung und Eigenschaften des Bitterstoffes des Salbeiblätter. *Dtsch. Apoth. Ztg.* 98:651–653.

Chipault, J.R., Mizuno, G.R., Hawkins, J.M. and Lundberg, O. 1952. The antioxidant properties of natural spices. *Food Res.* 17:46–55.

Cuvelier, M.-E., Berset, C. and Richard, H. 1994. Antioxidant constituents in sage (*Salvia officinalis*). *J. Agr. Food Chem.* 42:665–669.

Dapkevicius, A., Venskutonis, R., van Beck, T.A. and Linssen, J.P.H. 1998. Antioxidant activity of extracts obtained by different isolation procedures from some aromatic herbs grown in Lithuania. *J. Sci. Food Agric.* 77:140–146.

Dimayuga, R.E., Garcia, S.K., Nielsen, P.H. and Christophersen, C. 1991. Traditional medicine of Baja California Sur (Mexico) III. Carnosol: A diterpene antibiotic from *Lepechinia Hastata. J. Ethnopharmacol.* 31:43–48.

Eisenreich, W., Menhard, B., Hylands, P.J., Zenk, M.H. and Bacher, A. 1996. Studies on the biosynthesis of taxol: the taxane carbon skeleton is not of mevalonoid origin. *Proc. Natl. Acad. Sci. USA.* 93:6431–6436.

Fernandez, L., Duque, S., Sanchez, I., Quinones, D., Rodriguez, F. and Garcia Abujeta, J.L. 1997. Allergic contact dermatitis from rosemary (*Rosmarinus officinalis* L.). *Contact Dermatitis.* 37:248–249.

Fraga, B.M., Ganzalez, G., Herrera, J.R., Luis, J.G., Perales, A. and Ravelo, A.G. 1985. A revised structure of the diterpenoid rosmanol. *Phytochemistry.* 24:1853–1854.

Hanson, R. 1995. Diterpenoids. *Nat. Prod. Rep.* 12:207–218.

Hefnawy, Y.A., Moustafa, S.I. and Marth, E.H. 1993. Sensitivity of *Listeria* monocytogenes to selected spices. *J. Food Protection.* 56:876–878.

Hegnauer, R. 1996. *Chemotaxonomie der Pflanzen.* Basel: Birkhäuser Verlag, pp. 289–346.

Herrmann, K. 1961. Über die antioxidative Wirkung von Labiatendrogen und der ihnen enthaltenen Labiatensäure (Labiatengerbstoff). *Z. Lebensm.-unters. Forsch.* 116:224–228.

Hickey, M. and King, C. 1997. Labiatae Juss. (*Lamiaceae* Lindl.). In *100 Families of Flowering Plants.* Cambridge: Cambridge University Press, pp. 350–352.

Hjorther, A.B., Christophersen, C., Hausen, B.M. and Menne, T. 1997. Occupational allergic contact dermatitis from carnosol, a naturally-occuring compound present in rosemary. *Contact Dermatitis.* 37:99–100.

Hopia, A.I., Huang, S.-W., Schwarz, K., German, J.B. and Frankel, E.N. 1996. Effect of different lipid systems on antioxidant activity of rosemary constituents carnosol and carnosic acid with and without alpha-tocopherol. *J. Agric. Food Chem.* 44:2030–2036.

Houlihan, C.M., Ho, C.-T. and Chang, S.S. 1984 Elucidation of the chemical structure of a novel antioxidant, rosmaridiphenol, isolated from rosemary. *J. Am. Oil Chem. Soc.* 61:1036–1039.

Huang, S.-W., Frankel, E.N., Schwarz, K., Aeschbach, R. and German, J.B. 1996. Antioxidant activity of carnosic acid and methyl carnosate in bulk oils and oil-in-water emulsions. *J. Agric. Food Chem.* 44:2951–2956.

Huhtanen, C.N. 1980. *J. Food Protection.* 43:195–196, 200.

Ibáñez, E., Oca, A., de Murga, G., López-Sebastián, S., Tabera, J. and Reglero, G. 1999. Supercritical fluid extraction and fractionation of different preprocessed rosemary plants. *J. Agric. Food Chem.* 47:1400–1404.

Inatani, R., Nakatani, N. and Fuwa, H. 1983. Antioxidative effect of the constituents of rosemary (*Rosmarinus officinalis* L.) and their derivatives. *Agric. Biol. Chem.* 47:521–528.

Karakaya, S. and Kavas, A. 1999. Antimutagenic activities of some foods. *J. Food Sci. Agric.* 79:237–242.

Kleinig, H. 1989. The role of plastids in isoprenoid biosynthesis. *Ann. Rev. Plant Physiol. Plant Mol. Biol.* 40:39–59.

Krause, E.L. and Ternes, W. 2000. Bioavailability of the antioxidative *Rosmarinus officinalis* compound carnosic acid in eggs. *Eur. Food Res. Technol.* 210:161–164.

Krishnakumar, V. and Gordon, I. 1996. Antioxidants trends and developments. *Int. Food Ingredients.* 5:41–44.

Kunz, B. 1994. Gewürze zur Haltbarkeitsverlängerung von Brot. *Gordian.* 94:53.

Lemonica, I.P., Damasceno, D.C. and di Stasi, L.C. 1996. Study of the embryotoxic effects of an extract of rosemary (*Rosmarinus officinalis* L.). *Braz. J. Med. Biol. Res.* 29:223–227.

Leung, A.Y. and Foster, S. 1996. *Encyclopedia of Common Natural Ingredients Used in Food, Drugs, and Cosmetics.* New York: John Wiley and Sons, Inc., pp. 446–448, 457–459.

Lichtenthaler, H.K. 1999. The 1-deoxy-*d*-xylulose-5-phosphate pathway of isoprenoid biosynthesis in plants. *Ann. Rev. Plant Physiol. Plant Mol. Biol.* 50:47–65.

Liu, Z.-H. and Nakano, H. 1996. Antibacterial activity of spice extracts against food-related bacteria. *J. Facul. Appl. Biol. Sci., Hiroshima Univ.* 35:181–190.

Llewellyn, G.C., Burkett, M.L. and Eadie, T. 1981. *J. Assoc. Off. Analytical Chem.* 64:955–960.

Luis, J.G., Grillo, T.A. 1993. Abietane diterpenes from *Salvia munzii*. *Phytochemistry.* 34:863–864.

Luis, J.G., Grillo, T.A., Quinones, W. and Kishi, M.P. 1993. Columbaridione, a diterpene-quinone from *Salvia comlumbariae*. *Phytochemistry.* 36:251–252.

Mace, K., Offord, E.A., Harris, C.C. and Pfeifer, A.M. 1998. Development of *in vitro* models for cellular and molecular studies in toxicology and chemoprevention. *Arch. Toxicol. Suppl.* 20:227–236.

Madsen, H.L. and Bertelsen, G. 1995. Spices as antioxidants. *Trends Food Sci. and Technol.* 6:271–277.

McGarvey, D.J. and Croteau, R. 1995. Terpenoid metabolism. *Plant Cell.* 5:1015–1026.

Minnunni, M., Wolleb, U., Mueller, O., Pfeifer, A. and Aeschbacher, H.U. 1992. Natural antioxidants as inhibitors of oxygen species induced mutagenicity. *Mutat. Res.* 269:193–200.

Müller, M., Munné-Bosch, S., Alegre, L. and Schwarz, K. 2001. [Formation of phenolic Ditepenes in *Salvia officinalis* L. Plants]. DGQ (Deutsche Gesellschaft für Qualitäts-forschungs) XXXVI. Vortragstagung Gewürz- und Heilpflanzen, Jena, pp. 171–176.

Munné-Bosch, S., Schwarz, K. and Alegre, L. 1999. Enhanced formation of α-tocopherol and highly oxidized abietane diterpenes in water-stressed rosemary plants. *Plant Physiol.* 121:1047–1052.

Munné-Bosch, S., Alegre, L. and Schwarz, K. 2000. The formation of phenolic diterpenes in *Rosmarinus officinalis* L. under Mediterranean climate. *Eur. Food Res. Technol.* 210:263–267.

Murkovic, M., Steinberger, D. and Pfannhauser, W. 1998. Antioxidant spices reduce formation of heterocyclic amines in fried meat. *Z. Lebensm.-unters. Forsch.* 207:477–480.

Nakatani, N. and Inatani, R. 1981. Structure of rosmanol, a new antioxidant from rosemary (*Rosmarinus officinalis* L.). *Agric. Biol. Chem.* 45:2385–2386.

Nakatani, N.R. and Inatani, R. 1983. A new diterpene lactone, rosmadial from rosemary. *Agric. Biol. Chem.* 47:353–358.

Nakatani, N. and Inantani, R. 1984. Two antioxidative diterpenes from rosemary (*Rosamrinus officinalis* L.) and a revised structure for rosmanol. *Agric. Biol. Chem.* 48:2081–2085.

Nakatani, N.R., Inatani, T. and Koishi, T. 1984. Antioxidative compound, method of extracting same from rosemary, and use of same. U.S. Patent 4,450,097.

Nguyen, U., Frakman, G. and Evans, D.A. 1991. Process for extracting antioxidants from *Labiatae* herbs. U.S. Patent US5017397.

Nguyen, U., Evans, G. and Frakman, G. 1994. Natural antioxidants produced by supercritical extraction. In *Supercritical Fluid Processing of Food and Biomaterials*. Rizvi, S.S.H. (Ed.) Glasgow, UK: Blackie Academic and Professional, p. 103.

Offord, E.A., Mace, K., Ruffieux, C., Malnoe, A. and Pfeifer, A.M. 1995. Rosemary components inhibit benzo(a)pyrene-induced genotoxicity in human bronchial cells. *Carcinogenesis.* 16:2057–2062.

Offord, E.A., Mace, K., Avanti, O. and Pfeifer, A.M. 1997. Mechanisms involved in the chemoprotective effects of rosemary extract studied in human liver and bronchial cells. *Cancer Lett.* 114:275–281.

Pandit, V.A. and Shelef, L.A. 1994. Sensitivity of Listeria monocytogenes to rosemary (*Rosmarinus officinalis* L.). *Food Microbiol.* 11:57–63.

Paris, A., Strukelj, B., Renko, M. and Turk, V. 1993. Inhibitory effect of cornosolic acid on HIV-1 protease in cell-free assays. *J. Nat. Prod.* 56:1426–1430.

Petkov, V. 1986. *J. Ethnopharmacol.* 15:121–132, cited in Hänsel, R. Heller, K., Rimpler, H. and Schneider, G. 1994. *"Salvia" Hagers Handbuch der Pharmazeutischen Praxis*, Berlin: Springer Verlag, p. 498.

Pukl, M., Umek, A., Paris, A., Strukelj, B., Renko, M., Korant, B.D. and Turk, V. 1992. Inhibitory effect of carnosolic acid on HIV-1 protease. *Planta Medica.* 58:A632.

Richheimer, S.L., Bernart, M.W., King, G.A., Kent, M.C. and Bailey, D.T. 1996. Antioxidant activity of lipid-soluble phenolic diterpenes from rosemary. *J. Am. Oil Chem. Soc.* 73:507–514.

Richheimer, S.L., Bailey, D.T., Bernart, M.W., Kent, M., Viniski, J.V. and Anderson, L.D. 1999. Antioxidant activity and oxidative degradation of phenolic compounds isolated from rosemary. *Rec. Res. Develop. Oil. Chem.* 3:45–48

Romero, E., Tateo, F. and Debiaggi, M. 1989a. *Mitt Gebiete Lebensm.-lunters. Hyg.* 80:113–19.

Romero, E., Tateo, F. and Debiaggi, M. 1989b. *Mitt. Gebiete Lebensm. Hyg.* 80:113–119, cited in Hänsel, R., Heller, K., Rimpler, H. and Schneider, G. 1994. *"Rosmarinus" Hagers Handbuch der Pharmazeutischen Praxis*, Berlin: Springer Verlag, p. 556.

Samejima, K., Kanazawa, K., Ashida, H. and Danno, G.-I. 1995. Luteolin: a strong antimutagen against dietary carcinogen, Trp-P-2, in peppermint, sage, and thyme. *J. Agric. Food Chem.* 43:410–414.

Santamaria, A., Tateo, F., Bianchi, A. and Bianchi, L. 1987a. *Med. Biol. Env.* 15:97–101.

Santamaria, L., Tatea, F. and Bianchi, L. 1987b. Rosmarinus officinalis extract inhibits as antioxidant mutagenesis by 8-methoxypsoralen (8-MOP) and benzo(a)pyrene (BP) in Salmonella typhimurium. *Med. Biol. Env.* 15:92–101.

Schmitz, S., Weidenbörner, M. and Kunz, B. 1993. Herbs and spices as selective inhibitors of mould growth. *Chem. Mikrobiol. Technol. Lebensm.* 15:175–177.

Schneider, G. and Hiller, K. 1999. *Arzneidroge*. Heidelberg: Spektrum Akademischer Verlag, pp. 3–30.

Schreck, R., Albermann, K. and Baeuerle, P.A. 1992. Nuclear factor kappa-B: an oxidative stress-responsive transcription factor of eukaryotic cells. *Free Rad. Res. Comm.* 17:221–237.

Schwarz, K. and Ternes, W. 1992a. Antioxidative constituents of *Rosmarinus officinalis* and *Salvia officinalis* I. Determination of phenolic diterpenes with antioxidative activity amongst tocochromanols using HPLC. *Z. Lebensm.-unters. Forsch.* 195:95–98.

Schwarz, K. and Ternes, W. 1992b. Antioxidative constituents of *Rosmarinus officinalis* and *Salvia officinalis* II. Isolation of carnosic acid and formation of other phenolic diterpenes. *Z. Lebensm.-unters. Forsch.* 195:99–103.

Schwarz, K., Ternes, W. and Schmauderer, E. 1992. Antioxidative constituents of *Rosmarinus officinalis* and *Salvia officinalis*. III. Stability of phenolic diterpenes of rosemary extracts under thermal stress as required for technological processes. *Z. Lebensm. unters. Forsch.* 195:104–107.

Schwarz, K., Ernst, H. and Ternes, W. 1996. Evaluation of antioxidative constituents from thyme. *J. Sci. Food. Agric.* 70:217–223.

Shelef, L.A., Jyothi, E.K. and Bulgarelli, M.A. 1984. Growth of enteropathogenic and spoilage bacteria in sage-containing broth and foods. *J. Food Sci.* 49:737–740.

Singletary, K.W. 1996. Rosemary extract and carnosol stimulate rat liver glutathione-S-transferase and quinone reductase activities. *Cancer Lett.* 100:139–144.

Singletary, K.W. and Nelshoppen, J.M. 1991. Inhibition of 7,12-dimethylbenz(a)anthracene (DMBA)-induced mammary tumorigenesis and of *in vivo* formation of mammary DMBA-DNA adducts by rosemary extract. *Cancer Lett.* 60:169–175.

Singletary, K.W. and Rokusek, J.T. 1997. Tissue-specific enhancement of xenobiotic detoxification enzymes in mice by dietary rosemary extract. *Plant Foods Hum. Nutr.* 50:47–53.

Singletary, K., MacDonald, C. and Wallig, M. 1996. Inhibition by rosemary and carnosol of 7,12-dimethylbenz(a)anthracene (DMBA)-induced rat mammary tumorigenesis and *in vivo* DMBA-DNA adduct formation. *Cancer Lett.* 104:43–48.

Tateo, F., Fellin, M., Santamaria, L., Bianchi, A. and Bianchi, L. 1988. *Rosmarinus officinalis* L. extract production antioxidant and antimutagenic activity. *Perfumer Flavorist.* 13:48–55.

Tawfiq, N., Wanigatunga, S., Heaney, R.K., Musk, S.R., Williamson, G. and Fenwick, G.R. 1994. Induction of the anti-carcinogenic enzyme quinone reductase by food extracts using murine hepatoma cells. *Eur. J. Cancer Prev.* 3:285–292.

Tena, M.T., Valcárel, M., Hidalgo, P.J. and Ubera, J.L. 1997. Supercritical fluid extraction of natural antioxidants from rosemary: comparison with liquid solvent sonication. *Analytical Chem.* 69:521–526.

Ting, W.T.E. and Deibel, K.E. 1992. Sensitivity of Listeria monocytogenes to spices at two temperatures. *J. Food Safety.* 12:129–137.

Venkatachalam, K.V., Kjonaas, R. and Croteau, R. 1984. Development and essential oil content of secretory glands of sage (*Salvia officinalis*). *Plant Physiol.* 76:148–150.

Wagner, H. 1993. *Pharmazeutische Biologie, Drogen und ihre Inhaltsstoffe.* Stuttgart: Gustav Fischer Verlag.

Weinreich, B. 1989. Untersuchungen zur Struktur und zum Extrakrionsverhalten antioxidativ wirksamer Inhaltsstoffe aus *Rosmarinus-* und *Salvia officinalis* L. München: Technische Universität München, 182.

Wenkert, E., Fuchs, A. and McChesney, J.D. 1965. Chemical artifacts from the family Labiatae. *J. Org. Chem.* 30:2931–2940.

Zaika, L.L., Kissinger, J.C. and Wasserman, A.E. 1983. *J. Food Sci.* 48:1455–1459.

Zhu, B.T., Loder, D.P., Cai, M.X., Ho, C.-T., Huang, M.-T. and Conney, A.H. 1998. Dietary administration of an extract from rosemary leaves enhances the liver microsomal metabolism of endogenous estrogens and decreases their uterotropic action in CD-1 mice. *Carcinogenesis.* 19:1821–1827.

7 Organosulfur Compounds from Garlic

Bruce J. Holub, Karen Arnott, Jean-Paul Davis, Arun Nagpurkar and Jason Peschell

CONTENTS

7.1 INTRODUCTION

The tenet "Let food be thy medicine and medicine be thy food" espoused by Hippocrates nearly 2500 years ago is receiving renewed interest. In particular, increasing attention has been paid to the biological functions and medicinal value of naturally occurring compounds. Of these, garlic is perhaps the best example of the continuum between plants as food and plants as medicine. Occasionally called the "stinking rose," garlic has enjoyed a tremendous surge in popularity over the past two decades. The focus of more than 2000 research papers (Lawson, 1998a), garlic appears to exhibit a number of physiological actions that mimic several of today's pharmaceutical agents. With the possibility of fewer potential side effects and a reduction in the financial burden of prescription drug therapy, garlic may prove to be an effective alternative treatment in the prevention or progression of cardio-vascular disease (CVD) and cancer, as well as in combating various infections caused by microorganisms. At present, however, garlic's role in alleviating these ailments remains unclear because of methodological inconsistencies and recent conflicting reports in the biomedical literature. Researchers continue to address these contro-versial issues, which include the development of more reliable garlic products, standardized preparation protocols and determination of biologically active chemical constituents. Nevertheless, more dependable and reproducible clinical results are needed before the future of garlic as a medicinal aid can be ascertained.

7.1.1 HISTORY

Apart from being one of the most ancient remedies, garlic is a staple in the diets of an incredibly diverse number of cultures. Garlic's usage as both food and medicine pre-dates written history. The first recorded use of garlic occurred circa 3000 B.C. by both the Sumarians of Mesopotamia and the people of ancient India. Garlic was extremely important to the Egyptians, who recorded more than 22 therapeutic garlic formulae in their medical papyrus (Codex Ebers), which dates back to 1550 B.C. It was used extensively by slaves during the building of the pyramids and has also been found preserved in Pharaohs' tombs. Greek and Roman physicians, such as Hippocrates, Galen, Pliny the Elder and Dioscorides, recom-mended garlic for conditions including cancer, blood flow problems and bacterial infections (Block, 1985; Murry, 1995; Agarwal, 1996; Brown, 1996). Garlic was used in World War I to treat battle wounds and dysentery, as well as in World War II, when it earned the nickname "Russian Penicillin" because it was used when other antibiotics were not available (Fenwick and Hanley, 1985; Lawson, 1998a).

The word *garlic* originates from the Anglo-Saxon "*Garleac*" or "spear-plant" (Yutsis, 1994). Garlic is a member of the family Liliaceae (which contains more than 700 species including onions, leeks and chives) and has the botanical name *Allium sativum*. Garlic is a herbaceous perennial that is not found in the wild, having evolved to its present form after thousands of years of cultivation. Cultivated garlic does not produce true seeds, and therefore, no crossing occurs between strains. All garlic is propagated vegetatively from its cloves, and a complete new plant can grow within 2 years (Schumacher and Uyenaka, 1989). Garlic grows to approximately

30–100 cm in height in well-fertilized, sandy, loamy soil under warm, sunny conditions and flowers from June until August (Murry, 1995; Hahn, 1996).

7.1.2 PRODUCTION AND CONSUMPTION

In 1999, the world production of garlic was 9,280,188 metric tons (up 4.1% from
1998), with China as the leading producer (5,964,066 metric tons, up 2.5% from
1998), followed by India (517,700 metric tons, up 6.4% from 1998) and the Republic
of Korea (383,778 metric tons, down 2.6% from 1998). Argentina (150,000 metric
tons, up 1.3% from 1998) was the major South American producer, and the United
States (310,000 metric tons, up 19.4% from 1998) was the largest North American
producer (Food and Agricultural Organization, 2000). In Canada, the commercial
production of garlic is a relatively new industry that produced 6400 metric tons in
1999 (up 65% from 1998). Of this, Ontario accounted for 35%, or 2220 metric tons
in 1999 (up 13% from 1998) (Ontario Garlic Growers Association, 2000). In 1999,
the U.S. consumption of fresh garlic as food was 350 million kg, while in Canada,
the 1999 consumption was 12.1 million kg.

The herbal supplement industry is a dynamic and fast-growing industry. Global
herbal supplement sales totaled $10.4 billion in 1999, of which garlic supplements
accounted for $250 million (Nutrition Business Journal, 2000). The two major global
garlic supplement manufacturers are Lichtwer Pharma GmbH., in Germany (brand
name "Kwai"), and Wakunaga Pharmaceutical Co. in Japan (brand name "Kyolic").

7.2 CHEMICAL COMPOSITION AND CHEMISTRY

7.2.1 COMPOSITION

Garlic contains 33 sulfur compounds, 17 amino acids (including all the essential
amino acids), germanium, calcium, copper, iron, potassium, magnesium, selenium,
zinc and vitamins A, B, C, and E (Fenwick and Hanley, 1985; U.S. Department of
Agriculture, 1998).

Garlic is composed of 56–68% water and 26–30% carbohydrate. The most
medicinally significant components are thought to be the organosulfur-containing
compounds (11–35 mg/g fresh garlic). Garlic contains nearly four times as much
sulfur-containing compounds (per gram fresh weight) as onions, broccoli, cauli-
flower and apricots. The unusually high amounts of these compounds have led many
researchers to investigate the possibility that garlic may have potential pharmacological activities similar to other well-known sulfur-containing compounds, such
as probucol and penicillin (Lawson, 1996).

7.2.2 CHEMISTRY

One of the earliest chemical studies on garlic was conducted in 1844 by the German
chemist Theodor Wertheim. Wertheim was interested in garlic and employed steam
distillation to produce small amounts of garlic oil. Redistillation of the oil yielded
some odoriferous volatile substances. Wertheim proposed the name allyl (from

Allium) for the hydrocarbon group in the oil. In 1892, another German investigator, F.W. Semmler, applied steam distillation to cloves of garlic, producing one to two grams of an odoriferous oil per kilogram of garlic. In turn, the oil yielded diallyl disulfide, accompanied by lesser amounts of diallyl trisulfide and diallyl tetrasulfide. Chester J. Cavallito and colleagues made the next key discovery with regard to the chemistry of garlic in 1944. They established that methods less vigorous than steam distillation yielded different substances. Cavallito applied ethyl alcohol to garlic, which produced an oil that was the oxide of diallyl disulfides and subsequently named it allicin. In 1948, Arthur Stoll and Ewald Seebeck were able to demonstrate that allicin is generated in garlic when an enzyme initiates its formation from an odorless precursor molecule they identified as S-allyl-L-cysteine sulfoxide. Evidently, mechanical disruption of garlic enables the enzyme, alliinase, to come in contact with the precursor of allicin. They named this precursor alliin. Alliin was the first natural substance found to display optical isomerism at both the sulfur and carbon atoms (Block, 1985).

7.2.2.1 γ-Glutamyl-Cysteines and Cysteine Sulfoxides

The organosulfur-containing compounds of the mature, intact garlic cloves are composed of the cysteine sulfoxides (mainly alliin, with lesser amounts of methiin and isoalliin) and the γ-glutamylcysteines (mainly γ-glutamyl-S-*trans*-1-propenyl-cysteine, with lesser amounts of γ-glutamyl-S-allylcysteine, and γ-glutamyl-S-methylcysteine) (see Figure 7.1). Cysteine sulfoxides are white, crystalline, odorless solids that are highly soluble in water and insoluble in most organic solvents. The γ-glutamylcysteine compounds serve as reservoir compounds for the formation of the cysteine sulfoxides. During wintering, sprouting and storage, the γ-glutamyl-cysteines are gradually hydrolyzed, then oxidized to form cysteine sulfoxides by increasing levels of γ-glutamyltranspeptidase (formation increases with cooler temperatures). The cysteine sulfoxides (8–19 mg/g fresh weight) and γ-glutamyl-cysteines (5–16 mg/g) account for approximately 95% of the total sulfur in fresh garlic (Lawson, 1993). Approximately 85% of the cysteine sulfoxides are found in the bulb, with 12% in the leaves and 2% in the roots, whereas the γ-glutamylcysteines are found only in the bulbs (Lawson, 1998a).

7.2.2.2 Alliinase and Thiosulfinates

When garlic is mechanically disrupted, alliinase or alliin lyase (EC 4.4.1.4.) catalyzes the conversion of the cysteine sulfoxides to the biologically active diallyl thiosulfinates via sulfenic acid intermediates (Block, 1992). Alliinase is localized to a few vascular bundle sheath cells around the veins or phloem, whereas alliin and other cysteine sulfoxides are found in the clove mesophyll storage cells. This enzyme is approximately 10 times more abundant in the cloves than in the leaves and accounts for at least 10% of the total protein in the cloves (Ellmore and Feldberg, 1994). Alliinase is temperature and pH dependent; optimal activity is between pH 5.0–10.0, but allinase can be irreversibly deactivated at pH 1.5–3.0 (Krest and Keusgen, 1999).

The volatile and reactive thiosulfinates (2–9 mg/g fresh crushed garlic) are believed to be the physiologically and medicinally active compounds in garlic,

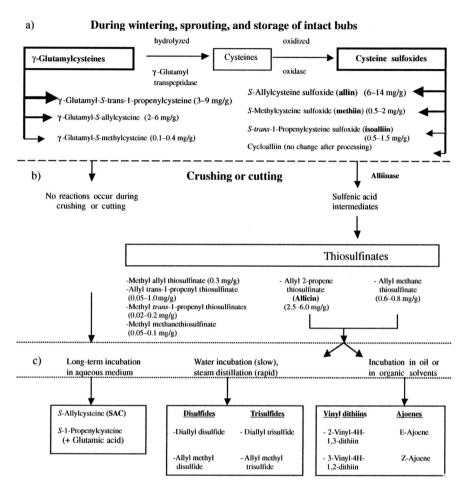

FIGURE 7.1 Chemical reactions of the main sulfur compounds found (a) in intact bulbs, (b) after crushing/cutting clove and (c) after processing. Source: Adapted from Lawson (1998a).

formed when garlic is cut, crushed or chewed (via alliinase activity) and are responsible for producing garlic's distinct odor. Allicin (allyl 2-propene thiosulfinate) is the most abundant thiosulfinate produced (70%), whereas allyl methyl thiosulfinate is the second most abundant (18%). Various other thiosulfinates are formed in lower concentrations. Conversion of the cysteine sulfoxides to thiosulfinates occurs within 0.2–0.5 min at room temperature for allicin and 1.5–5.0 min for allyl methyl thiosulfinates. The stability of the thiosulfinates is dependent upon their concentration and purity, as well as the solvent used and the temperature at which the reaction occurs. Allicin is less soluble in water and more soluble in organic solvents, especially polar solvents. The half-life of pure allicin in water is 30 days; the half-life of pure allicin in 1 mM citric acid is 60 days. Without a solvent, the half-life of allicin decreases to 16 h (Lawson, 1993). Thiosulfinates in garlic homogenates are less stable than their pure forms, possibly due to the presence of water-soluble substances. Refrigeration increases the stability greatly, but the only long-term storage option is freezing at –70°C (Lawson, 1996).

The thiosulfinates undergo various transformations to form more stable compounds such as di- and trisulfides, allyl sulfides, vinyl dithiins, ajoenes and mercaptocysteines. These reactions are also dependent upon temperature, pH and solvent conditions. Incubation of allicin or allyl methyl thiosulfinate in low polarity solvents produces mainly vinyl dithiins and smaller amounts of ajoenes. Incubation in alcohol gives variable results, forming ajoenes, diallyl trisulfides or vinyl dithiins, depending on whether pure allicin or cloves are incubated. Steam-distilled garlic oil consists of diallyl disulfide, diallyl trisulfide and mono- to hexasulfides (Lawson, 1993).

7.3 GROWING CONDITIONS AND SUPPLEMENTS

Although the effects of garlic have been investigated and reviewed considerably (Block, 1985; Fenwick and Hanley, 1985; Kendler, 1987; Kleijen et al., 1989; Block, 1992; Reuter and Sendl, 1994; Agarwal, 1996; Koch, 1996; Lawson, 1998b), the complex biological and pharmacological actions of garlic and its constituents are still not completely understood. One major problem is that the biological studies have proceeded at a far greater pace than the chemical and analytical processes used to identify the constitutive and active components found in the various garlic preparations employed in these biological studies (Lawson, 1993). This is of particular importance because the thiosulfinates released from fresh garlic are rapidly converted to a variety of other organosulfur compounds (the rapidity depending on the treatment condition). The variation in content of these compounds among the various types and sources of commercial garlic also creates a problem.

7.3.1 Variation in Organosulfur Compounds in Garlic

There is a large variation in the generation of allicin and other organosulfur compounds in garlic depending on variety/type, the growing soil composition and climate conditions, harvest dates and postharvest handling. The extent of these differences is important to monitor, because they may result in a reduction or increase in garlic's pharmacological efficacy.

There is natural variation in the alliin content (allicin-releasing potential) among the different varieties of garlic (approximately 40 types). Although the variation in allicin yield from bulb-to-bulb within a field (7%) and from farm-to-farm (25%) is relatively small, the variation among garlic varieties grown around the world is about fivefold (Lawson, 1998a). [According to Lawson (1998a), the most common type of garlic found in grocery stores (California strains) exhibits a variation of 1.8- to 2.7-fold in alliin and 1.5- to 4.2-fold in the γ-glutamylcysteines.]

Soil and climate conditions may have more of an effect on chemical variation than garlic variety. Garlic can be grown successfully in a wide range of soil types, least favorable being those that are prone to excessive frost heaving. Soils with high organic matter content are preferred growing media due to their increased moisture and nutrient-holding capacity accompanied by a pH range of 6.0–7.5 (Schumacher and Uyenaka, 1989). The addition of ammonium sulfate to the soil, as opposed to ammonium nitrate, causes a proportional increase in the sulfoxides in bulbs (Kosian, 1998). The addition of moderate amounts of inorganic selenium to crop soils has been shown to increase concentrations of selenium in garlic. The selenium-enriched garlic has exhibited anticarcinogenic properties, the ability to inactivate toxic metals and complement vitamin E in neutralizing free radicals (Irion, 1999).

Choice of harvest dates and postharvest handling can be very important. The amount of alliin and γ-glutamylcysteines present in the bulbs increases severalfold in the 4 weeks prior to harvest time. In addition, alliin increases by approximately 25% during the typical curing process (whole plants dried in the shade for at least 2 weeks). A delay in the normal harvest date by 2 weeks (until the plants are almost completely brown) increases the content of these compounds an additional 20% on a dry weight basis (Lawson, 1998a).

7.3.2 COMMERCIALLY AVAILABLE GARLIC SUPPLEMENTS

The organosulfur compounds present in garlic and compounds formed during the manufacturing of garlic preparations vary depending on the processing conditions (temperature, time, incubation media); hence, the various types of commercially available garlic supplements (see Table 7.1). Processing of garlic leads to a variety of chemical reactions that trigger the formation of a cascade of secondary organo-sulfur compounds that do not exist in raw garlic. These chemical reactions are critically important in determining which organosulfur compounds are present in a specific preparation. The uncertainty as to which of the chemical compounds is physiologically active in the body (and even whether a specific compound is actually present after ingestion) makes it impossible to claim that a specific preparation is

TABLE 7.1
Garlic Products and the Organosulfur Compounds Released

Products	Processing	Potential Active Principle(s)
Fresh garlic	Raw	Cysteines (γ-glutamylcysteines and S-allylcysteines)
		Thiosulfinates (allicin and allicin-derived sulfur compounds)
	Boiled	Mainly diallyl trisulfide, diallyl disulfide, allyl methyl trisulfide
	Fry/microwave	Mainly diallyl disulfide, allyl methyl disulfide, vinyl dithiins
Garlic powder	Spray dried	Varying amounts of cysteines and thiosulfinates
	Freeze dried	Varying amounts of cysteines and thiosulfinates
	Oven dried	Varying amounts of cysteines and thiosulfinates
Garlic oils	Steam-distilled	Mainly diallyl disulfide, diallyl trisulfide, allyl methyl trisulfide
	Macerated in oil	Thiosulfinate transformation compounds (vinyl dithiins, ajoenes, diallyl sulfide)
	Ether extraction	Thiosulfinate transformation compounds (vinyl dithiins, ajoenes, diallyl sulfide)
Aged garlic extract		S-allylcysteine and S-allylmercaptocysteine

responsible for the purported health benefits and presents a significant challenge to medical research.

In the area of garlic research, there are predominantly two schools of thought. One believes allicin and allicin-derived compounds to be responsible for the physiological effects of garlic, whereas the other considers non-allicin-derived compounds (e.g., *S*-allylcysteine) to be the physiologically active compounds. Several products on the market are known to contain a combination of allicin-derived and non-allicin-derived compounds.

7.3.2.1 Allicin-Derived Products

Most commercial products on the market either generate allicin or contain allicin-derived compounds. The potency of supplements that produce allicin is usually expressed as the "allicin-yield" or "allicin-releasing potential." Standardization of these supplements and of homogenized garlic usually involves chromatographic procedures including high-pressure liquid chromatography with atmospheric pressure chemical ionization mass spectrometric identification for the direct determination of alliin and for measuring the allicin release in an aqueous environment (Calvey et al., 1997) (see Table 7.2). The amount of allicin in a tablet/capsule can be used

---

TABLE 7.2
Example of Conditions Used to Run an HPLC Analysis on Various Organosulfur Compounds in Garlic

Detection	Column	Mobile Phase	Detector	Flow Rate
Allicin γ-glutamyl-*S*-allylcysteine	Supleco LC$_{18}$ (Bellefonte, PA) (5 μm, 250 × 4.6 mm id)	Methanol:H$_2$O: formic acid (v/v; 40:60:0.1)	Photodiode array (PDA)	1 ml/min (30 min run time)
Alliin	Spherisorb ODS 11 (3 μm, 100 × 4.0 mm, id)	THF:1,4-dioxane: acetonitrile: 45 mM potassium phosphate buffer (pH 7.0; v/v 2.2:2.9:25:69.9)	UV detector	1.2 ml/min (20 min run time)
S-allylcysteine	Capcelpak-C$_{18}$ (6 μm, 250 × 4.6 mm, id) (Shiseido, Japan)	A: 50 mM ammonium phosphate (pH 7.0) B: acetonitrile: 50 mM ammonium phosphate (pH 7.0; v/v 60:40)	UV detector	1 ml/min (30 min run time)

Source: S. De Grandis and C. Jackson, personal communications. University of Guelph.

as an index for evaluating and comparing the medicinal value of commercial garlic preparations. The phrase "allicin-releasing potential" refers to the amount of allicin generated from alliin when a tablet/capsule is exposed to an aqueous medium. Standardizing garlic products by using their potential for releasing allicin ensures the accuracy of dosage and effectiveness in long-term therapy. Because one molecule of allicin is chemically formed from two molecules of alliin, the numeric value of the allicin concentration is approximately half the numeric value of the specified amount of alliin [e.g., Kwai tablets are standardized to a minimum alliin content of 1.3% and allicin release of 0.6% (Jain et al., 1993)]. The variation in "allicin-releasing potential" among the major brands ranges from 0.1 to 8.9 mg/g powder (Han et al., 1995).

7.3.2.2 Garlic Powders

The most popular type of garlic supplement is in the powder form and is sold as either enteric-coated tablets or non-enteric-coated capsules/tablets. Garlic powder is simply dehydrated, pulverized garlic cloves. If garlic bulbs could be dried whole, their composition and enzyme activity would be similar to those of fresh garlic. However, the cloves are often chopped for efficient drying, which causes an immediate release of alliinase-generated thiosulfinates that are eventually lost through evaporation and chemical reactions. Therefore, it is very important that the slices have an optimum thickness so that a minimum of alliin is converted to allicin by released alliinase. Problems arise when different drying methods are used. Spray-drying homogenized garlic results in a loss of alliinase, while freeze-drying garlic (to avoid possible heat damage) destroys or denatures alliinase. Oven-drying sliced garlic in an airstream at low temperature (<60°C) has little effect on the yield of the main thiosulfinates (allicin and ally methyl thiosulfinate). Drying at higher temperatures, however, often results in an irreversible loss of activity of the enzyme alliinase, thereby preventing allicin generation. Garlic powder as a spice may or may not contain active alliinase, depending upon drying conditions (Lawson, 1993). Drying conditions and maintenance/storage of the final powder (dessication, etc.) need to be carefully controlled with monitoring to ensure freedom from microbiological contamination. Routine quality control measures to ensure minimal/acceptable levels of any trace contaminants are also critical.

7.3.2.3 Garlic Oils

Garlic supplements can also come in the form of oils: the oil of steam-distilled crushed garlic, the oil of crushed garlic macerate in vegetable oil and the oil produced by ether extraction of the thiosulfinates of crushed garlic followed by solvent removal. These oils do not generate allicin, but rather, contain allicin-derived compounds. They are often not standardized for organosulfur compounds and contain varying amounts of allicin-derived compounds such as polysulfides, ajoenes and vinyl dithiins.

The harshest technique is steam distillation, which involves boiling garlic followed by extracting compounds from condensed steam. Steam-distilled garlic oils consist of the diallyl (57%), allyl methyl (37%) and dimethyl (6%) mono- to hexasulfides. These oils are diluted (>99%) with vegetable oils to obtain a practical-sized pill and to stabilize the sulfides (Lawson, 1993).

Oil-macerated garlic products are produced by homogenizing chopped garlic in a vegetable oil, followed by separation from the oil-insoluble components. The thiosulfinates generated during chopping are very soluble in the oil and are subsequently transformed to vinyl dithiins (70%), sulfides (18%) and ajoenes (12%). These are the only commercially available garlic preparations that contain the vinyl-dithiins and ajoenes. Unfortunately, the commercial oil-macerates contain only about 20% of the total organosulfur compounds of the steam-distilled garlic oils or of the thiosulfinates of crushed garlic (Lawson, 1993).

The oil of ether-extracted garlic homogenate has a similar qualitative composition to the oil-macerates, but it has a major advantage in being highly concentrated. It contains up to nine times the amount of the vinyl dithiins and allyl sulfides and four times the amount of the ajoenes found in typical commercial oil-macerates (Lawson, 1993).

7.3.2.4 Non-Allicin-Derived Products

There are some products available that do not produce significant amounts of either allicin or allicin-derived compounds. Aged garlic extracts (AGE) remove the odor-forming compounds in garlic by incubation in 15–20% ethanol for 18–20 months prior to drying of the extract. They are sold in both a dry form and as a liquid containing 10% ethanol. These products do not contain cysteine sulfoxides or allicin-derived transformation products, because these volatile compounds are lost during the long incubation period. AGE does, however, contain γ-glutamylcysteine-derived compounds such as S-allylcysteine (SAC) and small amounts of S-allylmercaptocysteine (SAMC). SAC is a stable, odorless, water-soluble compound that is thought to be the active compound in these products. These commercially aged products claim to be standardized for SAC, but no specific or even minimum amount has ever been declared (Lawson, 1998a).

7.4 METABOLIC FATE OF GARLIC-DERIVED ORGANOSULFUR COMPOUNDS

There exists a large body of scientific evidence demonstrating the physiological effects of a variety of garlic products ranging from fresh and powdered garlic to garlic oil-macerates and aged extracts. Although these garlic preparations are composed of various organosulfur-containing compounds, the specific biologically active components responsible for garlic's physiological effects have not been elucidated. This is primarily because the metabolic fate of allicin, generally regarded to be the most important biologically active compound derived from garlic, is not well understood (Lawson and Wang, 1993; Reuter and Sendl, 1994; Lawson and Block, 1997; Freeman and Kodera, 1997; Lawson, 1998b).

The metabolic fate of garlic-derived organosulfur compounds begins in the mouth. The mechanical disruption of fresh garlic or the adequate hydration of dried garlic powder rapidly initiates the conversion of alliin to allicin via alliinase. However, once these compounds are swallowed and reach the acidic environment of the stomach (pH 1–3), alliinase is irreversibly deactivated, resulting in highly variable levels of allicin. This deactivation also occurs upon consumption of non-enteric-coated garlic capsules. Therefore, to ensure the intake of pharmacologically significant levels of allicin, it is often recommended that garlic capsules be enterically coated, thus bypassing the stomach intact and allowing for optimal transformation of alliin to allicin in the neutral pH of the intestine (Block, 1992).

At present, the role of allicin in garlic research is controversial (Freeman and Kodera, 1997; Lawson and Block, 1997). Although allicin and allicin-derived organosulfur compounds are regarded by most investigators to be responsible for the various physiological effects of garlic, neither allicin nor its common transformation products (diallyl sulfides, vinyl dithiins, ajoenes) can be detected in the blood, urine or stool upon consumption of large amounts of fresh garlic (up to 25 g) or pure allicin (60 mg) (Freeman and Kodera, 1997; Lawson, 1998b).

Another potential metabolic pathway for allicin and its transformation compounds is through its reaction with the amino acid cysteine in the intestine to form S-allylmercaptocysteine, which is then converted to allylmercaptan. This appears to be the primary metabolic route taken by allicin in isolated fresh human blood. However, it remains unknown whether allylmercaptan is the final end metabolite of allicin, because it has not been detected in measurable concentrations in blood, urine or stool after post-oral consumption of 150 mg (Lawson, 1998a). *In vitro* studies have shown physiological activity associated with allylmercaptan and S-allylmercaptocysteine, as well as their potential to be further metabolized to other organosulfur-containing compounds such as allyl sulfinic acid (2-propenesulfinic acid) and allyl sulfonic acid (2-propenesulfonic acid) (Lee et al., 1994; Gebhardt and Beck, 1996; Pinto et al., 1997; Sigounas et al., 1997a,b).

7.5 HEALTH BENEFITS

7.5.1 CHOLESTEROL-LOWERING PROPERTIES

Since 1900, approximately 170 clinical studies have been carried out on the potential hypocholesterolemic effect of garlic in humans (Reuter and Sendl, 1994; Han et al., 1995; Adler and Holub, 1997; Lawson, 1998c). On average, these trials have reported a lowering of total cholesterol (TC) by 10–12% (Kleijen et al., 1989; Warshafsky et al., 1993; Silagy and Neil, 1994a; Agarwal, 1996; Reuter et al., 1996). These results have been reviewed and supported by two meta-analyses (Warshafsky et al., 1993; Silagy and Neil, 1994a). One German study in particular demonstrated that a garlic powder supplement (900 mg/day) was as effective as the cholesterol-lowering drug, bezafibrate (Holzgartner et al., 1992).

Although it remains unknown as to which compounds in garlic are responsible for the hypocholesterolemic effect, there exists considerable animal and *in vitro* evidence to suggest that allicin or allicin-generated organosulfur compounds are the physiologically active compounds (Gebhardt et al., 1994; Yeh and Yeh, 1994; Gebhardt and Beck, 1996). Tissue culture studies have shown allyl disulfide and allylmercaptan to inhibit HMG-CoA reductase, the rate-limiting hepatic enzyme for the biosynthesis of cholesterol in the body. The mechanism of this inhibition is thought to involve the formation of sulfide bridges within the HMG-CoA reductase enzyme (the target of pharmaceutically active statins) by interaction with the allicin-derived compounds, rendering the enzyme inactive (Gebhardt et al., 1994). It should be noted that the vast majority of the clinical studies conducted on humans have utilized garlic preparations that contain either "allicin-releasing potential" or allicin-derived organosulfur compounds as their main ingredient.

Despite prior evidence of garlic's beneficial effects, new studies within the last 5 years have shown no statistically significant hypocholesterolemic effects in humans with garlic supplementation (Simons et al., 1995; Berthold et al., 1998; Isaacsohn et al., 1998; McCrindle et al., 1998;). A meta-analysis carried out in the United Kingdom (Neil et al., 1996) found no significant cholesterol-lowering in human subjects following garlic consumption. This finding has been recently confirmed by another meta-analysis of randomized clinical studies stating that the hypocholesterolemic effect of garlic is at best modest, and the robustness of the effect is debatable, therefore, its use for hypercholesterolemia is of questionable value (Stevinson et al., 2000).

Various explanations have been put forth regarding the lack of effect observed in the most recent garlic trials. Lawson (1998b) has suggested that recent negative studies employing enteric-coated, allicin-releasing garlic supplements are most likely due to inefficient and inconsistent *in vivo* production of allicin. Allicin-releasing potential is normally determined under conditions simulating those of the human intestine (where the enteric-coated garlic supplements are thought to disintegrate and generate allicin). These conditions may vary among individuals; thus, response to garlic supplements may be greater or less than expected. This could also lead to varying amounts of allicin and allicin-derived compounds being produced between individuals, resulting in varying therapeutic (cholesterol-lowering) effects, eventually affecting the statistical significance. Variability in both tablet quality (change in tablet coating may result in decreased time taken to disintegrate coating layer, consequently leading to a decreased amount of allicin available) and meal content (which can influence gastric emptying rate, pH), could very well account for differences observed between the positive and negative studies. Another possibility for the lack of effect may be related to the amount of fat in the diet. Several recent negative studies employed a diet restriction regime (NCEP step I/II diets) in contrast to the prior studies that had no dietary fat restrictions.

To date, no garlic supplementation trial has been conducted exclusively using women as subjects. Hormonal changes during a woman's menstrual cycle have been shown to elevate cholesterol levels 7–14 days after the start of menstruation. Therefore, fasting blood samples must be taken 3 to 7 days after the start of menstruation in order to ensure accurate blood lipid readings. Previous studies that included women as subjects made no mention of controlling for menstrual variation.

Questions regarding the apparent ineffectiveness of a particular brand-name product for cholesterol lowering in recent trials in contrast to earlier human studies have been raised and considered to possibly arise from compositional differences in newer vs. older versions of the product.

7.5.2 PLAQUE REGRESSION PROPERTIES

Atherosclerosis is a disease of the large- and intermediate-sized arteries in which fatty lesions called atheroma develop on the inside surfaces of the arterial wall. With time, the surrounding fibrous and smooth muscle tissues proliferate to form plaques and obstructions that can greatly reduce blood flow, sometimes to complete vessel occlusion. The fibroblasts of the plaque can eventually deposit such extensive amounts of dense connective tissue that arteries become stiff and unyielding. Still later, calcium salts often precipitate with cholesterol and other lipids of the plaque, leading to calcifications that make arteries rigid tubes. Both of these latter stages of the disease are called "hardening of the arteries."

Screening for the progression rate of atherosclerosis can be performed using intima-medial thickness (IMT), plaque volume measurements, or both. These diagnostics produce important information concerning arterial vessel wall morphology. IMT measurements correlate highly with other atherosclerotic risk factors such as hypertension and hypercholesterolemia. It has been demonstrated in men that an IMT of more than 1 mm was associated with a 2.1-fold increase in risk, and the presence of small or large plaques was associated with a 4.1-fold increase in risk of acute myocardial infarction as compared with men devoid of either of these manifestations (Salonen and Salonen, 1993). Plaque in the common carotid artery appears to be related to a 2.1-fold increased risk of myocardial infarction, and plaque found in the femoral arteries is associated with a 2.4-fold increased risk for ischemic coronary disease (Koscielny et al., 1998).

Recently, it has been hypothesized that garlic may prevent and possibly play a curative role in arteriosclerosis therapy by preventing plaque formation, increasing plaque regression, or both. A recent animal study demonstrated that treatment with allicin reduced the size of fatty streaks (atherosclerotic lesions) in mice, despite a lack of effect on lipid levels (Abramovitz et al., 1999). Another study demonstrated that 300 mg of garlic powder taken for more than 2 years protected the elastic properties of the human aorta, thus attenuating age-related increases in aortic stiffness (Breithaupt-Grögler et al., 1997). A recent human trial that provided continuous garlic powder supplementation of 900 mg/d for 4 years demonstrated a reduction in atherosclerotic plaque volume of 5–18%. An age-dependent increase in plaque volume between 50–80 years of age was also diminished by 6–13% under garlic treatment over a 4-year period (Koscielny et al., 1998).

In the face of these latest investigations, it appears that the beneficial effects of garlic on CVD may be attributable to its ability to delay or reduce the loss of elasticity in arterial vessels and the rise in atherosclerotic plaques, rather than its previously reported effects on blood lipid reduction.

7.5.3 ANTI-THROMBOTIC ACTIVITY

The relevance of platelet aggregation in the pathogenesis of atherosclerosis is well documented (Koscielny et al., 1998). The ability of garlic and its metabolites to prevent blood coagulation and platelet aggregation has been well established in numerous *in vitro* and *in vivo* (animals and humans) studies (Lawson et al., 1992; Han et al., 1995).

A number of garlic preparations have been used to study the effect of garlic on platelet aggregation (Lawson et al., 1992). It is thought that the effects of these commercial preparations are most likely attributable to allicin, ajoenes, vinyl dithiins and diallyl trisulfide, with allicin and diallyl trisulfide being the most active compounds for inhibiting platelet aggregation *in vitro*.

The exact mechanisms for the inhibition of platelet aggregation by garlic-derived compounds remain unknown. Adenosine, present in garlic cloves, is a strong competitive inhibitor of platelet clumping induced by adenosine diphosphate (ADP), thrombin and adrenaline, thus inhibiting platelet aggregation. However, this inhibition is seen only in platelet-rich plasma and is absent in whole blood (Han et al., 1995). These effects, therefore, may be of questionable physiological relevance. The most likely explanation is a reduction in eicosanoid formation, namely, the production of thromboxane (TXB_2) from arachidonic acid (Makheja and Bailey, 1990; Apitz-Castro et al., 1992; Lawson et al., 1992; Srivastava and Tyagi, 1993; Ali and Thomson, 1995; Batirel et al., 1996; Ali et al., 1999).

Garlic may prove effective in inhibiting plaque progression due to its ability to reduce platelet aggregation. The ability of garlic to reduce blood-clotting time is significant enough to recommend that patients discontinue taking any type of garlic at least 10 days prior to surgery (German et al., 1995).

7.5.4 BLOOD PRESSURE-LOWERING PROPERTIES

The effect of garlic preparations on blood pressure has been studied since 1921. A meta-analysis of controlled human trials using garlic preparations has concluded that garlic significantly lowers both the systolic and diastolic blood pressures although the reductions are generally greater for systolic pressure (decrease of 6.9% vs. 6.3%) (Silagy and Neil, 1994b). At present, evidence demonstrating the active compound(s) in garlic preparations responsible for lowering blood pressure is limited; however, several mechanisms of action have been proposed and are summarized in Table 7.3.

TABLE 7.3
Proposed Mechanisms of Action for Lowering Blood Pressure

Compound(s)	Proposed Mechanism
γ-Glutamylcysteines	These are thought to inhibit angiotensin I-converting enzyme in the renin-angiotensin-vasoconstriction system (blocks the formation of angiotensin II which is an extremely powerful vasoconstrictor of arterioles and to a lesser extent of veins), thus decreasing arterial blood pressure.
Aqueous garlic extract (AGE; e.g., Ajoene)	This is thought to increase the activity of nitric oxide synthase, an enzyme that produces nitric oxide, which is associated with a lowering of blood pressure; could have a direct relaxant effect on smooth muscle.
	AGE may act as a phytopharmacological K^+ channel opener, causing membrane hyperpolarization (decrease in intracellular Ca^{2+}), thus increasing vasodilation of vascular smooth muscle.
	AGE may have the ability to mimic prostaglandin E2 (vasodilator), which could ultimately decrease peripheral vascular resistance.
Fructans	These inhibit adenosine deaminase in isolated cells, which increase levels of adenosine, a compound that acts at purinergic receptors to relax and dilate blood vessel smooth muscle.

Recently, Kaye et al. (2000) determined that allicin had significant vasodilator activity in the pulmonary vascular bed of the rat, whereas allylmercaptan and diallyl disulfide produced no significant changes in pulmonary arterial perfusion pressure. Other indications for possible active compounds come from *in vitro* studies that show that γ-glutamylcysteines exert some effect on angiotensin (Sendl et al., 1992) and aqueous garlic extracts (ajoene and fructans) that may directly affect smooth muscle (Siegal et al., 1991; Das et al., 1995a,b; Dirsch et al., 1998; Pedraza-Chaverri et al., 1998; Siegal et al., 1999).

These findings are of potential clinical significance, because if the blood pressure-lowering effects found in these short periods of treatment can be sustained over a longer period of time, the incidence of stroke may be reduced by 30–40% and coronary heart disease by 20–25% (Collins et al., 1990).

As with the blood cholesterol trials, not all garlic studies have shown a blood pressure reducing effect. There is insufficient evidence at present to recommend garlic as an effective antihypertensive agent for routine clinical use. However, there is equally no evidence to suggest that it is harmful, and the currently available data strongly support the benefits of garlic as a mild hypertensive (Silagy and Neil, 1994b).

7.5.5 ANTIOXIDANT EFFECTS

Free radical damage is considered to be a causative factor in the development of cancer and inflammatory and chronic diseases. Therefore, free radical scavenging molecules (antioxidants) may play a beneficial role in these conditions. With repect to CVD, the oxidation of low-density lipoprotein (LDL) is believed to be a critical process in the development of atherosclerosis (Berliner et al., 1995; Navab et al., 1995). The presence of oxidized LDL in the intima of an artery leads to the production of macrophage-derived foam cells, the main cell type present in fatty streaks that are believed to be the earliest lesion of atherosclerosis (Fuster, 1994). Therefore, the use of antioxidants as dietary supplements to protect against LDL oxidation may reduce both the development and progression of atherosclerosis (Gey, 1995).

Garlic has been shown to prevent the oxidation of LDL to smaller, more dense atherogenic LDL species (i.e., LDL_3) in both animal and human subjects (Ide et al., 1996; Orekhov et al., 1996). Aged garlic extract has been shown to inhibit lipid peroxidation in rat liver microsomes (Horie et al., 1992), bovine pulmonary artery endothelial cells and murine macrophages (J774 cells) (Ide and Lau, 1999) in a dose-dependent manner. It has also been demonstrated that human subjects supplemented with AGE, but not raw garlic, have LDL particles that are significantly more resistant to oxidation (Munday et al., 1998). Two human clinical trials using garlic powder supplementation (300 or 600 mg/d) resulted in less susceptibility to copper-induced oxidation of LDL (Phelps and Harris, 1993). There exists, however, two recent studies using garlic powder tablets that do not support the hypothesis that dietary garlic supplementation decreases the susceptibility of LDL to oxidation (Simons et al., 1995; Byrne et al., 1999).

The exact compounds in garlic responsible for its antioxidant effects are undetermined. Some believe allicin to be the primary antioxidant compound in garlic and garlic powder supplements (Rekka and Kourounakis, 1994; Prasad et al., 1995; Rabinkov et al., 1998). Allylmercaptan, the most abundant allicin-derived metabolite found in the blood, also appears to have an antioxidant effect *in vivo* (Lawson and Wang, 1993). On the other hand, AGE does not contain allicin but is high in *S*-allylcysteine and can convert to *S*-allylmercaptocysteine, both of which have shown radical scavenging activities (Imai et al., 1994). Further studies are needed to determine the compounds primarily responsible for garlic's antioxidant effects.

Despite these shortcomings, garlic remains a promising antioxidant because of its powerful free radical scavenging ability, as well as its potential ability to decrease lipid peroxidation and increase glutathione levels (the most important scavenger of high-energy electrons inside the cell) (Ide and Lau, 1999; Balasenthil et al., 2000; Shobana and Naidu, 2000).

7.5.6 ANTI-CANCER PROPERTIES

Several epidemiological studies exist that document garlic's potential as an anticarcinogenic agent. Studies comparing regional populations in China have reported that the incidence of gastric cancer is significantly lower in regions where garlic consumption is high as compared with regions where garlic consumption is low (Haixiu et al., 1989; You et al., 1989). These findings are further supported by studies investigating stomach cancer in Italy, thyroid nodularity in Chinese women, and the incidence of colorectal cancer in Japanese-Hawaiians (Agarwal, 1996). In the United States, the "Iowa Women's Health Study" (41,837 women) (Steinmetz et al., 1994), determined an inverse relationship between colon cancer risk and garlic consumption, with an age- and energy-adjusted relative risk of 0.68. Most recently, a metaanalysis of 18 colorectal and stomach cancer investigations also reported a randomeffects relative risk estimate with garlic consumption (excluding supplements) of 0.69 and 0.53, respectively (Fleischauer et al., 2000).

Although several epidemiological studies have proposed an anticancer role for garlic, three epidemiological studies from the Netherlands have found no association between garlic supplement consumption and the prevention of breast, colon and lung cancers (Dorant et al., 1994, 1995, 1996).

Numerous investigations, *in vitro* and *in vivo*, have examined the antitumor effects of various constituents of garlic (Liu and Milner, 1990; Brady et al., 1991; Han et al., 1995; Milner, 1996; Schaffer et al., 1996; Hu and Singh, 1997; Pinto et al., 1997; Riggs et al., 1997; Sakamoto et al., 1997; Srivastava et al., 1997; Lawson, 1998b; Cohen et al., 1999; Lamm and Riggs, 2000). A recent study employing garlic powder and garlic extract preparations conducted on human tumor cell lines has found a clear dose-dependent inhibition of human hepatoma, colorectal carcinoma, and lymphatic leukemia cell lines (Siegers et al., 1999). The authors have suggested that the antiproliferative effects of garlic were due to allicin and the polysulfides. The allyl-containing organopolysulfides, such as diallyl di- and trisulfide and allyl methyl di- and trisulfide, have been shown to markedly suppress p34 (cdc2) kinase activity and induce a G2/M phase arrest in cultured human colon tumor (HCT-15) cells, and to significantly reduce the incidence of hepatic hyperplastic nodules, adenoma of the lung and thyroid, and hyperplasia of the urinary bladder in mice (Agarwal, 1996;

Knowles and Milner, 2000). Furthermore, these compounds are thought to be involved in the inhibition of certain cytochrome P-450 enzyme-dependent bioactivations of procarcinogens and protoxicants (Brady et al., 1991), as well as to increase levels of glutathione-S-transferase (GST), an enzyme of particular importance in the detoxification of xenobiotics in the body (Wilce and Parker, 1994). Most recently, the ability of various plants and plant extracts to influence apoptosis, or programmed cell death, in cancerous cells in an attempt to arrest their proliferation, has been the topic of much research. Allicin has been shown to induce apoptosis in a variety of cell lines, including human hepatocellular carcinoma cells (KIM-1) and human lymphoid leukemia (MOLT-4B) cells (Thatte et al., 2000).

SAC and SAMC have also been shown to have anticarcinogenic properties (Dorant et al., 1993; Koch, 1996; Heber, 1997; Cheng and Tung, 1981; Balasenthil and Nagini, 2000). Studies using SAC-rich aged garlic extract fed to rats at 2–4% of the diet, decreased carcinogen-induced tumors and DNA adducts (Liu and Milner, 1990). Recently, SAC was shown to inhibit proliferation of several human and murine melanoma cell lines in a dose-dependent manner (Agarwal, 1996). Likewise, cell proliferation of established erythroleukemia, breast and prostate cancer cell lines was inhibited by SAMC with the cell cycle being arrested in the G2/M phase (Sigounas et al., 1997a). SAMC has also been observed to induce apoptosis in a number of neoplasias, including breast and prostate cancer, by inhibiting the activation of interleukin-1β converting enzyme (Brady et al., 1991; Patel et al., 1996; Pinto et al., 1997; Srivastava et al., 1997).

7.5.7 Antimicrobial Effects

Traditionally, garlic has been most widely used for its reputed antibacterial and antifungal activities. Allicin is thought to be the main antibacterial agent of garlic, while the thiosulfinates (approximately 75% allicin) are widely held to be responsible for garlic's antimicrobial effects. Pure allicin was found to exhibit antibacterial activity against a wide range of gram-negative and gram-positive bacteria, including multidrug-resistant enterotoxicogenic strains of *Escherichia coli* (Ankri and Mirelman, 1999). The antibiotic activity of 1 mg of allicin has been equated to that of 15 IU of penicillin (Han et al., 1995).

Allicin has also demonstrated antifungal activity, particularly against *Candida albicans* and antiparasitic activity, including some major human intestinal protozoan parasites such as *Entamoeba histolytica* and *Giardia lamblia* (Ankri and Mirelman, 1999). It is well established that a main mechanism by which allicin disables dysentery-causing amebas is by its rapid reaction with the (SH)-group of cysteines in enzymes containing cysteine at their active sites, in particular, the cysteine proteinases and alcohol dehydrogenases. Cysteine protease enzymes are among the main culprits of infection, providing microorganisms with means to damage and invade tissues. Alcohol dehydrogenase enzymes play a major role in these harmful organisms' metabolism and survival. Because these groups of enzymes are found in a wide variety of infectious organisms such as bacteria, fungi and viruses, it is believed that allicin is a broad-spectrum antimicrobial, capable of warding off different types of infection (Ankri and Mirelman, 1999; Josling, 1999).

Ajoene, another garlic derivative, has exhibited antimicrobial activities (Yoshida et al., 1999) and, when used with chloroquine, has been shown to increase the antimalarial activity of chloroquine against the parasite *P. berghei* (Perez et al., 1994).

Recent studies have also demonstrated an inhibitory effect of garlic and AGE on *Helicobacter pylori*, a bacterium that causes gastric ulcers and is also implicated in the etiology of stomach cancer (Cellini et al., 1996; Sivam et al., 1997). Most recently, O'Gara et al. (2000) demonstrated a dose-dependent anti-*H. pylori* effect from garlic oil, garlic powder, allicin and diallyl disulfide *in vitro*. Other investigations conducted *in vitro* have been useful for infections of the skin and/or intestinal tract and are widely thought to be strong estimates of *in vivo* effects. Several studies in animals and people have demonstrated that orally consumed garlic and allicin-related compounds also have systemic antimicrobial effects in the lungs, kidney, blood, brain and cerebrospinal fluid (Lawson, 1998a).

The role of garlic in warding off infection may be particularly valuable in light of the growing resistance to antibiotics. It is unlikely that bacteria would develop resistance to allicin, because this would require modifying the very enzymes that make their activity possible (Josling, 1999).

7.6 POTENTIAL HEALTH CONCERNS

Garlic is widely used as a food and a spice and is "generally regarded as safe" (GRAS) by the U.S. FDA. Fresh, cooked garlic and garlic supplements are usually well tolerated when consumed at a reasonable level in or with meals. However, when consumed in the absence of food, or in high amounts, garlic will aggravate the throat and stomach (Sumiyoshi, 1997) and cause flatulence in some individuals. Odoriferous garlic breath and perspiration, which are the most common complaints by individuals, are caused mainly by the allyl sulfide compounds that are expelled from the blood via the lungs or by sweat (Block, 1992). Garlic may cause heartburn (an acid reflux effect) that can be attributed to a transient relaxation of the lower esophageal sphincter (Koch, 1996). A few individuals may experience mild allergic-type reactions, primarily in the form of skin irritation (e.g., contact dermatitis), while others may develop pemphigus, a group of chronic, relapsing skin diseases, if exposed to fresh garlic or diallyl disulfides (Brenner and Wolf, 1994; Asero et al., 1998; Jappe et al., 1999). Rhinoconjunctivitis and asthma have also been observed after inhalation or ingestion of garlic dust. One case has been reported in which an anaphylactic reaction occurred after ingestion of young garlic (Pérez-Pimiento et al., 1999). Postoperative bleeding and spontaneous epidural hematoma have been reported when garlic was used concurrently with the drug warfarin (Fugh-Berman, 2000). Therefore, individuals with clotting disorders, patients awaiting surgery, or those on anticoagulant therapy (e.g., warfarin), should be informed of garlic's anticoagulant activity (Evans, 2000; Heck et al., 2000).

Controversy exists about possible adverse effects of allicin. According to several *in vitro* investigations, synthetic allicin (0.2 mg/ml) is considered toxic because garlic oxidizes blood hemoglobin (Freeman and Kodera, 1997). The LD_{50} of allicin in mice is 120 mg/kg subcutaneously and 60 mg/kg intravenously, translating into 1200 g fresh garlic for a 70-kg person (Merck Index) (Block, 1992). However, these dosages are several hundredfold above any normal dose that would be generally ingested by humans. Moreover, garlic produces, on average, only 2–6 mg allicin/g which, at normal garlic consumption, is well below the toxic dose. Therefore, based on numerous human clinical trials, fresh garlic ingestion up to 10 g/day (one to two cloves), or 4–6 g garlic powder/day, is considered safe when taken with a meal (Koch, 1996).

One other potential health concern regarding the consumption of garlic arises when improper home canning and improper preparation and storage of garlic-in-oil mixtures takes place. Garlic is a low-acid vegetable with a typical clove pH ranging from 5.3 to 6.3. As with all low-acid vegetables, garlic can support the growth of the bacterium *Clostridium botulinum*. Environments of excessive moisture, high room temperature, lack of oxygen, and low-acid conditions all favor the growth of this bacterium and the subsequent production of an extremely potent toxin that causes botulism. Therefore, macerated garlic in oil, if not acidified, should be stored refrigerated for a limited time period (3 weeks maximum).

7.7 CONCLUSION

Several issues remain unresolved as to the physiological benefits of garlic and its derivatives due to recent conflicting reports in the biomedical literature. Results from studies using different garlic preparations are often grouped together and are not separated based on the type of preparation used. Consequently, the reported effects of garlic are often unpredictable and inconsistent. Therefore, all studies should be examined in relation to their active components and the specific garlic preparation employed.

In addition to the problem of pooled results, other factors, such as dose and duration, composition of the diet, variability in the levels of different organosulfur compounds in garlic supplements, and *in vivo* variability between individuals to generate the physiologically active garlic metabolites, may also contribute to the observed inconsistencies. Further research is needed to address these discrepancies and to produce consistent physiological effects for a given garlic preparation. With regard to garlic's influence on cardiovascular health and its anticancer, antioxidant and antimicrobial properties, the development of more reliable and defined garlic products, as well as a standard preparation protocol, will help to ensure more dependable and reproducible clinical results. Only then can it be ascertained whether garlic has a role in the future treatment of these ailments. Appropriate double-blind placebo-controlled clinical trials using specific doses/durations of brand-name products with defined compositions are needed before establishing whether nutraceuticals and functional foods containing garlic-derived material can be expected to qualify for evidence-based health claims.

7.8 ACKNOWLEDGMENTS

The authors acknowledge support from the Ontario Ministry of Agriculture, Food, and Rural Affairs (OMAFRA) and the Heart and Stroke Foundation of Ontario. We are also grateful to Patricia Swidinsky and Jeanette Arnott for their review of this document and helpful suggestions offered throughout its development.

REFERENCES

Abramovitz, D., Gavri, S., Harats, D., Levkovitz, H., Mirelman, D., Miron, T., Eilat-adar, S., Rabinkov, A., Wilchek, M., Eldar, M. and Vered, Z. (1999) Allicin-induced decrease in formation of fatty streaks (atherosclerosis) in mice fed a cholesterol-rich diet. *Coronary Artery Dis.* 10(7):515–519.

Adler, A.J. and Holub, B.J. (1997) Effect of garlic and fish-oil supplementation on serum lipid and lipoprotein concentrations in hypercholesterolemic men. *Am. J. Clin. Nutr.* 65:445–450.

Agarwal, K.C. (1996) Therapeutic actions of garlic constituents. *Med. Res. Rev.* 16:111–124.

Ali, M. and Thomson, M. (1995) Consumption of a garlic clove a day could be beneficial in preventing thrombosis. *Prostaglandins Leukot. Essent. Fatty Acids.* 53:211–212.

Ali, M., Bordia, T. and Mustafa, T. (1999) Effect of raw versus boiled aqueous extract of garlic and onion on platelet aggregation. *Prostaglandins Leukot. Essent. Fatty Acids.* 60(1):43–47.

Ankri, S. and Mirelman, D. (1999) Antimicrobial properties of allicin from garlic. *Microbes Infect.* 1(2):125–129.

Apitz-Castro, R., Badimon, J.J. and Badimon, L. (1992) Effect of ajoene, the major antiplatelet compound from garlic, on platelet thrombus formation. *Thromb. Res.* 68:145–155.

Asero, R., Mistrello, G., Roncarolo, D., Antoniotti, P.L. and Falagiani, P. (1998) A case of garlic allergy. *J. Allergy Clin. Immunol.* 101:427–428.

Balasenthil, S. and Nagini, S. (2000) Inhibition of 7,12-dimethylbenz[a]anthracene-induced hamster buccal pouch carcinogenesis by S-allylcysteine. *Oral Oncol.* 36(4):382–386.

Balasenthil, S., Arivazhagan, S. and Nagini, S. (2000) Garlic enhances circularory antioxidants during 7,12-dimethylbenz[a]anthracene-induced hamster buccal pouch carcinogenesis. *J. Ethnopharmacol.* 72(3):429–433.

Batirel, H.F., Aktan, S., Aykut, C., Yegen, B.C. and Coskun, T. (1996) The effect of aqueous garlic extract on the levels of arachidonic acid metabolites (leukotriene C4 and prostaglandin E2) in rat forebrain after ischemia-reperfusion injury. *Prostaglandins Leukot. Essent. Fatty Acids.* 54:289–292.

Berliner, J.A., Navab, M., Fogelman, A.M., Frank, J.S., Demer, L.L., Edwards, P.A., Watson, A.D. and Lusis, A.J. (1995) Atherosclerosis: basic mechanisms. Oxidation, inflammation, and genetics. *Circulation.* 91:2488.

Berthold, H.K., Sudhop, T. and von Bergmann, K. (1998) Effect of a garlic oil preparation on serum lipoproteins and cholesterol metabolism: a randomized controlled trial. *JAMA.* 279:1900–1902.

Block, E. (1985) The chemistry of garlic and onions. *Sci. Am.* 252:114–119.

Block, E. (1992) The organosulfur chemistry of the genus *Allium*—implications for the organic chemistry of sulfur. *Angew. Chem. Int. Ed. Engl.* 31:1135–1178.

Brady, J.F., Ishizaki, H., Fukuto, J.M., Lin, M.C., Fadel, A., Gapac, J.M. and Yang, C.S. (1991) Inhibition of cytochrome P-450 2E1 by diallyl sulfide and its metabolites. *Chem. Res. Toxicol.* 4:642–647.

Breithaupt-Grögler, K., Ling, M., Boudoulas, H. and Belz, G.G. (1997) Protective effect of chronic garlic intake on elastic properties of aorta in the elderly. *Circulation.* 96:2649–2655.

Brenner, S. and Wolf, R. (1994) Possible nutritional factors in induced pemphigus. *Dermatology.* 189:337–339.

Brown, D. (1996) Garlic. *Herbal Prescriptions for a Better Health.* Prima Publishing, Rocklin, CA, pp. 97–109.

Byrne, A.J., Neil, H.A.W. and Vallance, D.T. (1999) A pilot study of garlic consumption shows no significant effect on markers of oxidation or sub-fraction composition of low-density lipoprotein including lipoprotein(a) after allowance for non-compliance and the placebo effect. *Clinica. Chimica. Acta.* 285:21–33.

Calvey, E.M., Matusik, J.E., White, K.D., DeOrazio, R., Sha, D. and Block, E. (1997) Allium chemistry: supercritical fluid extraction and LC-APCI-MS of thiosulfinates and related compounds from homogenates of garlic, onion, and ramp. Identification in garlic and ramp and synthesis of 1-propanesulfinothioic acid *S*-allyl ester. *J. Agric. Food Chem.* 45(11):4405–4413.

Cellini, L., Di Campli, E., Masulli, M., Di Bartolomeo, S. and Allocati, N. (1996) Inhibition of *Helicobacter pylori* by garlic extract (*Allium sativum*). *FEMS Immunol. Med. Microbiol.* 13:273–277.

Cheng, H.H. and Tung, T.C. (1981) Effect of allithiamine on sarcoma-180 tumor growth in mice. *Chem. Abst.* 95:48.

Cohen, L.A., Ahao, A., Pittman, B. and Lubet, R. (1999) S-Allylcysteine, a garlic constituent fails to inhibit *N*-methylnitrosourea-induced rat mammory tumorigenesis. *Nutr. Cancer.* 35(1):58–63.

Collins, R., Peto, R., MacMahon, S., Hebert, P., Fiebach, N.H., Eberlein, K.A., Godwin, J., Qizilbash, N., Talyor, J.O. and Henneken, C.H. (1990) Blood pressure, stroke, and coronary heart disease. Part 2, Short-term reductions in blood pressure: overview of randomised drug trials in their epidemiological context. *Lancet.* 335(8693):827–838.

Das, I., Khan, N.S. and Sooranna, S.R. (1995a) Nitric oxide synthase activation is a unique mechanism of garlic action. *Biochem. Soc. Trans.* 23:136S.

Das, I., Khan, N.S. and Sooranna, S.R. (1995b) Potent activation of nitric oxide synthase by garlic: a basis for its therapeutic applications. *Curr. Med. Res. Opin.* 13:257–263.

Dirsch, V.M., Kiemer, A.K., Wager, H. and Vollmar, A.M. (1998) Effect of allicin and ajoene, two compounds of garlic, on inducible nitric oxide synthase. *Atherosclerosis.* 139(2):333–339.

Dorant, E., van den Brandt, P.A., Goldbohm, R.A., Hermus, R.J. and Sturmans, F. (1993) Garlic and its significance for the prevention of cancer in humans: a critical view. *Br. J. Cancer.* 67:424–429.

Dorant, E., van den Brandt, P.A. and Goldbohm, R.A. (1994) A prospective cohort study on *Allium* vegetable consumption, garlic supplement use, and the risk of lung carcinoma in The Netherlands. *Cancer Res.* 54:6148–6153.

Dorant, E., van den Brandt, P.A. and Goldbohm, R.A. (1995) *Allium* vegetable consumption, garlic supplement intake, and female breast carcinoma incidence. *Breast Cancer Res. Treat.* 33:163–170.

Dorant, E., van den Brandt, P.A. and Goldbohm, R.A. (1996) A prospective cohort study on the relationship between onion and leek consumption, garlic supplement use and the risk of colorectal carcinoma in The Netherlands. *Carcinogenesis.* 17:477–484.

Ellmore, G.S. and Feldberg, R.S. (1994) Alliin lyase localization in the bundle sheaths of the garlic clove (*Allium sativum*). *Am. J. Bot.* 81:89–94.

Evans, V. (2000) Herbs and the brain: fried or foe? The effects of ginko and garlic on warfarin use. *J. Neurosci. Nurs.* 32(4):229–232.

Fenwick, G.R. and Hanley, A.B. (1985) Genus *Allium*—Parts 1, 2 and 3. *C.R.C. Crit. Rev. Food Sci. Nutr.* 22:199–271, 273–377.

Fleischauer, A.T., Poole, C. and Arab, L. (2000) Garlic consumption and cancer prevention: meta-analyses of colorectal and stomach cancers. *Am. J. Clin. Nutr.* 72:1047–1052.

Food and Agricultural Organization (2000) FAOSTAT Database Results. www.fao.org

Freeman, F. and Kodera, Y. (1997) Rebuttal on garlic chemistry: Stability of *S*-(2-propenyl) 2-propene-1-sulfinothioate (allicin) in blood, solvents, and simulated physiological fluids. *J. Agric. Food Chem.* 45:3709–3710.

Fugh-Berman, A. (2000) Herb-drug interactions. *Lancet.* 355:134–138.

Fuster, V. (1994) Mechanisms leading to myocardial infarction: insights from studies of vascular biology. *Circulation.* 90:2126.

Gebhardt, R. and Beck, H. (1996) Differential inhibitory effects of garlic-derived organosulfur compounds on cholesterol biosynthesis in primary rat hepatocyte cultures. *Lipids.* 31:1269–1276.

Gebhardt, R., Beck, H. and Wagner, K.G. (1994) Inhibition of cholesterol biosynthesis by allicin and ajoene in rat hepatocytes and HepG2 cells. *Biochim. Biophys. Acta* 1213:57–62.

German, K., Kumar, U. and Blackford, H.N. (1995) Garlic and the risk of TURP bleeding. *Br. J. Urol.* 76:518.

Gey, K.F. (1995) Ten-year retrospective on the antioxidant hypothesis of atherosclerosis: Threshold plasma levels of antioxidant micronutrients related to minimum cardiovascular risk. *J. Nutr. Biochem.* 6:206.

Hahn, G. (1996) Botanical characterization and cultivation of garlic. Garlic: the science and therapeutic application of *Allium sativum* L. and related species (Koch, H.P. & Lawson, L.D., eds.) Williams and Wilkins, Baltimore, MD, pp. 25–36.

Haixiu, X., Rui, S., Hongzhou, N. and Aiguo, C. (1989) A five-year prospective study on the relationship between garlic and gastric cancer. *Chinese J. Cancer Res.* 1:60–62.

Han, J., Lawson, L., Han, G. and Han, P. (1995) A spectrophotometric method for quantitative determination of allicin and total garlic thiosulfinates. *Anal. Biochem.* 225:157–160.

Heber, D. (1997) The stinking rose: organosulfur compounds and cancer [editorial; comment]. *Am. J. Clin. Nutr.* 66:425–426.

Heck, A.M., Dewitt, B.A. and Lukes, A.L. (2000) Potential interactions between alternative therapies and warfarin. *Am. J. Health Syst. Pharm.* 57(13):1221–1227.

Holzgartner, H., Schmidt, U. and Kuhn, U. (1992) Comparison of the efficacy and tolerance of a garlic preparation vs. bezafibrate. *Arzneimittelforschung.* 42:1473–1477.

Horie, T., Awazu, S., Itakura, Y. and Fuwa, T. (1992) Identified diallyl polysulfides from an aged garlic extract which protects the membranes from lipid peroxidation. *Planta Med.* 58:468–469.

Hu, X. and Singh, S.V. (1997) Glutathione *S*-transferases of female A/J mouse lung and their induction by anticarcinogenic organosulfides from garlic. *Arch. Biochem. Biophys.* 340:279–286.

Ide, N. and Lau, B.H. (1999) Aged garlic extract attenuates intracellular oxidative stress. *Phytomedicine.* 6(2):125–131.

Ide, N., Nelson, A.B. and Lau, B.H. (1996) Aged garlic extract and its consituents inhibit copper-induced oxidative modification of low density lipoprotein. *Planta Med.* 63:263.

Imai, J., Ide, N., Nagae, S., Moriguchi, T., Matsuura, H. and Itakura, Y. (1994) Antioxidant and radical scavenging effects of aged garlic extract and its constituents. *Planta Med.* 60:417–420.

Irion, C.W. (1999) Growing allium and brassicas in selenium-enriched soils increases their anticarcinogenic potentials. *Med. Hypotheses.* 53(3):232–235.

Isaacsohn, J.L., Moser, M., Stein, E.A., Dudley, K., Davey, J.A., Liskov, E. and Black, H.R. (1998) Garlic powder and plasma lipids and lipoproteins: a multicenter, randomized, placebo-controlled trial. *Arch. Intern. Med.* 158:1189–1194.

Jain, A.K., Vargas, R., Gotzkowsky, S. and McMahon, F.G. (1993) Can garlic reduce levels of serum lipids? A controlled clinical study. *Am. J. Med.* 94(6):632–635.

Jappe, U., Bonnekoh, B., Hausen, B.M. and Gollnick, H. (1999) Case report: Garlic-related dermatoses: Case report and review of the literature. *Am. J. Contact. Dermatitis.* 10:37–39.

Josling, P. (1999) Allicin-Garlic's Magic Bullet. The Garlic Information Center. garlic@mistral. co.uk.

Kaye, A.D., De Witt, B.J., Anwar, M., Smith, D.E., Feng, C.J., Kadowitz, P.J. and Nossman, B.D. (2000) Analysis of responses of garlic derivatives in the pulmonary vascular bed of the rat. *J. Appl. Physiol.* 89(1):353–358.

Kendler, B.S. (1987) Garlic (*Allium sativum*) and onion (*Allium cepa*): A review of their relationship to cardiovascular disease. *Prev. Med.* 16:670–685.

Kleijen, J., Knipschild, P. and ter Riet, G. (1989) Garlic, onions and cardiovascular risk factors. A review of the evidence from human experiments with emphasis on commercially available preparations. *Br. J. Clin. Pharmacol.* 28:535–544.

Knowles, L.M. and Milner, J.A. (2000) Diallyl disulfide inhibits p34(cdc2) kinase activity through changes in complex formation and phosphorylation. *Carcinogenesis.* 21(6):1129–1134.

Koch, H.P. (1996) Toxicology, side effects and unwanted effects of garlic. In: *Garlic: The Science and Therapeutic Application of* Allium sativum *L. and Related Species.* (Koch, H.P. and Lawson, L.D., eds.), Williams and Wilkins, Baltimore, MD, pp. 221–228.

Koscielny, J., Klüßebdorf, D., Latza, R., Schmitt, R., Radtke, H., Siegel, G. and Kiesewetter, H. (1998) The antiatherosclerotic effect of *Allium sativum. Atherosclerosis.* 144:237–249.

Kosian, A.M. (1998) [Effect of sulfur nutrition for sulfoxide accumulation in garlic bulbs.] *Ukr. Biokhim. Zh.* 70:105–109.

Krest, I. and Keusgen, M. (1999) Quality of herbal remedies from *Allium sativum*: differences between garlic powder and fresh garlic. *Planta Med.* 65(2):139–143.

Lamm, D.L. and Riggs, D.R. (2000) The potential application of *Allium Sativum* (garlic) for the treatment of bladder cancer. *Urol. Clin. North Am.* 27(1):157–162.

Lawson, L.D. (1993) Bioactive organosulfur compounds of garlic and garlic products: Role in reducing blood lipids. In: *Human Medicinal Agents from Plants.* American Chemical Society, Washington, DC, pp. 306–330.

Lawson, L.D. (1996) The composition and chemistry of garlic cloves and processed garlic. In: *Garlic: The Science and Therapeutic Application of* Allium sativum *L. and Related Species.* (Koch, H.P. and Lawson, L.D., eds.), Williams and Wilkins, Baltimore, MD, pp. 37–107.

Lawson, L.D. (1998a) Garlic: A review of its medicinal effects and indicated active compounds. In: *Phytomedicines of Europe: Their Chemistry and Biological Activity.* (Lawson, L.D. and Bauer, R., eds.), American Chemical Society, Washington, DC, pp. 176–209.

Lawson, L.D. (1998b) Garlic powder for hyerlipideamia-analysis of recent negative results. *Quarterly Rev. Nat. Med.* Fall: 188–189.

Lawson, L.D. (1998c) Effect of garlic on serum lipids. *JAMA.* 280(18):1568.

Lawson, L.D. and Block, E. (1997) Comments on garlic chemistry: stability of *S*-(2-propenyl) 2-propene-1-sulfinothioate (allicin) in blood, solvents, and simulated physiological fluids. *J. Agric. Food Chem.* 45:542–542.

Lawson, L.D. and Wang, Z.J. (1993) Pre-hepatic fate of the organosulfur compounds derived from garlic (*Allium sativum*). *Planta Med.* 59:A688–A689.

Lawson, L.D., Ransom, D.K. and Hughes, B.G. (1992) Inhibition of whole blood platelet-aggregation by compounds in garlic clove extracts and commercial garlic products. *Thromb. Res.* 65:141–156.

Lee, E.S., Steiner, M. and Lin, R. (1994) Thioallyl compounds: potent inhibitors of cell proliferation. *Biochim. Biophys. Acta.* 1221:73–77.

Liu, J. and Milner, J. (1990) Influence of dietary garlic powder with and without selenium supplementation on mammary carcinogen adducts. *FASEB J.* 4:A1175.

Makheja, A.N. and Bailey, J.M. (1990) Antiplatelet constituents of garlic and onion. *Agents Actions.* 29:360–363.

McCrindle, B.W., Helden, E. and Conner, W.T. (1998) Garlic extract therapy in children with hypercholesterolemia. *Arch. Pediatr. Adolesc. Med.* 152:1089–1094.

Milner, J.A. (1996) Garlic: its anticarcinogenic and antitumorigenic properties. *Nutr. Rev.* 54:S82–S86.

Munday, J.S., James, K.A., Fray, L.M., Kirkwood, S.W. and Thompson, K.G. (1998) Daily supplementation with aged garlic extract, but not raw garlic, protects low density lipoprotein against *in vitro* oxidation. *Atherosclerosis.* 143:399–404.

Murry, M.T. (1995) *The Healing Power of Herbs*, 2nd Ed., Prima Publishing, Rocklin, CA.

Nagpurkar, A., Peschell, J. and Holub, B.J. (2000) Garlic constituents and disease prevention. In: *Herbs, Botanicals and Teas as Functional Foods and Nutraceuticals* (Mazza, G. and Oomah, B.D., eds.) Technomic Publishing Co. Inc., Lancaster, PA, pp. 1–22.

Navab, M., Fogelman, A.M., Berliner, J.A., Territo, M.C., Demer, L.L., Frank, J.S., Watson, A.D., Edwards, P.A. and Lusis, A.J. (1995) Pathogenesis of atherosclerosis. *Am. J. Cardiol.* 76:18C.

Neil, H.A., Silagy, C.A., Lancaster, T., Hodgeman, J., Vos, K., Moore, J.W., Jones, L., Cahill, J. and Fowler, G.H. (1996) Garlic powder in the treatment of moderate hyperlipidaemia: A controlled trial and meta-analysis. *J. R. Coll. Physicians. Lond.* 30:329–334.

Nutrition Business Journal. (2000).

O'Gara, E.A., Hill, D.J. and Maslin, D.J. (2000) Activities of garlic oil, garlic powder, and their diallyl constituents against *Helicobacter pylori*. *Appl. Environ. Microbiol.* 66(5):2269–2273.

Ontario Garlic Growers Association. (2000) Personal Communication.

Orekhov, A.N., Pivovarpva, E.M. and Tertov, V.V. (1996) Garlic powder tablets reduce atherogenicity of low-density lipoproteins. A placebo-controlled double-blind study. *Nutr. Metab. Cardiovasc. Dis.* 6:21.

Patel, T., Gores, G.J. and Kaufmann, S.H. (1996) The role of proteases during apoptosis. *FASEB J.* 10(5):587–597.

Pedraza-Chaverri, J., Tapia, E., Medina-Campos, O.N., de Los, A. and Franco, M. (1998) Garlic prevents hypertension induced by chronic inhibition of nitric oxide synthesis. *Life Sci.* 62:PL71–7.

Perez, H.A., De la Rosa, M. and Apitz, R. (1994) *In vivo* activity of ajoene against rodent malaria. *Antimicrob. Agents Chemother.* 38:337–339.

Pérez-Pimiento, A.J., Moneo, I., Santaolalla, M., Paz, S. DE., Fernández-Parra, B. and Dominguez-Lázaro, A.R. (1999) Anaphylactic reaction to young garlic. *Allergy.* 54(6):626–629.

Phelps, S. and Harris, W.S. (1993) Garlic supplementation and lipoprotein oxidation susceptibility. *Lipids.* 28:475–477.

Pinto, J.T., Qiao, C., Xing, J., Rivlin, R.S., Protomastro, M.L., Weissler, M.L., Tao, Y., Thaler, H. and Heston, W.D. (1997) Effects of garlic thioallyl derivatives on growth, glutathione concentration, and polyamine formation of human prostate carcinoma cells in culture. *Am. J. Clin. Nutr.* 66:398–405.

Prasad, K., Laxsal, V.A., Yu, M. and Raney, B.L. (1995) Antioxidant activity of allicin, an active principle in garlic. *Mol. Cell Biochem.* 148:183.

Rabinkov, A., Miron, T., Konnstantinovski, L., Wilchek, M., Mirelman, D. and Weiner, L. (1998) The mode of action of allicin: Trapping of radicals and interaction with thiol containing proteins. *Biochem. Biophys. Acta.* 1379(2):233–244.

Rekka, E.A. and Kourounakis, P.N. (1994) Investigation of the molecular mechanism of the antioxidant activity of some *Allium sativum* ingredients. *Pharmazie*. 49:539.

Reuter, H.D. and Sendl, A. (1994) *Allium sativum* and *Allium ursinum*: chemistry, pharmacology and medicinal applications. *Econ. Med. Plant Res.* 6:56–113.

Reuter, H.D., Koch, H.P. and Lawson, L.D. (1996) Therapeutic effects and applications of garlic and its preparations. In: *Garlic: The Science and Therapeutic Application of* Allium sativum *L. and Related Species*. (Koch, H.P. and Lawson, L.D., eds.), Williams and Wilkins, Baltimore, MD, pp. 135–212.

Riggs, D.R., DeHaven, J.I. and Lamm, D.L. (1997) *Allium sativum* (garlic) treatment for murine transitional cell carcinoma. *Cancer*. 79:1987–1994.

Sakamoto, K., Lawson, L.D. and Milner, J.A. (1997) Allyl sulfides from garlic suppress the in vitro proliferation of human A549 lung tumor cells. *Nutr. Cancer*. 29:152–156.

Salonen, J.T. and Salonen, R. (1993) Ultrasound B-mode imaging in observational studies of atherosclerotic progression. *Circulation*. 87(suppl II):56–65.

Schaffer, E.M., Liu, J.Z., Green, J., Dangler, C.A. and Milner, J.A. (1996) Garlic and associated allyl sulfur components inhibit *N*-methyl-*N*-nitrosourea-induced rat mammary carcinogenesis. *Cancer Lett.* 102:199–204.

Schumacher, B.R. and Uyenaka, J. (1989) Garlic Production. Ministry of Agriculture and Food, Ontario. Order No. 89-096.

Sendl, A., Elbl, G., Steinke, B., Redl, K., Breu, W. and Wagner, H. (1992) Comparative pharmacological investigations of *Allium ursinum* and *Allium sativum*. *Planta Med.* 58:1–7.

Shobana, S. and Naidu, K.A. (2000) Antioxidant activity of selected Indian spices. *Prostraglandins Leukot. Essent. Fatty Acids*. 62(2):107–110.

Siegal, G., Walter, A., Schnalke, F., Schmidt, A., Buddecke, E., Loirand, G. and Stock, G. (1991) Potassium channel activation, hyperpolarization and vascular relaxation. *Z. Kardiol.* 80(7):9–24.

Siegal, G., Walter, A., Engel, S., Walper, A. and Michel, F. (1999) Pleiotropic effects of garlic. *Wien Med. Wochenschr.* 149(8–10):217–224.

Siegers, C.P., Steffen, B., Robke, A. and Pentz, R. (1999) The effects of garlic preparations against human tumor cell proliferation. *Phytomedicine*. 6(1):7–11.

Sigounas, G., Hooker, J.L., Anagnostou, A. and Steiner, M. (1997a) *S*-allylmercaptocysteine inhibits cell proliferation and reduces the viability of erythroleukemia, breast, and prostate cancer cell lines. *Nutr. Cancer*. 27:186–191.

Sigounas, G., Hooker, J.L., Li, W., Anagnostou, A. and Steiner, M. (1997b) *S*-allylmercaptocysteine, a stable thioallyl compound, induces apoptosis in erythroleukemia cell lines. *Nutr. Cancer*. 28:153–159.

Silagy, C.A. and Neil, H. A. (1994a) Garlic as a lipid lowering agent—a meta-analysis. *J. R. Coll. Physicians. Lond.* 28:39–45.

Silagy, C.A. and Neil, H.A. (1994b) A meta-analysis of the effect of garlic on blood pressure. *J. Hypertens.* 12:463–468.

Simons, L.A., Balasubramaniam, S., von Konigsmark, M., Parfitt, A., Simons, J. and Peters, W. (1995) On the effect of garlic on plasma lipids and lipoproteins in mild hypercholesterolaemia. *Atherosclerosis*. 113:219–225.

Sivam, G.P., Lampe, J.W., Ulness, B., Swanzy, S.R. and Potter, J.D. (1997) *Helicobacter pylori*—in vitro susceptibility to garlic (*Allium sativum*) extract. *Nutr. Cancer*. 27:118–121.

Srivastava, K.C. and Tyagi, O.D. (1993) Effects of a garlic-derived principle (ajoene) on aggregation and arachidonic acid metabolism in human blood platelets. *Prostaglandins Leukot. Essent. Fatty Acids*. 49:587–595.

Srivastava, S.K., Hu, X., Xia, H., Zaren, H.A., Chatterjee, M.L., Agarwal, R. and Singh, S.V. (1997) Mechanism of differential efficacy of garlic organosulfides in preventing benzo(a)pyrene-induced cancer in mice. *Cancer Lett.* 118:61–67.

Steinmetz, K.A., Kushi, L.H., Bostick, R.M., Folsom, A.R. and Potter, J.D. (1994) Vegetables, fruit, and colon cancer in the Iowa Women's Health Study. *Am. J. Epidemiol.* 139:1–15.

Stevinson, C., Pittler, M.H. and Ernst, E. (2000) Garlic for treating hypercholesterolaemia: A meta-analysis of randomized clinical studies. *Ann. Intern. Med.* 133(6):420–429.

Sumiyoshi, H. (1997) New pharmacological activities of garlic and its constituents. *Nippon Yakurigaku Zasshi* 110(1):93–97.

Thatte, U., Bagadey, S. and Dahanukar, S. (2000) Modulation of programmed cell death by medicinal plants. *Cell. Mol. Biol.* 46(1):199–214.

U.S. Department of Agriculture. (1998) USDA Nutrient Database for Standard Reference, Release 12. Food Group 11 Vegetables and Vegetable Products. Nutrient Data Laboratory Home Page, http://www.nal.usda.gov/fnic/foodcomp, Agricultural Research Service.

Warshafsky, S., Kamer, R.S. and Sivak, S.L. (1993) Effect of garlic on total serum cholesterol. A meta-analysis. *Ann. Intern. Med.* 119:599–605.

Wilce, M.C. and Parker, M.W. (1994) Structure and function of glutathione *S*-transferase. *Biochem. Biophys. Acta.* 1205(1):1–18.

Yeh, Y.Y. and Yeh, S.M. (1994) Garlic reduces plasma lipids by inhibiting hepatic cholesterol and triacylglycerol synthesis. *Lipids.* 29:189–193.

Yoshida, H., Hirotaka, K., Ohta, R., Ishikawa, K., Fukuda, H., Fujino, T. and Suzuki, A. (1999) An organosulfur compound isolated from oil-macerater garlic-extract, and its antimicrobial effect. *Biosci. Biotechnol. Biochem.* 63(3):588–590.

You, W.C., Blot, W.J., Chang, Y.S., Ershow, A. and Yang Z.T. (1989) Allium vegetables and reduced risk of stomach cancer. *J. Natl. Cancer. Inst.* 81:162–164.

Yutsis, P.I. (1994) Wonderful "Stinking Rose." New Editions Health World. Mar/Apr. (1994).

8 Phytochemicals from *Echinacea*

Clifford Hall III and Jurgen Schwarz

CONTENTS

1-5667-6902-7/02/$0.00+$1.50
© 2002 by CRC Press LLC

8.1 INTRODUCTION

Echinacea (pronounced ek-a-NAY-sha), or purple coneflower, is a perennial plant in the Compositae or daisy family native to North America (Foster, 1985). The flower size and color varies between species of *Echinacea* but, in general, can be characterized by hues of purple ray florets that surround an orange-brown colored head containing numerous achenes, i.e., fruiting bodies that contain seeds (Foster, 1985; Schulthess et al., 1991). *Echinacea* is one of the best-selling herbal products in the United States (Brevoort, 1998). *Echinacea purpurea* (L) Moench., *E. angustifolia* DC., and *E. pallida* Nutt. are the most widely used *Echinacea* species for medicinal purposes. Other species, *E. laevigata, E. tennesseensis, E. atrorubens, E. sanguinea, E. simulata,* and *E. paradox* (white or yellow color)*,* are adapted to specific growing regions and are thus not cultivated commercially. In the western United States and Canada, *E. purpurea* (L) Moench. and *E. angustifolia* DC. account for 80% and 20%, respectively, of the cultivated *Echinacea* (Li, 1998).

The value of cultivated *Echinacea* is often dependent on the parts of the plant and the *Echinacea* species used in the preparation of medicinal or health products. The average retail cost per pound of bulk *Echinacea* was from $37 to $65, or $20 to $23 wholesale (Little, 1999) for roots and $11 to $14 for the herbal parts (informal survey of commercially available bulk *Echinacea* conducted by the authors). The upper cost values relate to the root and herbal parts of *E. angustifolia*, due primarily to its perceived image of having better medicinal activity, which is not supported by scientific literature (Little, 1999). The herbal parts of *E. purpurea* contain phytochemicals similar to *E. angustifolia*. Also, commercial preparations of *E. purpurea* have been used successfully in Germany as evidenced by the two million prescriptions filled by German physicians annually (Barrett et al., 1999). This suggests that harvesting the herbal parts of *E. purpurea* for medicinal purposes may be more economical due to the lower purchase cost of the herb.

Echinacea is a dietary supplement, in the United States, promoted as an immunostimulatory agent (Bauer, 1999a, 2000). The use of *Echinacea* as a component in functional foods, from a legal perspective, may be years away because *Echinacea* is not considered a generally recognized as safe (GRAS) ingredient. However, the economic importance of *Echinacea* in the dietary and medicinal markets is significant, and thorough studies need to be completed to assess the risk/benefits of consuming *Echinacea*. This chapter will cover the chemistry of *Echinacea,* including factors that affect the stability of the phytochemicals, potential uses and product regulations.

8.2 *ECHINACEA* PHYTOCHEMICALS

Echinacea is one of the best-selling herbal products in the United States (Brevoort, 1998) and is promoted as an immunostimulatory agent (Bauer, 1999a, 2000). The alkamides, caffeic acid derivatives (e.g., cichoric acid), glycoproteins, and polysaccharides are believed to be responsible for *Echinacea's* observed immunostimulatory activity. Bauer (1997) reported that the variation in alkamide concentrations of *Echinacea* products was due to a number of factors such as growing season, which part of the plant was utilized in the preparation of the commercial product, the species (i.e., *purpurea, angustifolia* DC., *pallida,* etc.) of *Echinacea* used, and method of harvest. Other components are expected to be influenced by similar factors. The data presented in this section are on a dry weight basis.

8.2.1 CAFFEIC ACID DERIVATIVES AND PHENOLIC ACIDS

Caffeic acid derivatives (CADs) (Figure 8.1) make up the largest percentage of the phenolic compounds, with flavonoids, phenolic acids, and anthocyanins contributing a smaller percentage. CADs can be classified as quinyl esters of caffeic acid, tartaric acid derivatives or phenylpropanoid glycosides (Cheminat et al., 1988). The tartaric acid derivatives are believed to be responsible for some of the immune-enhancing activity of *Echinacea,* with cichoric acid being the most important (Bauer et al., 1989). Chlorogenic acid, verbascoside, and echinacoside have no immune-enhancing activity but possess antioxidant activity and, thus, are potentially important phytochemicals.

A qualitative evaluation of *E. pallida* showed that cichoric acid was the predominant CAD in the flowers and leaves, while the roots also contain high echinacoside levels (Cheminat et al., 1988). The concentration of cichoric acid in commercial *Echinacea* products, i.e., dried roots and aerial parts, sold in Australia were determined by Wills and Stuart (1999). These authors report that the average cichoric acid was 13.2 mg/g (1.32%) and 12.9 mg/g (1.29%) for the *E. purpurea* roots and aerial parts, respectively. The cichoric acid concentration ranged from 0.4% to 2.14%, suggesting that the physiological activity of products made from roots and aerial parts could vary. Also, chromatographic separation indicated that cichoric acid accounted for 63% and 67% of the phenols in the root and aerial parts, respectively. Bauer et al. (1988a) found similar levels in German-grown *Echinacea.* They reported that cichoric acid levels in aerial parts were 1.2–3.1% and, in roots, 0.6–2.1%. Further investigation found that an average cichoric acid content of 2.1, 1.3, 1.0, and 0.4% in the flower, root, leaf, and stem segments, respectively. In addition, these authors noted that the *E. purpurea* had higher levels of cichoric acid than *E. angustifolia* DC. Wills and Stuart (1999) reported similar results (Table 8.1) for Australian grown *E. purpurea.* The average cichoric acid content of 3.4, 1.7, 0.9 and 1.0% in the flower, root, leaf and stem segments, respectively, suggests that geographic location may have a slight affect on cichoric acid concentration. However, the accumulation of cichoric acid in specific plant parts was not affected by location, because the flowers had the highest level of cichoric acid at both locations, followed by root, leaf, and stem segments (Table 8.1).

Pietta et al. (1998a) presented a qualitative evaluation on the CAD in *E. purpurea, E. pallida,* and *E. angustifolia* (Table 8.2) using micellar electrokinetic chromatography (MEKC). The presence of the CAD was consistent with previous reports (Becker and Hsieh, 1985; Bauer et al., 1988a,b; Cheminat et al., 1988). The only discrepancy was that cichoric acid was not found in *E. angustifolia* by MEKC. Although cichoric acid is considered the most important CAD, a brief discussion on the other CAD is warranted due to the antioxidant potential of phenolic materials.

A phenolic acid content of 190 and 870 g/g was reported for *E. angustifolia* and *E. purpurea,* respectively (Glowniak et al., 1996). *E. purpurea* had a high chlorogenic acid content at 125 µg/g, while 26 g/g was found in *E. angustifolia.*

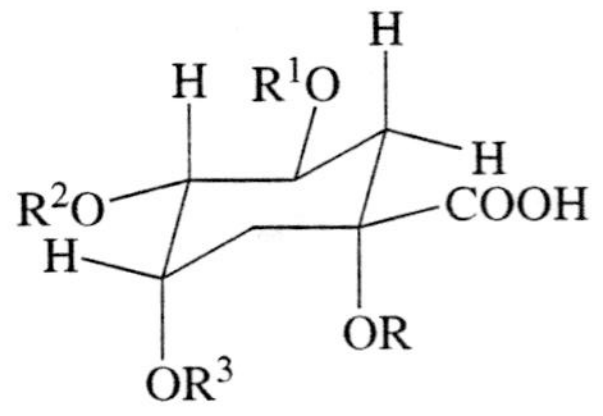

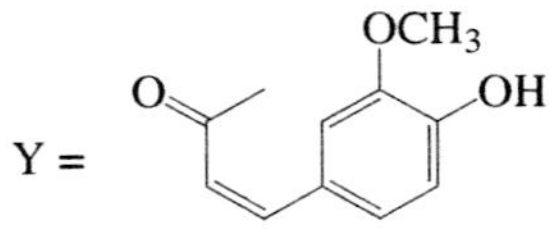

Feruloyl (i.e., ferulic acid)

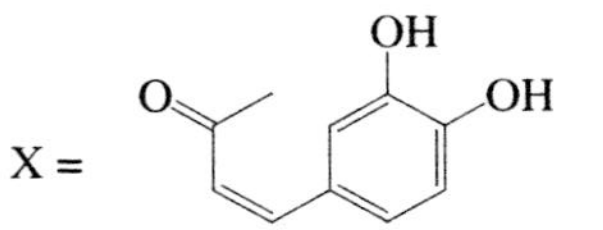

Caffeoyl (i.e., caffeic acid)

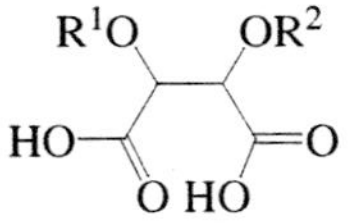

Quinic Acid

Quinyl Esters of Caffeic Acid

5-O-caffeoylquinic acid (Chlorogenic Acid): R^1 = x; R, R^2, R^3 = H
3,5-O-dicaffeoylquinic acid (isochlorogenic acid): R^1, R^1 = x; R, R^2 = H
1,3-O-dicaffeoylquinic acid (cynarine): R^1, R^3 = x; R^1, R^2 = H

Tartaric acid

Tartaric Acid Derivatives

2,3-O-dicaffeoyltartaric acid Cichoric Acid): R^1, R^2, = X
2-O-caffeoyl-3-O-feruloyltartaric acid: R^1 = X; R^2 = Y
2-O-caffeoyltartaric acid (Caftaric acid): R^1 = X; R^2 = H

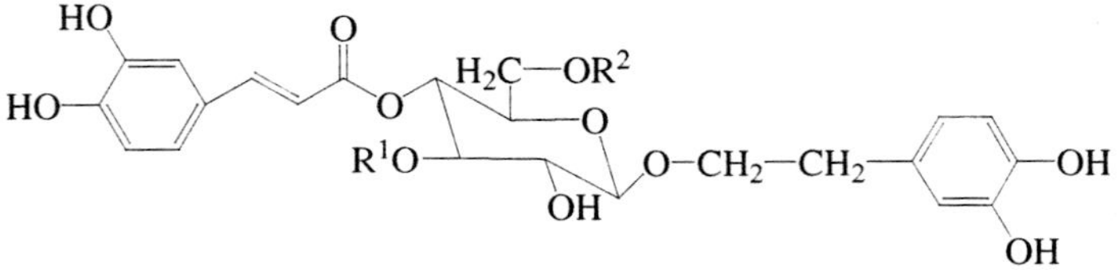

Phenylpropanoid Glycosides

β-(3,4-dihydroxyphenyl)ethyl-O-α-L-rhamnopyranosyl(1→3)-4-O-caffeoyl-β-D-glucopyranoside (Verbascoside): R^1 = rhamnose; R^2 = H
β-(3,4-dihydroxyphenyl)ethyl-O-α-L-rhamnopyranosyl(1→3)-β-D-glycopyranosyl(1→6)-4-O-caffeoyl-β-D-glucopyranoside (Echinacoside): R^1 = rhamnose; R^2 = glucose
β-(3,4-dihydroxyphenyl)ethyl-O-α-L-rhamnopyranosyl(1→3)-β-D-glycopyranosyl(1→6)-4-O-caffeoyl-β-D-glucopyranoside (6-O-Caffeoyl Echinacoside): R^1 = rhamnose; R^2 = 6-caffeoyl-glucose

FIGURE 8.1 Caffeic acid derivatives of *Echinacea*.

TABLE 8.1
Cichoric Acid Content of Various Plant Parts of
E. purpurea Grown in Germany and Australia

	Plant Part			
	Flower	**Roots**	**Leaf**	**Stem**
Germany[1]	13–30[3]	6–21	4–16	2–6
Australia[2]	29.5–38.3	10–23.8	4.1–15.3	3.8–10.9

[1] Bauer and colleagues (1988–1990).
[2] Wills and Stuart (1999).
[3] mg/g plant part dry weight.

TABLE 8.2
Qualitative Evaluation of the Caffeic Acid Derivatives in _Echinacea_ Species Using Micellar Electrokinetic Chromatography (Pietta et al., 1998)

Caffeic Acid Derivatives	Roots			Aerial		
	E. angustifolia	_E. pallida_	_E. purpurea_	_E. angustifolia_	_E. pallida_	_E. purpurea_
Caffeic acid	✔[1]	✔		✔	✔	
Caffeoyl-echinacoside		✔				
Caftaric			✔		✔	✔
Chlorgenic acid	✔	✔	✔	✔	✔	✔
Cichoric acid			✔		✔	✔
Cynarine	✔				✔	✔
Echinacoside	✔	✔		✔	✔	
Isochlorogenic acid	✔			✔		
Isochlorogenic acid (II)		✔				
Verbascoside				✔	✔	

[1] ✔ Denotes the presence of the CAD in the _Echinacea_ species.

Modified from Pietta et al., 1998a.

Protocatechuic, _p_-hydroxybenzoic, _p_-coumaric, and ferulic acids were common to both species, while vanillic and caffeic acids were specific to _E. angustifolia_ and _E. purpurea,_ respectively.

8.2.2 ALKAMIDES

Alkamides (Figure 8.2) are of interest because this group of compounds has immune-enhancing activity as well as anti-inflammatory activity (Müller-Jakic et al., 1994). Wills and Stuart (1999) reported the alkamide levels of 0.62% (6.2 mg/g) and 0.10% (1 mg/g) for _E. purpurea_ roots and aerial parts, respectively. Further investigation

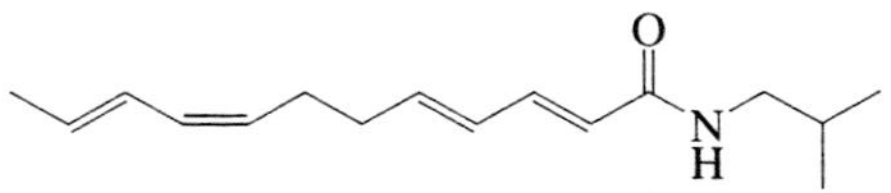

Dodeca-2*E*,4*E*,8*Z*,10*E*-tetraenoic acid isobutylamide

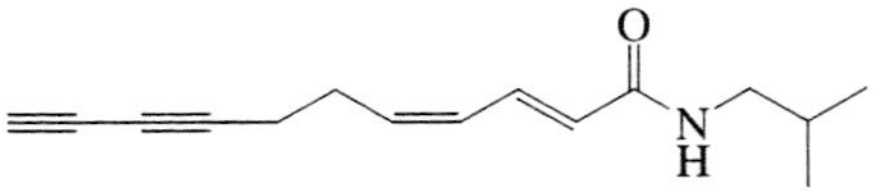

Undeca-2*E*,4*Z*-diene-8,10-diynoic acid isobutylamide

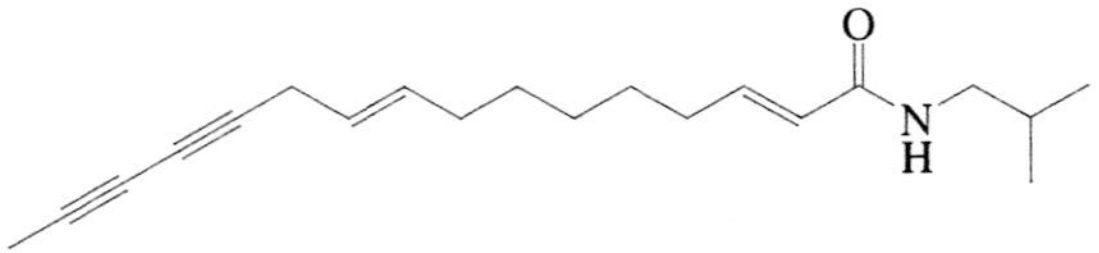

Hexadeca-2,9-diene-12,14-diynoic acid isobutylamide

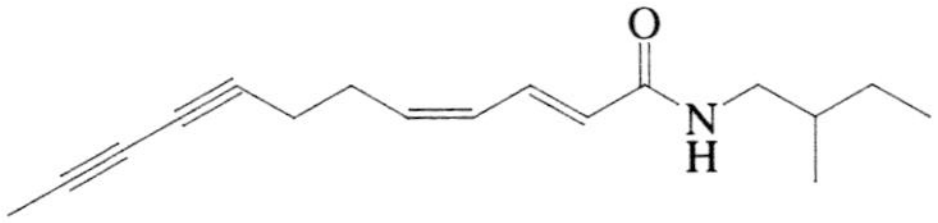

Dodeca-2*E*,4*Z*-diene-8,10-diynoic acid 2-methylbutylamide

FIGURE 8.2 Example of four alkamides isolated from *Echinacea*.

TABLE 8.3
Average Alkamide Content of Various Plant Parts of *E. purpurea*
Grown in Australia and New Zealand

			Plant Part			
Location	Roots	Rhizome	Flower	Leaf	Vegetative Stem	Productive Stem
Australia[3]	8.9[1]	N/A[2]	2.6	0.10	N/A	0.75
New Zealand[4]	6.2	8.0	2.9	0.24	19.0	1.52

[1] mg/g plant part dry weight.
[2] N/A = not applicable.
[3] Wills and Stuart (1999).
[4] Perry et al. (1997).

showed average alkamide levels of 0.89, 0.26, 0.01 and 0.08% in the root, flower, leaf, and stem segments, respectively. Perry et al. (1997) found an alkamide level in *E. purpurea* between 0.024% in the leaf to 1.9% in the vegetative stem (Table 8.3). Rogers et al. (1998) reported alkamide levels in *E. purpurea* between 0.024 and 0.11% for aerial parts. The vegetative stem and root portions gave 0.39% alkamide levels. The alkamide levels in the root of *E. angustifolia* grown in the United States and Australia was 0.12% and 0.059%, respectively (Rogers et al., 1998). Bauer and Remiger (1989) found that the dodec-2E,4E,8Z,10Z-tetraenoic acid isobutylamide accounted for 0.004–0.039% of the root of *E. purpurea* and 0.009–0.151% of the root of *E. angustifolia*, while the herbal part was less than 0.001–0.03%. These authors also reported that the alkamides in *E. purpurea* possess a 2,4-dienoic moiety, while the 2-monene moiety is more common in *E. angustifolia*. The lack of an isobutylamide moiety, to give 2-ketoalkenes and 2-alkynes, is common for *E. pallida* (Bauer et al., 1988c; Bauer and Remiger, 1989; Schulthess et al., 1991).

Nineteen alkamides have been identified (Bauer and Remiger, 1989; Perry et al., 1997). In *E. purpurea*, dodec-2E,4E,8Z,10Z-tetraenoic acid isobutylamide accounted for 45% and 76% of the alkamides from the roots and aerial parts, respectively. Undeca-2Z,4E-diene-8,10-diynoic acid isobutylamide and dodeca-2E,4Z-diene-8,10-diynoic acid isobutylamide accounted for an additional 18% and 15%, respectively, in the root. However, neither compound was found in the aerial parts. Undeca-2E,4Z-diene-8,10-diynoic acid isobutylamide accounted for 20% of the alkamides in the aerial parts and only 6% in the root of *E. purpurea* (Wills and Stuart, 1999). Perry et al. (1997) observed that dodec-2E,4E,8Z,10Z(E)-tetraenoic acid isobutylamide accounted for 27% of the alkamides in the roots, but that 6.6% of the total amount of this compound was derived from the roots. The majority (55%) of this compound was found in the vegetative stems, and 22% was found in the rhizome.

8.2.3 ESSENTIAL OILS

The alkamide component of the lipophilic fraction is believed to be partially responsible for the immune-enhancing response, whereas components such as

TABLE 8.4

Major Volatiles found in the Headspace of Various Plant Parts of *Echinacea* Species Using Headspace Gas Chromatography (Mazza and Cottrell, 1999)

	Myrcene	β-Pinene	α-Pinene	α-Phellandene	Dimethylsulfide
Flower					
E. angustifolia	33[1]	5.3	13.5	0	0.6
E. pallida	60	6.9	6.5	0	1.5
E. purpurea	43	7.4	22.6	0	1.0
Leaf					
E. angustifolia	38	4.0	3.5	0	<0.1
E. pallida	31	9.0	5.3	0	<0.1
E. purpurea	27	1.9	12.1	0	<0.1
Stem					
E. angustifolia	45	3.7	9.1	0	<0.1
E. pallida	49	9.7	7.6	0	<0.1
E. purpurea	45	4.4	33.7	0	<0.1
Root					
E. angustifolia	0	1.3	2.8	4.9	13.1
E. pallida	2.0	0.6	0	0	30.8
E. purpurea	0	0.2	0.6	16.7	14.7

[1] Denotes the percentage of volatile in the *Echinacea* species.

Data taken from Mazza and Cotrell, 1999.

essential oils can act as antioxidants. The essential oil content ranges from 0.1% to 2.0% (Schulte et al., 1967; Heinzer et al., 1988). Dodeca-2,4-diene-1-yl-iso-valerate and pentadeca-1,8Z-diene account for greater than 44% of the essential oil component of *E. angustifolia*. Pentadeca-1,8Z-diene, pentadeca-8Z-en-2-one and *E*-10-hydroxy-4,10-dimethyl-4,11-dodecadien-2-one (echinolon) are essential oil components of *E. pallida* (Schulte et al., 1967; Voaden and Jacobson, 1972; Heinzer et al., 1988). *E. purpurea* has an essential oil content of 0.2%, which is predominantly β-myrcene.

Mazza and Cottrell (1999) identified 70 compounds in the headspace of *Echinacea* species, 50 of which had not been reported previously. The distribution of the essential oils is, as one would expect, significantly higher in the flower, leaf, and stem tissue. The volatile content of *E. purpurea* root accounted for 10.4% of the total headspace volatiles, while 2.6% and 3.2% of the volatiles of *E. angustifolia* and *E. pallida*, respectively, were found in the roots. However, the volatile contents of the flower, stem and leaf tissues were similar between species.

A grouping of the compounds showed that aldehydes and terpenoids, respectively, made up 41–57% and 6–21% of the volatiles in root tissue, 19–24% and 46–58% of leaf volatiles and 6–14% and 81–91% of the stem volatiles (Mazza and Cottrell, 1999). β-Myrcene was the predominant aerial part volatile among *Echinacea* species (Table 8.4), but in roots, it was found only in *E. pallida*. An ethanol extract of the achene followed by steam distillation resulted in a β-myrcene content of 6.8% and

3.6% for *E. purpurea* and *E. angustifolia*, respectively, while *E. pallida* had only trace amounts (Schulthess et al., 1991). These values are far below those reported by Mazza and Cottrell (1999), in which the β-myrcene content ranged from 33% to 60% of the volatiles. Possible reasons for the difference in β-myrcene content could be the age of the plant materials or the analytical method used in the evaluations. Dimethyl sulfide was a major root volatile in all *Echinacea* species, while α-phellandrene accounted for 16.7% of the volatiles in *E. purpurea*, 4.9% in *E. angustifolia*, but was not found in *E. pallida*.

8.2.4 GLYCOPROTEINS AND POLYSACCHARIDES

Glycoproteins and polysaccharides are the final class of phytochemicals compounds important to *Echinacea*. The glycoproteins isolated from *E. purpurea* and *E. angustifolia* roots range from 17 to 30 kDa and contain 64–84% arabinose (Bauer, 1999a, 2000). Limited research has been completed on the glycoproteins, thus the reader is directed to the review by Bauer (1999a, 2000) for further information.

The polysaccharide 4-*O*-methyl-glucuronoarabinoxylan isolated from the hemicellulosic fraction of *E. purpurea* was characterized by Wagner et al. (1984) and linked to immune-stimulating activity using an *in vitro* granulocyte test. A second immune-stimulating polysaccharide, arabinorhamnogalactan, was identified in *E. purpurea* by Proksch and Wagner (1987) and later in *E. purpurea* cell cultures (Wagner et al., 1988). Also, two fucogalactoxyloglucans have been isolated from *E. purpurea* and cell cultures of *E. purpurea* (Wagner et al., 1988; Roesler et al., 1991a). For additional discussion on the polysaccharides, see Bauer (1999a, 2000) and Emmendörffer et al. (1999). The variety of phytochemical compounds in *Echinacea* demonstrates the difficulty in developing a single method for analyzing the physiologically active components for standardization purposes. Many analytical methods have been developed to analyze the alkamides and CAD and a few target the polysaccharides.

8.3 ANALYTICAL METHODS: ISOLATION, STRUCTURE VERIFICATION AND PHYTOCHEMICAL ASSESSMENT

The analysis of CAD and alkamides has been completed using a variety of analytical techniques that include high-performance liquid chromatography (HPLC), capillary electrophoresis, gas chromatography (GC), GC-mass spectrometry (GC-MS), liquid chromatography-mass spectrometry (LC-MS), and nuclear magnetic resonance (NMR). Sample preparation methodologies utilize hexane, methanol and ethanol as the primary extraction solvents.

8.3.1 CAD ISOLATION, VERIFICATION AND ASSESSMENT

The extraction of CAD can be completed using methanol or methanol-water mixtures (Bauer et al., 1988b; Cheminat et al., 1988; Bauer and Foster, 1991; Wills and Stuart, 1999). A common extraction approach uses a Soxhlet apparatus, methanol and an extraction time of 12–24 h. Cheminat et al. (1988) completed an in-depth structural evaluation of 12 CADs (Figure 8.1) isolated from *Echinacea* using NMR. Cheminat et al. (1988) grouped the CADs as quinyl esters of caffeic acid, tartaric acid derivatives and phenylpropanoid glycosides. This grouping can be advantageous to individuals interested in obtaining standards, because preliminary fractionation can be based on groups of similar compounds instead of single components.

8.3.1.1 CAD Isolation

The CAD extraction protocol of Cheminat et al. (1988) included a 12-h methanol-water (4:1) extraction followed by methanol removal and extraction of the remaining aqueous phase with petroleum ether and chloroform. The aqueous phase was acidified and then extracted with ethyl acetate and then *n*-butanol. After concentration of the ethyl acetate and *n*-butanol fractions, CADs were fractionated over a Sephadex LH20 column using methanol-water (4:1) followed by separation over silica gel, thin-layer chromatography, or HPLC. Unlike other CADs, cichoric acid was isolated in a purified form after separation on Sephadex LH20 column. Echinacoside isolation can be achieved using a polyamide column and eluding with water followed by ascending methanol concentrations. The purified CAD are powders and range in color from beige to white. Nuclear magnetic resonance (NMR) data for CADs are similar in that the molecules share a caffeic acid moiety. The variation in NMR data is due to the noncaffeic acid moiety.

8.3.1.2 CAD Structure Verification

In general, the caffeic acid moiety will have chemical shifts between 6.22–7.65 ppm with the aromatic proton ranging from 6.81 to 7.08 ppm. The olefinic protons have chemical shifts around 6.25 ppm for the proton located on the carbon next to the carbonyl, whereas the protons on the carbons neighboring the aromatic ring have chemical shifts near 7.59 ppm. The olefinic proton location can be readily observed by the large associated coupling constants (~16 Hz) (Becker and Hsieh, 1982, 1985; Cheminat et al., 1988). The tartaric acid moiety of cichoric acid and other tartaric acid derivatives is characterized by a proton resonance at 5.67 ppm, which integrates two protons. The elimination of one caffeic acid molecule from the tartaric acid moiety (i.e., 2-caffeoyltartaric acid or caftaric acid) results in two chemical shifts at 4.57 and 5.34 ppm.

Echinacoside (Figure 8.1) is a CAD that contains caffeic acid, a β-(3,4-dihydroxy-phenyl)-ethoxy substitution, two glucose groups, and a rhamnose group. Again, the caffeic acid moiety has chemical shifts similar to those listed above, while the aromatic proton chemical shifts for the β-(3,4-dihydroxyphenyl)-ethoxy substitution are found at 6.54, 6.66, and 6.67 ppm. The ethyl protons are found at 2.73 ppm for the protons on the carbon attached to the aromatic ring, while shifts of 3.67 and 3.92 ppm are characteristic of protons on the carbon attached to the glucose moiety. For the glucose and rhamnose moieties, the majority of the proton resonances are in the neighborhood of 3.14–3.75 ppm. Deuterated dimethyl sulfoxide (DMSO-d$_6$) is commonly used as the solvent when obtaining a proton NMR spectrum. In addition, trifluoroacetic acid can be added to eliminate the influence of hydroxyl protons (i.e., hydrogen bonding).

The carbon-13 NMR (^{13}CNMR) data can be used to support the proton data. Chemical shifts between 115 and 149 ppm are associated with aromatic carbons, whereas the aromatic carbons bearing hydroxyl groups have chemical shifts around 146–149 ppm. The carbonyl carbon involved in the ester linkage to tartaric acid or glucose is found around 169 ppm. In cichoric acid, the carbons associated with the tartaric acid display two resonances near 170 ppm, which correspond to the two carbonyl groups, and two aliphatic chemical shifts near 73 ppm. The chemical shifts of the β-(3,4-dihydroxyphenyl)-ethoxy substitution in echinacoside are similar to those of caffeic acid. The only difference is that the ethyl carbons of the β-(3,4-dihydroxyphenyl)-ethoxy are found ca. 35 and 70 ppm. The sugar moieties have chemical shifts predominantly in the range of 68–78 ppm. Deuterochloroform can be used as the solvent for the tartaric acid esters, while DMSO-d$_6$ should be used for the phenylpropanoid glycosides (e.g., echinacoside).

The use of mass spectrometry to support the NMR structural assignments has been reported by Cheminat et al. (1988) and Facino et al. (1993). Facino et al. (1993) used fast atom bombardment mass spectroscopy (FAB-MS) in the negative ion mode to elucidate CAD in *Echinacea* preparations. They found that a butyl acetate fraction had negative ions ([M-H]-) of 473, 353, and 311 m/z, which correspond to cichoric acid (MW 474), chlorogenic acid (MW 354), and caftaric acid (MW 312), respectively. Subsequent FAB-MS mass spectrometry (FAB-MS/MS) of cichoric acid produced [M-H]- daughter ions of 311, 293, 179, 149 and 113 m/z. A loss of a caffeoyl residue (160 m/z) via cleavage of the ester bond results in the formation of caftaric acid. The ion at m/z 293 depicts the loss of caffeic acid residue (180 m/z) from the cichoric acid structure. In addition, the m/z of 179 is due to the caffeic acid residue. The subsequent loss of a second caffeoyl or caffeic acid results in the 149 and 113 m/z ions, which correspond to the tartaric acid moiety.

The FAB-MS/MS of chlorogenic acid resulted in daughter ions of 191, 179, 161, and 135 m/z. Chlorogenic acid is a quinic acid ester of caffeic acid, thus, one would expect the loss of a caffeoyl unit or caffeic acid. The ion at 191 m/z represents the loss of caffeic acid from chlorogenic acid to give quinic acid. The presence of the 179, 161, and 135 m/z ions are related to the caffeic residue. Minor ions at 309 and 147 m/z are indicative of a loss of carbon dioxide (44 m/z units) from chlorogenic acid and quinic acid, respectively. For further discussion on MS analysis of CAD, see Facino et al. (1993). Using the same rationale, MS data from other CAD can be interpreted and used to support other analytical data.

8.3.1.3 CAD Assessment

Cichoric acid has been the target of many analytical methods as a means to standardize *Echinacea* preparations. The method of Bauer (1999b) illustrates the most common method used to evaluate cichoric acid and other CADs via HPLC. A reversed-phase (LiChroCART 125-4 column) system with gradient elution was used to complete the separation within 12 min. The gradient system included water plus 0.1% orthophosphoric acid (85%) as eluent A and acetonitrile plus 0.1% orthophosphoric acid (85%) as eluent B. A linear gradient from 10% to 30% of eluent B was completed within 20 min at 1 ml/min and detection at 330 nm. Hu and Kitts (2000) used a similar reversed-phase system to separate chlorogenic acid, caffeic acid, echinacoside, and cichoric acid. In addition, CADs can be assessed using TLC using silica gel as the stationary phase and a mobile phase of ethyl acetate-formic acid-acetic acid-water (100:11:11:27 v/v) with detection at 360 nm.

A promising alternative to HPLC and TLC is micellar electrokinetic chromatography (MEKC). Pietta et al. (1998a) used a 25 mM tetraborate buffer containing 30 mM sodium dodecyl sulfate (SDS) at pH 8.6 for the separation of CADs. The separation was completed within 20 min with good resolution. The advantage to this method is that the cost of operation is less than HPLC, because organic solvents are not needed.

8.3.1.4 Alkamide Isolation

Two of the most common methods for the extraction of alkamides involve the use of hexane (Bauer et al., 1989; He et al., 1998) or acetonitrile (Perry et al., 1997). Fractionation of crude extract can be completed using column chromatography in which silica gel serves as the stationary phase and mixtures of hexane-ethyl acetate as the mobile phase. Perry et al. (1997) used a reverse-phase extraction column and acetonitrile-water solvent system. Further purification can be completed using reverse-phase HPLC.

The extraction and isolation protocol of Bauer et al. (1989) involves the use of hexane, in a Soxhlet extraction apparatus, over a 3-day period. The resulting yellow extract can be fractionated over silica gel using ethyl acetate-hexane (1:5) or by thin-layer chromatography (TLC) (Bauer and Remiger, 1989) using ethyl acetate-hexane (1:2). The use of TLC can prove valuable as a prescreening method prior to column chromatography over silica gel. Bauer and Remiger (1989) used silica gel plates containing fluorescence indicator (F_{254}) and anisaldehyde/sulfuric acid spray for detection. They found that the 2-monoenamide structures gave a yellow color, while a violet color was formed when 2,4-dienamides were present. In addition, the 2,4-dienamides fluoresce at 254 nm, providing a second means for separating the various alkamides into classes. Purification of the alkamides can be completed on a reverse-phase semipreparative column using water and acetonitrile (40–80%) gradient and detection at 254 nm (Bauer et al., 1988c, 1989). The NMR data for the alkamides are similar, because they all contain an isobutylamide moiety and an unsaturated carbon chain. The variation in NMR data is reflected in the position and type of chemical bonds present in the alkamide.

8.3.1.5 Alkamide Structure Verification

The NMR data of alkamides have been reported by Bohlmann and Hoffman (1983), Bauer et al. (1988c, 1989), and Perry et al. (1997). The isobutylamide region of the alkamides (Figure 8.2) is characterized by a broad singlet at ca. 5.5 ppm, a double doublet at 3.18 ppm, a multiplet at 1.80 ppm and a doublet at 0.93 ppm. The signal at 5.5 ppm represents the proton on the nitrogen (N-H), while the doublet at 0.93 ppm represents the methyl protons (2 CH_3). The chemical shift at 3.18 ppm represents the methylene proton next to the nitrogen (-NH-CH$_2$-) and the multiplet signal at 1.80 ppm represents the methine protons (-NH-CH$_2$-CH-). If the isobutylamide is replaced by a 2-methyl-butylamide substitution (Figure 8.2), the most obvious change would be the additional triplet signal at 0.91 ppm, which represents the protons of the methyl group adjacent to the methylene group. The presence of a doublet at 0.91 ppm is indicative of a methyl substitution on a methine group. The presence of a multiplet signal at 1.41 and 1.16 ppm represents the methylene protons adjacent to methyl and methine groups. The proton signal at 1.58 ppm represents the methine group next to the nitrogen.

The unsaturated carbon moiety will produce signals in two primary regions. The signal from 5.87–7.51 ppm represents the olefenic protons, while the allylic methylenes resonate at 2.18–2.55 ppm. A terminal acetylene (i.e., triple bond) would have a signal near 1.98 ppm. The stereochemistry (E and Z configurations) can be assessed by determining the coupling constants. Coupling constants of 15 and 11 Hz correspond to the E and Z configurations, respectively. For additional discussion on proton NMR for the alkamides, see Bauer et al. (1988c, 1989).

The ^{13}CNMR data show a characteristic carbonyl carbon at 166 ppm, which corresponds to the carbon attached to the nitrogen via an amide bond. The aliphatic carbon signals can be found between 19–47 ppm for the isobutylamides and 11–45 ppm for the 2-methyl-butylamides. The carbon signals associated with the double and triple bonds of the unsaturated chain can be found between 119–136 and 64–77 ppm, respectively. For additional information regarding ^{13}CNMR, see Perry et al. (1997).

Mass spectral data for the alkamides was reported by Bauer et al. (1988c, 1989) and He et al. (1998) using an HPLC-MS. The mass-to-charge (m/z) ratios ranged from 229 to 261 m/z for the alkamides isolated from *E. purpurea* roots. In addition, He et al. (1998) reported a range in m/z of 229–278 for alkamides isolated from *E. purpurea* achenes. The daughter ion fragmentation patterns showed that the isobutylamides and the 2-methyl-butylamides can be differentiated. Fragments characteristic of isobutylamides include [M-57]$^+$, [M-72]$^+$ and [M-100]$^+$, while 2-methyl-butylamides have [M-29]$^+$, [M-86]$^+$ and [M-114]$^+$. The loss of 57 mass units is indicative of a loss of the isobutyl group [C$_4$H$_9$]$^+$, while the loss of 72 and 100 mass units represents [C$_4$H$_{10}$N]$^+$ and [C$_4$H$_{10}$NCO]$^+$, respectively. The loss of 29 mass units from 2-methyl-butylamides represents a [C$_2$H$_5$]$^+$ or ethyl (-CH$_2$-CH$_3$) unit. The loss of [C$_5$H$_{12}$N]$^+$ and [C$_5$H$_{12}$NCO]$^+$ are indicated by a loss of 86 and 144 m/z units from the parent ion.

8.3.1.6 Alkamide Assessment

The evaluation of alkamides, as an indicator for standardization, in *Echinacea* preparations is commonly completed using HPLC. The method of Bauer (1999b) illustrates the most common method to evaluate alkamides using reverse-phase HPLC. In this method, the lipophilic components are separated by gradient elution using water (eluent A) and acetonitrile (eluent B) linearly from 40% to 80% eluent B at 1 ml/min. The separation was completed on a C18 reverse-phase column and detection was at 254 nm. Thin-layer chromatography using silica 60 plates with indicator (F254 nm) can be used as an alternative to HPLC. The solvent system includes *n*-hexane and ethyl acetate (2:1 v/v), with detection at 254 nm or at visual inspection after spraying the developed plate with anisaldehyde/sulfuric acid. The 2,4-dienamides fluoresce at 254 nm, providing a means for separating the various alkamides into classes. In addition, 2-monoenamide structures produce a yellow color, while 2,4-dienamides are visualized by a violet color.

8.3.1.7 Glycoprotein and Polysaccharide Isolation and Characterization

Of the phytochemical components in *Echinacea*, the glycoproteins and polysaccharides have been the least characterized. Three glycoproteins have been isolated from *E. purpurea* and *E. angustifolia* and characterized as having molecular weights of 17, 21 and 30 kDa (Bauer, 1999a). Arabinose, galactose, and glucosamine make up 64–84, 1.9–5.3 and 6% of the sugars, respectively.

Five polysaccharides have been isolated from *E. purpurea*. 4-*O*-methyl-glucuronoarabinoxylan (35 kDa; PS1) was isolated from *E. purpurea* and characterized by Proksch and Wagner (1987). A second immune-stimulating polysaccharide, arabinorhamnogalactan (45 kDa; PS2), was identified in *E. purpurea* by Proksch and Wagner (1987) and later in *E. purpurea* cell cultures (Wagner et al., 1988). Also, two fucogalactoxyloglucans (MW 10 and 25 kDa) and an acidic arabinogalactan (MW 75 kDa) have been isolated from *E. purpurea* and cell cultures of *E. purpurea* (Wagner et al., 1988; Roeslet et al., 1991a). The complexity of isolating the polysaccharides has limited the advancement of characterizing *Echinacea* polysaccharides.

A general protocol (Proksch and Wagner, 1987; Wagner et al., 1988) for separating the polysaccharides can be found in Figure 8.3. After separation, acid hydrolysis, methanolysis, and reduction and oxidation reactions were completed to aid in structural determination. For complete NMR and sugar distribution data, see Proksch and Wagner (1987) and Wagner et al. (1988).

8.4 STABILITY OF *ECHINACEA* PHYTOCHEMICALS

8.4.1 CICHORIC ACID STABILITY

Cichoric acid is believed to contribute to the immunostimulatory activity, but the compound tends to decompose through enzymatic degradation during extractions (Bauer, 1997); thus, the variability in cichoric acid levels found by many researchers may be due to the extraction methodology and not to the plant species or origin. Bauer (1999b) evaluated the cichoric acid content of six commercially available expressed juice preparations of *E. purpurea*. The thermally treated preparations had higher cichoric acid than ethanol-preserved preparations. The inactivation of polyphenol oxidase by heat may account for the difference found between heated and nonheated preparations.

Wills and Stuart (2000) evaluated cichoric acid levels in *E. purpurea* to assess the effect of postharvesting handling methods on *Echinacea* phytochemicals. These authors found that cichoric concentration was not significantly altered when *Echinacea*

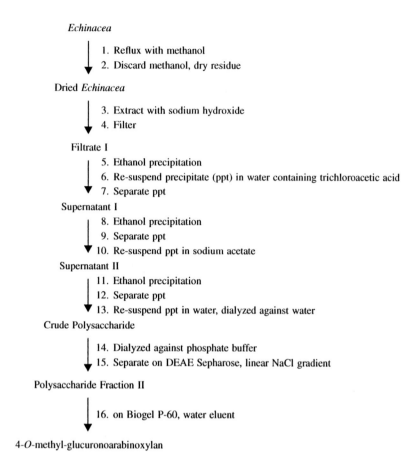

FIGURE 8.3 Isolation protocol for polysaccharides from *Echinacea*.

was subject to various physical treatments that included cutting, compression, or crushing. Further investigation showed that storage of freshly harvested *Echinacea* at 20°C and 60% relative humidity (RH) did not significantly affect the concentration of cichoric acid over a 30-day storage period (Wills and Stuart, 2000). However, storage at 5°C and 80% RH resulted in a significant reduction (80%) in cichoric acid levels. Further experimentation showed that cichoric acid levels were not reduced significantly when the *E. purpurea* roots were blanched prior to storage at 5°C and 80% RH. These data imply that enzymatic, i.e., polyphenol oxidase, reactions may be responsible for the degradation of cichoric acid (Bauer, 1997, 1999b; Wills and Stuart, 2000). Nüsslein et al. (2000) reported that polyphenol oxidase activity decreased with the addition of ascorbic acid or ethanol. However, a complete inhibition of polyphenol oxidase was noted when both ascorbic acid (50 nM) and ethanol (40%) were added during the extraction process.

An assessment of CAD retention in *E. purpurea* processed using various drying methods showed that freeze-drying and vacuum-microwaved drying of the aerial (i.e., flower) parts had significantly higher total cichoric (also reported as chicoric) and caftaric acid levels compared with air drying (Kim et al., 2000a). Air drying at 40°C had significantly higher CAD levels than products air dried at 25 and 70°C. Also, storage of a 9.3% moisture *Echinacea* product had a significantly lower cichoric acid content compared with one dried to 6.1% prior to storage. Again, polyphenol oxidase may contribute to the degradation of cichoric acid. In addition, these authors noted a slight reduction in CAD in samples harvested between August and October.

8.4.2 ALKAMIDE STABILITY

Perry et al. (2000) found that drying *E. purpurea* roots (69.2% to 6.8% moisture) at 32°C for 48 h did not significantly reduce the alkamide levels. A significant reduction in the concentration of two alkamides, but not total alkamide levels, was found when *E. purpurea* roots were chopped. However, time and temperature storage had a significant effect on alkamide levels. A 40% and 80% loss in alkamide levels were found in the chopped *E. purpurea* stored at –18 or 24°C for 64 weeks, respectively. The reduction in alkamides was hypothesized as an oxidation problem as noted by Bauer et al. (1988b) during storage of *E. pallida*. The storage of herbal medicines containing *Echinacea*, in dry form, would be expected to have similar alkamide reductions. In contrast, no significant reduction of alkamides was found when alkamides were stored under solvent, i.e., acetonitrile (Perry et al., 2000). Although acetonitrile would not be used in tinctures, the alkamides may behave in a similar manner under ethanol-preserved tinctures.

Wills and Stuart (2000) found that alkamide concentrations increased when *Echinacea* was subject to various physical treatments that included cutting, compressing or crushing. These authors hypothesized that the higher alkamide concentrations observed in the physically abused samples was due to the faster drying time as opposed to a biosynthetic production of alkamides.

The physically abused samples were dried, at 40°C, to less than 12% moisture in 12 h, whereas whole plants required 48 h to reach the same moisture level. Storage of freshly harvested *Echinacea* at 20°C and 60% RH did not significantly affect the concentration of alkamide levels over a 30-day storage period (Wills and Stuart, 2000). However, a significant reduction in alkamide concentrations was found when dried, crushed *Echinacea* was stored at 20°C in the presence of light or at 30°C in the dark over 60 days. These data support the findings of Rogers et al. (1998), who found that storage of powdered *E. angustifolia* roots in a sealed bag at room temperature, under desiccation, resulted in a reduction in alkamide levels by 13%.

Kim et al. (2000b) assessed alkamide retention in *Echinacea* processed using various drying methods. These authors found that freeze-dried *Echinacea* roots retained significantly higher total alkamide levels compared to vacuum (50 mmHg)-microwave-dried, air-dried at 70°C, and vacuum-microwave-dried at partial vacuum (200 mmHg) products. However, no significant differences in alkamide concentrations in *Echinacea* leaves were found between the freeze-dried, vacuum-microwave-dried, or air-dried at 50°C leaves. Again, the products that were vacuum-microwave-dried using partial vacuum treatment had the lowest alkamide concentrations. The lower alkamide levels were thought to be the result of an oxidation reaction.

Bauer (1999b) found that the alkamide, dodeca-2E,4E,8Z,10E/Z-tetraenoic acid isobutylamide, level was influenced by preparation method. The nonthermal preparations appeared to have slightly higher levels of the tested alkamide than the thermally treated products. Bauer et al. (1988b) noted that 8-hydroxy-9-ene derivatives were formed during the storage of *E. pallida*. They noted that the polyacetylenes and polyenes undergo oxidative reactions and suggest that the root remains whole and that the extract remain in solution to prevent oxidative degradation.

8.5 *ECHINACEA*: POTENTIAL USES AND HEALTH BENEFITS

E. purpurea, E. angustifolia DC., and *E. pallida* are the most widely used *Echinacea* species for medicinal or health purposes. However, species such as *E. tennesseensis* are often harvested along with *E. purpurea* and *E. angustifolia* to the point where the exploitation by wildcrafters has threatened the *E. tennesseensis* species in their original range. *Echinacea* is promoted as an immune-enhancing herbal product but could be easily incorporated into cereal-based products (Wilson, 1998), provided that GRAS status is granted to *Echinacea* for use in foods by the U.S. Food and Drug Administration (FDA). *Echinacea* addition to a food system, to create a functional food, has merit that includes a body of over 300 research articles that focus on medicinal benefits. Of the 300 research articles, 34 deal with clinical trials. In addition, over two million prescriptions are filled by German physicians annually (Barrett et al., 1999), thus providing some evidence of the safety of *Echinacea* products. However, additional research is needed to support the potential health benefits and determine adverse effects of consuming *Echinacea* before it can be considered a component in functional foods. The following discussion highlights only a fraction of studies completed to date. The authors suggest the reviews of Bauer (1999a, 2000), Emmendörffer et al. (1999), and Melchart and Linde (1999) as additional reading material for more detailed information.

8.5.1 HEALTH BENEFITS

The benefits of consuming *Echinacea* concentrates range from improving immune response to prolonging the life span of cancer patients. Improving immune response has been the most widely documented benefit of *Echinacea* (Vömel, 1985; Bauer et al., 1988a; Erhard et al., 1994; Burger et al., 1997; See et al., 1997; Rehman et al., 1999). Vömel (1985) first reported that phagocytosis of erythrocytes was significantly enhanced, over nontreatment, upon treatment with *Echinacea*, while Bauer et al. (1988a) found that the lipophilic fraction of *Echinacea* was more active than the polar fraction at enhancing phagocytosis (i.e., first immune reaction against invading foreign substances). The enhanced immune response was believed, but not proven, to be due to the isobutylamides and polyacetylene components in the lipophilic fraction. Ethanol extracts of *Echinacea* roots enhanced phagocytosis (Erhard et al., 1994) and the rate of immunoglobulin G (IgG) production (Rehman et al., 1999). Rehman et al. (1999) data support previous research that the *Echinacea* enhanced-immune response, but they noted that the maximal IgG level was not affected when compared to a placebo group. Burger et al. (1997) reported that cytokine production was enhanced by the addition of an ethanol extract from the aerial parts of *Echinacea*. Glycoprotein-polysaccharide complex stimulated the release of interleukin 1, tumor necrosis factor (TNF) and interferon (IFN) in macrophages, *in vitro* (Wagner et al., 1988; Roesler et al., 1991b). However, the same immune responses were not significantly enhanced after human subjects were injected intravenously with a polysaccharide solution, although nonspecific immune functions could be shown (Roesler et al., 1991a). For additional information regarding the immunological activity and clinical investigations of *Echinacea*, see the reviews of Bauer (1999a, 2000), Emmendörffer et al. (1999), and Melchart and Linde (1999).

8.5.2 ANTIMICROBIAL AND ANTIVIRAL ACTIVITIES

Echinacea has been found to promote resistence to viruses. Wacker and Hilbig (1978) found that L 929 cells were 50–80 times more resistant to the influenza, herpes, and vesicular stomatitis viruses for 24 h after treatment with methanol and aqueous extracts of *Echinacea*. These authors noted that the cell became sensitive to the viruses after 48 h, suggesting that repeat treatment would be required. Macrophages eliminate *Listeria monocytogens* and *Candida albicans* from internal organs of humans when polysaccharides isolated from cell cultures of *Echinacea* were added to a model test system (Roesler et al., 1991b; Steinmüller et al., 1993). Cichoric and caftaric acids inhibited the hyaluronidase enzyme; an enzyme produced by pathogenic organisms to penetrate tissue and cause infection (Facino et al., 1995). The exact mode by which *Echinacea* phytochemicals enhance immune functions is not fully elucidated, but the inhibition of the growth of microorganisms may play a role in immune-enhancing activity.

Preliminary results, in the authors' laboratory, using a paper disc assay, have shown that methanol-water extracts of *Echinacea* were effective at inhibiting two strains of *Escherichia coli*, *Bacillus cereus*, *Salmonella thyphimurium*, and *Staphylococcus aureus*, while hexane extracts were less effective. *B. cereus* was the most susceptible, and *E. coli* was the least susceptible. In addition, extracts of the herbal portion of *E. purpurea* were more effective than *E. angustifolia*. CADs may be responsible for the microbial inhibition, similar to that found with phenolic antioxidants (Gailani and Fung, 1984). The preliminary antimicrobial activity is promising, and research continues to determine the inhibitory concentrations for the extracts against various microorganisms.

8.5.3 ANTIOXIDANT AND ANTI-INFLAMMATORY ACTIVITIES

Cichoric acid and other CADs such as echinacoside, cynarine, and chlorogenic acid, were found to act as radical scavengers (Facino et al., 1995). Although not tested, verbascoside and 2-caffeoyl-cichoric acid would be expected to have similar radical scavenging activity. Facino et al. (1995) reported that *Echinacea* protected collagen from free radical damage by scavenging reactive oxygen species. Echinacoside and cichoric acid were found to prevent collagen degradation best followed by cynarine

= caffeic acid and chlorogenic acid. These authors noted that skin preparations containing *Echinacea* may prevent skin damage caused by UVA and UVB radiation.

Contrary to Facino et al. (1995), research completed by Pietta et al. (1998b) showed that *Echinacea* was an ineffective antioxidant. Work completed by authors (Hall III and Schwarz, unpublished data) showed that commercial *Echinacea* products had antioxidant potential. The test systems included the determination of induction time, using the Rancimat, as a way to evaluate the oxidative stability of a cold-pressed sunflower oil. Three commercial preparations included *Echinacea* dried aerial parts (i.e., herb), an ethanol-preserved *Echinacea* tincture, and a root extract. The addition of 200 ppm *Echinacea* preparation significantly improved the oxidative stability of the sunflower oil, which had an induction time of 23.3 h at 110°C. The induction time of the ethanol-preserved *Echinacea* tincture was 31.6 h, while the root extract and herb inhibited oxidation for 36.6 and 40 h, respectively. Further investigation using bulk *E. purpurea* and *angustifolia* showed that the herb part of *E. purpurea* was the most effective at reducing oxidation followed by *E. purpurea* root and *E. angustifolia* root and herb. A qualitative assessment of the CADs by reverse-phase HPLC showed that *E. purpurea* appeared to have the highest concentration of phenolic compounds. A quantitative characterization of the phenolic compounds is underway to establish a relationship between *Echinacea* species and antioxidant activity. The results from this investigation follow similar trends observed by Hu and Kitts (2000). They found that a methanol extract of *E. pallida* had a higher antioxidant activity than *E. purpurea* and *angustifolia*. In addition, the relationship between CAD and antioxidant activity was established. The CAD concentrations of 2.61, 1.11, and 0.49% were found in *E. pallida, E. angustifolia*, and *E. purpurea*, respectively; which correlated to antioxidant activity. Hu and Kitts (2000) hypothesized that echinacoside may account for the antioxidant activity based on the observation that the *E. purpurea* lacked echinacoside and had the lowest antioxidant potential.

The anti-inflammatory property of *Echinacea* has been associated with alkamides and polysaccharides. Tubaro et al. (1987) found that an *E. angustifolia* root polysaccharide fraction (EPF) effectively inhibited a croton oil-induced edema when applied topically, while an intravenous injection (0.5 mg/kg) inhibited a carrageenan-induced edema. The intravenous injection was slightly more effective than topical applications. Water extracts of *E. angustifolia* root (EAE) were less effective than the polysaccharide fraction at inhibiting carrageenan-induced edema. To achieve similar inhibition rates, 100 times more EAE was required compared to EPF. The application of *Echinacea* alkamides was shown to inhibit 5-lipoxygenase, a key enzyme in arachidonic acid metabolism to prostagladins (a component identified as promoting inflammation), via a competitive inhibition of the enzyme mechanism (Müller-Jakic et al., 1994). Alternatively, radical scavenging activity of the highly unsaturated alkamides may reduce oxidative reactions that may occur during arachidonic acid metabolism. Dodeca-2*E*,4*E*,8*Z*,10*E*/*Z*-tetraenoic acid isobutylamides had the highest inhibitory activity (62.6%) against 5-lipoxygenase. However, a hexane extract of *E. angustifolia* roots inhibited 81.8% of the 5-lipoxygenase activity, suggesting that additional components are involved in the enzyme inhibition (Müller-Jakic et al., 1994).

8.6 COMMERCIAL PRODUCTS: STANDARDIZATION AND PRODUCT REGULATION

Echinacea is sold as a dietary supplement in the United States and as natural health products in Canada, while in Germany and many European countries, *Echinacea* products are sold as drugs in pharmacies (Bauer, 2000). There are a number of products on the market, which include dried herbal and root, alcohol tinctures and extracts, and expressed juice products, thus, standardization would be a difficult task. However, regulation would be less difficult because all of these products would fall under a dietary supplements category.

8.6.1 PRODUCT REGULATION

The regulation of *Echinacea* products in the United States would fall under the Dietary Supplement Health and Education Act of 1994 (DSHEA). Under DSHEA, structure/function claims are permissible, while claims targeting specific diseases would not be permissible, unless reviewed by the FDA. A structure/function claim such as "support of the immune system" would be acceptable, while "alleviates the common cold or flu" would not be acceptable for *Echinacea*, because the claim would target a specific disease state (i.e., cold or flu). In the United States, statute 403 (a)(1) of the Federal Food, Drug, and Cosmetic Act "prohibits labeling that is false or misleading"; thus, a structure/function claim must meet the criteria set forth in 403(r)(6).

Echinacea must first be approved as a GRAS ingredient before it can legally be added to food. In many European countries, health claims are not permitted on food products, while in other countries, such as Japan, claims are allowed under the Foods for Specified Health Use (FOSHU) system. The benefits of *Echinacea* in laboratory studies show promise and potential that functional foods may come from this research, provided that the studies support the safety of *Echinacea*.

8.6.2 PRODUCT STANDARDIZATION

One issue not resolved by the DSHEA, specifically for botanical or herbal dietary products, is the lack of standardization of the active component. In general, the standardization is based on the level of plant material rather than on one specific compound. In many cases, the specific compound responsible for the health benefit has not been fully characterized. In addition, many phytochemicals constituents may participate in an observed health benefit.

The standardization of *Echinacea* products based on cichoric acid and alkamides has been proposed (Bauer 1999a,b; Perry et al., 2000, 2001). The CADs are also commonly used as marker compounds and labeled as "phenolics" on many dietary supplements. The main reason for standardization is that the level of the active components varies based on plant material, age of the *Echinacea*, and method of preparation. However, the specific compounds responsible for the immune-enhancing activity of *Echinacea* are not fully understood. Echinacoside has been used to standardize many preparations, but the lack of immune-enhancing activity of this component suggests that it should not be used as the component for standardization. However, this compound could be used as a marker compound for authenticating the species of *Echinacea* used in the preparation.

The use of GC-MS coupled with multivariate statistical analysis proved valuable in verifying the authenticity of *Echinacea* species (Lienert et al., 1998). The principal components analysis and cluster analysis were used as a way to group similar root extracts based on the identified compounds from the GC run. The correct grouping of the *Echinacea* species (i.e., *purpurea, angustifolia* and *pallida*) was not influenced by extraction method or by the aging process of the roots. In addition, TLC is often used to verify the authenticity of *Echinacea* preparations.

Echinacea has been used for centuries as a medicinal plant. Research from the last two decades shows that *Echinacea* can enhance the immune system when *in vitro* and *in vivo* indicators are used in the model investigations. Further investigations, e.g., bioavailability and clinical studies, are needed to thoroughly understand the immune-enhancing activity of *Echinacea*. In addition, further research is needed to characterize processing methods that do not significantly reduce the levels of phytochemicals.

REFERENCES

Barrett, B., Kiefer, D. and Rabago, D. 1999. "Assessing the risks and benefits of herbal medicine: an overview of scientific evidence," *Altern. Therap. Health Med.* 5(4):40–9.

Bauer, R. 1997. "Standardisierung von *Echinacea purpurea*-Preßsaft auf cichoriensäure und alkamide," *Z. Phytother.* 18:270–276.

Bauer, R. 1999a. "Clinical investigations of *Echinacea* phytopharmaceuticals," in *Immunomodulatory Agents from Plants*, ed., H. Wagner, Basel, Switzerland: Birkhäuser Verlag, pp. 41–88.

Bauer, R. 1999b. "Standardization of *Echinacea purpurea* expressed juice with reference to cichoric acid and alkamides," *J. Herbs Spices Med. Plants.* 6:51–61.

Bauer, R. 2000. "Chemistry, pharmacology, and clinical applications of *Echinacea* products," in *Herbs, Botanicals and Teas as Functional Foods and Nutraceuticals*, eds., G. Mazza and B.D. Oomah, Lancaster, PA: Technomic Publishing Co., Inc., pp. 45–73.

Bauer, R. and Foster, S. 1991. "Analysis of alkamides and caffeic acid derivatives from *Echinacea simulata* and *E. paradoxa* roots," *Planta Med.* 57(5):447–449.

Bauer, R. and Remiger, P. 1989. "TLC and HPLC analysis of alkamides in *Echinacea* drugs," *Planta Medica.* 55:367–371.

Bauer, R., Jurcic, K., Puhlmann, J. and Wagner, H. 1988a. "Immunologische *in-vivo* und *in-vitro* untersuchungen mit *Echinacea*-Extrakten," *Arzneim.-Forsch.* 38:276–281.

Bauer, R., Khan, A. and Wagner, H. 1988b. "TLC and HPLC analysis of *Echinacea pallida* and *E. angustifolia* roots," *Planta Med.* 54(5):426–430.

Bauer, R., Remiger, P. and Wagner, H. 1988c. "Alkamides from the roots of *Echinacea purpurea*," *Phytochemistry.* 27(7):2339–2342.

Bauer, R., Remiger, P. and Wagner, H. 1989. "Alkamides from the roots of *Echinacea angustifolia*," *Phytochemistry.* 28(2):505–508.

Becker, H. and Hsieh, W. 1982. "Structure of Echinacoside," *Zeitschriff Fur Naturfuschung*, Section C. 37:351–353.

Becker, H. and Hsieh, W. 1985. "Chocoree-Saüre und deren Derivate aus *Echinacea* arten," *Zeitschriff Fur Naturfuschung*, Section C. 40:585–587.

Bohlmann, F. and Hoffman, H. 1983. "Further Amides form *Echinacea purpurea*," *Phytochemistry.* 22(5):1173–1175.

Brevoort, P. 1998. "The booming U.S. botanical market," *Herbalgram.* 44:33–46.

Burger, R.A., Torres, A.R., Warren, R.P., Caldwell, V.D. and Hughes, B.G. 1997. "*Echinacea*-Induced cytokine production by human macrophages," *Int. J. Immunopharmacol.* 19(7):371–9.

Cheminat, A., Zawatzky, R., Becher, H. and Brouillard, R. 1988. "Caffeoyl conjugates from *Echinacea* species: structures and biological activity," *Phytochemistry.* 27:2787–2794.

Emmendörffer, A., Wagner, H., and Lohmann-Matthes, M.L. 1999. "Clinical investigations of *Echinacea* phytopharmaceuticals," in *Immunomodulatory Agents From Plants*, ed., H. Wagner, Basel, Switzerland: Birkhäuser Verlag, pp. 89–104.

Erhard, M., Kellner, J., Wild, J., Losch, U. and Hatiboglu, F.S. 1994. "Effect of *Echinacea, Aconitum, Lachesis* and *Apis* extracts, and their combinations on phagocytosis of human granulocytes," *Phytother. Res.* 8(1):14–17.

Facino, R.M., Carini, M., Aldini, G., Marinello, C., Arlandini, E., Franzoi, L., Colombo, M., Pietta, P. and Mauri, P. 1993. "Direct characterization of caffeoyl esters with antihyaluronidase activity in crude extracts from *Echinacea angustifolia* roots by fast atom bombardment tandem mass spectrometry," *Farmaco.* 48(10):1447–61.

Facino, R.M., Carini, M., Aldini, G., Saibene, L., Pietta, P. and Mauri, P. 1995. "Echinacoside and caffeoyl conjugates protect collagen from free radical-induced degradation: a potential use of *Echinacea* extracts in the prevention of skin photodamage," *Planta Med.* 61(6):510–4.

Foster, S. 1985. "Echinacea: the purple coneflower," *Amer. Horticult.* August: 14–17.

Gailani, M. and Fung, D. 1984. "Antimicrobial effects of selected antioxidants in laboratory media and in ground pork," *J. Food Protection.* 47:428–33.

Glowniak, K., Zgorka, G. and Kozyra, M. 1996. "Solid-phase extraction and reversed-phase high-performance liquid chromatography of free phenolic acids in some *Echinacea* species," *J. Chromatogr. A.* 730(1/2):25–29.

He, X., Lin, L., Bernart, M.W. and Lian, L. 1998. "Analysis of alkamides in roots and achenes of *Echinacea purpurea* by liquid chromatography-electrospray mass spectrometry," *J. Chromatogr. A.* 815(2):205–211.

Heinzer, F., Chavanne, M., Meusy, J.P., Maitre, H.P., Giger, E. and Baumann, T.W. 1988. "The classification of therapeutically used species of the genus *Echinacea.*" *Pharm. Acta Helv.* 63(4–5):132–6.

Hu, C. and Kitts, D. 2000. "Studies on the antioxidant activity of *Echinacea* root extract." *J. Agric. Food Chem.* 48:1466–72.

Kim, H., Durance, R., Scaman, C. and Kitts, D. 2000a. "Retention of caffeic acid derivatives in dried *Echinacea purpurea.*" *J. Agric. Food Chem.* 48:4182–86.

Kim, H., Durance, T., Scaman, C. and Kitts, D. 2000b. "Retention of alkamides in dried *Echinacea purpurea.*" *J. Agric. Food Chem.* 48:4187–92.

Li, T.S.C. 1998. "*Echinacea*: cultivation and medicinal value," *HortTechnol.* 8(2):122–129.

Lienert, D., Anklam, E. and Panne, U. 1998. "Gas chromatography-mass spectral analysis of roots of *Echinacea* species and classification by multivariate data analysis," *Phytochem. Anal.* 9:88–98.

Little, R. 1999. "Taming *Echinacea* angustifolia: research at SDSU and insights from a grower," http://www.abs.sdstate.edu/bio/reesen/Echinaca/newsletter.htm.

Mazza, G. and Cottrell, T. 1999. "Volatile components of roots, stems, leaves, and flowers of *Echinacea* species," *J. Agric. Food Chem.* 47:3081–3085.

Melchart, D. and Linde, K. 1999. "Clinical investigations of *Echinacea* phytopharmaceuticals," in *Immunomodulatory Agents from Plants*, ed., H. Wagner, Basel, Switzerland: Birkhäuser Verlag, pp. 105–118.

Müller-Jakic, B., Breu, W., Probstle, A., Redl, K., Greger, H. and Bauer, R. 1994. "*In Vitro* inhibition of cyclooxygenase and 5-lipoxygenase by alkamides from *Echinacea* and *Achillea* species," *Planta Med.* 60(1):37–40.

Nüsslein, B., Kurzmann, M., Bauer, R. and Kreis, W. 2000. "Enzymatic degradation of cichoric acid in *Echinacea purpurea* preparations," *J. Natural Products.* 63(12):1615–1618.

Perry, N.B., Van Klink, J.W., Burgess, E.J. and Parmenter, G.A. 1997. "Alkamide levels in *Echinacea purpurea*: a rapid analytical method revealing differences among roots, rhizomes, stems, leaves and flowers," *Planta Med.* 63(1):58–62.

Perry, N.B., Van Klink, J.W., Burgess, E.J. and Parmenter, G.A. 2000. "Alkamide levels in *Echinacea purpurea*: effects of processing, drying, and storage," *Planta Med.* 66:54–56.

Perry, N., Burgess, E. and Glennie, V. 2001. "*Echinacea* standardization: analytical methods for phenolic compounds and typical levels in medicinal species," *J. Agric. Food Chem.* 49(4):1702–1706.

Pietta, P., Mauri, P. and Bauer, R. 1998a. "MEKC analysis of different *Echinacea* species," *Planta Med.* 64(7):649–652.

Pietta, P., Simonetti, P. and Mauri, P. 1998b. "Antioxidant activity of selected medicinal plants," *J. Agric. Food Chem.* 46:4487–4490.

Proksch, A. and Wagner, H. 1987. "Structural analysis of a 4-*O*-methyl-glucuronoarabinox-ylan with immuno-stimulating activity from *Echinacea purpurea*," *Phytochemistry.* 26(7):1989–1993.

Rehman, J., Dillow, J.M., Carter, S.M., Chou, J., Le, B. and Maisel, A.S. 1999. "Increased production of antigen-specific immunoglobulins G and M following *in vivo* treatment with the medicinal plants *Echinacea angustifolia* and *Hydrastis canadensis*." *Immunol. Lett.* 68(2–3):391–5.

Roesler, J., Emmendorffer, A., Steinmuller, C., Luttig, B., Wagner, H. and Lohmann-Matthes, M.L. 1991a. "Application of purified polysaccharides from cell cultures of the plant *Echinacea purpurea* to test subjects mediates activation of the phagocyte system," *Int. J. Immuno-pharmacol.* 13(7):831–41.

Roesler, J., Steinmuller, C., Kiderlen, A., Emmendorffer, A., Wagner, H. and Lohmann-Matthes, M.L. 1991b. "Application of purified polysaccharides from cell cultures of the plant *Echinacea purpurea* to mice mediates protection against systemic infections with *Listeria monocytogenes* and *Candida albicans*." *Int. J. Immunopharmacol.* 13(1):27–37.

Rogers, K.L., Grice, I.D., Mitchell, C.J. and Griffiths, L.R. 1998. "High performance liquid chromatography determined alkamide levels in Australian-grown *Echinacea* spp." *Aust. J. Exp. Agric.* 38(4):403–408.

Schulte, K.E., Rucker, G. and Perlick, J. 1967. "The presence of polyacetylene compounds in *Echinacea purpurea* Munch and *Echinacea angustifolia* DC." *Arzneimittelforschung.* 17(7):825–9.

Schulthess, B.H., Giger, E. and Baumann, T.W. 1991. "*Echinacea*: anatomy, phytochemical pattern, and germination of the achene." *Planta Med.* 57(4):384–8.

See, D.M., Broumand, N., Sahl, L. and Tilles, J.G. 1997. "*In vitro* effects of *Echinacea* and ginseng on natural killer and antibody-dependent cell cytotoxicity in healthy subjects and chronic fatigue syndrome or acquired immunodeficiency syndrome patients." *Immunopharmacology.* 35(3):229–35.

Steinmüller, C., Roesler, J., Grottrup, E., Franke, G., Wagner, H. and Lohmann Matthes, M.L. 1993. "Polysaccharides isolated from plant cell cultures of *Echinacea purpurea* enhance the resistance of immunosuppressed mice against systemic infections with *Candida albicans* and *Listeria monocytogenes*." *Int. J. Immunopharmacol.* 15(5):605–14.

Tubaro, A., Tragni, E., Del Negro, P., Galli, C. and Loggia, R. 1987. "Anti-inflammatory activity of a polysaccharidic fraction of *Echinacea angustifolia*." *J. Pharm. Pharmacol.* 39:567–9.

Voaden, D. and Jacobson, M. 1972. "Tumor inhibitors 3. Identification and synthesis of an oncolytic hydrocarbon from American coneflower roots." *J. Med. Chem.* 15:619–623.

Vömel, T. 1985. "Der einflußeines pflanzlichen immunostimulans auf die phagozytose von erythrozyten durch das retikulohistozytäre ststem der isoliert perfundierten ratten-leber." *Arzneim. Forsch./Drug Res.* 35:1437–1439.

Wacker, A. and Hilbig, W. 1978. "Virus-Inhibition by *Echinacea purpurea* (Author's Translation)," *Planta Med.* 33(1):89–102.

Wagner, H., Proksch, A., Riess Maurer, I., Vollmar, A., Odenthal, S., Stuppner, H., Jurcic, K., Le Turdu, M. and Heur, Y.H. 1984. "Immunostimulant action of polysaccharides (heteroglycans) from higher plants. Preliminary communication." *Arzneimittelforschung.* 34(6):659–61.

Wagner, H., Stuppner, H., Schafer, W. and Zenk, M. 1988. "Immunologically active polysaccharides of *Echinacea purpurea* cell cultures." *Phytochemistry.* 27(1):119–126.

Wills, R.B.H. and Stuart, D.L. 1999. "Alkylamide and cichoric acid levels in *Echinacea purpurea* grown in Australia." *Food Chem.* 67:385–388.

Wills, R.B.H. and Stuart, D.L. 2000. "Effect of handling and storage on alkylamide and cichoric acid levels in *Echinacea purpurea*," *J. Sci. Food Agric.* 80:1402–6.

Wilson, D. 1998. "Nutraceuticals: a natural choice for grain-based products." *Cereal Foods World.* Sept., 43(9):718–719.

9 Pectin from Fruits

Qi Wang, Jordi Pagán and John Shi

CONTENTS

1-5667-6902-7/02/$0.00+$1.50
© 2002 by CRC Press LLC

9.1 INTRODUCTION

Vauquelin was the first to report, in 1790, pectin as a soluble substance in fruit juices. It was not truly identified until 1824, when French scientist Braconnot took up Vauquelin's work and described a "substance widely available in plant life and already occasionally observed in the past has jellifying properties which gel when mixed with acid in its solution" (Braconnot, 1825). He called this substance "Pectin-acid" from the Greek word "pectos," which means solidify, congeal or curdle. Pectin has since come to be the general term for a complex group of polysaccharides that occur as structural materials in all higher plants. The term pectic substances is commonly used to encompass methoxyl esterified pectin, deesterified pectic acid and its salts (pectates), and certain neutral polysaccharides often found in association with pectin (Voragen et al., 1995). In this chapter, pectin is used to include both pectin and pectic substance. Another associated term, protopectin, is used to describe the native pectin fraction within the cell walls that cannot be extracted without some degradation.

Pectin makes up about one-third of the cell wall dry matter of dicotyledoneous and some monocotyledoneous plants (Jarvis et. al., 1988). Much smaller proportions of these substances are found in the cell walls of grasses (Wada and Ray, 1978). Pectin contributes both to adhesion between the cells and to the mechanical strength of the cell wall, behaving in the manner of stabilized gels (Jarvis, 1984). Physical changes in the plant tissue, such as softening, are frequently accompanied by changes in the properties of the pectic substances.

Over the years, there has been a significant amount of research on pectin because of the wide and long-established application of this material in the food and nonfood industries (Endress, 1991). Due to the wide occurance and potentially high content of pectin in fruits and vegetables, it has been recognized as a valuable source of dietary fiber (DF). The health benefits of DF and specifically of pectins, both as an isolate form and as a natural component of cell walls, are well documented and remain a priority research area (Truswell and Beynen, 1992). There is also growing evidence of the pharmacological activities of certain types of pectins.

This chapter reviews the physicochemical and functional properties of pectins, their biological role in plants and physiological effects either as naturally occuring or as intent plant foods.

9.2 FUNCTIONS OF PECTINS IN PLANT TISSUES

9.2.1 PECTIN IN CELL WALLS

Pectin is a major cell wall component with a variety of important biological functions in plants. It plays a role in the control of cell growth, in defense against invasions of microorganisms and in maintaining the physical and sensory properties of fresh fruits and their processing characteristics.

Plant cell walls consist of a series of layers, from outer to inner, respectively, are the middle lamella, primary cell wall, secondary cell wall (in some of the plants) and plasma membrane. Cell walls contain approximately 60% water and 40% polymers, of which pectins make up 20–35% (Jarvis, 1982). The highest concentration of pectin is seen in the middle lamella, with a gradual decrease in passing through the primary cell wall toward the plasma membrane (Kertesz, 1951; Darvill et al., 1980; Brett and Waldron, 1996). The concentration of pectin in the middle lamella is estimated to be in the order of 10–30%. At these concentrations, pectins form large numbers of both nonspecific entanglements and cross-links through Ca^{2+}, which lead to strong cohesiveness in the structure. One of the functions of the middle lamella pectin is to bind adjacent cell walls together through cell-cell adhesion, which requires the middle lamella pectin to be effectively interconnected with components embedded in the primary cell walls. These embedded components may be pectins and/or hemicelluloses (mainly xyloglucan). A primary cell wall is a complex structure that allows growth of the cell, while acting as a skeleton for the plant. A recent model of primary cell wall structure is shown in Figure 9.1 (Carpita and Gibeaut, 1993). The proposed structure consists of three independent domains: a cellulose-xyloglucan framework, a matrix of pectic polysaccharides and a domain of structural proteins. Pectin chains can form a gel-like network contributing to the firmness and structure of plant tissue. In some cell walls, the network structure can also be stabilized by cross-links between feruloyl groups that are attached to the pectin chain (Carpita and Gibeaut, 1993; Goldberg et al., 1996).

9.2.2 CHANGES DURING RIPENING AND STORAGE

The role of pectin in ripening of fruits has been the subject of many studies over the years (Knee, 1978; Poovaiah et al., 1988). Evaluations have varied greatly depending on species, stage of ripening and the approach of the study. In general,

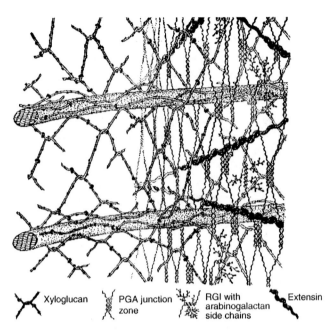

Xyloglucan   PGA junction   RGI with   Extensin
              zone   arabinogalactan
                             side chains

FIGURE 9.1 A model for the primary cell wall structure in plants, where PGA stands for polygalacturonan and RG I stands for rhamnogalacturonan. With permission from Carpita and Gibeaut (1993).

ripening of fruits is characterized by softening, loss of cell cohesion and by an accumulation of water-soluble pectins at the expense of protopectin (Bartley and Knee, 1982; Redgwell et al., 1997). The increase in pectin solubility is considered to result from the actions of a number of cell-wall-degrading enzymes, which include *endo*- and *exo*-polygalacturonases (Gross and Wallner, 1979), methylesterases (Castaldo et al., 1989; Seymour et al., 1987) and various glycosidases (Wallner and Walker, 1975). In most fruits, *endo*-polygalacturonases play a crucial role in converting protopectin to a soluble form acting by depolymerization of the pectic backbones, although, in some fruits (apple and freestone peach), only *exo*-polygalacturonases are detected during ripening (Pressey and Avants, 1978). In the latter case, as the fruit ripens, the pectic fraction is replaced by newly synthesized pectins that are highly esterified, of high molecular weight and contain a lower proportion of neutral residues (Knee, 1978; Knee and Bartley, 1981). However, a decreased molecular weight with a broad molecular weight distribution of the water-soluble fractions of pectin is seen in most of the fruits during ripening. The role of methylesterase in the softening of the cell wall is not very clear. Although a decrease in degree of esterification (DE) of the middle lamella pectins was observed by some researchers (Hobson, 1963; Burns and Pressey, 1987), other researchers found no change of DE during ripening (de Vries et al., 1984). Reduced DE may render the pectin more susceptible to *endo*-polygalacturonases, because both *endo*- and *exo*-polygalacturonases act only on de-esterified residues (Pressey and Avants, 1982). The loss of neutral sugars during ripening, mainly galactose and arabinose, implicates the action of various glycosidases (Gross and Sams, 1984). It has been suggested that the increase in pectin solubility may also result from cleavage of linkages between side chains of pectin and hemicelluloses (de Vries et al., 1982; Nogata et al., 1996).

9.3 NUTRITIONAL AND PHYSIOLOGICAL EFFECTS

Pectins display a wide range of physiological and nutritional effects important to human nutrition and health. Some of these functions are determined by chemical structures, while others are more related to physical properties (albeit controlled by structure). Pectin is a dietary fiber, because it is not digested by enzymes produced by humans. Although not digested and absorbed in the upper gastrointestinal tract, pectin can be fermented by colonic microflora in the colon to CO_2, CH_4, H_2 and short-chain fatty acids (SCFA), mainly, acetate, propionate and butyrate. These fatty acids are a potential energy source for the mucosal cells of the large intestine, and some may be absorbed from the colon, providing energy and having additional metabolic effects.

9.3.1 EFFECT OF PECTIN ON CARBOHYDRATE METABOLISM

In the mid 1970s, Jenkins and colleagues reported that pectin attenuated the postprandial rise in blood glucose and insulin concentrations in both insulin-dependent and non-insulin-dependent diabetic people after a carbohydrate meal (Jenkins et al., 1976). Many have since confirmed these results, showing that pectin either mixed into carbohydrate-containing foods or in glucose drinks, significantly decreases postprandial glycemia and insulinemia in diabetic patients (Monnier et al., 1978; Williams et al., 1980; Poynard et al., 1982; Schwartz et al., 1988; Tunali et al., 1990). A similar response is observed in healthy volunteers (Jenkins et al., 1977; Holt et al., 1979; Sahi et al., 1985). It is well documented that an improvement in glycemic control in diabetic patients can reduce the risk of microvascular complications (e.g., nephropathy, retinopathy) (DCCT Research Group, 1993; UKPDS Group, 1998). Guar gum, another water-soluble dietary fiber with a glycemia-lowering effect, elicited significant long-term improvement in diabetic control (Jenkins et al., 1980; Aro et al., 1981; Groop et al., 1993). It is reasonable to suppose that pectin and other viscous fibers may similarly have long-term beneficial effects. Evidence in the literature suggests that a dose of 15–30 g/day of pectin and adequate mixing with carbohydrate foods in a hydrated form are necessary for efficacy. Some studies have shown that lower doses of pectin or consumption before the carbohydrate test meal reduces or eliminates the effect on the postprandial plasma metabolites (Wahlqvist et al., 1979; Gold et al., 1980; Frape and Jones, 1995).

The mechanisms by which pectin and other viscous fibers affect carbohydrate metabolism have been attributed mainly to the ability of these materials to decrease the rate of carbohydrate absorption within the gastrointestinal tract. This seems to be a result of inhibition of a series of processes associated with lumenal digestion of carbohydrate, including gastric function, intestinal transit and mixing, α-amylase-starch interactions and movement of the hydrolyzed products of starch. A number of physicochemical mechanisms may be involved; which of these predominate will depend on various factors, particularly on the type of pectin and the form in which the fiber is ingested. For nongelling polysaccharide gums, such as guar gum, the ability to effectively increase the viscosity of digesta is a critical factor (Ellis et al., 1996; Gallaher et al., 1999). For pectins, particularly high methoxyl pectins, which readily gel at low pH, gelation may also contribute to the glycemia-lowering effects so long as there remains sufficient co-solutes, such as starch and its hydrolysis products.

9.3.2 EFFECT OF PECTIN ON LIPID METABOLISM

The lowering of serum cholesterol by pectin was reported in rats and humans as early as 1960s (Keys et al., 1961; Wells and Ershoff, 1961; Palmer and Dixon, 1966). Since then, the short-term cholesterol-lowering effect of pectin has been demonstrated repeatedly in a wide variety of subjects and experimental conditions (Jenkins et al., 1975; Kay and Truswell, 1977; Judd and Truswell, 1982; Vargo et al., 1985; Schuderer, 1986; Haskell et al., 1992). Most studies have shown that 10–50 g/day pectin substances reduces total cholesterol, low-density cholesterol (LDL-c) and very low-density cholesterol, and the low-density to high-density cholesterol (HDL-c) ratio; HDL cholesterol itself is only minimally affected. Long-term effects of pectin on serum cholesterol and consequences of this need further investigation. Some earlier studies with guar gum failed to show long-term reduction in cholesterol levels (Jenkins et al., 1980; Simons et al., 1982). However, a recent study using subjects with mild to moderate hypercholesterolemia demonstrated that 15 g/day of a pectin and guar gum mixture produced significant reductions in total cholesterol and LDL-c sustained over a period of 51 weeks (Knopp et al., 1999); HDL-c and triglycerides were not significantly changed. Similar cholesterol-lowering effects were observed with a 15 g/day mixture of psyllium, pectin, guar gum and locust bean gum for a period of 24 weeks (Jensen et al., 1997). Because elevated LDL-c has been identified as a risk factor for cardiovascular diseases, consumption of pectin may benefit those who are at increased risk of cardiovascular diseases, including individuals with hyperlipidemia and diabetes.

There are suggestions that the hypolipidemic efficacy of high-methoxyl pectin is higher than that of low-methoxyl pectin (Schuderer, 1986; Furda, 1990). *In vitro*, high-methoxyl pectin bound more LDL than the same amount of low-methoxyl pectin (Falk and Nagyvary, 1982). Despite this, DE of pectin does not seem to be a determinant factor for the cholesterol-lowering effect, as long as sufficient quantity is taken. Using 15 g/day high-methoxyl (DE 71%), low-methoxyl (DE 37%) and amidated pectins, Judd and Truswell (1982) obtained similar reductions in serum cholesterol. Because different qualities and grades of pectin exist, it is possibly a redundant exercise to attempt to evaluate relative efficacy in the absence of physico-chemical data and a clear idea of mechanism. In clinical trials, there needs to be a minimum amount to achieve a statistically significant result, and also, there must be a plateau reached in response. Current evidence suggests that at least 10–15 g/day is necessary for cholesterol-lowering effects in both healthy individuals and hyper-lipidemic patients (Schwandt et al., 1982; Endress, 1991; Truswell and Beynen, 1992). Studies by Palmer and Dixon (1966) and Delbarre et al. (1977) demonstrated that little effect could be seen in individuals taking 6 g/day or less of pectin. However, a substantial lowering of serum total and LDL cholesterol levels was observed when 12 g/day was taken (Schwandt et al., 1982).

The mechanisms by which pectin affects lipid metabolism remain uncertain and may differ with the types and forms of pectin consumed. However, it is likely that rheological changes in the gastrointestinal contents, i.e., increased viscosity or gelation, are important determinants of the hypocholesterolemic effects. As for a number of other dietary fibers, the ingestion of pectin enhances excretion of bile acids and steroids (Kay and Truswell, 1977; Miettinen and Tarpila, 1977). To compensate for these bile acid losses, more cholesterol in the liver is converted to bile acids, which eventually leads to a decrease in total serum cholesterol and low-density lipoproteins. It has also been suggested that pectin reduces the rate and the quantity of lipid absorption by entrapping or binding the lipids and by interfering with micelle formation (Furda, 1990). Lewinska et al. (1994) added pectin in a granular form to human plasma in *in vitro* batch sorption experiments and found that 40% of total cholesterol, 45% of LDL-c and 36% of HDL-c were removed. It has also been demonstrated that SCFA (acetate) may be absorbed into the portal vein and ultimately impair hepatic cholesterol synthesis in hypercholesterolemic subjects (Veldman et al., 1999). The lipid-lowering effects were also related to the reduced insulin secretion and, hence, reduced stimulus to lipid synthesis.

9.3.3 PECTIN AND COLORECTAL CANCER

The possible association between dietary fiber intake and reduced risk of colorectal cancer has received considerable attention (Klurfeld, 1992) but remains controversial. Considerable evidence has accumulated to support the connection between reduced risk of colon cancer and a diet rich in water-insoluble fiber. Such a connection has not been seen consistently for diets rich in water-soluble fiber, including pectin. Results from studies of carcinogenesis in animals are variable; indeed, some dietary fibers appear to promote carcinogenesis. Jacobs (1990) evaluated eight studies in animals with chemically induced colon cancer, the feeding of pectin enhanced tumor development in half of the studies and showed protective effect in only one study. A more recent review (Ma et al., 1996) similarly found that of seven studies, dietary pectin showed protective effect in two, promotion of tumor development in three and no effect in the remaining two. The apparently conflicting results may arise from experimental design such as variability in animal models, the choice and dose of carcinogen, diet composition (particularly the type of fat), duration of the experiment and the choice of control groups. A recent study suggested that pectin was protective against colon cancer when fish oil, but not corn oil, was the lipid source (Lupton, 2000). The combination of pectin and fish oil was even more protective against tumor development than the combination of cellulose and fish oil. There is increasing evidence that results may depend on the stage of the tumorigenic process (Ma et al., 1996; Lupton, 2000). Taper and Roberfroid (1999) showed that dietary pectin significantly inhibited the growth of intramuscularly transplanted mouse tumous. Heitman et al. (1992) reported significant suppression of incidence of chemically induced colon cancer when rats were fed a 10% pectin diet at the promotion stage of carcinogenesis.

The high fermentability of pectin by colonic bacterial is perhaps the most important characteristic related to the development of colon cancer. Some epidemiological data suggest that acidic stools are negatively associated with colon cancer (Walker et al., 1986; Bruce, 1987). Because the fermentation of pectin lowers colonic pH, this may be protective against colon cancer (Thornton, 1981). However, animal models do not always support this hypothesis (Jacobs and Lupton, 1986; Lupton et al., 1988). *In vitro* experiments have shown that butyrate promotes differentiation and apoptosis in a variety of colon tumor cell lines (Kruh et al., 1995; Smith et al., 1998), which is likely to be beneficial with respect to colon cancer. However, there are again some discrepancies in the data on the effects of butyrate and other SCFA (Lupton, 1995). Thus, infusion of SCFA into the rat colon increased cell proliferation (Sakata, 1987; Kripke et al., 1989), a possible cancer promoting factor. Many other hypotheses on the role of pectin and other fermentable fiber in colon-tumor prevention have been proposed (Hsieh and Wu, 1995; Ohkami et al., 1995; Jiang et al., 1997) and add to our understanding of the question, but it is far from resolved.

9.3.4 Effect of Pectin on the Absorption of Other Nutrients

The possible chelating effect of pectin on the bioavailability of nutrients, other than carbohydrates and lipids, is a concern. Despite extensive investigations in both animals and humans, there are no final conclusions because of the conflicting results.

Experiments with animals generally tend to demonstrate a negative influence of pectin on protein digestibility and utilization (Shah et al., 1982; Mosenthin et al., 1994). Only a slight increase in ileal output of nitrogen, due to either the decreased digestion and absorption of protein or the increased losses of endogenous nitrogen was observed in ileostomy patients receiving 15 g/day citrus pectin (Sandberg et al., 1983). This negative effect of pectin was probably caused by a decreased accessibility of protein molecules to digestive enzymes and decreased transport of the products of digestion to the absorptive sites. Pectin may also inhibit enzyme activity and increase proteolytic enzyme secretion (Foreman and Schneeman, 1980; Eggum, 1992).

The effect of pectin on mineral absorption and utilization has been the subject of much controversy. Although many studies have shown that neither high nor low DE pectin have an influence on the absorption of zinc, calcium, magnesium and iron (Sandberg et al., 1983; Gillooly et al., 1984; Rossander, 1987), other studies have suggested a detrimental effect on the utilization of these minerals (Drews et al., 1979; Kelsay et al., 1979a,b). Pectin can bind Ca^{++}, which may reduce its availability for absorption in the small intestine. However, after fermentation of pectin, the bound Ca^{++} would be available for potential absorption in the colon (Trinidad et al., 1996). Some investigators have observed that pectin increased the absorption of ferrous iron in rats, probably through enhanced iron solubility (Gordon and Chao, 1984; Kim, 1998).

A number of studies using animal models demonstrated that dietary pectin affected the absorption and bioavailability of β-carotene (Rock and Swendseid,

1992), a precursor for vitamin A. In a recent study with Mongolian gerbils, Deming et al. (2000) found that consumption of citrus pectin (7 g/100 g) depressed the bioavailability of β-carotene, resulting in less conversion of β-carotene to vitamin A. Similar results were reported in humans, in whom the bioavailability of β-carotene given in a mixed supplement was markedly reduced by pectin in healthy young women (Riedl et al., 1999).

9.3.5 BIOMEDICAL PROPERTIES OF PECTIN

There are numerous reports on the biomedical effects of pectin. Antidiarrheic action is the most widely known property, and pectin has been used to treat diarrhea for more than half a century (Malyoth, 1934; Tompkins, 1938). This effect is often followed by an antivomitive action, which in babies improves assimilation and toleration, particularly of dairy products. This is a consequence of some protection and regulation of the gastrointestinal system (Pilnick and Zwicker, 1970). High-methoxyl pectins are often used to treat gastritis and ulcers. Because the stomach walls become covered with a gel-like film after swallowing pectin, a protective effect on the stomach from gastric and biliar hypersecretions occurs (Navarro and Navarro, 1985).

Pectin is a very efficient antireflux remedy. Patients with heartburn as their main symptom but normal esophageal mucosa at endoscopy are classified as having endoscopy-negative gastroesophageal reflux disease (GORD). GORD may be treated with a pectin-based raft-forming antireflux agent (Aflurax) (Havelund and Aalykke, 1997; Havelund et al., 1997). After 6 months of treatment, Aflurax significantly delayed recurrence of moderate or severe heartburn and erosive esophagitis.

C_3 and C_4 components are natural antibacterial agents present in human colostrum. Sepehri et al. (1998) reported that pectin-rich plant extracts enhanced transfer of C_3 and C_4 from blood to colostrum by an unknown mechanism. This observation suggests that these plant extracts may be used to reinforce the antibacterial activity of human colostrum.

In addition to the general health benefits mentioned above, pectin and pectic polysaccharides isolated from some plants have been reported to have a number of pharmacological activities as listed in Table 9.1. Not all pectins are biologically active, and not all studies are reliable. Results suggest that each bioactivity of pectin may depend on fine chemical structure and molecular weight. Several studies have indicated that the anticomplementary activity of pectic polysaccharides depends on the detailed structures of the galactan side chains that are attached to the rhamnogalacturonan backbones (Yamada et al., 1989; Wagner, 1990). The high molecular weight polysaccharides have more potent anticomplementary activity than low molecular weight polysaccharides. Samuelsen et al. (1996) also found that the "hairy regions" with 1,3,6-linked galactose side chains were the most active fraction with respect to the anticomplementary activity. Detailed information on the pharmacological properties of pectins can be found in a recent review by Yamada (1996).

It seems that pectin may produce a synergistic effect, increasing the activity of principal compounds (Negrevergne, 1974a,b; Nie et al., 1999). This characteristic

TABLE 9.1
Pharmacological Activity of Pectins Isolated from Plants

Immunostimulating activity
- Anti-complement activity
- Anti-complementary activity
- Fc receptor up-regulation on macrophages
  (enhancing activity of immune complex clearance)
- stimulation of macrophage phagocytosis

Antiulcer activity
Antimetastasis activity
Antinephritis activity and antinephrosis activity
Hypoglycemic activity
Cholesterol decreasing effect

Source: From Yamada (1996).

has recently resurfaced in the justification of the use of pectin as raw material in the manufacture of different pharmaceutical preparations (Kertesz, 1951).

9.3.6 REGULATORY STATUS

Commercial pectin is regulated as a food additive in most countries and is not considered a problem from a toxicological point of view. Both pectin and amidated pectin have been granted an acceptable daily intake value (ADI) "not specified" by the international expert committees JECFA (Joint Food and Agriculture Organization and World Health Organization Experts Committee on Food Additives) (FAO, 1984) and the European Union Scientific Committee for Food (SCF, 1975). It has been approved for a wide range of uses under the EC Directive (95/2/EC) on Food Additives other than colors and sweeteners. The U.S. Food and Drug Administration (FDA) has also affirmed the generally recognized as safe status of pectin (FDA, 1986). Nevertheless, most authorities proscribe criteria of purity for commercial pectins from time to time.

9.4 STRUCTURE, EXTRACTION AND ANALYSIS OF PECTIN

9.4.1 STRUCTURE OF ISOLATED PECTIN

9.4.1.1 Chemical Structure

A representative structure of pectin is illustrated in Figure 9.2. Pectins are composed of a α-(1→4) linked D-galacturonic acid backbone interrupted by single α-(1→2) linked L-rhamnose residues (BeMiller, 1986). Depending on the pectin source and the extraction mode, the carboxyl groups of galacturonic acid units are partly esterified by methanol and, in certain pectins, are partially acetylated. The mole-percent of rhamnose may be different depending on the source of pectins (Ryden and

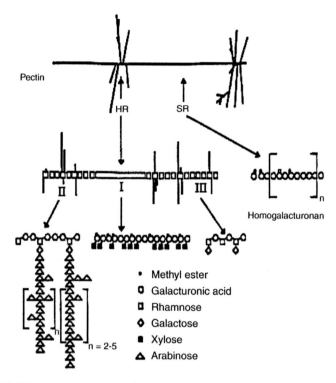

FIGURE 9.2 Hypothetical structure of apple pectin, where I is the xylogalacturonan region, II is the region with arabinan side chains, III is the rhamnogalacturonan region, HR is the hairy region and SR is the smooth region. With permission from Schols et al. (1989).

Selvendran, 1990). Often, arabinan, galactan or arabinogalactan side chains are attached to the C-4 position of the rhamnose residues, although attachment to C-2 and C-3 of the galacturonic acid have also been reported (Talmadge et al., 1973). In these side chains, the arabinosyl units have (1→5) linkages, while galactoses are mutually joined mainly by (1→4) linkages, but (1→3) and (1→6) linkages also occur. Other sugars such as glucuronic acid, L-fucose, D-glucose, D-mannose and D-xylose are sometimes found in side chains. For large parts of the chains, the rhamnose may be distributed in a fairly regular fashion, because acid hydrolysis gives segments 25–35 units long (Powell et al., 1982). In other regions, the rhamnose units are close together; there are two to three moles of neutral sugar per mole of galacturonic acid residue (de Vries et al., 1982). Pectins from the middle lamella appear to have fewer rhamnose insertions, fewer and shorter branches and a higher DE than those from the primary cell walls (Brett and Waldron, 1996). The size of the neutral sugar side chains appears to differ between the sparsely rhamnosylated regions (smooth region) and the densely rhamnosylated regions (hairy region). Using the assumptions that all the rhamnose units have neutral sugar chains, and these chains attach only to rhamnose, it is estimated that the lengths of the side chains are 4–10 residues long in the sparse regions and 8–20 residues long in the dense regions (Selvendran, 1985; de Vries et al., 1982).

9.4.1.2 Physical Structure

9.4.1.2.1 Backbone Conformation

The axial-axial type, α-(1→4) linkages between galacturonic acid units impose a severe conformational constraint on the adjacent monomers, which results in an intrinsic stiffness to the pectin backbone. Two types of conformation, a right-handed 3_1 helix (Di Nola et al., 1994; Cros et al., 1992; Walkinshaw and Arnott, 1981a) or a right-handed 2_1 helix (Di Nola et al., 1994; Cros et al., 1992) have been considered the most probable conformations for pectin molecules using conformational analysis and X-ray studies. These helices may be stabilized by intramolecular hydrogen bonding and/or intermolecular calcium ions forming an ordered and fairly stiff structure in solid state. Whether this ordered tertiary structure persists in solution is not clear, but it has generally been accepted that the backbone adopts a worm-like conformation in solution. Results from small-angle neutron and light scattering on pectins of various DE gave persistence length (IP, a parameter of chain flexibility) in the range of 2–31 nm (Rinaudo, 1996). This suggests that pectin is much more flexible than xanthan (IP ~ 120 nm, Sato et al., 1984) but more rigid than amylose (IP ~ 3 nm, Ring et al., 1985). Several studies on the solution properties of pectins suggested that pectin molecules are further aggregated into either rods or segmented rods in solution and held together by noncovalent interactions (Fishman et al., 1992).

The Mark-Houwink-Sakurada (MHS) relation relates the intrinsic viscosity [η] to the average molecular weight (MW) in the following form:

$$[\eta] = k \, MW^{\alpha} \tag{9.1}$$

The MHS exponent α is a measure of the chain stiffness for a polymer-solvent pair. Values from 0.7 to 2.0 were reported for pectins in aqueous solutions, but most of these converged to approximately 0.8 ~ 0.9 (Voragen et al., 1995). This indicates a slightly stiff conformation. The large discrepancy of α values is presumably a result of aggregation phenomena in addition to the differences in the fine structures. Many studies have shown that clarification methods used to prepare the solution have great influences on the molecular weight (MW) measurements (Berth et al., 1994), indicating the existence of aggregates.

The insertion of rhamnose in the galacturonan chain introduces a "kink" to the polymer chain, which is believed to reduce the stiffness of the molecule (Rees and Wight, 1971). Molecular modeling has shown that the chain flexibility increases monotonically with each insertion of rhamnose randomly into the galacturonan backbone (Whittington et al., 1973); however, this has not yet been clearly demonstrated experimentally. Molecular modeling also suggests that methoxyl groups have no effect on the flexibility of the linkages between the galacturonic acids (Cros et al., 1992), but, a recent study by Morris et al. (2000) demonstrated clearly that the stiffness of the pectin chain decreased with increasing DE. They suggest that both steric and electrostatic interactions are important in these conformational changes. An earlier study also reported that the flexibility was at a maximum for a DE of around 50% and that amidation induced greater rigidity (Axelos and Thibault, 1991).

9.4.1.2.2 Molecular Weight and Molecular Weight Distribution

Molecular weight is one of the most important characteristics in determining the functional behavior of pectins. Pectins are highly heterogeneous with respect to molecular weight. The MW and MW distribution of pectins in situ can be expected to vary with plant source and stage of ripening. For isolated and purified pectins, the molecular weight is largely determined by the extraction modes and conditions. Kar and Arslan (1999b) compared the molecular weight of pectins extracted from orange peels by extraction with HCl (pH 2.5, 90°C and 90 min), ammonium oxalate (0.25%, pH 3.5, 75°C and 90 min) and EDTA (0.5%, 90°C and 90 min). The weight average MWs as measured by light scattering were reported as 84,500, 91,400 and 102,800 g/mol, respectively, for the samples obtained from HCl, ammonium oxalate and EDTA extraction. The tendency to aggregate in aqueous medium made MW measurement itself a major challenge, because complete molecular dispersion is seldom the prevailing physical state of pectin in water. All of these factors contribute to the large discrepancies in the reported MW values of pectin in the literature. From most of the studies, the weight average molecular weight from various fruit sources is typically in the order of $10^4 \sim 10^5$ g/mol (Kontominas and Kokini, 1990; Fishman, 1991; Corredig et al., 2000). However, values as high as 4.2×10^6 and 1.2×10^6 g/mol were reported for pectins extracted from cider apple pomace by CDTA and Na_2CO_3, respectively (Chapman et al., 1987).

9.4.2 PECTIN EXTRACTION

9.4.2.1 Laboratory Extraction

Pectin extraction is a complex physicochemical process in which solubilization, extraction and depolymerization of pectin macromolecules from plant tissues may take place (Kertesz, 1951). High-esterified pectins are readily extracted by hot water, but low-esterified pectins usually are not readily extracted under such conditions because they are physically bound in situ via metallic cations, especially divalent cations. Sequestering agents, such as sodium hexametaphosphate, ammonium oxalate, ethylene diamine tetraacetate (EDTA) and cyclohexane diamine tetraacetate (CDTA), which readily bind cations, are added to the extractants for efficient extraction of this type of pectin. The presence of an acid or base and elevated temperature help cell wall disruption, protopectin hydrolysis and solubilization of pectic substances (Kirtchev et al., 1989). In such processes, degradation of pectin macromolecules probably takes place (Thibault and Rombouts, 1985). Extraction with dilute alkali generally yields a pectin with a low degree of esterification as a result of saponification of ester group, whereas the acid extraction yields a pectin with a relatively high DE.

In order to study the composition and structure of pectins and their changes during ripening, storage and processing, various procedures have been developed for fractional extraction of pectins. A typical laboratory scheme for sequential extraction of pectin from plant cell wall materials is summarized in Figure 9.3. (Selvendran and O'Neill, 1987; Voragen et al., 1995, Vierhuis et al., 2000). Not every step in the scheme is used by all of the researchers, and decisions depend on the source materials.

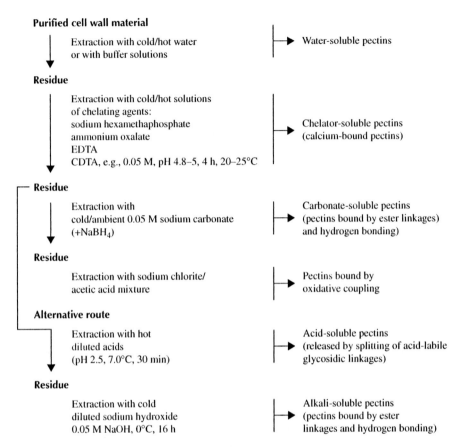

Purified cell wall material

Extraction with cold/hot water
or with buffer solutions → Water-soluble pectins

Residue

Extraction with cold/hot solutions
of chelating agents:
sodium hexamethaphosphate
ammonium oxalate
EDTA
CDTA, e.g., 0.05 M, pH 4.8–5, 4 h, 20–25°C → Chelator-soluble pectins
(calcium-bound pectins)

Residue

Extraction with
cold/ambient 0.05 M sodium carbonate
(+NaBH₄) → Carbonate-soluble pectins
(pectins bound by ester linkages)
and hydrogen bonding)

Residue

Extraction with sodium chlorite/
acetic acid mixture → Pectins bound by
oxidative coupling

Alternative route

Extraction with hot
diluted acids
(pH 2.5, 7.0°C, 30 min) → Acid-soluble pectins
(released by splitting of acid-labile
glycosidic linkages)

Residue

Extraction with cold
diluted sodium hydroxide
0.05 M NaOH, 0°C, 16 h → Alkali-soluble pectins
(pectins bound by ester
linkages and hydrogen bonding)

FIGURE 9.3 Scheme for laboratory extraction of pectins. With permission from Voragen et al. (1995).

Some nonconventional approaches have also been applied to pectin extraction. These include use of various enzymes (Sakai, 1992; Shkodina et al., 1998; Matthew et al., 1990; Williamson et al., 1990), extrusion (Thibault et al., 1996), heat and pressure treatments (Fishman, 2000) and ultrasonic treatment (Panchev et al., 1988). These studies indicate that among many factors that affect the yields and properties of pectin extracts, pH, temperature and, in the case of low-DE pectin, the presence of chelating agents are the most important. (Renard and Thibault, 1993; Eriksson, 1997). In general, severe conditions increase the yield but also accelerate degradation, resulting in lower MW and lower DE pectins.

9.4.2.2 Industrial Extraction

Most commercial pectins are produced from apple pomace and citrus peels, although other raw materials are also used, including sugar beet pulp (in Sweden, Germany and England), sunflower seed head (in Bulgaria), and potato pulp. The industrial process for pectin extraction was described in detail by May (1990) and Voragen

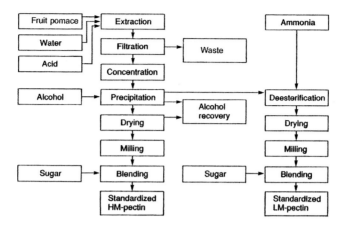

FIGURE 9.4 Industrial production of pectins. Adapted from Copenhagen Pectin (1985).

et al. (1995), as summarized in Figure 9.4. The source materials are refluxed with hot dilute mineral acid (hydrochloric, nitric and sulfuric acids) at ~ pH 2 and ~ 60–100°C for 0.5–10 h. The hot pectin extract is separated from the solid residue as efficiently as possible using different approaches, such as vibrating screens, decanter centrifuges, vacuum drum filters and various types of presses. When the substrate materials are apple or peach pomace, pectinase-free α-amylase is usually added to hydrolyze starch. The clarified extract is then concentrated under vacuum to 3–4% pectin content and precipitated with isopropanol (methanol and ethanol are also permitted). The precipitate obtained on adding alcohol is then washed to remove contaminants, pressed to remove as much as possible of the liquid, dried and ground to a powder form. The extraction conditions are usually optimized with respect to yield, gelling capacity and desired DE. Fast-gelling pectins (DM >70%) are typically extracted at pH 2.5 and 100°C for 45 min; medium-, fast- and slow-set pectins (DM 60–70%) are extracted at lower temperatures for longer periods of time, because at lower temperatures, de-esterification proceeds faster than depolymerization (Voragen et al., 1995). To produce low-DE types of pectins, acid treatment is commonly applied to remove some of the ester groups in different stages of the extraction process. A range of low-DE pectin can be produced in this way. Alkaline hydrolysis can also be used for this purpose, but at a low temperature (<10°C) to prevent pectin degradation by β-elimination reaction. When ammonia is used for alkaline hydrolysis, a proportion of the ester groups is replaced by amide, and amidated pectins are obtained.

To get a product that is consistent over a range of properties, it is customary to blend a number of production batches and dilute them with sugars or dextrose, to obtain a standardized quality (May, 1990).

9.4.2.3 Production of Fruit Dietary Fiber

Fruit dietary fiber is the product obtained by drying the washed wet pulp (the main by-product of juice industries) that remains after fruit juice extraction. Because fruit dietary fibers contain a high level of pectins as the major water-soluble fibers, they have similar technological functions as isolated pectins (Diepenmat-Wolters, 1993) and may also be used as a potential source of dietary fiber supplements. An example of such an application is low-fat, high-dietary fiber frankfurters containing peach dietary fiber (PDF) (Grigelmo-Miguel et al., 1999).

9.4.3 Physical and Chemical Characterization Methods

9.4.3.1 Determination of End-Reducing Sugars

The end-reducing sugars can be determined by the Cu^{++} reactions (Nelson, 1944), neocuproine ferricianur method (Dygert et al., 1965), dinitrosalicilic acid method (White and Kennedy, 1981) and 2-2′ bicinchoninic acid (BCA) photometric method (Waffenschmidt and Jaenicke, 1987). Alkaline pH and extended boiling period are the main inconveniences of these methods. These methods are used to quantify the reducing end units resulting from the depolymerization of pectin. For pectins of low esterified or low MW, the BCA method gives the most reliable results. This method is based on the reaction between BCA and Cu^+, which forms a deep-blue complex in alkaline solution; the cuprous ions are produced from the reduction of Cu^{++} by the reducing sugars and reducing end units in the solution. The absorbency given by this complex can be measured at 560 nm.

9.4.3.2 Pectin Quantitation in Samples

Because galacturonic acid is the fundamental unit of pectin, quantitation of this acid is a primary method for determination of the amounts of pectin present in a sample. The pectin is usually expressed as the anhydrouronic acid content (AUC). Colorimetric methods are the most commonly used, based on the application of different chromophores: carbazole (Dische, 1947), meta-hydroxydiphenyl (Blumenkrantz and Asboe-Hansen, 1973) and 3,5-dimetylphenol (Scott, 1979; McFeeters and Armstrong, 1984). These reagents react with 5-formyl-2-furancarboxylic acid, a chromogen formed from sulfuric acid dehydration of the galacturonic acid residues of pectin, producing a solution with an absorbance in the visible light range, which can be measured colorimetrically. Among these methods, the meta-hydroxydiphenyl method, which has been automated by Thibault (1979), has the advantages of greater sensitivity, higher specificity and faster execution (Ahmed and Labavitch, 1977; Robertson, 1979). The carbazole method and its modified forms (Bitter and Muir, 1962; Dekker et al., 1972) are subject to interference from the neutral sugars associated with pectin samples (Blumenkrantz and Asboe-Hansen, 1973).

Decarboxylation, another traditional method for pectin quantitation, is achieved by treatment of pectins with 19% hydrochloric acid—evolved carbon dioxide is collected in a standard solution of sodium hydroxide and back-titrated with acid (Schultz, 1965). This method is considered most reliable but is time-consuming and needs a relatively large quantity of samples.

Chromatographic methods have also been developed for quantitative determination of the AUC of pectin. Uronic acids have been analyzed using gas-liquid chromatography (GLC) after conversion to L-galactono-1,4-lactone via potassium borohydride reduction and lactonization using methanolic hydrogen chloride. The derived lactone was then analyzed by GLC as the trimethylsilyl derivative (Perry and Hulyalkar, 1965; Ford, 1982). Ion exchange has been used in high-performance liquid chromatography (HPLC) with pulsed-amperometric or refractive index detection of uronic acids without derivatization (Forni et al., 1982; Voragen et al., 1982; Hatanaka et al., 1986; Matsuhashi et al., 1989; Schols et al., 1989). These early methods suffer from the difficulty of obtaining complete release of the uronic acid units from pectin during the hydrolysis step (Garleb et al., 1991; Leitao, 1995). Hydrolytic techniques have been significantly improved by the use of multifunction enzymes (Leitao, 1995) or methanolysis combined with trifluoroacetic acid (TFA) hydrolysis (de Ruiter et al., 1992). The HPLC method is now widely used in researches and food industries for pectin analysis.

9.4.3.3 Determination of Degree of Esterification

Degree of esterification is an important molecular index for pectin classification that describes the extent to which carboxyl groups in pectin molecules exist as the methyl ester. Depending on the methods used for determination, DE can be expressed either by the ratio of methylated carboxyl groups to a total of carboxyl groups or by the percentage of methoxyl content. The theoretical maximum value for each expression is 100% for the former and 16.3% for the latter (Doessburg, 1965).

Direct titrimetric method is a classic method for DE determination (National Research Council, 1972). In this method, pectin is first washed by acidic alcohol to convert the free carboxyl groups to the protonated form and is then washed with alcohol and dried. The washed sample is then dissolved in water and titrated with standard base and acid. The uronic content and DE are calculated from the neutralization and saponification equivalents of the pectic carboxyl groups. The existence of acetyl groups in some pectin samples leads to an overestimate of DE in the titration method. In this case, the copper binding method is an alternative approach (Keijbets and Pilnik, 1974). In the copper binding method, the amounts of copper ions that bind to the pectin molecules before and after saponification are measured by the complexometry, the ratio of which gives the DE.

The measurement of methoxyl content in pectin is another way to determine DE. Total pectin content is determined by the methods described in the previous section, and the methoxyl content is found by measuring the amount of methanol released by alkaline or enzymatic demethylation (Kujawski and Tuszynski, 1987). The measurement of methanol can be achieved by a number of methods. In the method of Wood and Siddiqui (1971), methanol is oxidized to formaldehyde with potassium permanganate and is then condensed with pentane-2,4-dione to yield a colored product (3,5-diacetyl-1,4-dihydro-2,6-dimethylpiridine) that can be measured spectrophotometrically. Klavons and Bennett (1986) modified this method by using alcohol oxidase instead of potassium permanganate that increased the sensitivity and speed of the analysis. Methanol can be converted to methyl nitrite by reacting with nitrous acid in a closed tube and analyzed by GLC (Hoff and Feit, 1964; Bartolomé and Hoff, 1972). HPLC has also been used for methanol quantification, usually using a cation exchange resin column and refractive index detection (Gandelman and Birks, 1982; Voragen et al., 1986). The HPLC method developed by Voragen et al. (1986) also allows for the simultaneous measurement of acetyl content.

9.4.3.4 Analysis of Neutral Sugars

The neutral monosaccharide constituents of pectin are commonly analyzed by total acid hydrolysis of the samples, usually the procedure of Saeman hydrolysis, which uses 12 M sulfuric acid for solvolysis and 1 M sulfuric acid for subsequent hydrolysis (Selvendran et al., 1979; Englyst and Cummings, 1984). de Ruiter et al. (1992) compared five hydrolysis methods using different acids and concluded that methanolysis with 2 M HCl prior to TFA hydrolysis provided the best results. The monosaccharides are then converted to the alditol acetate derivatives and analyzed by GLC. HPLC has also been used, with the advantage that the sugars can be analyzed without further derivatization. Cation exchange columns with refractive index detection have been used, and more recently, anion exchange systems with pulsed amperometric detection are gaining in popularity (Ball, 1990).

9.4.3.5 Molecular Weight

Measurement of MW of pectin and other natural polysaccharides is not an easy task because of the highly polydisperse nature of the polysaccharides and their tendency to aggregate in aqueous solution. This may partly explain the large diversity of data reported for pectin in the literature. Methods for MW determination may be classified into three categories: absolute techniques, relative techniques and combined techniques. Absolute techniques include static light scattering, sedimentation equilibrium and membrane osmometry. These techniques require no assumptions about molecular conformation and do not require calibration using standards of known MW. Light scattering and sedimentation equilibrium give weight-average MW, while membrane osmometry yields number-average MW. Light scattering has been widely applied for pectin characterization (Jordan and Brant, 1978; Plashchina et al., 1985; Chapman et al., 1987). The method is very sensitive to molecular aggregation, and great care has to be taken in the preparation and clarification of sample solutions. Application of membrane osmometry was reported by Fishman et al. (1986) and Berth et al. (1980) and sedimentation equilibrium was reported by Morris et al. (2000).

Relative techniques include gel permeation chromatography (GPC) (Anger and Berth, 1986; Deckers et al., 1986; Berth and Lexow, 1991; Hourdet and Muller, 1991), sedimentation velocity and viscometry (Christensen, 1953). These methods require assumptions of molecular conformation or calibration using standards of known MW.

In the last two decades, methods for MW determination have been improved by combining two or more techniques, for instance, GPC combined with a laser light scattering detector (Fishman et al., 1986; Kontominas and Kokini, 1990) and GPC combined with sedimentation equilibrium (Harding et al., 1991). These methods not only allow absolute measurement of average MW but also provide information on the MW distribution and molecular conformations.

9.4.3.6 Rheological and Textural Characterization

The diverse biological functions of pectins are reflected in a range of rheological behaviors that will be described in the next section. Aqueous pectin dispersions may exist in the form of a viscous flow or a gel at certain conditions. Accordingly, rheological characterization of the pectin system is a fundamental requirement when considering both the production and utilization of pectins. The information obtained from rheological measurement identifies parameters essential for process design, and quality control and optimization and helps in the design of new textural characteristics for products. There are various types of viscometers and rheometers that operate on different principles (Rolin and de Vries, 1990; Lapasin and Pricl, 1995). Only a few of the most commonly used types are described here.

9.4.3.6.1 Characterization of Pectin Dispersions

9.4.3.6.1.1 Dilute Solution

Capillary viscometers have been widely used in determining the viscosity of Newtonian fluids. In these viscometers, the driving force is usually the hydrostatic head of the test liquid itself, although, application of external pressure is also used in order to increase the range of measurement and allow non-Newtonian behavior to be studied. In operation, the efflux time of a fixed volume of test liquid is measured, from which the kinematic viscosity is calculated.

Intrinsic viscosity $[\eta]$ is a characteristic property of an isolated polymer molecule in a given solvent and it represents a measure of its hydrodynamic volume. Capillary viscosity is often used for the measurement of $[\eta]$ of polymer solutions, although for some uses, a more conventional concentric cylinder viscometer is preferred (Ross-Murphy, 1994). Two viscosity values are measured using a capillary viscometer: one is the viscosity of the solvent, η_s, and the other is the viscosity of the polymer solution, η. From η_s and η, relative viscosity (η_r), which is the viscosity enhancement due to the contribution of the polymer, is obtained:

$$\eta_r = \eta/\eta_s \tag{9.2}$$

η_r of a series polymer concentrations need to be measured and usually treated according to the Huggins (1942) [Equation 9.3] and Kraemer (1938) [Equation 9.4] equations:

$$\eta_{sp}/c = [\eta] + K'[\eta]^2c \tag{9.3}$$

$$\ln(\eta_r)/c = [\eta] + (K' - 0.5)[\eta]^2c \tag{9.4}$$

where η_{sp} $(= \eta_r - 1)$ is called specific viscosity, c is polymer concentration and K' is the Huggins coefficient. Intercepts from either of the two plots η_{sp}/c vs. c and ln $(\eta_r)/c$ vs. c extrapolated to c = 0 give $[\eta]$, which is best estimated from the average of the two.

Because of the polyelectrolytic nature, pectin solutions need to be made in excess of salt, usually in 0.05 ~ 0.1 M sodium chloride or phosphate, and use the same solvent for dilution (isoionic dilution) (Pals and Hermans, 1952). This is because, unlike neutrol polymers, the viscosity of dilute solution of polyelectrolytes displays unique dependence on concentration. As shown in Figure 9.5, the η_{sp} of sodium pectate exhibits a maximum in pure water and low concentration of salt, a phenomenon caused by the so-called "electroviscous effect." When the salt concentration is

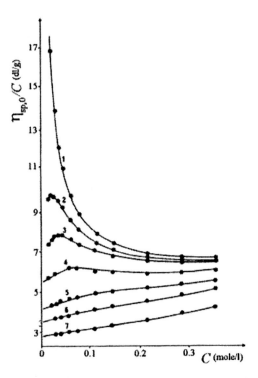

FIGURE 9.5 Zero shear rate reduced viscosity ($\eta_{sp,0}/c$) vs. concentration of sodium pectate at various NaCl concentrations; 1–7:0, 0.2, 0.4, 1.0, 2.5, 5.0 and 50 mM. From Pals and Hermans (1952), with permission of Elsevier Applied Science.

higher than ~5 mM, the viscosity decreases with concentration monotonically, as with neutrol polymers.

[η] is often used for estimation of MW through the Mark Houwink equation [Equation 9.1] (Deckers et al., 1986). Thus, the capillary viscometer is often used for monitoring the depolymerization (Kar and Arslan, 1999a) and oxidative cross-linking of pectins (Oosterveld et al., 2001). It has also been used to study the effects of salts and ionic strength (Smidsrød and Haug, 1971; Michel et al., 1982), sugars and other agents (Chen and Joslyn, 1967; Kar and Arslan, 1999a), temperature and concentration (Kar and Arslan, 1999b) on the viscosity of pectin solutions.

9.4.3.6.1.2 Semi-Dilute Solutions

Both strain- and stress-controlled rotational rheometers are widely employed to study the flow properties of non-Newtonian fluids. Different measuring geometries can be used, but coaxial cylinder, cone-plate and plate-plate are the most common choices. Using rotational rheometers, two experimental modes are mostly used to study the behavior of semi-dilute pectin solutions: steady shear measurements and dynamic measurements. In the former, samples are sheared at a constant direction of shear, whereas in the latter, an oscillatory shear is used.

The viscosity of pectin solutions is influenced by the conditions of measurement. Steady shear is used to study the dependence of shear stress (or shear viscosity) on the shear rate and also to study how intrinsic factors (structure, MW and concentration) and extrinsic factors (pH, ionic strength and temperature) influence the viscosity (Chou and Kokini, 1987; Basak and Ramaswamy, 1994). It has been shown that the viscosity of many non-Newtonian fluids, including pectin dispersions, is time dependent (thixotropy or antithixotropy). This arises as a result of the structural reorganization within the dispersion. To characterize the thixotropic behavior, measurements are carried out at a constant shear rate, and the variation of the shear stress with time is monitored until a constant value is reached (Tung et al., 1970). To eliminate time-dependent effects, viscosity measurement should be carried out at "equilibrium conditions" by such means as allowing the sample to rest for a considerable time before the application of shear or by waiting until the shearing force or the torque have stabilized before taking a measurement.

Although creep-compliance (Kawabata, 1977; Dahme, 1985) and stress-relaxation techniques (Comby et al., 1986) have been used to study the viscoelestic properties of pectin solutions and gels, the most common technique is small-deformation dynamic measurement, in which the sample is subjected to a low-amplitude, sinusoidal shear deformation. The resultant stress response may be resolved into an in-phase and 90° out–of–phase components; the ratio of these stress components to applied strain gives the storage and loss moduli (G′ and G″), which can be related by the following expression:

$$G^* = G' + iG'' \qquad (9.5)$$

where G^* is the complex modulus (Axelos et al., 1991). The storage modulus G' represents ideal elastic solid behavior and is a measure of the energy stored in the material during deformation. For a gel, it is directly related to the cross-link density of the network. The loss modulus G'' represents Newtonian fluid viscous behavior and is a measure of the frictional energy generated between liquid layers during deformation and dissipated as heat (Ferry, 1980). The variations of G' and G'' with oscillatory frequency ϖ allow a qualitative determination of the nature of a material (Ferry, 1980) and can essentially identify a pectin system as a viscoelastic solution or a gel.

9.4.3.6.2 Characterization of Pectin Gels

The most important textural characteristic of a pectin gel is gel strength, and there are a number of good reviews on methods and instruments used to determine this (Mitchell, 1976; Crandall and Wicker, 1986; Alexos et al., 1991). Both nondestructive approaches, which measure the elastic properties of the gel, and destructive approaches, which measure the inelastic properties or breaking strength of a gel, are used.

9.4.3.6.2.1 Nondestructive Tests

For many years, the pectin industry has measured gel strength using the IFT-SAG method (Cox and Higby, 1944; IFT, 1959). In this method, pectin gels are prepared under standard conditions, and the amount of sag under the force of gravity is measured with a micrometer called a Ridgelimeter. The method is precise, reproducible and simple to operate, but it is incapable of a comprehensive evaluation of gel structure.

Different types of concentric cylinder and parallel plate instruments have been developed for determining gel strength (Mitchell, 1976). In these instruments, pectin is allowed to set between two corrugated concentric cylinders (or parallel plates), then one-half of the geometry is twisted by a torsion wire and the extent of torsion produced in the gel is measured through the other half of the geometry. The corrugation prevents slippage of the gel. These instruments allow fundamental measurements on gels, such as creep compliance, stress relaxation and rigidity modulus (Mitchell and Blanshard, 1976; Plashchina et al., 1979).

Small–deformation dynamic measurements have become a popular technique for characterization of food gels. Because it is nondestructive, it is often used to monitor the gelation process and study the gelation kinetics of pectins under various conditions (Rao and Cooley, 1993; Grosso and Rao, 1998). By comparing the data obtained from such small deformation oscillatory measurement with the theory of Rouse and Zimm, Chou et al. (1991) proposed a compact random coil conformation for citrus pectin. In conjunction with other techniques, such as DSC and NMR, small-deformation dynamic methods have greatly extended our understanding of gelation mechanisms of pectin and pectin-based products.

9.4.3.6.2.2 Destructive Tests

Both destructive and nondestructive measurements can be done on an Instron Material Tester. In this system, the sample is loaded in a test cell, and the compression or tension force is measured when the upper part of the cell is moved over a given distance (time). Within the elastic limit of the gel, the elastic modulus E (or gel strength) is obtained from the initial slope of the nondestructive stress/strain curve; additional deformation results in the breakage of the sample, giving the characteristic parameters—yield stress and breaking strain.

Using similar principles, several types of texture analyzers have been developed and found extensive use in the characterization of food gels (Rolin and de Vries, 1990). In some, a plunger is applied to the gel at a constant speed, and the stress vs. deformation graph is recorded, or in some, simply the force required to press a plunger of specified dimension into the gel is measured (Bloom Gelometer). A more sophisticated texture-profile analyzer uses a double-penetration action to imitate the biting action of the teeth, and thus, provides force-deformation data required for the evaluation of the seven texture parameters: fracturability, hardness, cohesiveness, adhesiveness, springiness, gumminess and chewiness (Bourne, 1982). A typical such profile is shown in Figure 9.6.

9.5 FUNCTIONAL PROPERTIES OF PECTIN IN FOOD PRODUCTS

9.5.1 SOLUTION PROPERTIES OF PECTIN DISPERSIONS

9.5.1.1 Solubility and Dispersibility

Pectins are generally soluble in water and insoluble in most organic solvents. Pectic acids are only soluble as the sodium or potassium salt; divalent salts of pectic acids must first be converted to their sodium or potassium forms for dissolution. The

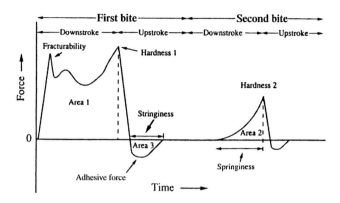

FIGURE 9.6 A typical texture profile of food gels obtained through a texturimeter. With permission from Bourne (1982).

solubility of pectins usually decreases with increasing ionic strength and MW and with decreasing DE. Pectin will not dissolve in media in which gelling conditions exist.

Commercial pectins are usually sold as a powder, but in most applications, the preparation of a uniform pectin solution is required. The addition of a pectin powder to water tends to produce lumps, and subsequent dissolution of these lumps is very difficult, leading to nonuniform solutions. There are several approaches for overcoming this problem. In the laboratory, pectin solutions can be easily prepared by first wetting the powder with alcohol or glycerol. In industrial applications, dispersing agents, such as finely powdered sucrose and dextrose, can be used. However, pectin becomes increasingly difficult to dissolve as the soluble solids in the medium increase. It is recommended that high-methoxyl (HM) pectin be dissolved at solid concentrations below 20% (Copenhagen Pectin, 1985). By incorporating a mixture of organic acid and inorganic carbonate into the pectin powder, effervescence during mixing serves to agitate and disperse the particles throughout the medium (Towle and Christensen, 1973). Special treatments have also been developed to produce pectins with improved dispersibility (Leo and Taylor, 1955), but these have limited uses in practice because of cost and taste considerations.

9.5.1.2 Flow Properties

Aqueous pectin dispersions show flow behavior similar to many other polysaccharide solutions. Flow curves of specific viscosity η_{sp} vs. shear rate have a Newtonian plateau (constant η_{sp}) at low shear rates, followed by a shear thinning region at moderate shear rates (Morris et al., 1981). Most pectin solutions have relatively low viscosity compared to some other commercial polysaccharides, such as guar gum, mainly because of the lower MW. Consequently, pectin has limited use as a thickener.

Similar to solutions of other random coil polysaccharides, the viscosity of pectin dispersion decreases with increasing temperature but increases with increasing concentration, and the effect of temperature is stronger at higher concentrations (Kar and Arslan, 1999b). The effect of temperature on apparent viscosity is usually analyzed by an Arrhenius-type relationship to calculate activation energy of flow

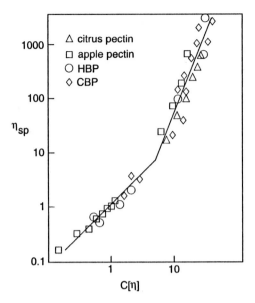

FIGURE 9.7 The dependence of specific viscosity [η_{sp}] on the overlapping parameter c[η] for citrus pectin ($\triangle$), apple pectin ($\square$), hot buffer extracted tomato pectin (HBP) (○) and cold buffer extracted pectin (CBP) (◇). With permission from Chou and Kokini (1987).

over a selected range of temperature and concentration. Kar and Arslan (1999b) found the activation energies of flow for orange peel pectin dispersions to be 19.53–27.16 kJ/mol (concentration 2.5–20 kg/m³, temperature 20–60°C).

The dependence of viscosity on concentration is usually described through a nondimensional parameter c[η], which, as a measure of volume occupancy of polymer molecules, may be used to evaluate the extent of overlapping of polymer domains. Chou and Kokini (1987) showed that pectins from citrus, apple and tomato in citrate phosphate buffer solution (pH 4.6) are characterized by overlapping concentrations (c*) that separate the dilute and concentrate regions at a c[η] value of about 6 (Figure 9.7). Up to c*, viscosity depends on concentration to the power 1.4 and beyond c*, to the power 3.3.

In addition to concentration and temperature, the viscosity of aqueous solutions of pectin also depends on MW, DE, electrolyte concentration, type and concentration of co-solute and pH (Oakenfull, 1991). Pectin behaves as a typical polyelectrolyte, and viscosity is influenced by pH and ionic strength (Fuoss and Strauss, 1949). The addition of salts of monovalent cations, such as sodium chloride, to pectin dispersions reduces viscosity (Michel et al., 1982); the effect is more pronounced with decreasing DE. This decrease is caused by the shielding of the carboxyl groups' charges by the cations, which lessens chain-chain repulsion and suppresses chain expansion (Delahay, 1965; Cesaro et al., 1982). By this mechanism, salt can affect firmness in vegetable tissues (Van Buren, 1984). In contrast, the addition of calcium or other polyvalent cations increases the viscosity of pectin solutions (Lotzkar et al., 1946). This effect has been attributed to bridging between suitably positioned carboxyl groups, by forming an "egg box structure," although other mechanisms may also be involved. In a calcium-free pectin solution, viscosity is inversely related to pH (Copenhagen Pectin, 1985).

Concentrations and types of sugars or oligosaccharides also affect the viscosity of pectin solutions. Chen and Joslyn (1967) and Kar and Arslan (1999a) found that sucrose, dextrose and maltose increased the viscosity of aqueous pectin solutions whereas dextrins reduced it. The viscosity-enhancing effect of the sugars was interpreted in terms of the decrease in dielectric constant of the solvent, dehydration action and hydrogen bonding formation. However, the effect of dextrins on the viscosity of pectin was apparently an artifact due to ionic impurities in the dextrin.

9.5.2 PECTIN GELS

The best known property of pectin is that it can gel under suitable conditions. A gel may be regarded as a system in which the polymer is in a state between fully dissolved and precipitated. In a gel system, the polymer molecules are cross-linked to form a tangled, interconnected three-dimensional network that is immersed in a liquid medium (Flory, 1953). In pectin and most other food gels, the cross-linkages in the network are not point interactions as in covalently linked synthetic polymer gels, but involve extended segments, called junction zones, from two or more pectin molecules that are stabilized by the additive effect of weak intermolecular forces.

On cooling a hot pectin solution, the thermal motions of the molecules decrease, increasing their tendency to associate and form a gel network. Any system containing pectins with the potential to gel has an upper temperature limit above which gelation will never occur. Similarly, there is a minimum concentration and MW for specific structural types of pectin, below which gel formation is not possible. For both HM and LM (low-methoxyl) pectins, high MW and high concentration generally favor gelation, and the higher the concentration, the higher the strength of the gel. Other intrinsic and extrinsic factors such as pH, amount and nature of co-solute and DE are also important with different effects for HM and LM pectins. The mechanism of formation and properties of HM and LM pectin gels are very different and described separately as follows.

9.5.2.1 Gelation of High-Methoxyl Pectins

9.5.2.1.1 Gelation Mechanism

At sufficiently low pH and in the presence of sugars or other co-solutes, HM pectins form gels, typically, at a pH below ~3.6 and in the presence of sucrose at a concentration greater than 55% by weight. The co-solutes lower water activity and "dehydrate" the pectin, while the acid suppresses ionization of the carboxyl groups, decreasing the electrostatic repulsion between the pectin molecules. Both effects promote chain-chain association, consequently leading to partial precipitation of pectin, i.e., gelation. X-ray diffraction studies suggest that the junction zones in HM pectin gels have a structure as shown in Figure 9.8 (Walkinshaw and Arnott, 1981b; Oakenfull and Scott, 1984). The structure is stabilized by hydrogen bonds (Jenchs, 1969) and hydrophobic interactions of the ester groups (Chen and Joslyn, 1967). These weak intermolecular forces, while insufficient to maintain the structural integrity of the

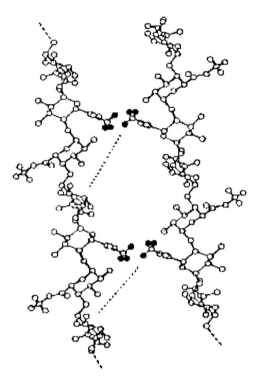

FIGURE 9.8 Structure of the junction zone in high-methoxyl (HM) pectin gels as inferred from x-ray diffraction studies. The hydrogen atoms of the hydrophobic methyl groups are represented by filled circles, and hydrogen bonds are indicated by dotted lines. With permission from Oakenfull and Scott (1984).

junction zones individually, cumulatively are able to confer thermodynamic stability to the gel network (Oakenfull, 1991). In contrast to LM pectin, HM pectin gels are not temperature reversible; that is, they do not melt at elevated temperatures.

9.5.2.1.2 Factors Affecting Gelation

Pectin is a polycarboxylic acid with a pK_a value of approximately 3.5. Decreasing the ratio of dissociated to nondissociated acid group by lowering pH renders the pectin molecule less hydrophilic, increasing the tendency to form gels. There is a specific pH value at which gelation is optimum for a particular HM pectin and a pH range within which gelation can be obtained in practice. If the sugar and pectin contents are held constant, as would be normal in preserves manufacture, the effect of a change in pH is seen as a loss in gel strength above a certain critical pH [Figure 9.9(a)] and a gradual lowering of the setting temperature [Figure 9.9(b)] as pH increases toward the pK_a value.

The degree of methylation and overall distribution of hydrophilic and hydrophobic groups have major effects on the gel formation of a particular pectin. The ester group is less hydrophilic than the acid group, and consequently, at a given pH, HM pectins with increasingly greater DE can gel at a somewhat higher pH under the same cooling gradient, while the requirement for a dehydrating agent necessary

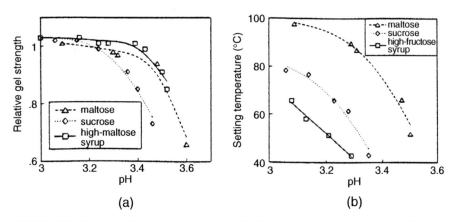

FIGURE 9.9 Effect of pH on the relative strength (a) and setting temperature (b) of pectin gels (0.5%) in the presence of different sugars (68% w/w total solids). With permission from May and Stainsby (1986).

TABLE 9.2
Gelation of High Methoxyl (HM) Pectins with
Various Degrees of Esterification (DE) at pH 3.0,
Total Soluble Solids 65% and Pectin Concentration 0.43%

HM Pectin Type	DE (%)	Gelling Time when Cooled at Different Temperatures			
		95°C	85°C	75°C	65°C
Rapid set	73.5	60 min	10 min	Pre-gel	Pre-gel
Medium set	69.5	No gel	40 min	5 min	Pre-gel
Slow set	64.5	No gel	No gel	No gel	30 min

Source: Copenhagen Pectin, 1985).

---

for gelation increases (Towle and Christensen, 1973). Gelation rate declines with decreasing DE. In the pectin industry, the difference in setting rate is used to classify HM pectin to rapid-set, medium-set and slow-set types, as illustrated in Table 9.2 (Copenhagen Pectin, 1985). The rate of gelation influences the product's texture.

Sucrose and other co-solutes also affect the gelation of HM pectin. At high solid contents, there is less water available to act as a solvent for the pectin, favoring the tendency to gel. A typical range of soluble solids content of various products is 50–80%, and for a particular pectin, gelling temperature decreases with decreasing soluble solids. The effect of type of co-solute on promotion of gelation varies, as demonstrated in Figure 9.9(b) (May and Stainsby, 1986). At the same levels of soluble solids and pectin, the setting temperature is highest when maltose is the co-solute, followed by sucrose and high-fructose corn syrup. Above levels of 85% soluble solids, the hydration effect is so strong that in practice, gelation of any commercial pectin can hardly be controlled.

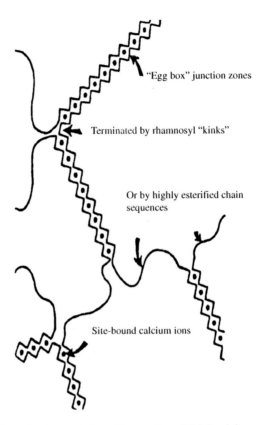

"Egg box" junction zones

Terminated by rhamnosyl "kinks"

Or by highly esterified chain
sequences

Site-bound calcium ions

FIGURE 9.10 Schematic representation of low-methoxyl (LM) calcium pectate gel. Redrawn from Lapasin and Pricl (1995).

9.5.2.2 Gelation of LM Pectin

9.5.2.2.1 Gelation Mechanism

The solubility of the calcium salt of completely de-esterified pectin, i.e., poly-galacturonic acid, is extremely low, and the similar tendency to precipitate in the presence of calcium ions leads to gel formation with LM pectin. The lower the DE of LM pectin, the greater the reactivity with calcium, as reflected in the higher gelling temperatures (Anyas-Weisz and Deuel, 1950). The ability of calcium to form insoluble complexes with pectins is associated with the free carboxyl groups on the pectin chains. Calcium complexes with neutral as well as acidic carbohydrates involve coordination bonds utilizing the unfilled orbitals of the calcium ions (Angyal, 1989). This is largely because its ionic radius (0.1 nm) is large enough that it can coordinate with the spatial arrangement of oxygen atoms found in many sugars and also because of flexibility in the directions of its coordinate bonds (Van Buren, 1991). For the calcium-induced coagulation and gelation of pectin, a so-called egg box structure has been proposed, as shown in Figure 9.10 (Rees et al., 1982). Calcium ions ionically interact and coordinate with the oxygens of two adjacent chains, giving rise to cross-linking. Individual calcium cross-linkages become

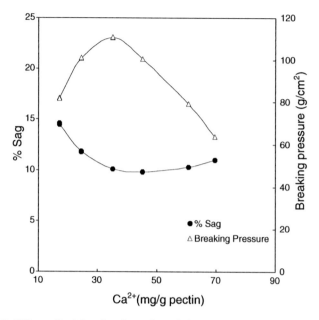

FIGURE 9.11 Effects of calcium levels on the gelation characteristics for LM pectin gels at pH = 3.6, 1% pectin and 30% sugar. Redrawn from El-Nawawi and Heikel (1995).

stabilized by cooperative neighboring cross-linkages (junction zones) with maximal stability being reached when 7–14 consecutive cross-links are present (Kohn and Luknar, 1977). Thus, LM pectins with a blockwise distribution of free carboxyl groups are very sensitive to low calcium levels (Thibault and Rinaudo, 1986).

9.5.2.2.2 Factors Affecting Gelation

The presence in the chain of both neutral sugars and methoxyl groups hinders the formation of junction zones; thus, the ability to gel increases with decreasing DE—the lower the pectin DE, the lower the calcium requirement to reach a given texture of acid-precipitated pectinates (Owens et al., 1949). Calcium chloride did not coagulate pectins with DE >60%, and the concentration needed to coagulate low DE pectins increased as the MW decreased (Anyas-Weisz and Deuel, 1950). The insertion of rhamnose introduces a severe kink into the backbone, which terminates the junction zone (Figure 9.10), resulting in decreased gelling properties. The inclusion of acetyl groups, as for sugar beet pectins, imparts poor gelation characteristics (Pippen et al., 1950), whereas amidation promotes gel formation (the amide groups possibly allow other types of chain associations through hydrogen bonding) (Racapé et al., 1989).

The amount of calcium considerably affects the gelling properties of LM pectins. In general, when the calcium:pectin ratio (R) is low, an increase in R results in improvements in the gel properties until an optimum, after which there is a decline and finally, syneresis occurs (Figure 9.11) (Sosulski et al., 1978; Chang and Miyamoto, 1992; El Nawawi and Heikel, 1995). The calcium sensitivity of LM pectin is pH related. Sosulski et al. (1978) reported that citrus pectin was insensitive to calcium changes in the pH range of 2.6–3.2; but above this range, typically in the range of pH 3.5–7.0, pectins become more sensitive to calcium; less calcium is needed at neutral pH than at low pH to induce gelation (Axelos and Thibault, 1991; Grosso and Rao, 1998). This occurs because at a pH above the pK_a value, the carboxyl groups are ionized, resulting in greater repulsion between chains and facilitating an easier calcium link. In addition, the presence of monovalent salt reduces the amount of calcium needed for inducing gelation of LM pectin, with the gel strength increasing with ionic strength until syneresis occurs.

Although LM pectins form gels in the absence of a co-solute, the presence of sugars affects the gelation, although much less than calcium and pH (Axelos and Thibault, 1991; El Nawawi and Heikel, 1995; Grosso and Rao, 1998). Depending on the type of the sugars and the pH, gelation properties are affected differently (May and Stainsby, 1986). Generally, the addition of sugars increases the setting temperature and gel strength, and syneresis is reduced (Christensen, 1986). The addition of sucrose and glucose up to 30% enhanced the gelation of LM pectin (DE 31.5) at pH 4 and 0.1% calcium, whereas the addition of fructose and sorbitol hindered gelation (Grosso and Rao, 1998). The detrimental effects of fructose and sorbitol were attributed to competition with pectin for calcium ions.

9.5.3 STABILITY OF PECTINS

9.5.3.1 Chemical and Physical Degradation

Pectins are most stable at pH 3–4. At lower or higher pH, the removal of methoxyl, acetyl and neutral sugar groups, as well as the cleavage of the polymer backbone may occur. The susceptibility to acid hydrolysis of different glycosidic linkages varies, and the linkages between two uronic acid residues are most resistant. Thus, acid hydrolysis may be used for selective degradation of pectins. At lower temperatures, the hydrolysis of the methoxyl, acetyl groups and neutral side chains is predominant. Elevated temperatures accelerate degradation rates, and cleavage of the glycosidic bonds of the galacturonan backbone progressively increases.

At pH 5–6, pectin is stable at room temperature, but as the temperature or pH is increased, molecular degradation starts by a β-elimination reaction as shown in Figure 9.12 (Kenner, 1955; Whistler and BeMiller, 1958). The cleavage only takes place at a glycosidic linkage next to an esterified galacturonic acid; thus, LM pectin and pectic acid show better stability under these conditions. The effect of previous heat treatment at different pH levels on LM and HM pectin gels (Figure 9.13) shows that at pH 5.5, the gel strength of HM pectin gels decreased dramatically compared to LM pectin, mainly as a result of the β-elimination degradation. Increased temperature promotes the β-elimination process more than the competitive de-estirification, while increased pH favored the de-esterification (Kirtchev et al., 1989). An enhancing

FIGURE 9.12 β-elimination reaction of pectin molecules.

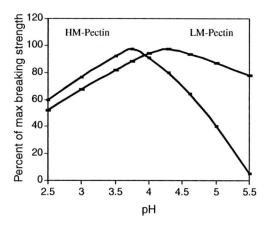

FIGURE 9.13 Breaking strength of pectin gels as a function of previous heat treatment of pectin solutions (90°C, 15 min) at various pH values. Redrawn from Copenhagen Pectin (1985).

effect of cations on the β-elimination depolymerization has been observed in several studies (Keijbets, 1974; Sajjaanantakul et al., 1993). At alkaline pH, pectin is rapidly de-esterified and degraded, even at room temperature (Launer and Tomimatsu, 1961).

Pectins are also readily degraded by oxidation (Neukom, 1963), γ-irradiation (Sjoberg, 1987), mechanical treatments such as grinding and extrusion (Bock et al., 1977; Ralet and Thibault, 1994), and dehydration (Ben-Shalom et al., 1992).

9.5.3.2 Enzyme Degradation

Pectic substances can be degraded by many enzymes produced by higher plants or microorganisms, which produce many textural changes in fruits and vegetables during ripening, storage and processing (as detailed in Section 2.3). The enzymes have specific functions and target sites and are usually divided into two groups: pectin esterases, such as methylesterase and acetyl esterase, which split off methoxyl groups and acetyl groups, respectively, and pectin depolymerases, such as polygalacturonase and lyase, which degrade pectin backbones. Recently, Daas et al. (1999) have suggested a classification that divides pectin-active enzymes into those that act on the homogalacturonan region and those that act on the rhamnogalacturonan region.

Commercially available pectinases, used on an industrial scale in fruit and vegetable processing, are of fungal origin and generally contain pectin methylesterases, polygalacturonases and pectin-lyase, proteases, hemicellulases, cellulases and glycosidases (Voragen et al., 1995).

9.5.4 INTERACTIONS OF PECTINS WITH OTHER BIOPOLYMERS

In most food systems, pectin is present together with other biopolymers, such as protein and other polysaccharides. The mixture of these polymers may produce synergistic effects and generate novel rheological and textural properties. Consequently,

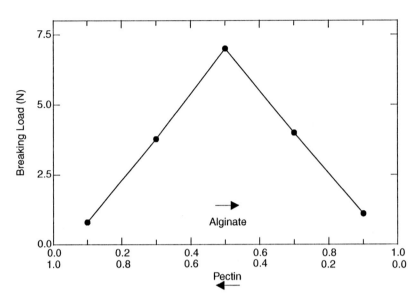

FIGURE 9.14 Stoichiometry of interaction in alginate-pectin mixed gels. (Redrawn from Toft et al., 1986).

interactions between pectin and other biopolymers have generated much interest over the last two decades or so.

When two biopolymers that differ in structure and conformation are mixed together, enthalpically favorable interactions will promote association between the two polymers; but if the heterotypic interactions are enthalpically unfavorable, thermodynamic incompatibility occurs with a tendency to phase separation (Tolstoguzov, 1991).

A mixture of pectin and alginate provides a good example of heterotypic interactions between two polymers. In the presence of alginate at acidic pH, HM pectin can form gels without sugar, a useful requirement for low–calorie food production (Steinnes, 1975; Morris and Chilvers, 1984; Thakur et al., 1997). A mixture of two polymers formed gels under conditions in which neither would gel in isolation. Maximum gel strength was attained when the poly-D-galacturonate content of pectin and the poly-L-guluronate content of alginate were present in equal amounts (Figure 9.14) (Toft et al., 1986). Gel formation was attributed to the heterotypic association between the poly-L-guluronate sequences and the near mirror-image poly-D-galacturonate sequences (Thom et al., 1982). Co-gelation is promoted at lower pH and by an increase in DE of the pectin.

The interactions between pectin and gelatin or other proteins in a mixture are more complex; both segregative and associative interactions may exist depending on the pH and ionic strength. By manipulating conditions such as pH, ionic strength and relative amounts of pectin and protein, different types of gels can be obtained. Normally, associative interactions involve electrostatic attraction between polyanionic pectin and polycationic gelatin or other proteins below their isoelectric points (IP). The interaction may result in the formation of soluble or insoluble complexes (Grinberg and Tolstoguzov, 1997; Wang and Qvist, 2000). Complex formation is suppressed by high salt concentration. For example, at ~ pH 3, LM pectin and gelatin (type B, IP = ~ 4.8–4.9) form monophasic co-gels in the absence of salt; while at a high salt concentration, they form phase-separated co-gels (Gilsenan et al., 2001). At a pH above the IP of the protein, because both pectin and protein then bear negative net charges, both segregative interactions (thermodynamic incompatibility) between the two polymers and self-association of each polymer tend to promote the formation of phase-separated systems. The formation of phase-separated co-gels at near neutral pH has been reported for several mixed systems, for example, HM pectin/gelatin in the presence of a co-solute (Chronakis et al., 1997), LM pectin/β-lactoglobulin with or without calcium (Dumay et al., 1999) and pectin/bovine serum albumin (Cai and Arntfield, 1997). Although there have been reports that pectin may also form soluble interpolymer complexes with some proteins by hydrogen bonding plus local electrostatic interactions, these complexes do not usually resist high temperatures or high salt concentrations (Antonov et al., 1996; Wang and Qvist, 2000).

9.6 CONCLUSION

The present review indicates that while pectin products continue to serve as a well-established food additive for technological purposes, the application of pectins for health benefits has many potential opportunities. In the last few decades, considerable research has been done on the physiological effects of pectin in humans as a water-soluble dietary fiber supplement. Some novel pharmacological activities of pectins and pectic polysaccharides are gaining more attention. Although most natural pectins from fruits and vegetables have not been shown to have these activities, chemical and enzymatic modifications may provide useful tools for producing such products for human health care.

REFERENCES

Ahmed, A.E. and Labavitch, J.M. 1977. A simplified method for accurate determination of cell wall uronide content [Pears, kiwi fruits]. *J. Food Biochem.*, 1(4):361–365.

Andoh, A., Bamba, T. and Sasaki, M. 1999. Physiological and anti-inflammatory roles of dietary fiber and butyrate in intestinal functions. *J. Parenteral Enteral Nutr.*, 23(5 Suppl): S70–S73.

Anger, H., and Berth, G. 1986. Gel permeation chromatography and the Mark-Houwink relations for pectins with different degrees of esterification. *Carbohydr. Polym.*, 6 (3):193–202.

Angyal, S.J. 1989. Complexes of metal cations with carbohydrates in solution, *Adv. in Carbohydr. Chem. Biochem.*, 47:1–43.

Antonov, Y.A., Lashko, N.P., Glotova, Y.K., Malovikova, A. and Markovich, O. 1996. Effect of the structural features of pectins and alginates on their thermodynamic compatibility with gelatin in aqueous media, *Food Hydrocolloids*, 10(1):1–9.

Anyas-Weisz, L. and Deuel, H. 1950. The coagulation of sodium pectinates, *Helv. Chim. Acta*, 33:559–562.

Aravantinos-Zafiris, G. and Oreopoulou, V. 1992. The effect of nitric acid extraction variables of orange pectin, *J. Sci. Food Agric.*, 60:127–129.

Aro, A., Uusitupa, M., Voutilainen, E., Hersio, K., Korhonen, T. and Siitonen, O. 1981. Improved diabetic control and hypocholesterolaemic effect induced by long-term dietary supplementation with guar gum in type 2 (insulin independent) diabetes, *Diabetologia*, 21:29–33.

Axelos, M.A.V. and Thibault, J.F. 1991. The chemistry of low-methoxyl pectin gelation in: *The Chemistry and Technology of Pectin*, ed., R.H. Walter, New York: Academic Press, pp. 109–118.

Axelos, M.A.V., Lefebvre, J., Qiu, C.G. and Rao, M.A. 1991. Rheology of pectin dispersions and gels in: *The Chemistry and Technology of Pectin*, ed., R.H. Walter, New York: Academic Press, pp. 228–250.

Ball, G.F.M. 1990. The application of HPLC to the determination of low molecular weight sugars and polyhydric alcohols in foods: a review, *Food Chem.*, 35:117–152.

Bartley, I.M. and Knee, M. 1982. The chemistry of textural changes in fruit during storage, *Food Chem.*, 9:47–58.

Bartolomé, L.G. and Hoff, J.E. 1972. Gas chromatographic methods for the assay of pectin methyesterase, free methanol and methoxy groups in plant tissues, *J. Agric. Food Chem.*, 20(2):262–266.

Basak, S. and Ramaswamy, H.S. 1994. Simultaneous evaluation of shear rate and time dependency of stirred yogurt rheology as influenced by added pectin and strawberry concentrate, *J. Food Eng.*, 21:385–393.

BeMiller, J.N. 1986. An introduction to pectins: structure and properties, in: *Chemistry and Functions of Pectins*, eds., M.L. Fishman and J.J. Jen. Washington DC: American Chemical Society, pp. 2–12.

Ben-Shalom, N., Plat, D., Levi, A. and Pinto, R. 1992. Changes in molecular weight of water-soluble and EDTA-soluble pectin fractions from carrot after heat treatments, *Food Chem.*, 45(4):243–245.

Berth, G., Anger, H. and Lexow, D. 1980. The determination of the molecular weights of pectins by means of membrane osmometry in aqueous solution, *Nahrung*, 24(6):529–534.

Berth, G., Dautzenberg, H., Lexow, D. and Rother, G. 1990. The determination of the molecular weight distribution of pectins by calibrated GPC, Part I. Calibration by light scattering and membrane osmometry, *Carbohydr. Polym.*, 12(1):39–59.

Berth, G., Dautzenberg, H. and Rother, G. 1994. Static light scattering technique applied to pectin in dilute solution. Part II. The effects of clarification, *Carbohydr. Polym.*, 25:187–195.

Bitter, T. and Muir, H.M. 1962. A modified uronic acid carbazole reaction, *Anal. Biochem.*, 4:330–334.

Blumenkrantz, N. and Asboe-Hansen, G. 1973. New method for quantitative determination of uronic acids, *Anal. Biochem.*, 54:484–489.

Bock, W., Anger, H., Kohn, H., Malovikova, A., Dongowski, G. and Friebe, R. 1977. Charakterisierung mechanolytisch abgebauter Pektinpraparate, *Angew Makromol Chem*, 64:133–146.

Bourne, M.C. 1982. *Food Texture and Viscosity: Concept and Measurement*, New York: Academic Press.

Braconnot, H. 1825. Recherches sur un novel acide universellement rependu dans tous les vegetaux, *Ann. Chim. Phys. Ser. II*, 28:173–178.

Brett, C.T. and Waldron, K. 1996. *Physiology and Biochemistry of Plant Cell Walls*, 2nd ed., London: Chapman and Hall.

Bruce, W.R. 1987. Recent hypotheses for the origin of colon cancer, *Cancer Res.*, 47(16):4237–4242.

Burns, J.K. and Pressey, R. 1987. Ca²⁺ in cell walls of ripening tomato and peach, *J. Am. Soc. Hortic. Sci.*, 112(5):783–787.

Cai, R. and Arntfield, S.D. 1997. Thermal gelation in relation to binding of bovine serum albumin-polysaccharide systems, *J. Food Sci.*, 62(6):1129–1134.

Carpita, N.C. and Gibeaut, D.M. 1993. Structural models of primary cell walls in flowering plants: consistency of molecular structure with the physical properties of the walls during growth, *Plant J.*, 3(1):1–30.

Castaldo, D., Quagliuolo, L., Servillo, L., Balestrieri, C. and Giovane, A. 1989. Isolation and characterization of pectin methylesterase from apple fruit, *J. Food Sci.*, 54(3):653–655, 673.

Cesaro, A., Ciana, A., Delben, F., Manzini, G. and Paoletti, 1982. Thermodynamic evidence of a pH-induced conformational transition in aqueous solution, *Biopolymers*, 21:431–449.

Chapman, H.D., Morris, V.J., Selvendran, R.R. and O'Neill, M.A. 1987. Static and dynamic light scattering studies of pectic polysaccharides from the middle lamellae and primary cell walls of cider apples, *Carbohydr. Res.*, 165:53–68.

Chang, K.C. and Miyamoto, A. 1992. Gelling characteristics of pectin from sunflower head residues, *J. Food Sci. Off. Publ. Inst. Food Technol.*, 57(6):1435–1438, 1443.

Chen, T.S. and Joslyn, M.A. 1967. The effect of sugars on viscosity of pectin solutions II. Comparison of dextrose, maltose and dextrins, *J. Colloid Interface Sci.*, 25:346–352.

Chou, T.C. and Kokini, J L. 1987. Rheological properties and conformation of tomato paste pectins, citrus and apple pectins, *J. Food Sci.*, 52:1658–1664.

Chou, T.C., Pintauro, N. and Kokini, J.L. 1991. Conformation of citrus pectin using amplitude oscillatory rheometry, *J. Food Sci.*, 56(5):1365–1368, 1371.

Christensen, P.E. 1953. Methods of grading pectin in relation to the molecular weight (Intrinsic Viscosity) of pectin, *Pectin Symposium of the New York Preservers Association*, pp. 163–172.

Christensen, S.H. 1986. Pectins in: *Food Hydrocolloids*, vol. 3, ed., M. Glicksman, Boca Raton: CRC Press, pp. 205–230.

Chronakis, I.S., Kasapis, S. and Abeysekara, R. 1997. Structural properties of gelatin-pectin gels. Part I: Effect of ethylene glycol, *Food Hydrocolloids*, 11(3):271–279.

Comby, S., Doublier, J.L. and Lefebvre, J. 1986. Stress-relaxation study of high-methoxyl pectin gels, in: *Gums and Stabilizers for the Food Industry 3*, eds., G.O. Phillips, D.J. Wedlock and P.A. Williams, New York: Elsevier Science Publishers, pp. 203–212.

Copenhagen Pectin. 1985. *Pectin. General Description.* Lille Skensved, Denmark.

Corredig, M., Kerr, W. and Wicker, L. 2000. Molecular characterisation of commercial pectins by separation with linear mix gel permeation columns in-line with multi-angle light scattering detection, *Food Hydrocolloids*, 14:4–47.

Cox, R.E. and Higby, R.H. 1944. An improved method for determining the grade of commercial pectins, *Food Manuf.*, 19:199–202.

Crandall, P.G. and Wicker, L. 1986. Pectin internal gel strength: theory, measurement, and methodology, in: *Chemistry and Function of Pectins*, eds., M.L. Fishman and J.J. Jen, Washington, DC: American Chemical Society, pp. 88–102.

Cros, S., Bouchemal, N., Ohassan, H., Imberty, A. and Perez, S. 1992. Solution conformation of a pectin fragment dissaccharide using molecular modeling and nuclear magnetic resonance, *Int. J. Biol. Macromol.*, 14(6):313–320.

Daas, P.J.H., van-Alebeek, G.J.W.M., Voragen, A.G.J. and Schols, H.A. 1999. Determination of the distribution of non-esterified galacturonic acid in pectin with *endo*-polygalacturonase, *Gums and Stabilisers for the Food Industry 10, Proceedings of the 10th Conference*, Wrexham, Cambridge: RSC, pp. 3–18.

Dahme, A. 1985. Characterization of linear and nonlinear deformation behaviour of a weak standard gel made from high-methoxyl citrus pectin, *J. Texture Stud.*, 16:227–239.

Darvill, A., McNeil, M., Albersheim, P. and Delmer, D.P. 1980. The primary cell walls of flowering plants, in *The Biochemistry of Plants*, Vol. 1, eds., P.K. Stumpf and E.E. Conn, New York: Academic Press, pp. 91–161.

DCCT (Diabetes Control and Complications Trial) Research Group. 1993. The effect of intensive treatment of diabetes on the development and progression of long-term complications in insulin-dependent diabetes mellitus, *N. Engl. J. Med.*, 329:977–986.

de Vries, J.A. Voragen, A.J., Rombouts, F.M. and Pilnick W. 1982. Enzymatic degradation of apple pectins, *Carbohydr. Polym.*, 2:25–33.

de Vries, J.A., Voragen, A.G.J., Rombouts, F.M. and Pilnick, W. 1984. Changes in the structure of apple pectic substances during ripening and storage, *Carbohydr. Polym.*, 4:3–13.

de Vries, J.A., Voragen, A.G.J., Rombouts, F.M. and Pilnick, W. 1986. Structural studies of apple pectins with pectolytic enzymes, in: *Chemistry and Function of Pectins*, eds., M.L. Fishman and J.J. Jen, Washington: ACS Symposium Series No. 310, pp. 38–48,

Deckers, H.A., Olieman, C., Rombouts, F.M. and Pilnik, W. 1986. Calibration and application of high-performance size exclusion columns for molecular weight distribution of pectins, *Carbohydr. Polym.*, 6:361–378.

de Ruiter, G.A., Schols, H.A., Voragen, A.G.J. and Rombouts, F.M. 1992. Carbohydrate analysis of water-soluble uronic acid-containing polysaccharides with high-performance anion-exchange chromatography using methanolysis combined with TFA hydrolysis is superior to four other methods, *Anal. Biochem.*, 207:176–185.

Delahay, P. 1965. *Double Layer and Electrode Kinetics*, New York: Wiley.

Delbarre, F., Rondier, J. and Gery, A. 1977. Lack of effect of two pectins in idiopathic or gout-associated hyperdyslipidemia hypercholesterolaemia, *Am. J. Clin. Nutr.*, 30(4):463–465.

Deming, D.M., Boileau, C., Lee, C.M. and Erdman, J.W. 2000. Amount of dietary fat and type of soluble fiber independently modulate postabsorptive conversion of beta-carotene to vitamin A in Mongolian gerbils, *J. Nutr.*, 130(11):2789–2796.

Di Nola, A., Fabrizi, G., Lamba, D. and Segre, A.L. 1994. Solution conformation of a pectic acid fragment by 1 H-NMR and molecular dynamics, *Biopolymers*, 34(4):457–462.

Diepenmaat-Wolters, M.G.E. 1993. Food Ingredients Europe, in: *Functional Properties of Dietary Fibre in Foods*, Maarsen, The Netherlands, Expoconsult, pp. 162–164.

Dische, Z. 1947. A new specific color reaction of hexuronic acids, *J. Biol. Chem.*, 167:189–198.

Drews, L.M., Kies, C. and Fox, H.M. 1979. Effect of dietary fiber on copper, zinc, and magnesium utilisation by adolescent boys, *Am. J. Clin. Nutr.*, 32:1893–1897.

Dumay, E., Laligant, A., Zasypkin, D. and Cheftel, J.C. 1999. Pressure- and heat-induced gelation of mixed beta-lactoglobulin/polysaccharide solutions: Scanning electron microscopy of gels, *Food Hydrocolloids*, 13(4):339–351.

Dygert, S., Li, L.H., Florida, D. and Thoma, J.A. 1965. Determination of reducing sugar with improved precision, *Anal. Biochem.*, 13:367–374.

Eggum, B.O. 1992. The influence of dietary fiber on protein digestion and utilisation, in: *Dietary Fiber—A Component of Food, Nutritional Function in Health and Disease*, eds., T.F. Schweizer and C.A. Edwards, New York: Springer-Verlag.

El-Nawawi, S.A. and Heikel, Y. A. 1995. Factors affecting the production of low-ester pectin gels, *Carbohydr. Polym.*, 26:189–193.

Ellis, P.R., Rayment, P. and Wang, Q. 1996. A physico-chemical perspective of plant polysaccharides in relation to glucose absorption, insulin secretion and the entero-insular axis, *Proc. Nutr. Soc.*, 55:881–898.

Endress, H.-U. 1991. Nonfood uses of pectin, in: *The Chemistry and Technology of Pectin*, ed., R. Walter, New York: Academic Press.

Englyst, H.N. and Cummings, J.H. 1984. Simplified method for the measurement of total nonstarch polysaccharides by gas liquid chromatography of constituent sugars as alditol acetates, *Analyst*, 109:937–942.

Falk, J.D. and Nagyvary, J.J. 1982. Exploratory studies of lipid-pectin interactions, *J. Nutr.*, 112(1):182–188.

FAO. 1984. Food and Nutrition Paper 31/2, p. 75, Rome.

FDA. 1986. GRAS regulation 21; CFR paragraph 18–1588.

Ferry, J.D. 1980. *Viscoelastic Properties of Polymers*, New York: Wiley.

Fishman, M.L., Pepper, L., Damert, W.C., Phillips, J.G. and Barford, R.A. 1986. A critical reexamination of molecular weight and dimensions of citrus pectins, in *Chemistry and Functions of Pectins*, eds., M.L. Fishman and J.J. Jen, Washington: ACS Symposium Series No. 310, pp. 22–37.

Fishman, M.L., Cooke, P., Levaj, B., Gillespie, D.T., Sondey, S.M. and Scorza, R. 1992. Pectin microgels and their subunit structure, *Arch. Biochem. Biophys.*, 294(1):253–260.

Flory, P.J. 1953. *Principles of Polymer Chemistry*. Ithaca, New York: Cornell University Press.

Ford, C.W. 1982. A routine method for identification and quantitative determination by gas-liquid chromatography of galacturonic acid in pectic substances, *J. Sci. Food Agric.*, 33:318–324.

Foreman, L.P. and Schneeman, B.O. 1980. Effects of dietary pectin and fat on the small intestinal contents and exocrine pancreas of rats, *J. Nutr.*, 110:1992–1999.

Frape, D.L. and Jones, A.M. 1995. Chronic and postprandial responses of plasma insulin, glucose and lipids in volunteers given dietary fibre supplements, *Br. J. Nutr.*, 73(5):733–751.

Fuoss, R. M. and Strauss, V.P. 1949. The viscosity of mixtures of polyelectrolytes and simple electrolytes, *Ann. N.Y. Acad. Sci.*, 51:836–851.

Furda, I. 1990. Interaction of dietary fiber with lipids-mechanistic theories and their limitations, *Adv. Exp. Med. Biol.*, 270:67–82.

Gallaher, D.D., Wood, K.J., Gallaher, C.M., Marquart, L.F. and Engstrom, A.M. 1999. Intestinal contents supernatant viscosity of rats fed oat-based muffins and cereal products, *Cereal Chem.* 76(1):21–24.

Gandelman, M.S. and Birks, J.W. 1982. Photooxygenation-chemiluminescence high-performance liquid chromatographic detector for the determination of apliphatic alcohols, aldehydes, ethers and saccharides, *J. Chromatogr.*, 242:21–31.

Garleb, K.A., Bourquin, L.D. and Fakey, G.C. 1991. Galacturonate in pectic substances from fruits and vegetables: comparison of anion-exchange HPLC with pulsed amperometric detection to standard colorimetric procedure, *J. Food Sci.*, 56:423–426.

Gillooly, M., Bothwell, T.H., Charlton, R.W., Torrance, J.D., Bezwoda, W.R., MacPhail, A.P., Derman, D.P., Novelli, L., Morrall, P. and Mayet, F. 1984. Factors affecting the absorption of iron from cereals, *Br. J. Nutr.*, 51(1):37–46.

Gilsenan, P.M., Richardson, R.K. and Morris, E.R. 2001. Rheologoy of pectin-gelatin co-gels, *Annu. Trans. Nordic Rheol. Soc.*, 9:5–10.

Gold, L.A., McCourt, J.P. and Merimee, T.J. 1980. Pectin: An examination in normal subjects, *Diabetes Care*, 3:50–52.

Goldberg, R., Morvan, C., Jauneau, A. and Javis, M.C. 1996. Methyl-esterfication, de-esterfication and gelation of pectins in the primary cell wall. In: *Pectins and Pectinases*, J. Visser and A.G.J. Voragen, Amsterdam: Elsevier, pp. 151–172.

Goto, A., Araki, C. and Uzumi, Y. 1981. Molecular weight distribution of water soluble pectins in citrus juices, *Proceedings of the International Society of Citriculture*, pp. 931–934.

Grigelmo-Miguel, N., Abadias-Seros, M.I. and Martin-Belloso, O. 1999. Characterization of low-fat high-dietary fibre frankfurters, *Meat Sci.*, 52:247–256.

Grinberg, V.Ya. and Tolstoguzov, V.B. 1997. Thermodynamic incompatibility of proteins and polysaccharides in solutions, *Food Hydrocolloids*, 11(2):145–158.

Groop, P.H., Aro, A., Stenman, S. and Groop, L. 1993. Long-term effects of guar gum in subjects with non-insulin-dependent diabetes mellitus, *Am. J. Clin. Nutr.*, 58:513–518.

Gross, K.C. and Wallner, S.J. 1979. Degradation of cell wall polysaccharides during tomato fruit ripening, *Plant Physiol.*, 63:117–120.

Gross, K.C. and Sams, C.E. 1984. Changes in cell wall neutral sugar composition during fruit ripening, *Phytochemistry*, 23(11):2457–2461.

Grosso, C.R.F. and Rao, M.A. 1998. Dynamic rheology of structure development in low-methoxyl pectin + Ca^{2+} sugar gels, *Food Hydrocolloids*, 12(3):357–363.

Grudeva-Popova, J. and Sirakova, I. 1998. Effect of pectin on some electrolytes and trace elements in patients with hyperlipoproteinemia, *Folia Medica (Plovdiv)*, 40(1):41–45.

Harding, S.E., Berth, G., Ball, A., Mitchell, J.R. and de la Torre, J.G. 1991. The molecular weight distribution and conformation of citrus pectins in solution studied by hydrodynamics, *Carbohydr. Polym.*, 16(1):1–16.

Harris, P.J. and Fergusson, L.R. 1993. Dietary fibre: its composition and role in protection against colorectal cancer, *Mutation Res.*, 290(1):97–110.

Haskell, W.L., Spiller, G.A., Jensen, C.D., Ellis, B.K. and Gates, J.E. 1992. Role of water soluble dietary fiber in the management of elevated plasma cholesterol in healthy subjects, *Am. J. Cardiol.*, 69(5):433–439.

Hatanaka, C., Yokohiki, K. and Matsuhashi, S. 1986. Analysis of galacturonic acid and oligogalacturonic acids by high-performance liquid chromatography, *J. Fac. Appl. Biol. Sci., Hiroshima Univ.*, 25:41–48.

Havelund, T., Aalykke, C. and Rasmussen, L. 1997. Efficacy of a pectin-based anti-reflux agent on acid reflux and recurrence of symptoms and oesophagitis in gastro-oesophageal reflux disease, *Eur. J. Gastroenterol. Hepatol.*, 9(5):509–514.

Havelund, T. and Aalykke, C. 1997. The efficacy of a pectin-based raft-forming anti-reflux agent in endoscopy-negative reflux disease, *Scand. J. Gastroenterol.*, 32(8):773–777.

Heitman, D.W., Hardman, W.E. and Cameron, I.L. 1992. Dietary supplementation with pectin and guar gum on 1,2-dimethylhydrazine-induced colon carcinogenesis in rats, *Carcinogenesis (Eynsham)*, 13(5):815–818.

Hobson, G.E. 1963. Pectinesterase in normal and abnormal tomato fruit, *Biochem. J.*, 86:358–365.

Hoff, J.E. and Feit, E.D. 1964. New technique for functional group analysis in gas chromatography-syringe reactions, *Anal. Chem.*, 36:1002–1008.

Holt, S., Heading, R.C., Carter, D.C., Prescott, L.F. and Tothill, P. 1979. Effect of gel fibre on gastric emptying and absorption of glucose and paracetamol, *Lancet*, 1:636–639.

Hourdet, D. and Muller, G. 1991. Solution properties of pectin polysaccharides—III: molecular size of heterogeneous pectin chains. Calibration and application of SEC to pectin analysis, *Carbohydrate-Polymers*, 16(4):409–432.

Hsieh, T.C. and Wu, J.M. 1995. Changes in cell growth, cyclin/kinase, endogenous phosphoproteins and nm23 gene expression in human prostatic JCA-1 cells treated with modified citrus pectin, *Biochem. Mol. Biol. Int.*, 37(5):833–841.

Huang, J.M.G. 1973. Improved method for the extraction of pectin, *Fla. Agric. Exp. St. J. Ser. No. 5101*.

Huggins, M.L. 1942. The viscosity of dilute solutions of long-chain molecules. IV. Dependence on concentration, *J. Am. Chem. Soc.*, 64:2716–2718.

IFT. 1959. Committee on Pectin Standardization, Final Report of the IFT Committee, *Food Technol.*, 13:496–500.

Jacobs, L.R. 1990. Influence of soluble fibers on experimental colon carcinogenesis, in: *Dietary Fiber Chemistry, Physiology, and Health Effects*, eds., D. Kritchevsky, C. Bonfield and J.W. Anderson, New York: Plenum, pp. 389–401.

Jacobs, L.R. and Lupton, J.R. 1986. Relationship between colonic luminal pH, cell proliferation, and colon carcinogenesis in 1,2-dimethylhydrazine treated rats fed high fiber diets, *Cancer Res*, 46:1727–1734.

Jarvis, M.C. 1982. The proportion of calcium-bound pectin in plant cell walls, *Planta*, 154:344–346.

Jarvis, M.C. 1984. Structure and properties of pectin gels in plant cell walls, *Plant Cell Environ.*, 7:153–164.

Jarvis, M.C., Forsyth, W. and Duncan, H.J. 1988. A survey of the pectin content of nonlignified monocot cell walls, *Plant Physiol.*, 88(2):309–314.

Jenchs, W.P. 1969. *Catalysis in Chemistry and Enzymology*, New York: McGraw-Hill.

Jenkins, D.J.A., Leeds, A.R., Newton, C. and Cummings, J.H. 1975. Effect of pectin, guar gum, and wheat fiber on serum cholesterol, *Lancet*, 1:1116–1117.

Jenkins, D.J.A., Leeds, A.R., Gassull, M.A., Wolever, T.M.S., Goff, D.V., Alberti, K.G.M.M. and Hockaday, T.D.R. 1976. Unabsorbable carbohydrates and diabetes: decreased postprandial hyperglycemia, *Lancet*, 2:172–174.

Jenkins, D.J.A., Leeds, A.R., Gassull, M.A., Cochet, B. and Alberti, K.G.M.M. 1977. Decrease in postprandial insulin and glucose concentrations by guar and pectin, *Ann. Intern. Med.*, 86:20–23.

Jenkins, D.J.A., Wolever, T.M.S., Taylor, R.H., Reynolds, D., Nineham, R. and Hockaday, T.D.K. 1980. Diabetic glucose control, lipids and trace elements on long term guar, *Brit. Med. J.*, 280:1353–1354.

Jensen, C.D., Haskell, W. and Whittam, J.H. 1997. Long-term effects of water-soluble dietary fiber in the management of hypercholesterolemia in healthy men and women, *Am. J. Cardiol.*, 79(1):34–37.

Jiang, Y.H., Lupton J.R. and Chapkin, R.S. 1997. Dietary fat and fiber modulate the effect of cancinogen on colonic protein kinase C lambda expression in rats, *J. Nutr.*, 127(10):1938–1943.

Jordan, R.C. and Brant, D.C. 1978. An investigation of pectin and pectic acid in dilute aqueous solution, *Biopolymers*, 17:2885–2895.

Joseph, G.H. 1953. Better Pectins, *Food Eng.*, 25:71–73.

Joseph, G.H. and Havighorst, C.R. 1952. Engineering quality pectins, *Food Eng.*, 24:87–89, 160–162.

Judd, P.A. and Truswell, A.S. 1982. Comparison of the effects of high and low methoxyl pectins on blood and fecal lipids in man, *Br. J. Nutr.*, 48:451–458.

Kar, F. and Arslan, N. 1999a. Characterisation of orange peel pectin and effect of sugars, L-ascorbic acid, ammonium persulfate, salts on viscosity of orange peel pectin solutions, *Carbohydr. Polym.*, 40:285–291.

Kar, F. and Arslan, N. 1999b. Effect of temperature and concentration on viscosity of orange peel pectin solutions and intrinsic viscosity-molecular weight relationship, *Carbohydr. Polym.*, 40:277–284.

Kawabata, A. 1977. Studies on chemical and physical properties of pectin substances from fruits, viscoelasticity of fruit pectin gels, *Memoirs Tokyo Univ. Agric.*, 19:166–178.

Kay, R.M. and Truswell, A.S. 1977. Effect of citrus pectin on blood lipids and fecal steroid excretion in man, *Am. J. Clin. Nutr.*, 30:171–175.

Keijbets, M.J.H. 1974. β-elimination of pectin in the presence of anions and cations, *Carbohydr. Res.*, 33:359–362.

Keijbets, M.J.H. and Pilnik, W. 1974. Some problems in the analysis of pectin in potato tuber tissue, *Potato Res.*, 17:169–177.

Kelsay, J.L., Jacob, R.A. and Prather, E.S. 1979a. Effect of fiber from fruits and vegetables on metabolic responses of human subjects, III. Zinc, copper, and phosphorus balances, *Am. J. Clin. Nutr.*, 32(11):2307–2311.

Kelsay, J.L., Jacob, R.A. and Prather, E.S. 1979b. Effect of fiber from fruits and vegetables on metabolic responses of human subjects, II. Calcium, magnesium, iron, and silicon balances, *Am. J. Clin. Nutr.*, 32(9):1876–1880.

Kenner, J. 1955. The alkaline degradation of carbonyl oxycelluloses and the significance of saccharinic acids for the chemistry of carbohydrates, *Chem. Ind.*, pp. 727–730.

Kertesz, Z.I. 1951. *The Pectic Substances*. New York: Interscience.

Keys, A., Grande, F. and Anderson, J.T. 1961. Fiber and pectin in the diet and serum cholesterol concentration in man, *Proc. Soc. Exp. Biol. Med.*, 106:555–558.

Kim, M. 1998. Highly esterified pectin with low molecular weight enhances intestinal solubility and absorption of ferric iron in rats, *Nutr. Res.*, 18(12):1981–1994.

Kirtchev, N., Panchev, I. and Kratchanov, C. 1989. Kinetics of acid-catalyzed deesterification of pectin in a heterogeneous medium, *Int. J. Food Sci. Technol.*, 24:479–486.

Klavons, J.A. and Bennett, R.D. 1986. Determination of methanol using alcohol oxidase and its applications to methyl ester content of pectins, *J. Agric. Food Chem.*, 34(4):597–599.

Klurfeld, D.M. 1992. Dietary fiber-mediated mechanisms in carcinogenesis, *Cancer Res.*, 52(7):2055s–2059s.

Knee, M. 1978. Metabolism of polymethylgalacturonate in apple fruit cortical tissue during ripening, *Phytochemistry*, 17:1261–1264.

Knee, M. and Bartley, J.M. 1981. Composition and metabolism of cell wall polysaccharides in ripening fruits, in: *Recent Advances in the Biochemistry of Fruits and Vegetables*, eds., J. Friend and M.J.C. Rhodes, New York: Academic Press, pp. 133–148.

Knopp, R.H., Superko, H.R., Davidson, M., Insull, W., Dujovne, C.A., Kwiterovich, P.O., Zavoral, J.H., Graham, K., O'Connor, R.R. and Edelman, D.A. 1999. Long-term blood cholesterol-lowering effects of a dietary fiber supplement, *Am. J. Prev. Med.*, 17(1):18–23.

Kohn, R. and Luknar, O. 1977. Intermolecular calcium ion binding on polyuronates-polygalacturonate and polyguluronate, *Coll. Czech. Commun.*, 42:731–744.

Kontominas, M.G. and Kokini, J.L. 1990. Measurement of molecular parameters of water soluble apple pectin using low angle laser light scattering, *Lebensmittel Wiss und Technologie*, 23:174–177.

Kraemer, E.O. 1938. Molecular weights of celluloses and cellulose derivatives, *Ind. Eng. Chem.*, 30:1200–1203.

Kripke, S.A., Fox, A.D., Berman, J.M., Settle, R.G. and Rombeau, J.L. 1989. Stimulation of intestinal mucosal growth with intracolonic infusion of short-chain fatty acids, *J. Parenteral Enteral Nutr.*, 13(2):109–116.

Kujawski, M. and Tuszynski, T. 1987. A comparison of some methods for the determination of methoxyl groups in commercial pectin preparations, *Nahrung*, 31(3):233–238.

Lapasin, R. and Pricl, S. (1995). *Rheology of Industrial Polysaccharides, Theory and Applications*, London: Blackie Academic and Professional, pp. 495–578.

Launer, H. and Tomimatsu, Y. 1961. Alkali sensitivity of polysaccharides: periodate starches, periodate dextran and a polygalacturonide, *J. Org. Chem.*, 26:541–545.

Leo, H.T. and Taylor, C.C. 1955. Jelly-making composition, U.S. Patents 2,703,757, 2,703,758 and 2,703,759.

Lewinska, D., Rosinski, S. and Piatkiewicz, W. 1994. A new pectin-based material for selective LDL-cholesterol removel, *Artificial Organs*, 18(3):217–222.

Lotzkar, H., Schultz, T.H., Owens, H.S. and Maclay, W.D. 1946. Effect of salts on the viscosity of pectinic acid solutions, *J. Phys. Chem.*, 50:200–210.

Lupton, J.R. 2000. Is fiber protective against colon cancer? Where the research is leading us, *Nutrition*, 16:558–561.

Lupton, J.R., Coder, D.M. and Jacobs, L.R. 1988. Long-term effects of fementable fibers on rat colonic pH and epithelial cell cycle, *J. Nutr.*, 118:840–845.

Ma, Q., Hoper, M., Halliday, I. and Roelands, B.J. 1996. Diet and experimental colorectal cancer, *Nutr. Res.*, 16(3):413–426.

Macleod, G.S., Fell, J.T., Collett, J.H., Sharma, H.L. and Smith, A.M. 1999. Selective drug delivery to the colon using pectin: chitoson:hydroxypropyl methylcellulose film coated tablets, *Int. J. Pharm.*, 187(2):251–257.

Malyoth, G. 1934. Pectin, the most effective constituent of the apple diet, *Klin. Wochschr.*, 13:51–54.

Matsuhashi, S., Yokohiki, K. and Hatanaka, C. 1989. High performance liquid chromatography measurement of the degradation limits of pectate by exopectinases, *Agric. Biol. Chem.*, 53(5):1417–1418.

Matthew, J.A., Howson, S.J., Keenan, M.H.J. and Belton, P.S. 1990. Improvement of the gelation properties of sugar-beet pectin following treatment with an enzyme preparation derived from *Aspergillus niger*: comparison with a chemical modification, *Carbohydr. Polym.*, 12(3):295–306.

May, C.D. 1990. Industrial pectins: sources, production and applications, *Carbohydr. Polym.*, 12:79–99.

May, C.D. and Stainsby, G. 1986. Factors affecting pectin gelation, in: *Gums and Stabilisers for the Food Industry 3*, eds., G.O. Phillips, D.J. Wedlock and P.A. Williams, London: Elsevier Applied Science Publishers, pp. 515–523.

McFeeters, R.F. and Armstrong, S.A. 1984. Measurement of pectin methylation in plant cell walls, *Anal. Biochem.*, 139:212–217.

Michel, F., Doublier, J.L. and Thibault, J.F. 1982. Investigations of high-methoxyl pectins by potentiometry and viscometry, *Prog. Food Nutr. Sci.*, 6:367–372.

Michel, F., Thibault, J.-F., Mercier, C., Heitz, F. and Pouillade, F. 1985. Extraction and characterization of pectins from sugar beet pulp, *J. Food Sci.*, 50:1499–1502.

Miettinen, T.A. and Tarpila, S. 1977. Effect of pectin on serum cholesterol, fecal bile acids and biliary lipids in normolipidemic and hyperlipidemic individuals, *Clin. Chim. Acta.*, 79:471–477.

Mitchell, J.R. 1976. Rheology of gels, *J. Texture Stud.*, 7:313–339.

Mitchell, J.R. and Blanshard, J.M.V. 1976. Rheological properties of citrus sodium pectate gels, *J. Texture Stud.*, 7(3):341–351.

Monnier, L., Pham, T.C., Aguirre, L., Orsetti, A. and Mirouze, J. 1978. Influence of indigestible fibers on glucose tolerance, *Diabetes Care*, 1:83–88.

Morris, V.J. and Chilvers, G.R. 1984. Cold setting alginate-pectin mixed gels, *J. Sci. Food Agric.*, 35(12):1370–1376.

Morris, E.R., Cutler, A.N., Ross-Murphy, S.B., Rees, D.A. and Price, J. 1981. Concentration and shear-rate dependence of viscosity in random coil polysaccharide solutions, *Carbohydrate Polym.*, 1:5–21.

Morris, G.A., Foster, T.J. and Harding, S.E. 2000. The effect of the degree of esterification on the hydrodynamic properties of citrus pectin, *Food Hydrocolloids*, 14(3):227–235.

Mosenthin, R., Sauer, W.C. and Ahrens, F. 1994. Dietary pectin's effect on ileal and fecal amino acid digestibility and exocrin pancreatic secretions in growing pigs, *J. Nutr.*, 124 (8):1222–1229.

National Research Council. 1972. Pectins, in: *Food Chemicals Codex*, 2nd edition, Washington, DC: National Academy of Sciences, pp. 577–581.

Nelson, N. 1944. A photometric adaptation of the Somogyi Method for the determination of glucose, *J. Biol. Chem.*, 153:375–380.

Nie, Y., Li, Y., Wu, H., Du, H., Dai, S., Wang, H. and Li, Q. 1999. Colloidal bismuth pectin: an alternative to bismuth subcitrate for the treatment of *Helicobacter pylori*-positive duodenal ulcer, *Helicobacter*, 4(2):128–134.

Nogata, Y., Yoza, K., Kusumoto, K. and Ohta, H. 1995. Changes in molecular weight and carbohydrate composition of cell wall polyuronide and hemicellulose during ripening in strawberry fruit, in: *Pectins and Pectinases*, eds., J. Visser and A.G.J. Voragen, Amsterdam: Elsevier, pp. 591–596.

Oakenfull, D.G. 1991. The chemistry of high-methoxyl pectins, in: *The Chemistry and Technology of Pectin*, ed., R. Walter, New York: Academic Press, pp. 87–108.

Oakenfull, D. and Scott, A. 1984. Hydrophobic interaction in the gelation of high methoxyl pectins, *J. Food Sci.*, 49:1093–1098.

Ohkami, H., Tazawa, K., Yamashita, I., Shimizu, T., Murai, K., Kobashi, K. and Fujimaki, M. 1995. Effects of apple pectin on fecal bacterial enzymes in azoxymethane-induced rat colon carcinogenesis, *Jpn. J. Cancer Res.*, 86(6):523–529.

Oosterveld, A., Pol, I.E., Beldman, G. and Voragen, A.G.J. 2001. Isolation of feruloylated arabinans and rhamnogalacturonans from sugar beet pulp and their gel forming ability by oxidative cross-linking, *Carbohydr. Polym.*, 44:9–17.

Owens, H.S., McReady, R.M. and Maclay, W.D. 1949. Gelation and characteristics of acid-precipitated pectinates, *Food Technol.*, 3:77–82.

Palmer, G.H. and Dixon, D.G. 1966. Effect of pectin dose on serum cholesterol levels, *Am. J. Clin. Nutr.*, 18:437–442.

Pals, D.T.F. and Hermans, J.J. 1952. Sodium salts of pectin and of carboxy methyl cellulose in aqueous sodium chloride, *Rec. Trav. Chim. Pays-Bas*, 71:433–457.

Panchev, I., Kirchev, N. and Kratchanov, C. 1988. Improving pectin technology. II. Extraction using ultrasonic treatment, *Int. J. Food Sci. Technol.*, 23(4):337–341.

Perry, M.B. and Hulyalkar, R.K. 1965. The analysis of hexuronic acids in biological materials by gas-liquid partition chromatography, *Can. J. Biochem.*, 43:573–584.

Pilgrim, G.W., Walter, R.H. and Oakenfull, D.G. 1991. Jams, jellies and preserves, in: *The Chemistry and Technology of Pectin*, ed., R. Walter, New York: Academic Press.

Pilnick, W. and Zwicker, P. 1970. Pectins, *Gordian*, 70:202–204.

Pippen, E.L., McCready, R.M. and Owens, H.S. 1950. Gelation properties of partially acetylated pectins, *J. Am. Chem. Soc.*, 72:813–816.

Plashchina, I.G., Formina, O.A., Braudo, E.E. and Tolstoguzov, V.B. 1979. Creep study of highly esterified pectin gels. I. The creep of saccharose-containing gels, *Colloid Polym. Sci.*, 257:1180–1187.

Plashchina, I.G., Semenova, M.G., Bravdo, E.E. and Tolstoguzov, V.B. 1985. Structural studies of the solutions of anionic polysaccharides. IV. Study of pectin solutions by light-scattering, *Carbohydr. Polym.*, 5(3):159–179.

Poovaiah, B.W., Glenn, G.M. and Reddy, A.S.N. 1988. Calcium and fruit softening: physiology and biochemistry, *Hortic. Rev. Portland*, 10:107–152.

Powell, D.A., Morris, E.R., Gidley, M.J. and Rees, D.A. 1982. Conformation and interaction of pectins II. Influences of residue sequence on chain association in calcium pectate, *J. Mol. Biol.*, 155:517–531.

Poynard, T., Slama, G. and Tchobroutsky, G. 1982. Reduction of post-prandial insulin needs by pectin as assessed by the artificial pancreas in insulin-dependent diabetics, *Diabete Metabolisme (Paris)*, 8:187–189.

Pressey, R. and Avants, J.K. 1978. Difference in polygalacturonase composition of clingstone and freestone peaches, *J. Food Sci.*, 43:1415–1423.

Racapé, E., Thibault, J.F., Reitsma, J.C.E. and Pilnik, W. 1989. Properties of amidated pectins II. Polyelectrolyte behaviour and calcium binding of amidated pectins and amidated pectic acids, *Biopolymers*, 28:1435–1448.

Ralet, M.-C. and Thibault, J.-F. 1994. Effect of extrusion-cooking on plant cell-walls from lemon. I. Characterisation of the water soluble pectins and comparison with acid-extracted pectins, *Carbohydr. Res.*, 260:283–296.

Rao, M.A. and Cooley, H.J. 1993. Dynamic rheological measurement of structure development in high-methoxyl pectin/fructose gels, *J. Food Sci.*, 58:876–879.

Raymond, W.R. and Nagel, C.W. 1969. Gas-liquid chromatographic determination of oligo-galacturonic acids, *Anal. Chem.*, 41:1700–1703.

Redgwell, R.J., MacRae, E., Hallett, I., Fischer, M., Perry, J. and Harker, R. 1997. *In vivo* and *in vitro* swelling of cell walls during fruit ripening, *Planta*, 203(2):162–173.

Rees, D.A. and Wight, A.W. 1971. Polysaccharide conformation. Part VII. Model building computations for α-1,4-galacturonan and the kinking function of L-rhamnose residues in pectic substances, *J. Chem. Soc. B*, 1366–1372.

Rees, D.A., Morris, E.R., Thom, D. and Madden, J.K. 1982. Shapes and interactions of carbo-hydrate chains, in: *The Polysaccharides*, ed., G.O. Aspinall, New York: Academic Press, pp. 195–290.

Renard, C.M.G.C. and Thibault, J.F. 1993. Structure and properties of apple and sugar-beet pectins extracted by chelating agents, *Carbohydr. Res.*, 244(1):99–114.

Riedl, J., Linseisen, J., Hoffmann, J. and Wolfram, G. 1999. Some dietary fibers reduce the absorption of carotenoids in women, *J. Nutr.*, 129(12):2170–2176.

Rinaudo, M. 1996. Physicochemical properties of pectins in solution and gel states, *Pectins and Pectinases*, eds., J. Visser and A.G.J. Voragen, New York: Elsevier, pp. 21–34.

Ring, S.G., I'Anson, K.J. and Morris, V.J. 1985. Static and dynamic light scattering studies of amylose solutions, *Macromolecules*, 18:182–188.

Robertson, G.L. 1979. The fractional extraction and quantitative determination of pectic substances in grapes and musts, *Am. J. Enol. Vitic.*, 30(3):182–186.

Rock, C.L. and Swendseid, M.E. 1992. Plasma beta-carotene response in humans after meals supplemented with dietary pectin, *Am. J. Clin. Nutr.*, 55(1):96–99.

Rolin, C. and de Vries, J. 1990. *Food Gels*. London: Elsevier Applied Science Publishers Ltd., pp. 401–434.

Rossander, L. 1987. Effect of dietary fiber on iron absorption in man, *Scand. J. Gastroenterol.*, 22 (suppl 129):68–72.

Ross-Murphy, S.B. 1994. Rheological methods, in: *Physical Techniques for the Study of Food Biopolymers*, ed., S.B. Ross-Murphy, Glasgow: Blackie Academic and Professional, pp. 343–392.

Ryden, P. and Selvendran, R.R. 1990. Structural features of cell-wall polysaccharides of potato (solanum tuberosum), *Carbohydr. Res.*, 195:257–272.

Sahi, A., Bijlani, R.L., Karmarkar, M.G. and Nayar, U. 1985. Modulation of glycemic response by protein, fat and dietary fibre, *Nutr. Res.*, 5(12):1431–1435.

Sajjaanantakul, T., Van Buren, J.P. and Downing, D.L. 1993. Effect of cations on heat degradation of chelator-soluble carrot pectin, *Carbohydr. Polym.*, 20:207–214.

Sakai, T. 1992. Degradation of pectins, in: *Microbial Degradation of Natural Products*, ed., G. Winkelmann, Weinheim: VCH Publishers, pp. 57–81.

Sakata, T. 1987. Stimulatory effect of short-chain fatty acids on epithelial cell proliferation in the rat intestine: a possible explanation for trophic effects of fermentable fiber, gut microbes and luminal trophic factors, *Br. J. Nutr.*, 58(1):95–104.

Samuelsen, A.B., Paulsen, B.S., Wold, J.K., Otsuka, H., Kiyohara, H., Yamada, H. and Knutsen, S.H. 1996. Characterization of a biologically active pectin from *Plantago major* L., *Carbohydr. Polym.*, 30(1):37–44.

Sandberg, A.S., Ahderinne, R., Andersson, H., Hallgren, B. and Hulten, L. 1983. The effect of citrus pectin on the absorption of nutrients in the small intestine, *Hum. Nutr. Clin. Nutr.*, 37C(3):171–183.

Saravacos, G.D. 1970. Effect of temperature on viscosity of fruit juices and purees, *J. Food Sci.*, 35:122–125.

Sato, T., Norisuye, T. and Fujita, H. 1984. Double-stranded helix of xanthan in dilute solution: evidence from light scattering, *Polym. J.*, 16:341–350.

Schols, H.A., Reitsma, J.C.E., Voragen, A.G.J. and Pilnik, W. 1989. High-performance ion exchange chromatography of pectins, *Food Hydrocolloids*, 3:115–121.

Schuderer, U. 1986. Wirkung von Apfelpektin auf die Cholesterin-und Lipoprotein konzentration bei Hypercholesterinämie, Dissertation, University of Giessen, Germany.

Schultz, T.H. 1965. Determination of the degree of esterification of pectin. Determination of the ester methoxyl content of pectin by saponification and titration. Determination of the anhydrouronic acid content by decarboxylation and titration of the liberated carbon dioxide, in: *Methods in Carbohydrate Chemistry 5*, New York: Academic Press, pp. 189–194.

Schwandt, P., Richter, W.O., Weisweiler, P. and Neureuther, G. 1982. Cholestyramine plus pectin in treatment of patients with familial hypercholesterolemia, *Atherosclerosis*, 44:379–383.

Schwartz, S.E., Levine, R.A., Weinstock, R.S., Petokas, S., Mills, C.A. and Thomas, F.D. 1988. Sustained pectin ingestion: effect on gastric emptying and glucose tolerance in non-insulin-dependent diabetic patients, *Am. J. Clin. Nutr.*, 48(6):1413–1417.

Scientific Committee for Foods of the European Communities. 1975. SCF Reports, 15th Series.

Scott, R.W. 1979. Colorimetric determination of hexuronic acids in plant materials, *Anal. Chem.*, 51(7):936–941.

Selvendran, R.R. 1985. Developments in the chemistry and biochemistry of pectic and hemicellulosic polymers, in: *The Cell Surface and Plant Growth and Development*, eds., K. Roberts, A.W.B. Johnston, C.W. Lloyd, P. Shaw and H.W. Woolhouse, Cambridge, England: Company of Biologists, pp. 51–88.

Selvendran, R.R. and O'Neill, M.A. 1987. Isolation and analysis of cell walls from plant materials, *Methods Biochem. Anal.*, 32:25–153.

Selvendran, R.R., March, J.F. and Ring, S.G. 1979. Determination of aldoses and uronic acid content of vegetable fiber potatoes, *Anal. Biochem.*, 96(2):282–292.

Sepehri, H., Roghani, M. and Houdebine, M.L. 1998. Oral administration of pectin-rich plant extract enhances C3 and C4 complement concentration in woman colostrums. *Reprod. Nutr. Dev.*, 38(3):225–260.

Seymour, G.B., Harding, S.E., Taylor, A.J., Hobson, G.E. and Tucker, G.A. 1987. Polyuronide solubilization during ripening of normal and mutant tomato fruit, *Phytochemistry*, 26:1871–1875.

Shah, N., Atallah, M.T., Mahoney, R.R. and Pellett, P. 1982. Effect of dietary fiber components on fecal nitrogen excretion and protein utilization in growing rats, *J. Nutr.*, 112:658–666.

Shkodina, O.G., Zeltser, O.A., Selivanov, N.Y. and Ignatov, V.V. 1998. Enzymic extraction of pectin preparations from pumpkin, *Food Hydrocolloids*, 12(3):313–316.

Simons L.A., Gayst, S., Balasubramaniam, S. and Ruys, J. 1982. Long-term treatment of hypercholesterolemia with a new palatable formulation of guar gum, *Atherosclerosis*, 45:101–108.

Sjoberg, A.M. 1987. The effects of gamma irradiation on the structure of apple pectin, *Food Hydrocolloids*, 1(4):271–276.

Smidsrød, O. and Haug, A. 1971. Estimation of the relative stiffness of the molecular chain in polyelectrolytes from measurements of viscosity at different ionic strengths, *Biopolymers*, 10:1213–1227.

Smith, J.G., Yokoyama, W.H. and German, J.B. 1998. Butyric acid from the diet: actions at the level of gene expression, *CRC Crit. Rev. Food Sci. Nutr.*, 38(4):259–297.

Sosulski, F., Lin, M.J.Y. and Humbert, E.S. 1978. Gelatin characteristics of acid-precipitated pectin from sunflower heads, *J. Can. Inst. Food Sci. Technol.*, 11(3):113–116.

Steinnes, A. 1975. Alginate in foods, *Gordian*, 75(7–8):228–230.

Stoddart, R.W. 1984. *The Biosynthesis of Polysaccharides*, London: Croom Helm.

Talmadge, K.W., Keegstra, K., Bauer, W.D. and Albersheim, P. 1973. The structure of plant cell walls. I. the macromolecular components of the walls of suspension-cultured sycamore cells with a detailed analysis of the pectic polysaccharides, *Plant Physiol.*, 51(1):158–173.

Taper H.S. and Roberfroid, M. 1999. Influence of insulin and oligofructose on breast cancer and tumor growth, *J. Nutr.*, 129(7):1488S–1491S.

Thakur, B.R., Singh, R.K. and Handa, A.K. 1997. Chemistry and uses of pectin—a review, *CRC Crit. Rev. Food Sci Nutr.*, 37(1):47–73.

Thibault, J.F. 1979. Automatisation du dosage des substances pectiques par la methode au metanhydroxydiphenyl, *Lebensm. Wiss. u. Technol.*, 12:247–.

Thibault, J.F. and Rinaudo, M. 1986. Chain association of pectic molecules during calcium-induced gelation, *Biopolymers*, 25:455–468.

Thibault, J.F. and Rombouts, F.M. 1985. Effects on some oxidizing agents, especially ammonium peroxosulfate on sugar beet pectins, *Carbohydr. Res.*, 154:205–215.

Thibault, J.F., Ralet, M.C., Axelos, M.A.V. and Della, V.G. 1996. Effects of extrusion-cooking on pectin-rich materials, in: *Pectins and Pectinases*, eds., J. Visser and A.G.J. Voragen, Amsterdam: Elsevier, pp. 425–437.

Thom, D., Dea, I.C.M., Morris, E.R. and Powell, D.A. 1982. Interchain associations of alginate and pectins, in: *Gums and Stabilisers for the Food Industry 1*, ed., G.O. Phillips, Clwyd: Pergamon Press, pp. 97–108.

Thornton, J.R. 1981. High colonic pH promotes colorectal cancer, *Lancet*, 1:1081–1082.

Tiwary, C.M., Ward, J.A. and Jackson, B.A. 1997. Effect of pectin on satiety in healthy US Army Adults. *J. Am. Coll. Nutr.*, 16(5):423–428.

Toft, K., Grasdalen, H. and Smidsrød, O. 1986. Synergistic gelation of alginates and pectins, in: *Chemistry and Function of Pectins*, eds., M.L. Fishman and J.J. Jen, Washington, DC: ACS, pp. 117–132.

Tolstoguzov, V.B. 1991. Functional properties of food proteins and role of protein-polysaccharide interaction, *Food Hydrocolloids*, 4(6):429–468.

Towle, G.A. and Christensen, T. 1973. Pectin, in: *Industrial Gums, Polysaccharides and Their Derivatives*, eds., R.L. Whistler and J.N. BeMiller, New York: Academic Press, pp. 429–461.

Trinidad, T.P., Wolever, T.M. and Thompson, L.U. 1996. Availability of calcium for absorption in the small intestine and colon from diets containing available and unavailable carbohydrates: an *in vitro* assessment, *Int. J. Food Sci. Nutr.*, 47(1):83–88.

Truswell, A.S. and Beynen, A.C. 1992. Dietary fiber and plasma lipids: potential for prevention and treatment of hyperlipidaemias, in: *Dietary Fiber—A Component of Food, Nutritional Function in Health and Disease*, eds., T.F. Schweizer and C.A. Edwards, New York: Springer-Verlag, pp. 295–332.

Tsuji, K., Shimizu, M., Nishimura, Y., Nakagawa, Y. and Ichikawa, T. 1992. Simultaneous determination of hydrogen, methane and carbon dioxide of breath using gas-solid chromatography, *J. Nutr. Sci. Vitaminol.*, 38(1):103–109.

Tung, M.A., Richards, J.F., Morrison, B.C. and Watson, E.L. 1970. Rheology of fresh, aged and gamma-irradiated egg white, *J. Food Sci.*, 35:872–874.

UKPDS (United Kingdom Prospective Diabetes Study) Group. 1998. Intensive blood glucose control with sulphonylureas or insulin compared with conventional treatment and risk of complications in patients with type 2 diabetes (UKPDS 33), *Lancet*, 352:837–853.

Van Buren, J.P. 1984. Effects of salts added after cooking on the texture of canned snap beans, *J. Food Sci.*, 49:910–912.

Van Buren, J.P. 1991. Function of pectin in plant tissue structure and firmness, in: *The Chemistry and Technology of Pectin*, ed., R.H. Walter, New York: Academic Press, Inc., pp. 1–22.

Vargo, D., Doyle, R. and Floch, M.H. 1985. Colonic bacterial flora and serum cholesterol: alterations induced by dietary citrus pectin, *Am. J. Gastroenterol.*, 80(5):361–364.

Veldman, F.J., Nair, C.H., Vorster, H.H., Vermaak, W.J., Jerling, J.C., Oosthuisen, N. and Venter, C.S. 1999. Possible mechanisms through which dietary pectin influences fibrin network architecture in hypercholesterolemic subjects, *Thrombosis Res.*, 93(6):253–264.

Vierhuis, E., Schols, H.A., Beldman, G. and Voragen, A.G.J. 2000. Isolation and characterisation of cell wall material from olive fruit (*Olea europaea* cv koroneiki) at different ripening stages, *Carbohydr. Polym.*, 43(1):11–21.

Voragen, A.J.G., Schols, H.A., de Vries, J.A. and Pilnik, W. 1982. High performance liquid chromatographic analysis of uronic acids and oligogalacturonic and oligogalacturonic acids, *J. Chromatogr.*, 244:327–336.

Voragen, A.J.G., Schols, H.A. and Pilnik, W. 1986. Determination of the degree of methylation and acetylation of pectins by HPLC, *Food Hydrocolloids*, 1(1):65–70.

Voragen, A.G.J., Pilnik, W., Thibault, J.F., Axelos, M.A.V. and Renard, C.M.G.C. 1995. Pectins, in: *Food Polysaccharides and Their Applications*, ed., A.M. Stephen. New York: Marcel Dekker, Inc., pp. 287–339.

Wada, S. and Ray, P.M. 1978. Matrix polysaccharides of oat celeoptile cell walls, *Phytochemistry*, 17:923–931.

Waffenschmidt, S. and Jaenicke, L. 1987. Assay on reducing sugars in the nanomole range with 2-2×-bicinchoninate, *Anal. Biochem.*, 165:337–340.

Wahlqvist, M.L., Morris, M.J., Littlejohn, G.O., Bond, A. and Jackson, R.V.J. 1979. The effect of dietary fiber on glucose tolerance in healthy males, *Aust. N. Z. J. Med.*, 9:154–158.

Walker, A.R.P., Walker, B.F. and Walker, A.J. 1986. Faecal pH, dietary fiber intake, and proneness to colon cancer in four South African populations, *Br. J. Cancer*, 53:489–495.

Walkinshaw, M.D. and Arnott, S. 1981a. Conformations and interactions of pectins. I. X-ray diffraction analyses of sodium pectate in neutral and acidified forms, *J. Mol. Biol.*, 153:1055–1073.

Walkinshaw, M.D. and Arnott, S. 1981b. Conformations and interactions of pectins. II. Models for junction zones in pectinic acid and calcium pectate gels, *J. Mol. Biol.*, 153:1075–1085.

Wallner, S.J. and Walker, J.E. 1975. Glycosidases in cell wall-degrading extracts of ripening tomato fruits, *Plant Physiol.*, 55(1):94–98.

Wang, Q. and Qvist, K.B. 2000. Investigation of the composite system of β-lactoglobulin and pectin in aqueous solutions, *Food Res. Int.*, 33:683–690.

Wells, A.F. and Ershoff, B.H. 1961. Beneficial effects of pectin in prevention of hypercholesterolemia and increase in liver cholesterol in cholesterol-fed rats, *J. Nutr.*, 74:87–92.

Whistler, R.L. and BeMiller, J.N. 1958. Alkaline degradation of polysaccharides, *Adv. Carbohydr. Chem.*, 13:289–329.

White, C.A. and Kennedy, J.F. 1981. Manual and automated spectrophotometric techniques for the detection and assay of carbohydrates and related molecules, *Tech. Carbohydr. Metab.*, B312:1–64.

Whittington, S.G., Glover, R.M. and Hallman, G.M. 1973. The conformational statistics of pectic acid, *J. Polym. Sci.*, 42:1493–1497.

Williams, D.R.R., James, W.P.T. and Evans, I.E. 1980. Dietary fiber supplementation of a normal breakfast administered to diabetics, *Diabetologia*, 18:379–383.

Williamson, G., Faulds, C.B., Matthew, J.A., Archer, D.B., Morris, V.J., Brownsey, G.J. and Ridout, M.J. 1990. Gelation of sugarbeet and citrus pectins using enzymes extracted from orange peel, *Carbohydr. Polym.*, 13:387–397.

Wood, P.J. and Siddiqui, I.R. 1971. Determination of methanol and its application to measurement of pectin ester content and pectin methyl esterase activity, *Anal. Biochem.*, 39:418–428.

Yamada, H. 1996. Contribution of pectins on health care, in: *Pectins and Pectinases*, eds., J. Visser and A.G.J. Voragen, Amsterdam: Elsevier, pp. 173–190.

Yamada, H., Ra, K.S., Kiyohara, H., Cyong, J.C. and Otsuka, Y. 1989. Structural characterisation of an anti-complementary pectic polysaccharide from the roots of *Bupleurum falcatum* L., *Carbohydr. Res.*, 189:209–226.

10 Human Health Effects of Docosahexaenoic Acid

Julie Conquer and Bruce J. Holub

CONTENTS

10.1 INTRODUCTION

There has been a marked surge in interest during the past decade for n-3 poly-unsaturated fatty acids (n-3 PUFA, omega-3 fatty acids) to serve as essential nutrients for human health as well as to play a potential role in the prevention and management of various chronic disorders, including cardiovascular disease (CVD), neurological disorders, and others. The present brief review will focus upon docosahexaenoic acid (DHA, 22:6n-3) with respect to dietary intakes, metabolism and functioning, as well as various clinical conditions in which an insufficiency of DHA has been implicated in the specific disorder. While DHA is consumed primarily in the form of fish and fish oils, which contain varying mixtures of DHA plus eicosapentaenoic

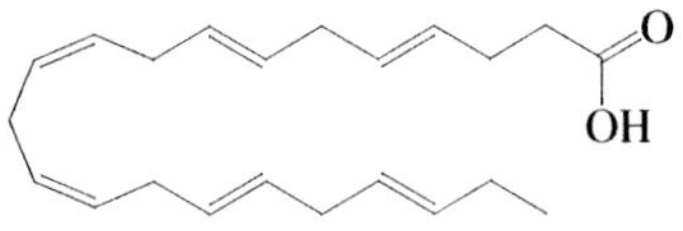

FIGURE 10.1 Docosahexaenoic acid (all-*cis*, 22:6*n*-3).

acid (EPA, 20:5n-3), this review will focus on DHA. Plant-based food sources, including vegetable oils, are important sources of the n-3 PUFA known as alpha-linolenic acid (LNA, 18:3n-3); however, these sources are lacking in DHA. The present review will restrict discussions on LNA with respect to its potential to give rise to some DHA in mammalian tissues via metabolic conversion.

10.2 DIETARY INTAKES, SOURCES, AND DIGESTION OF DHA

DHA has a very low melting point (approaching –50°C), rendering it highly fluid at both room and body temperatures. Through accumulation via the aquatic food chain (Arts et al., 2001), DHA accumulates to varying levels in aquatic organisms including fish, marine mammals, and the corresponding edible oils derived there-from. The fluidity of DHA, despite cold aquatic temperatures (approaching 0°C), is considered to play an important role in the biochemical adaptation of aquatic ectothermic organisms to colder environmental temperatures via its enrichment in membrane/cellular phospholipid components. The structure for DHA is depicted in Figure 10.1.

Current North American intakes of total n-3 PUFA have been estimated at approx-imately 1.6 g/day (approximately 0.7% of total dietary energy), of which close to 90% is LNA from plant sources, including various oils (selected vegetable oils such as canola oil, soybean oil, flax/flaxseed oil, etc.). The estimated dietary ratio of n-6 PUFA (primarily as linoleic acid, 18:2n-6) to LNA (n-6:n-3 ratio) is approximately 10:1. DHA plus EPA are consumed at a level of approximately 0.1–0.2 g/day in a typical North American diet (Kris-Etherton et al., 2000). Raper et al. (1992) have estimated the mean daily intake of DHA to be 78 mg, and the mean daily intake of EPA to be 46 mg, for a total of 124 mg of combined DHA plus EPA per day. The vast majority of DHA is currently consumed in the form of fish or fish oils with much smaller amounts being consumed in egg yolk and organ meats. The levels of DHA (USDA data) in selected fish, crustaceans, and mollusks are shown in Table 10.1. Standard shell eggs typically contain 30–50 mg of DHA per large egg, which approaches considerably higher levels with flax (10–20% by weight of diet) feeding (80–100 mg/egg) or DHA feeding at 300–600 mg per hen daily (200–300 mg/egg). Currently, there is much interest and activity in developing novel nutraceutical sources of DHA for inclusion in functional foods from large-scale production and use of single-cell oils, highly enriched in DHA from microalgae (Kyle, 2001; Zeller et al., 2001) and other sources through bioengineering technologies. Furthermore, active attempts at genetic manipulation of existing plants that lack DHA is under investi-gation with respect to the introduction of the necessary elongation/desaturation

TABLE 10.1
DHA contents (mg/100 g edible portion, raw) of selected fish, crustaceans, and mollusks

Food	mg/100 g	Food	mg/100 g
Bass, freshwater	200	Swordfish	100
Cod, Atlantic	200	Arctic Char	500
Flounder, unspecified	100	Trout, brook	200
Haddock	100	Trout, lake	1100
Halibut, Pacific	300	Trout, rainbow	400
Herring, Atlantic	900	Tuna, bluefin	1200
Herring, Pacific	700	Tuna, skipjack	300
Mackerel, Atlantic	1600	Whitefish, lake	1000
Mackerel, king	1200	Crab, Alaska king	100
Ocean perch	100	Shrimp, Atlantic white	200
Pike, northern	100	Shrimp, northern	200
Pickerel, walleye	200	Spiny lobster, southern rock	100
Salmon, Atlantic	900	Clam, softshell	200
Salmon, coho	500	Oyster, Pacific	200
Salmon, sockeye	700	Scallop, unspecified	100
Snapper, red	200	Squid, unspecified	200

enzymes to allow for the generation of unique vegetable oils enriched in DHA for wide food applications.

Most of the DHA found in aquatic food sources, including fish and fish oils, is found esterified to storage fat in the form of triglyceride and, to a much lesser extent, esterified to membrane/cellular phospholipid (PL) components. The physiological/enzymic processes involved in the digestibility (including hydrolysis and absorption) of EPA/DHA as found in dietary triglycerides have been reviewed (Nelson and Ackman, 1988). Pancreatic lipase action in the small intestine on dietary triglyceride provides for the release of the 1- and 3-positioned DHA as free fatty acid with the concomitant release of a 2-monoglyceride. Many marine mammal oils (e.g., seal and whale oils) contain significant amounts of DHA in the 1- and 3-positions. In fish oils, where DHA is commonly esterified in the 2-position, pancreatic lipase activity will yield a 2-monoglyceride containing DHA. The DHA (as free fatty acid) and/or as 2-monoglyceride are incorporated into micellar particles that provide for the intestinal absorption of DHA and its subsequent incorporation into intestinally-derived chylomicron particles as DHA-containing triglyceride molecules for subsequent storage in adipose tissue or uptake and delivery to peripheral tissues including liver, heart, etc.

The preparation of DHA-enriched concentrates from fish oil triglycerides often involves the formation/processing of ethyl ester derivatives as well as free fatty acid forms. Comparative studies have been reported in the literature on the digestibility and bioavailability of DHA when consumed in the form of triglyceride, free fatty acid, or ethyl ester in human volunteers. These studies have indicated that DHA is readily absorbed as a free fatty acid form with higher apparent digestibility/absorption

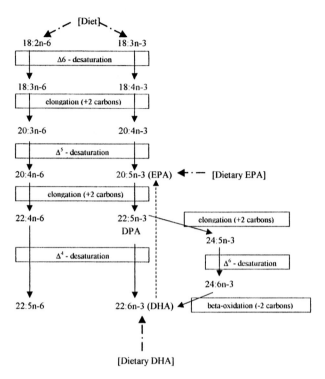

FIGURE 10.2 Desaturation, elongation, and retroconversion of PUFA.

being exhibited for DHA in free fatty acid form as compared to triglyceride or ethyl ester forms (Beckermann et al., 1990; Lawson and Hughes, 1988). Subsequent studies (Krokan et al., 1993; Nordoy et al., 1991) have indicated that test meals containing DHA in triglyceride or ethyl ester forms yielded near-equivalent bioavailabilities in human studies. So-called "natural" forms of dietary DHA as a food ingredient may be differentially defined by regulatory agencies in different countries according to whether or not the DHA is in a triglyceride ("natural") form or in an ethyl ester form. With the increasing interest in DHA as a nutraceutical ingredient for functional foods, microencapsulated forms of DHA concentrates are becoming available using differing microencapsulation technologies that require studies in humans to evaluate and compare the digestibility coefficients (bioavailability) of DHA in foods enriched with microencapsulated DHA vs. other delivery forms (Wallace et al., 2000).

10.3 METABOLISM AND BIOCHEMICAL ASPECTS OF DHA

Until recently, the commonly accepted pathway for the metabolic conversion of LNA (18:3n-3) to DHA (22:6n-3), and the corresponding conversion of dietary linoleic acid (18:2n-6) to 22:5n-6, involved the sequential utilization of delta 6-, 5-, and 4-desaturases along with elongation reactions (2-carbon additions) as depicted in Figure 10.2. The more recent and pioneering work of Dr. Howard

Sprecher and colleagues (Voss et al., 1991) established that the metabolism of docosapentaenoic acid (DPA), (22:5n-3) to DHA in liver is independent of a 4-desaturase (see Figure 10.2). This latter pathway involves the microsomal chain elongation of 22:5n-3 to 24:5n-3, followed by its desaturation to 24:6n-3 and conversion to DHA via a peroxisomal retroconversion of tetracosahexaenoic acid (24:6n-3). The conversion efficiency of LNA to DHA is limited in human adults, with conversion efficiencies estimated to be approximately 3.8% (Emken et al., 1994), and even more limited in infants, where the conversion efficiency appears to be less than 1% (Salem et al., 1996). It is apparent, therefore, that the most direct means of maintaining optimal/high levels of DHA in the human body is by direct consumption of preformed dietary DHA. It has been pointed out (Moore et al., 1995) that the limited ability to synthesize DHA may underlie some of the pathology that occurs in genetic diseases involving peroxisomal beta-oxidation. Some question regarding the existence of two separate delta 6-desaturase activities or a single enzyme performing both delta 6-desaturase reactions in human PUFA metabolism exists (Marzo et al., 1996; Willard et al., 2001).

The feeding of an EPA-free source of supplementary DHA (from algae) to human volunteers indicated a metabolic retroconversion of DHA to EPA (Conquer and Holub, 1996, 1997). Previous animal and *in vitro* studies in isolated rat liver cells have demonstrated that DHA can be retroconverted to EPA, and that this retro-conversion is a peroxisomal function (Schlenk et al., 1969; Gronn et al., 1991). Studies in isolated rat liver cells by Schlenk et al., 1969 have also indicated that the resultant EPA can be chain-elongated to DPA (22:5n-3) for subsequent esterification into cellular lipids. An acyl-CoA oxidase has been identified as the enzyme responsible for the chain shortening of DHA in the peroxisomal beta-oxidation of PUFA in human fibroblasts (Christensen et al., 1993). The aforementioned *in vivo* human studies have estimated the extent of retroconversion of DHA to EPA to be approximately 10% (Conquer and Holub, 1996, 1997).

Various techniques including stable isotope tracer and gas chromatography combustion and isotope ratio mass spectrometry have been utilized to study the *in vivo* compartmental metabolism of DHA. Whereas DHA is naturally occurring in various lipid/lipoprotein classes as well as circulating cells, there is evidence to suggest that lysophosphatidylcholine enriched in DHA (LysoPC-DHA) may be an important intermediate when bound to plasma albumin for the delivery of DHA to various tissues including the brain (Thies et al., 1994; Polette et al., 1999).

The aforementioned evidence for the limited conversion efficiency (less than 1% overall) of dietary LNA to DHA in human infants (Salem et al., 1996) has led to concerns regarding the efficiency of dietary LNA alone (in the absence of concomitant DHA) to supply sufficient levels of cellular DHA for optimal brain (neuronal) and retinal functioning for learning ability and visual acuity, respectively (Cunnane et al., 2000). Studies in neonatal baboons consuming commercial human infant formula have indicated that preformed DHA is sevenfold more efficient than LNA-derived DHA as a source for DHA accretion in the brain (Su et al., 1999). In the case of retinal pigment epithelium, preformed DHA gave levels 12-fold and 15-fold greater than for LNA-derived DHA, whereas the corresponding ratios for liver, plasma, and erythrocytes were 27, 29, and 51, respectively. Estimations of the actual accumulation of DHA in the human infant brain and other tissues, and considerations of dietary intakes of LNA in infant formula devoid of DHA, have led to the conclusion that preformed dietary DHA (in addition to LNA) should likely be provided during the first 6 months of life in order to provide sufficient levels of DHA for optimal physiological functioning (Cunnane et al., 2000).

The biochemical functioning of DHA in cellular membranes (including its essentiality for nervous system function) has been attributed to the unique physicochemical properties of DHA-containing membrane phospholipids. The potential mechanisms underlying the functional essentiality of DHA at the cellular/membrane level have been reviewed (Mitchell et al., 1998) and include its impact on membrane fluidity, its capacity to modify cell signaling processes and gene expression, as well as its utilization as a substrate for the formation of DHA metabolites and the attenuation of eicosanoid synthesis from arachidonic acid. Many of the physicochemical studies on DHA have been based on the fact that DHA represents approximately 50% of the lipid chain in the retinal rod outer segment disk membranes and a large fraction of the lipid chains in the membranes of neuronal tissues. Nuclear magnetic resonance (NMR) spectra have revealed unique properties of multilamellar dispersions of DHA-containing phospholipids (Everts and Davis, 2000). It has been demonstrated that DHA modulates interactions of the interphotoreceptor retinoid-binding protein with 11-*cis*-retinal, which may account, in part, for the biochemical basis of membrane DHA in photoreceptor cells and associated functioning (Chen et al., 1996). Recently, Suh et al. (2000) demonstrated that small manipulations of the dietary level of DHA can modify the fatty acid composition of membrane lipid and visual pigment content and kinetics in the developing photoreceptor cell. Published studies by Gaudette and Holub (1991) have shown the potential for DHA (when added in free fatty acid form) to modify agonist-induced cell signaling processes (including polyphosphoinositide turnover) in human platelets. More recently, it has been demonstrated that DHA can alter inositol phosphate metabolism and protein kinase C activity in adult porcine cardiac myocytes (Nair et al., 2001). Such modulation of intracellular cell signaling may underlie the recognized antiarrhythmic potential of DHA.

10.4 DHA AND CARDIOVASCULAR DISEASE: EPIDEMIOLOGY AND INTERVENTION TRIALS

10.4.1 LIPIDS AND LIPOPROTEINS

Various epidemiological trials have suggested that blood levels of DHA are inversely associated with risk factors for CVD. In a study by Leng et al. (1994a), the plasma level of DHA was reported to be negatively associated with levels of triglyceride and hemostatic factors such as fibrinogen and blood viscosity. In a separate study by the same group (Leng et al., 1994b), DHA levels were found to be significantly lower in blood from men and women with peripheral arterial disease as compared with a group of age- and sex-matched controls. In 1995, Simon et al. (1995) determined that the levels of DHA in the blood are inversely correlated with CVD risk.

Most trials involving supplementation with DHA in humans use DHA from a marine source, which usually contains substantial amounts of EPA. Recently, the availability of purified sources of DHA (meaning EPA-free in this review, although not necessarily free of other fatty acids) has made it possible to investigate the effects of supplementation with DHA, in the absence of EPA, on various disorders. There are very few studies on the effect of purified DHA on lipid and lipoprotein levels in humans. DHA supplementation has been shown to increase DHA levels in serum and membrane phospholipid and nonesterified fatty acids (Agren et al., 1996; Conquer and Holub, 1996, 1997, 1998; Rambjor, 1996; Davidson et al., 1997; Grimsgaard et al., 1997; Nelson et al., 1997; Vidgren, 1997). Furthermore, supplementation with EPA-free DHA capsules has also been shown to increase EPA levels in serum and membrane phospholipid (Conquer and Holub, 1996, 1997, 1998; Rambjor, 1996; Grimsgaard et al., 1997; Nelson et al., 1997; Vidgren, 1997) and decrease the omega-3 fatty acid as DPA. Although there is still a discrepancy as to whether DHA modifies CVD risk factors to the same extent as EPA, various studies have shown a fasting triglyceride-lowering effect with DHA supplementation in the absence of EPA (Agren et al., 1996, 1997; Conquer and Holub, 1996; Davidson et al., 1997; Grimsgaard et al., 1997; Nelson et al., 1997; Hirai et al., 1989; Mori et al., 2000a). Decreases in fasting triglyceride levels in published studies range from 16% to 26%, with supplementation of 1.25–6 g DHA per day for 6–15 weeks. Postprandial triglyceridemia also appears to be suppressed by DHA supplementation. In 1998, Hansen et al. (1998) gave 4 g per day of highly purified EPA or DHA, for five weeks, to male volunteers. The results suggested that prolonged intake of DHA may be more efficient than that of EPA in lowering postprandial triglyceridemia.

DHA supplementation has also been shown to decrease the ratios of total cholesterol:HDL cholesterol and LDL cholesterol:HDL cholesterol (Conquer and Holub, 1996). Although this may be partially due to an increase in HDL levels as suggested in some studies (Agren et al., 1996; Conquer and Holub, 1996; Davidson et al., 1997; Grimsgaard et al., 1997; Nelson et al., 1997; Hirai et al., 1989), most studies suggest that there is not a direct effect of DHA on total or LDL cholesterol (Nelson et al., 1997; Conquer and Holub, 1998; Mori et al., 2000a). Furthermore, DHA supplementation does not appear to modify levels of apo A1, apo B, or lipoprotein(a) (Nelson et al., 1997; Conquer and Holub, 1998).

10.4.2 THROMBOGENIC RISK FACTORS

At least one study (Conquer and Holub, 1998) suggests that DHA supplementation in humans results in an increase in serum DHA as nonesterified fatty acid to levels that are potentially antithrombotic. Whether DHA supplementation modifies thrombotic risk factors is not clear. It has been suggested that DHA supplementation results in reduced platelet aggregation in response to collagen and adenosine diphosphate (ADP) (Hirai et al., 1989). In contrast, (Conquer and Holub, 1996) and others (Agren et al., 1997) have shown no effect of DHA supplementation on platelet aggregation. These authors (Conquer and Holub, 1996; Agren et al., 1997) also found no effects of DHA supplementation on thromboxane, fibrinogen, antithrombin III, factor VII, or factor X levels or on prothrombin and activated partial prothrombin time. A more recent study suggests that DHA has no unfavorable effect on the activity of plasminogen activator inhibitor-1 (PAI-1) activity after 7 weeks of supplementation (3.6 g DHA per day) (Hansen et al., 2000). These findings suggest that there are no observable changes in blood coagulation, platelet function, or thrombotic tendencies in healthy, adult men with DHA supplementation.

Recently, Mori et al. (1999) showed that 4 g/day of purified DHA reduced the 24-hour and daytime ambulatory blood pressure, and the 24-hour daytime and nighttime ambulatory heart rate in overweight, mildly hyperlipidemic men. The same group found that DHA supplementation enhances vasodilator mechanisms and attenuates constrictor responses in the forearm microcirculation (Mori et al., 2000b).

10.5 DHA AND NEUROLOGICAL DISORDERS

10.5.1 ATTENTION DEFICIT DISORDER AND HYPERACTIVITY

Attention deficit hyperactive disorder (ADHD) is the diagnosis used to describe children who are inattentive, impulsive, and hyperactive. At least a subgroup of these patients, with symptoms of essential fatty acid (EFA) deficiency has been shown to have significantly lower proportions of plasma arachidonic acid and DHA (Stevens et al., 1995). Another study suggests that subjects with lower compositions of total n-3 fatty acids have significantly more behavioral problems; temper tantrums; and learning, health, and sleep problems than those with a higher proportion of n-3 fatty acids (Stevens et al., 1996). Currently, a double-blind, placebo-controlled study with long-chain n-3 fatty acids is being undertaken in children who have been clinically diagnosed with ADHD and who exhibit symptoms of EFA deficiency (Burgess et al., 2000). There are no studies in the literature investigating the effect of purified DHA on ADHD symptoms.

Levels of serum EFA in control children have also been compared with hyperactive children, and levels of DHA were found to be significantly lower in the hyperactive children (Mitchell et al., 1987). It is not clear whether this is due to dietary differences in omega-3 fatty acid intakes or some other mechanism.

10.5.2 SCHIZOPHRENIA

There is a growing body of evidence that suggests that abnormalities of the red blood cell membrane are found in patients suffering from schizophrenia. Decreased levels of DHA and arachidonic acid (AA) have been shown in red blood cell and fibroblast membranes from patients with schizophrenia (Horrobin, 1996; Peet et al., 1996) as well as their close relatives (Peet et al., 1996). A hypothesis has been made that the abnormality of photoreceptor function observed in patients with schizophrenia (Warner et al., 1999) may be a result of decreased membrane omega-3 levels in these patients. The decrease in DHA levels in these patients appears to be the result of both decreased synthesis from omega-3 precursors (Mahadik et al., 1996) and increased breakdown and loss (Horrobin, 1996; Peet et al., 1995). Preliminary studies suggest that supplementation with omega-3 fatty acids (DHA and EPA), as well as an increase in omega-3 fatty acid intake in the diet, results in less severe symptoms of schizophrenia (Peet et al., 1996; Assies et al., 2001), and this appears to be related to the increased level of omega-3 fatty acids in red cell membranes. Significantly reduced DHA concentrations in erythrocyte membranes from patients with schizophrenia compared with a carefully matched control group have recently been reported that could not be attributed to differences in dietary DHA intakes (Assies et al., 2001). To our knowledge, there are no studies in the literature investigating the effect of purified DHA on schizophrenia symptoms.

10.5.3 DEMENTIA

Expanding knowledge in the area of lipid biochemistry suggests that Alzheimer's Disease (AD) is associated with brain lipid defects (Nitsch et al., 1991; Söderburg et al., 1991, 1992; Mulder et al., 1998). Decreased levels of AA and DHA have been shown in membrane phospholipids of brains from patients with AD by some (Söderburg et al., 1991; Prasad et al., 1998) but not all authors (Brooksbank and Martinez, 1989). Furthermore, the levels of the DHA metabolites, neuroprostanes, have been shown to be significantly greater in AD brains (Reich et al., 2001). Aging itself has no influence on the fatty acid composition of brain phospholipids (Söderburg et al., 1991; Prasad et al., 1998). Although there is little information on differences in blood levels of DHA between subjects with AD vs. control subjects, it has been hypothesized that a decreased level of DHA is a risk factor for the development of AD (Kyle et al., 1999). Recently, Conquer et al. (2000a) found decreased levels of DHA in the plasma of patients with AD and other dementias as compared with healthy age-matched individuals. There are no studies in which patients with AD have been supplemented with purified DHA. However, in 1999, Terano et al. published an article describing increased scores on the mini-mental state examination (MMSE) and the Hasegawa's Dementia rating scale (HDS-R) in elderly subjects with moderately severe dementia resulting from thrombotic cerebrovascular diseases who had received DHA supplementation for at least three months.

10.5.4 PEROXISOMAL DISORDERS

Patients with Zellweger syndrome (a neurological disorder) and related peroxisomal disorders have been shown to have extremely low levels of DHA in their plasma and erythrocytes (Martinez et al., 1993, 1994; Martinez, 1994, 1995), brain, liver, kidney, and retina (Martinez, 1990, 1991, 1992a, 1995). A recent study suggests that DHA is drastically reduced in all brain glycerophospholipid fractions (Martinez and Mougan, 1999) in patients with peroxisomal disorders. The retroconversion of DHA to EPA (Gronn et al., 1990) and the synthesis of DHA (Petroni et al., 1998) were also deficient in cultured fibroblasts from patients with Zellweger syndrome.

Trials with small numbers of patients have been conducted on the effect of DHA supplementation in patients with generalized peroxisomal disorders. Treatment with DHA ethyl ester has been shown to favorably modify the fatty acid composition of erythrocytes and plasma in patients with peroxisomal disorders (Martinez, 1992b, 1995; Martinez et al., 1993; Sovik et al., 1998; Moser et al., 1999). In terms of clinical effects, the results are mixed and may depend upon the specific type of peroxisomal disorder. At least some studies suggest that a clear clinical improvement parallels the increase in blood DHA (Martinez, 1995, 1996; Martinez and Vazquez, 1998; Martinez et al., 2000). Patient improvements that have been noted include improvements in vision, alertness, social contact, and muscle control (Martinez, 1996; Martinez and Vazquez, 1998; Martinez et al., 2000).

10.5.5 DEPRESSION

Decreased levels of omega-3 fatty acids, including DHA, are hypothesized to be of importance in depression (Hibbeln and Salem, 1995). In a recent study, the red blood cell membrane fatty acids from a group of patients with depression were compared with those of a group of healthy controls (Edwards et al., 1998). There was a reduced level of total omega-3 fatty acids, as well as DHA, in the patients. Furthermore, reduced red blood cell membrane DHA emerged as a significant predictor of depression in the multiple regression analysis of the study population, which included patients with depression and controls (Edwards et al., 1998). There do not appear to be any studies in which patients with depression have been supplemented with purified DHA.

10.5.6 OTHER

There is preliminary evidence to suggest DHA involvement with other neurological disorders. Many publications have appeared in the past linking reduced levels of DHA in the plasma with retinitis pigmentosa (Anderson et al., 1987; Converse et al., 1987; Gong et al., 1992; Hoffman et al., 1993; Holman et al., 1994; Hoffman and Birch, 1995). There do not appear to be any studies in which retinitis pigmentosa patients have been supplemented with purified DHA devoid of EPA.

Preliminary evidence suggests that dietary essential fatty acids, including DHA, may change neurotransmitter concentrations. Plasma DHA concentrations were shown to be negatively correlated with cerebrospinal fluid 5-hydroxyindoleacetic acid in violent subjects (Hibbeln et al., 1998).

10.6 DHA AND OTHER DISORDERS

10.6.1 ASTHENOZOOSPERMIA

DHA is found in extremely high levels in human ejaculate (Nissen and Kreysel, 1983; Gulaya et al., 1993), and the concentrations of DHA in both ejaculate and spermatozoa have been suggested to be positively associated with sperm motility in humans (Nissen and Kreysel, 1983; Gulaya et al., 1993; Conner et al., 1997; Zalata et al., 1998; Conquer et al., 1999). DHA levels have been shown to be lower in complete ejaculates from asthenozoospermic individuals vs. normozoospermic individuals (Nissen and Kreysel, 1983; Gulaya et al., 1993), as well as in certain sperm fractions (Zalata et al., 1998), total sperm (Conquer et al., 1999), and seminal plasma (Conquer et al., 1999) from asthenozoospermic vs. normozoospermic individuals. Furthermore, there is a net decrease in DHA content in sperm during the process of sperm maturation (Ollero et al., 2000). DHA may contribute to the membrane fluidity necessary for the motility of sperm tails (Connor et al., 1998), and as such, it makes up approximately 14% of PL fatty acids in the tails of human sperm (Zalata et al., 1998; Conquer et al., 1999). The role of DHA in the motility of the spermatozoa is not completely clear, although it has been suggested that DHA is involved in the regulation of free fatty acid utilization by sperm (Jones and Plymate, 1988, 1993). DHA can also be slowly metabolized to 19,20-dihydroxy-4,7,10,13,16-DPA by seminal vesicles (Oliw and Sprecher, 1991).

Even though the level of DHA in human sperm is correlated with sperm motility, and DHA levels are lower in the sperm and seminal plasma of asthenozoospermic individuals, there are very few studies in which human males are supplemented with DHA. In 1990, Knapp (1990) fed 50 ml menhaden oil (containing DHA plus EPA) per day to 10 normozoospermic men for 4 weeks. DHA was increased in both phosphatidylcholine (PC) and phosphatidylethanolamine (PE) fractions, yet there was no effect on sperm motility. In 2000, Conquer et al. (2000b) fed 375 mg or 750 mg of DHA per day to asthenozoospermic men for 3 months. DHA levels increased in the serum and seminal plasma, but there was no increase in the sperm. Furthermore, there was no effect on sperm motility. In asthenozoospermic men, the inability of DHA to further enrich sperm phospholipid may be responsible for the lack of effect of DHA supplementation on sperm motility. It is unclear why DHA did not increase in the sperm phospholipid of the asthenozoospermic individuals. A lack of DHA precursors in the preparation and/or the presence or absence of certain compounds of the seminal plasma of these individuals and/or membrane transport deficits and/or insufficient supplementation durations may have been responsible.

10.6.2 IMMUNE DEFENSE

There is very little information on the effect of supplementation with DHA, in the absence of EPA, on various immune parameters in humans. Recently, Kelley et al. (1998, 1999) suggested that DHA supplementation has no effect on the proliferation of peripheral blood mononuclear cells and the delayed hypersensitivity skin response. Also, there were no changes in the number of T cells producing interleukin 2, the ratio between helper/suppressor T cells in the circulation, or serum concentrations of immunoglobulin G, C3, or interleukin 2 receptor. DHA supplementation caused a significant decrease in the number of circulating white blood cells, mainly due to a decrease in circulating granulocytes. Furthermore, natural killer cell activity and *in vitro* secretion of interleukin-1 beta (IL-1β) and tumor necrosis factor alpha (TNFα) were significantly reduced by DHA supplementation. This was accompanied by reductions in the lipopolysaccharide-induced production of prostaglandin E_2 and leukotriene B_4 by peripheral blood mononuclear cells.

10.6.3 OTHER

Plasma levels of DHA, in cholesterol esters and phospholipids as well as erythrocyte phosphatidyl-choline and -ethanolamine, have been shown to be lower in patients with cystic fibrosis as compared with age-matched controls (Biggemann et al., 1988). Plasma phospholipid DHA has also been shown to be lower in well-nourished patients with cystic fibrosis than in gender-, age- and height-matched controls, suggesting that the deficiency may be related to a specific defect in fatty acid metabolism (Roulet et al., 1997).

Children with classical phenylketonuria (PKU) have reduced levels of DHA in plasma cholesterol esters (0.25 vs. 0.54 wt%) and membrane phospholipids compared with nonaffected controls (Sanjuro et al., 1994; Poge et al., 1998). It appears, however, that the reduced levels of DHA in PKU patients occur only in patients with strict diet therapy and are most probably caused by a reduced intake of omega-3 fatty acids (Poge et al., 1998). Studies are underway and planned on the effects of DHA supplementation in pregnant women with PKU and children with PKU (Giovannini et al., 1995).

DHA also appears to play a role in aggression. Supplementation with DHA has been shown to prevent aggression enhancement at times of mental stress (academic examinations) (Hamazaki et al., 1996). However, in a separate study, changes in aggression were investigated under nonstressful conditions (Hamazaki et al., 1998). Aggression levels remained constant after supplementation with 1.5 g DHA per day for 3 months.

10.7 CONCLUDING COMMENTS

DHA has become recognized as a physiologically-essential nutrient in the brain and retina for optimal neuronal functioning and visual performance, respectively. In addition to playing an important role in human health, there are a number of chronic disorders (cardiovascular, neurological, genetic, others) in which apparently insufficient levels of DHA have been implicated in the associated pathophysiology. In some cases, placebo-controlled intervention trials with dietary/supplemental DHA have provided significant benefit. However, there is an urgent need for carefully controlled intervention trials to determine whether supplementation with DHA (at various doses and durations) can provide clinical benefit for the various pathophysiologies, where reduced physiological levels of DHA have been reported in patients with selected disorders as compared to healthy controls. Current consumption levels of DHA in North America and most countries of the world per capita are well below intakes that are considered to be optimal for human health, including the prevention and/or attenuation of various chronic disorders. For example, the average per capita intake of DHA in the adult population of North America is approximately 80 mg/day (with a wide standard deviation including intakes of essentially zero in vegan vegetarians). Recent recommendations for dietary intakes of DHA (Simopoulos et al., 1999) have advised ensuring intakes of at least 300 mg/day of DHA for pregnant and lactating women, which is approximately four times that of current mean North American intakes. Insufficient exposure to DHA during development of the infant during gestation and often during lactation due to low breast milk levels (and formula feeding with DHA-deficient products in North America and elsewhere) is of concern. For adults in general, it has been recommended (Simopoulos et al., 1999) that DHA plus EPA intakes be at least 650 mg/day with DHA counting 220 to 435 mg of the summed total daily. The large gap in current intakes of DHA in most countries relative to ideal/optimal intakes for health and disease prevention/modification is extremely wide. Furthermore, many clinical conditions may prove to require much higher levels of DHA than for healthy groups in order to yield beneficial effects on selected disease-related parameters and risk factors because of metabolic abnormalities in the patient groups. Thus, the development of various DHA concentrates from conventional (aquatic) and alternative sources and their inclusion in various forms (oils, powders, emulsions, microencapsulated preparations, etc.) as nutraceuticals in a plethora of functional foods will undergo a dramatic surge in activity/availability in the coming decade. Such population increases in DHA intakes from a wide selection of processed and natural food sources (e.g., by DHA feeding to domestic animals), as well as increased fish/seafood consumption, will hopefully enhance human health, including brain and visual functioning, and increase the capacity for preventing and managing various chronic disorders.

REFERENCES

Agren, J.J., Hanninen, O., Julkunen, A., Fogelholm, L., Vidgren, H., Schwab, U., Pynnonen, O. and Uusitupa, M. (1996) Fish diet, fish oil and docosahexaenoic acid rich oil lower fasting and postprandial plasma lipid levels. *Eur. J. Clin. Nutr.* 50:765–771.

Agren, J.J., Vaisanen, S., Hanninen, O., Muller, A.D. and Hornstra, G. (1997) Hemostatic factors and platelet aggregation after a fish-enriched diet or fish oil or docosahexaenoic acid supplementation. *Prostaglandins Leukot. Essent. Fatty Acids.* 57:419–421.

Anderson, R., Maude, M., Lewis, R., Newsome, D. and Fishman, G. (1987) Abnormal plasma levels of polyunsaturated fatty acid in autosomal dominant retinitis pigmentosa. *Exp. Eye Res.* 44:155–159.

Arts, M.T., Ackman, R.G. and Holub, B.J. (2001) "Essential fatty acids" in aquatic ecosystems: a crucial link between diet and human health and evolution. *Can. J. Fish. Aquat. Sci.* 58:122–137.

Assies, J., Lieverse, R., Vreken, P., Wanders, R.J., Dingemans, P.M. and Linszen, D.H. (2001) Significantly reduced docosahexaenoic and docosapentaenoic acid concentrations in erythrocyte membranes from schizophrenic patients compared with a carefully matched control group. *Biol. Psychiatry.* 49:510–522.

Beckermann, B., Beneke, M. and Seitz, I. (1990) Comparative bioavailability of eicosapentaenoic acid and docosahexaenoic acid from triglycerides, free fatty acids and ethyl esters in volunteers. *Arzneimittelforschung.* 40:700–704.

Biggemann, B., Laryea, M.D., Schuster, A., Griese, M., Reinhardt, D. and Bremer, H.J. (1988) Status of plasma and erythrocyte fatty acids and vitamin A and E in young children with cystic fibrosis. *Scand. J. Gastroenterol. Suppl.* 143:135–141.

Brooksbank, B.W.L. and Martinez, M. (1989) Lipid abnormalities in the brain in adult Down's syndrome and Alzheimer's disease. *Mol. Chem. Neuropathol.* 11:157–185.

Burgess, J.R., Stevens, L., Zhang, W. and Peck, L. (2000) Long-chain polyunsaturated fatty acids in children with attention-deficit hyperactivity disorder, *Am. J. Clin. Nutr.* 71S:327–330.

Chen, Y., Houghton, L.A., Brenna, J.T. and Noy, N. (1996) Docosahexaenoic acid modulates the interactions of the interphotoreceptor retinoid-binding protein with 11-*cis*-retinal. *J. Biol. Chem.* 271:20507–20515.

Christensen, E., Woldseth, B., Hagve, T.A., Poll-The, B.T., Wanders, R.J., Sprecher, H., Stokke, O. and Christophersen, B.O. (1993) Peroxisomal beta-oxidation of polyunsaturated long chain fatty acids in human fibroblasts. The polyunsaturated and the saturated long chain fatty acids are retroconverted by the same acyl-CoA oxidase. *Scand. J. Clin. Lab. Invest.* 215:S61–S74.

Connor, W.E., Weleber, R.G., DeFrancesco, C., Lin, D.S. and Wolf, D.P. (1997) Sperm abnormalities in retinitis pigmentosa. *Invest. Opthmol. Vis. Sci.* 38:2619–2628.

Connor, W.E., Lin, D.S., Wolf, D.P. and Alexander, M. (1998) Uneven distribution of desmosterol and docosahexaenoic acid in the heads and tails of monkey sperm. *J. Lipid Res.* 39:1404–1411.

Conquer, J.A. and Holub, B.J. (1996) Supplementation with an algae source of docosahexaenoic acid increases (n-3) fatty acid status and alters selected risk factors for heart disease in vegetarian subjects. *J. Nutr.* 126:3032–3039.

Conquer, J.A. and Holub, B.J. (1997) Dietary docosahexaenoic acid as a source of eicosapentaenoic acid in vegetarians and omnivores. *Lipids.* 32:341–345.

Conquer, J.A. and Holub, B.J. (1998) Effect of supplementation with different doses of DHA on the levels of circulating DHA as non-esterified fatty acid in subjects of Asian Indian background. *J. Lipid Res.* 39:286–292.

Conquer J.A., Martin, J.B., Tummon, I., Watson, L. and Tekpetey, F. (1999) Fatty acid analysis of blood serum, seminal plasma and spermatozoa of normozoospermic vs. asthenozoospermic males. *Lipids.* 34:793–799.

Conquer, J.A., Tierney, M.C., Zecevic, J., Bettger, W.J. and Fisher, R.H. (2000a) Fatty acid analysis of blood plasma of patients with Alzheimer's disease, other types of dementia and cognitive impairment. *Lipids.* 35:1305–1311.

Conquer, J.A., Martin, J.B., Tummon, I., Watson, L. and Tekpetey, F. (2000b) Effect of DHA supplementation on DHA status and sperm motility in asthenozoospermic males. *Lipids.* 35:149–154.

Converse, C., McLachlan, T., Bow, A., Packard, C. and Chepherd, J. (1987) Lipid metabolism in retinitis pigmentosa. In: *Degenerative Retinal Disoders, Clinical and Laboratory Investigations.* (Hollyfield, J., Anderson, R., and LaVail, M., eds.) New York: Alan R. Liss Inc., pp. 93–101.

Cunnane, S.C., Francescutti, V., Brenna, J.T. and Crawford, M.A. (2000) Breast-fed infants achieve a higher rate of brain and whole body docosahexaenoate accumulation than formula-fed infants not consuming dietary docosahexaenoate. *Lipids.* 35:105–111.

Davidson, M.H., Maki, K.C., Kalkowski, J., Schaefer, E.J., Torri, S.A. and Drennan, K.B. (1997) Effects of docosahexaenoic acid on serum lipoproteins in patients with combined hyperlipidemia: a randomized, double-blind, placebo-controlled trial. *J. Am. Coll. Nutr.* 16:236–243.

Edwards, R., Peet, M., Shay, J. and Horrobin, D. (1998) Omega-3 polyunsaturated fatty acid levels in the diet and in red blood cell membranes of depressed patients. *J. Affect. Disord.* 48:149–155.

Emken, E.A., Adolf, R.O. and Gulley, R.M. (1994) Dietary linoleic acid influences desaturation and acylation of deuterium-labeled linoleic and linolenic acids in young adult males. *Biochim. Biophys. Acta.* 1213:277–288.

Everts, S. and Davis, J.H. (2000) 1H and (13)C NMR of multilamellar dispersions of polyunsaturated (22:6) phospholipids. *Biophys. J.* 79(2):885–897.

Gaudette, D.C. and Holub, B.J. (1991) Docosahexanoic acid (DHA) and human platelet reactivity. *J. Nutr. Biochem.* 2:116–121.

Giovannini, M., Biasucci, G., Agostoni, C., Luotti, D. and Riva, E. (1995) Lipid status and fatty acid metabolism in phenylketonuria. *J. Inherited. Metab. Dis.* 18:265–272.

Gong, J., Rosner, B., Rees, D., Berson, E., Weigel-DiFranco, C. and Schaefer, E. (1992) Plasma docosahexaenoic acid levels in various genetic forms of retinitis pigmentosa. *Invest. Opthamol. Vis. Sci.* 33:2596–2602.

Grimsgaard, S., Bonaa, K.H., Hansen, J.B. and Nordoy, A. (1997) Highly purified eicosapentaenoic acid and docosahexaenoic acid in humans have similar triacylglycerol-lowering effects but divergent effect on serum fatty acids. *Am. J. Clin. Nutr.* 66:649–659.

Gronn, M., Christensen, E., Hagve, T.A. and Christophersn, B.O. (1990) The Zellweger syndrome: deficient conversion of docosahexaenoic acid (22:6(n-3)) to eicosapentaenoic acid (20:5(n-3)) and normal delta 4-desaturase activity in cultured skin fibroblasts. *Biochim. Biophys. Acta.* 1044:249–254.

Gronn, M., Christensen, E., Hagve, T.A. and Christophersen, B.O. (1991) Peroxisomal retroconversion of docosahexaenoic acid (22:6(n-3)) to eicosapentaenoic acid (20:5(n-3)) studied in isolated rat liver cells. *Biochim. Biophys. Acta.* 1081:85–91.

Gulaya, N.M., Tronko, M.D., Volkov, G.L. and Margitich, M. (1993) Lipid composition and fertile ability of human ejaculate. *Ukrainian J. Biochemistry.* 65:64–70.

Hamazaki, T., Sawazaki, S., Itomura, M., Asaoka, E., Nagao, Y., Nishimura, N., Yazawa, K., Kuwamori, T. and Kobayashi, M. (1996). The effects of docosahexaenoic acid on aggression in young adults: a placebo-controlled, double-blind study. *J. Clin. Invest.* 97:1129–1133.

Hamazaki, T., Sawazaki, S., Nagao, Y., Kuwamori, T., Yazawa, K., Mizushima, Y. and Kobayashi, M. (1998). Docosahexaenoic acid does not affect aggression of normal volunteers under nonstressful conditions. A randomized, placebo-controlled, double-blind study. *Lipids.* 33:663–667.

Hansen, J.B., Grimsgaard, S., Nilsen, H., Nordoy, A. and Bonaa, K.H. (1998) Effects of highly purified eicosapentaenoic acid and docosahexaenoic acid on fatty acid absorption, incorporation into serum phospholipids and postprandial triglyceridemia. *Lipids.* 33:131–138.

Hansen, J., Grimsgaard, S., Nordoy, A. and Bonaa, K.H. (2000) Dietary supplementation with highly purified eicosapentaenoic acid and docosahexaenoic acid does not influence PAI-1 activity. *Thromb. Res.* 98:123–132.

Hibbeln, J.R. and Salem, Jr., N. (1995) Dietary polyunsaturated fatty acids and depression: when cholesterol does not satisfy. *Am. J. Clin. Nutr.* 62:1–9.

Hirai, A., Terano, T., Makuta, H., Ozawa, A., Fujita, T., Tamura, Y. and Yoshida, S. (1989) Effect of oral administration of highly purified eicosapentaenoic acid and docosahexaenoic acid on platelet function and serum lipids in hyperlipidemic patients. In: *Advances In Prostaglandin, Thromboxane, and Leukotriene Research*, vol. 19, New York: Raven Press, pp. 627–630.

Hoffman, D. and Birch, D. (1995) Docosahexaenoic acid in red blood cells of patients with X-linked retinitis pigmentosa. *Invest. Opthalmol. Vis. Sci.* 36:1009–1018.

Hoffman, D., Uauy, R. and Birch, D. (1993) Red blood cell fatty acid levels in patients with autosomal dominant retinitis pigmenosa. *Exp. Eye Res.* 57:359–368.

Holman, R., Bibus, D., Jeffrey, G., Smethurst, P. and Crofts, J. (1994) Abnormal plasma lipids of patients with retinitis pigmentosa. *Lipids.* 29:61–65.

Horrobin, D.F. (1996) Schizophrenia as a membrane lipid disorder which is expressed throughout the body. *Prostaglandins Leukot. Essent. Fatty Acids.* 55:3–7.

Jones, R.E. and Plymate, S.R. (1988) Evidence for the regulation of fatty acid utilization in human sperm by docosahexaenoic acid. *Biol. Reprod.* 39:76–80.

Jones, R.E. and Plymate, S.R. (1993) Synthesis of docosahexaenoyl coenzyme A in human spermatozoa. *J. Androl.* 14:428–432.

Kelley, D.S., Taylor, P.C., Nelson, G.J. and Mackey, B.E. (1998) Dietary docosahexaenoic acid and immunocompetence in young healthy men. *Lipids.* 33:559–566.

Kelley, D.S., Taylor, P.C., Nelson, G.J., Schmidt, P.C., Ferretti, A., Erickson, K.L., Yu, R., Chandra, R.K. and Mackey, B.E. (1999) Docosahexaenoic acid ingestion inhibits natural killer cell activity and production of inflammatory mediators in young healthy men. *Lipids.* 34:317–324.

Knapp, H.R. (1990) Prostaglandins in human semen during fish oil ingestion: evidence for *in vivo* cyclooxygenase inhibition and appearance of novel trienoic compounds. *Prostaglandins.* 39:407–423.

Kris-Etherton, P.M., Taylor, D.S., Yu-Poth, S., Huth, P., Moriarty, K., Fishell, V., Hargrove, R.L., Zhao, G. and Etherton, T.D. (2000) Polyunsaturated fatty acids in the food chain in the United States. *Am. J. Clin. Nutr.* 71:S179–S188.

Krokan, H.E., Bjerve, K.S. and Mork, E. (1993) The enteral bioavailability of eicosapentaenoic acid and docosahexaenoic acids is as good from ethyl ester as from glyceryl ester in spite of lower hydrolytic rates by pancreatic lipase in vitro. *Biochim. Biophys. Acta.* 1168:59–67.

Kyle, D. (2001) The large-scale production and use of a single-cell oil highly enriched in docosahexaenoic acid. In: *Omega-3 Fatty Acids: Chemistry, Nutrition and Health Effects.* (Shahidi, F. and Finley, J.W., eds.) Washington, D.C.: American Chemical Society, Oxford University Press, pp. 92–107.

Kyle, D.J., Schaefer, E., Patton, G. and Beiser, A. (1999) Low serum docosahexaenoic acid is a significant risk factor for Alzheimer's dementia. *Lipids.* 34 (Suppl):S245.

Lawson, L.D. and Hughes, B.G. (1988) Human absorption of fish oil fatty acids as triacylglycerols, free acids or ethyl esters. *Biochem. Biophys. Res. Commun.* 152:328–335.

Leng, G.C., Smith, F.B., Fowkes, F.G., Horrobin, D.F., Ells, K., Morse-Fisher, N. and Lowe, G.D. (1994a) Relationship between plasma essential fatty acids and smoking, serum lipids, blood pressure and haemostatic and rheological factors. *Prostaglandins Leukot. Essent. Fatty Acids.* 51:101–108.

Leng, G.C., Horrobin, D.F., Fowkes, F.G., Smith, F.B., Lowe, G.D., Donnan, P.T. and Ells, K. (1994b) Plasma essential fatty acids, cigarette smoking, and dietary antioxidants in peripheral arterial disease. A population-based case-control study. *Arterioscler. Thromb.* 14:471–478.

Mahadik, S.P., Shendarkar, N.S., Scheffer, R.E., Mukherjee, S. and Correnti, E.E. (1996) Utilization of precursor essential fatty acids in culture by skin fibroblasts from schizophrenic patients and normal controls. *Prostaglandins Leukot. Essent. Fatty Acids.* 55:65–70.

Martinez, M. (1990) Severe deficiency of docosahexaenoic acid in peroxisomal disorders: a defect of delta 4 desaturation? *Neurology.* 40:1292–1298.

Martinez, M. (1991) Developmental profiles of polyunsaturated fatty acids in the brain of normal infants and patients with peroxisomal diseases: severe deficiency of docosahexaenoic acid in Zellweger's and pseudo-Zellweger's syndomes. *World Rev. Nutr. Diet.* 66:87–102.

Martinez, M. (1992a) Abnormal profiles of polyunsaturated fatty acids in the brain, liver, kidney, and retina of patients with peroxisomal disorders. *Brain Res.* 583:171–182.

Martinez, M. (1992b) Treatment with docosahexaenoic acid favourably modifies the fatty acid composition of erythrocytes in peroxisomal patients. *Prog. Clin. Biol. Res.* 375:389–397.

Martinez, M. (1994) Polyunsaturated fatty acids in the developing human brain, red cells and plasma: influence of nutrition and peroxisomal disease. Fatty acids and lipids: biological aspects. (Galli, C., Simpolous, A.P. and Tremoli, E., eds.). *World Rev. Nutr. Diet.* Karger, vol. 75, pp. 70–78.

Martinez, M. (1995) Polyunsaturated fatty acids in the developing human brain, erythrocytes and plasma in peroxisomal disease: therapeutic implications. *J. Inherit. Metab. Dis.* 18:S61–S75.

Martinez, M. (1996) Docosahexaenoic acid therapy in docosahexaenoic acid-deficient patients with disorders of peroxisomal biogenesis. *Lipids.* 31:S145–S152.

Martinez, M. and Mougan, I. (1999) Fatty acid composition of brain glycerophospholipids in peroxisomal disorders. *Lipids.* 34:733–740.

Martinez, M. and Vazquez, E. (1998) MRI evidence that docosahexaenoic acid ethyl ester improves myelination in generalized peroxisomal disorders. *Neurology.* 51:26–32.

Martinez, M., Pineda, M., Vidal, R., Conill, J. and Martin, B. (1993) Docosahexaenoic acid— a new therapeutic approach to peroxisomal disorder patients: experience with two cases. *Neurology.* 43:1389–1397.

Martinez, M., Mougan, I., Roig, M. and Ballabriga, A. (1994) Blood polyunsaturated fatty acids in patients with peroxisomal disorders. A multicenter study. *Lipids.* 29:273–280.

Martinez, M., Vazquez, E., Garcia-Silva, M.T., Manzanares, J., Bertran, J.M., Castello, F. and Mougan, I. (2000) Therapeutic effects of docosahexaenoic acid ethyl ester in patients with generalized peroxisomal disorders. *Am. J. Clin. Nutr.* 71:376S–385S.

Marzo, I., Alava, M.A., Pineiro, A. and Naval, J. (1996) Biosynthesis of docosahexaenoic acid in human cells: evidence that two different delta 6-desaturase activities may exist. *Biochim. Biophys. Acta.* 1301:263–272.

Mitchell, E.A., Aman, M.G., Turbott, S.H. and Manku, M. (1987) Clinical characteristics and serum essential fatty acid levels in hyperactive children. *Clin. Pediatr.* 26:406–411.

Mitchell, D.C., Gawrisch, K., Litman, B.J. and Salem, Jr., N. (1998) Why is docosahexaenoic acid essential for nervous system function? *Biochem. Soc. Trans.* 26:365–370.

Moore, S.A., Hurt, E., Yoder, E., Sprecher, H. and Spector, A.A. (1995) Docosahexaenoic acid synthesis in human skin fibroblasts involves peroxisomal retroconversion of tetracosahexaenoic acid. *J. Lipid Res.* 36:2433–2443.

Mori, T.A., Bao, D.Q., Burke, V., Puddey, I.B. and Beilin, L.J. (1999) Docosahexaenoic acid but not eicosapentaenoic acid lowers ambulatory blood pressure and heart rate in humans. *Hypertension.* 34:253–260.

Mori, T.A., Burke, V., Puddey, I.B., Watts, G.F., O'Neal, D.N., Best, J.D. and Beilin, L.J. (2000a) Purified eicosapentaenoic and docosahexaenoic acids have differential effects on serum lipids and lipoproteins, LDL particle size, glucose, and insulin in mildly hyperlipidemic men. *Am. J. Clin. Nutr.* 71:1085–1094.

Mori, T.A., Watts, G.F., Burke, V., Hilme, E., Puddey, I.B. and Beilin, L.J. (2000b) Differential effects of eicosapentaenoic acid and docosahexaenoic acid on vascular reactivity of the forearm microcirculation in hyperlipidemic, overweight men, *Circulation.* 102:1264–1269.

Moser, A.B., Jones, D.S., Raymond, G.V. and Moser, H.W. (1999) Plasma and red blood cell fatty acids in peroxisomal disorders. *Neurochem. Res.* 24:187–197.

Mulder, M., Ravid, R., Swaab, D.F., deLoet, E.R., Haasdijk, E.D., Julk, J., van der Bloom, J. and Havekes, L.M. (1998) Reduced levels of cholesterol, phospholipids, and fatty acids in CSF of Alzheimer's disease patients are not related to Apo E4. *Alz. Dis. Assoc. Dis.* 12:198–203.

Nair, S.S., Leitch, J. and Garg, M.L. (2001) N-3 polyunsaturated fatty acid supplementation alters inositol phosphate metabolism and protein kinase C activity in adult porcine cardiac myocytes. *J. Nutr. Biochem.* 12:7–13.

Nelson, G.J. and Ackman, R.G. (1988) Absorption and transport of fat in mammals with emphasis on n-3 polyunsaturated fatty acids. *Lipids.* 23:1005–1014.

Nelson, G.J., Schmidt, P.C., Bartolini, G.L., Kelley, D.S. and Kyle, D. (1997) The effect of dietary docosahexaenoic acid on plasma lipoproteins and tissue fatty acid composition in humans. *Lipids.* 32:1137–1146.

Nissen, H.P. and Kreysel, H.W. (1983) Polyunsaturated fatty acids in relation to sperm motility. *Andrologia.* 15:264–269.

Nitsch, R., Pittas, A., Blusztajn, J.K., Slock, B.E., and Growdon, J. (1991) Alterations of phospholipid metabolites in post-mortem brains from patients with Alzheimer's disease. *Ann. N.Y. Acad. Sci.* 640: 110–113.

Nordoy, A., Barstad, L., Connor, W.E. and Hatcher, L. (1991) Absorption of the n-3 eicosapentaenoic and docosahexaenoic acids as ethyl esters and triglycerides by humans. *Am. J. Clin. Nutr.* 53:1185–1190.

Oliw, E.H. and Sprecher, H.W. (1991) Metabolism of polyunsaturated (n-3) fatty acids by monkey seminal vesicles: isolation and biosynthesis of omega-3 epoxides. *Biochim. Biophys. Acta.* 1086:287–294.

Ollero, M., Powers, R.D. and Alvarez, J.G. (2000) Variation of docosahexaenoic acid content in subsets of human spermatozoa at different stages of maturation: implications for sperm lipoperoxidative damage. *Mol. Reprod. Dev.* 55:326–334.

Peet, M., Laugharne, J., Rangarajan, N., Horrobin, D. and Reynolds, G. (1995) Depleted red cell membrane essential fatty acids in drug-treated schizophrenic patients. *J. Psychiatr. Res.* 29:227–232.

Peet, M., Laugharne, J.D., Mellor, J. and Ramchand, C.N. (1996) Essential fatty acid deficiency in erythrocyte membranes from chronic schizophrenic patients, and the clinical effects of dietary supplementation. *Prostaglandins Leukot. Essent. Fatty Acids.* 55:71–75.

Petroni, A., Bertagnolio, B., LaSpada, P., Blasevich, M., Papini, N., Govoni, S., Rimoldi, M. and Galli, C. (1998) The beta-oxidation of arachidonic acid and the synthesis of docosahexaenoic acid are selectively and consistently altered in skin fibroblasts from three Zellweger patients versus X-adrenoleukodystrophy, Alzheimer and control subjects. *Neurosci.* 250:145–148.

Poge, A.P., Baumann, K., Muller, E., Leichsenring, M., Schmidt, H., Bremer, H.J. (1998) Long-chain polyunsaturated fatty acids in plasma and erythrocyte membrane lipids of children with phenylketonuria after controlled linoleic intake. *J. Inherit. Metab. Dis.* 21:373–381.

Polette, A., Deshayes, C., Chantegrel, B., Croset, M., Armstrong, J.M. and Lagarde. M. (1999) Synthesis of acetyl,docosahexaenoyl-glycerophosphocholine and its characterization using nuclear magnetic resonance. *Lipids.* 34:1333–1337.

Prasad, M.R., Lovell, M.A., Yatin, M., Dhillon, H., and Markesbery, W.R. (1998) Regional membrane phospholipid alterations in Alzheimer's disease. *Neurochem. Res.* 23:81–88.

Rambjor, G.S., Walen, A.I., Windsor, S.L. and Harris, W.S. (1996) Eicosapentaenoic acid is primarily responsible for hypotriglyceridemic effect of fish oil in humans. *Lipids.* 31:S45–S49.

Raper, N.R., Cronin, J.F. and Exler, J. (1992) Omega-3 fatty acids content of the U.S. food supply. *J. Am. Coll. Nutr.* 11:304–308.

Reich, E.E., Markesbery, W.R., Roberts, L.J., Swift, L.L., Morrow, J.D. and Montine, T.J. (2001) Brain regional quantification of F-ring and D-/E-ring isoprostanes and neuroprostanes in Alzheimer's disease. *Am. J. Pathol.* 158:293–297.

Roulet, M., Frascarolo, P., Rappaz, I. and Pilet, M. (1997) Essential fatty acid deficiency in well nourished young cystic fibrosis patients. *Eur. J. Pediatr.* 156:952–956.

Salem, N.Jr., Wegher, B., Mena, P. and Uauy, R. (1996) Arachidonic and docosahexaenoic acids are biosynthesized from their 18-carbon precursors in human infants. *Proc. Natl. Acad. Sci. USA.* 93:49–54.

Sanjuro, P., Perteagudo, L., Rodriguez Soriano, J., Vilaseca, A. and Campistol, J. (1994) Polyunsaturated fatty acid status in patients with phenylketonuria. *J. Inherit. Metab. Dis.* 17:704–709.

Schlenk, H., Sand, D.M. and Gellerman, J.L. (1969) Retroconversion of docosahexaenoic acid in the rat. *Biochim. Biophys. Acta.* 187:201–207.

Simon, J.A., Hodgkins, M.L., Browner, W.S., Neuhaus, J.M., Bernert, Jr., J.T. and Hulley, S.B. (1995) Serum fatty acids and the risk of coronary heart disease. *Am. J. Epidemiol.* 142:469–476.

Simopoulos, A.P., Leaf, A. and Salem, Jr., N. (1999) Essentiality of and recommended dietary intakes for omega-6 and omega-3 fatty acids. *Ann. Nutr. Metab.* 43:127–130.

Söderburg, M., Edlund, C., Kristensson, K. and Dallner, G. (1991) Fatty acid composition of brain phospholipids in aging and in Alzheimer's disease. *Lipids.* 26:421–428.

Söderburg, M., Edlund, C., Kristensson, K., Alafuzoff, I., and Dallner, G. (1992) Lipid composition in different regions of the brain in Alzheimer's disease/senile dementia of Alzheimer's type. *J. Neurochem.* 59:1646–1653.

Sovik, O., Mansson, J.E., Bjorke Monsen, A.L., Jellum, E. and Berge, R.K. (1998) Generalized peroxisomal disorder in male twins: fatty acid composition of serum lipids and response to n-3 fatty acids. *J. Inherit. Metab. Dis.* 21:662–670.

Stevens, L.J., Zentall, S.S., Abate, M.L., Kuczek, T., and Burgess, J.R. (1995) Essential fatty acid metabolism in boys with attention-deficit hyperactivity disorder. *Am. J. Clin. Nutr.* 62:761–768 (Abstract).

Stevens, L.J., Zentall, S.S., Abate, M.L., Kuczek, T. and Burgess, J.R. (1996) Omega-3 fatty acids in boys with behaviour, learning, and health problems. *Physiol. Behav.* 59:915–920.

Su, H.M., Bernardo, L., Mirmiran, M., Ma, X.H., Corso, T.N., Nathaneilsz, P.W. and Brenna, J.T. (1999) Bioequivalence of dietary alpha-linolenic and docosahexaenoic acids as sources of docosahexaenoate in brain and associated organs in neonatal baboons. *Pediatr. Res.* 45:87–93.

Suh, M., Wierzbicki, A.A., Lien, E.L. and Clandinin, M.T. (2000) Dietary 20:4n-6 and 22:6n-3 modulates the profile of long- and very-long-chain fatty acids, rhodopsin content and kinetics in developing photoreceptor cells. *Pediatr. Res.* 48:524–530.

Terano, T., Fujishiro, S., Ban, T., Yamamoto, K., Tanaka, T., Noguchi, Y., Tamura, Y., Yazawa, K. and Hirayama, T. (1999) Docosahexaenoic acid supplementation improves the moderately severe dementia from thrombotic cerebrovascular diseases. *Lipids.* 34:S345–S346.

Thies, F., Pillon, C., Moliere, P., Lagarde, M. and Lecerf, J. (1994) Preferential incorportation of sn-2 lysoPC DHA over unesterified DHA in the young rat brain. *Am. J. Physiol.* 267:R1273–1279.

Vidgren, H.M., Agren, J.J., Schwab, U., Rissanen, T., Hanninen, O. and Uusitupa, M.I. (1997) Incorporation of n-3 fatty acids into plasma lipid fractions, and erythrocyte membranes and platelets during dietary supplementation with fish, fish oil, and docosahexaenoic acid-rich oil among healthy young men. *Lipids.* 32:697–705.

Voss, A., Reinhart, M., Sankarappa, S. and Sprecher, H. (1991) The metabolism of 7,10,13,16,19-docosapentaenoic acid to 4,7,10,13,16,19-docosahexaenoic acid in rat liver is independent of a 4-desaturase. *J. Biol. Chem.* 266:19995–20000.

Wallace, J.M., McCabe, A.J., Robson, P.J., Keogh, M.K., Murray, C.A., Kelly, P.M., Marquez-Ruiz, G., McGlynn, H., Glimore, W.S. and Strain, J.J. (2000) Bioavailability of n-3 polyunsaturated fatty acids (PUFA) in foods enriched with microencapsulated fish oil. *Ann. Nutr. Metab.* 44:157–162.

Warner, R., Laugharne, J., Peet, M., Brown, L. and Rogers, N. (1999) Retinal function as a marker for cell membrane omega-3 fatty acid depletion in schizophrenia: a pilot study. *Biol. Psychiatry.* 45:1138–1142.

Willard, D.E., Nwankwo, J.O., Kaduce, T.L., Harmon, S.D., Irons, M., Moser, H.W., Raymond, G.V. and Spector, A.A. (2001) Identification of a fatty acid delta(6)-desaturase deficiency in human skin fibroblasts. *J. Lipid Res.* 42:501–508.

Zalata, A.A., Christophe, A.B., Depuydt, C.E., Schoonjans, F. and Comhaire, F.H. (1998) The fatty acid composition of phospholipids of spermatozoa from infertile patients. *Mol. Hum. Reprod.* 4:111–118.

Zeller, S., Barclay, W. and Abril, R. (2001) Production of docosahexaenoic acid from microalgae In: *Omega-3 Fatty Acids: Chemistry, Nutrition and Health Effects.* (Shahidi, F. and Finley, J.W., eds.) Washington, D.C.: American Chemical Society, Oxford University Press, pp. 108–124.

11 Solid-Liquid Extraction Technologies for Manufacturing Nutraceuticals

Dennis D. Gertenbach

CONTENTS

1-5667-6902-7/02/$0.00+$1.50
© 2002 by CRC Press LLC

Many phytochemicals and nutraceutical ingredients are derived from botanicals. In the manufacture of many of these nutraceuticals, processes begin with the extraction of plant materials using a suitable solvent. Many technologies and types of equipment exist to achieve this solid-liquid extraction. To successfully choose and operate the proper equipment for producing the desired product in an economic manner, the fundamentals of equilibrium and mass transfer must be understood. Once these fundamentals are understood, they can be applied to the botanical raw material of interest and the chemical properties of the desired phytochemical to select and operate the most cost-effective extraction equipment.

It should be noted that the term "extraction" is used throughout this chapter to signify solid-liquid extraction. Although supercritical extraction is governed by the same mass transfer principles as those describing extraction with liquid solvents, it is not discussed in this chapter.

11.1 DEFINITIONS FOR SOLID-LIQUID EXTRACTION

It is important to define what is meant by "solid-liquid extraction."

> *Definition:* Solid-liquid extraction is the use of a solvent to dissolve and remove a soluble fraction (called the solute) from an insoluble, permeable solid.

Fortunately, botanicals contain a network of passageways throughout the plant that allow nutrients and water to be distributed while the plant is alive. These same passageways provide a raw material for extraction that is quite porous, allowing the solvent to readily penetrate and remove the solute.

In many instances, the extraction of a specific compound or family of compounds is desired to make the final nutraceutical product. Extraction removes these phytochemicals of interest along with many other compounds, depending on the conditions chosen for extraction. These compounds of interest are commonly referred to as "markers" in the nutraceutical industry. It should be noted that for some widely used nutraceuticals, the marker is the active ingredient, providing a medicinal benefit. For many other nutraceutical products, the active ingredients are not known, but the beneficial properties of the nutraceutical product are known to be associated with the marker compounds. Thus, by standardizing to a stated marker content, the manufacturer and user are assured that the product will provide the desired benefit, as the unknown active ingredients will be found in the product with the marker compounds.

11.2 FUNDAMENTALS OF SOLID-LIQUID EXTRACTION

As with many mass transfer operations, two fundamental concepts need to be realized to fully understand extraction: equilibrium and mass transfer rate. These control how much and how fast the marker compounds extract from the botanical raw material.

11.2.1 EQUILIBRIUM

When a botanical is contacted with a solvent, an equilibrium relationship develops that describes the concentration of each compound within the botanical that will dissolve into the solvent. This relationship can be described as follows:

$$K = \frac{C_e}{C_{dm}} \tag{11.1}$$

where K is the equilibrium constant, C_e is the concentration of a given compound in the solvent, and C_{dm} is the concentration of a given compound in the dry (extract free) marc. From this relationship, the larger the value of K, the more of a given compound will dissolve into the solvent. Also, the more of a given compound that was originally found in the botanical, the higher the concentration will be in the liquid. The amount of a compound that will dissolve is a function of the solvent used and the temperature of the solvent. Thus, solvent and temperature will dictate how large K is. In many instances, the solvent and temperature can be chosen so that K is quite large, and nearly all of the marker will dissolve into the solvent.

11.2.2 MASS TRANSFER

The equilibrium relationship in Equation 11.1 describes how much of a given compound will dissolve into a solvent at a given temperature but does not tell how fast it will dissolve. To determine how long it will take for a compound to reach the equilibrium concentration in the liquid, mass transfer concepts must be employed.

To best understand how mass transfer takes place in solid-liquid extraction, consider a particle of a botanical, as shown in Figure 11.1. As mentioned earlier, botanicals are very porous, containing a network of passageways and internal pores. To extract the solute from the particle to the bulk solvent surrounding the particle, four steps are needed:

1. Diffusion of the solvent into the solid particle through its internal pore structure
2. Dissolution of the solute, both the marker compounds and other soluble compounds, from the solid into the solvent within the particle

Solvent

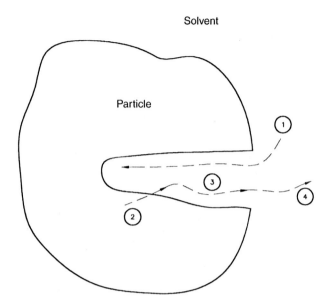

FIGURE 11.1 Four steps for mass transfer: (1) solvent soaks into the pores within the particle, (2) solute dissolves into the solvent within the pores of the particle, (3) dissolved solute migrates to the surface of the particle, and (4) dissolved solute at the particle surface diffuses into the bulk solvent.

3. Diffusion of the dissolved solute from within the particle to the particle surface through the internal pores
4. Washing the solute-rich solvent from the boundary layer at the particle surface into the bulk solvent surrounding the particle

Experience has shown that for most botanicals, step 3, the diffusion of solvent through the pores within the particle to the surface, is the rate limiting step. The rate of diffusion in step 3 can be described by Fick's Second Law, which states the following:

$$\frac{\delta C}{\delta t} = D\frac{\delta^2 C}{\delta x^2} \tag{11.2}$$

where C is the concentration of the solute or marker compound, t is time, D is the diffusion coefficient or diffusivity, and x is the particle diameter.

Fick's Second Law shows that the driving force for extraction is the concentration gradient within the particles, i.e., the difference between the phytochemical concentration inside the particle and the concentration at the particle surface. The rate of extraction, $\delta C/\delta t$, increases with a larger concentration gradient. Equation 11.2 also shows that the rate of extraction can be altered by increasing the diffusion coefficient, D, or reducing the particle diameter, x.

Typical diffusion coefficients for extracting various components from plant raw materials are listed in Table 11.1.

TABLE 11.1
Typical Diffusion Coefficients

Plant Raw Material	Solute	Solvent	Particle Size mm	Temperature °C	Diffusion Coefficient $m^2/sec \times 10^{10}$
Sugar beets[1]	Sucrose	Water	5	75	7.2
Sugar beets[1]	Sucrose	Water	1	75	4.3
Sugar beets[1]	Sucrose	Water	1	23–25	1.6
Coffee[1]	Caffeine	CH_2Cl_2	Not reported	30	0.47
Soybean flakes[1]	Soybean oil	Hexane	0.56	69	1.13
Soybean flakes[1]	Soybean oil	Trichloroethylene	0.51	69	0.74–1.27
St. John's wort[2]	Hypericins	Ethanol-water	0.9 (weighted mean)	55	0.093
Echinacea herb[2]	Phenolic compounds	Ethanol-water	1.8 (weighted mean)	50	0.21

[1] From Schwartzberg and Chao (1982).
[2] Data generated by Botanicals International Extracts, Boulder, Colorado, 1999.

The solution of Equation 11.2 is dependent on the geometry of the particle and on the type of equipment used for extraction. Several authors, including Crank (1956), Schwartzberg (1987), and Plachco and Lago (1975, 1976, 1978), provide solutions for Equation 11.2 for several particle geometries and a number of equipment types. However, these solutions have found little use in choosing and operating the appropriate extraction equipment and are not further discussed in this chapter.

11.2.3 VARIABLES AFFECTING EXTRACTION

With the goal of removing as much of the phytochemical of interest from the plant raw material as quickly as possible, knowledge of the concepts of equilibrium and mass transfer is essential for economically operating extraction equipment. Extraction variables and operation are chosen to maximize the value of the equilibrium constant, K, in Equation 11.1 and the rate of extraction, $\delta C/\delta t$, in Equation 11.2. These two equations show that only four factors affect extraction operation:

1. Solvent composition
2. Temperature
3. Particle size
4. Liquid-to-solid ratio

11.2.3.1 Solvent Composition

The choice of solvent can significantly change both the amount of phytochemical that extracts into the liquid and the rate at which the phytochemical is extracted. Generally,

TABLE 11.2
Robbins Chart of Solute/Solvent Group Interactions

Solute Class	Solvent Class											
	1	2	3	4	5	6	7	8	9	10	11	12
H Donor Groups												
1 Phenol	0	0	–	0	–	–	–	–	–	–	+	+
2 Acid, thiol	0	0	–	0	–	–	0	0	0	0	+	+
3 Alcohol, water	–	–	0	+	+	0	–	–	+	+	+	+
4 Active H on multihalogen paraffin	0	0	+	0	–	–	–	–	–	–	0	+
H Acceptor Groups												
5 Ketone, amide with no H on N, sulfone, phosphine oxide	–	–	+	–	0	+	+	+	+	+	+	+
6 Tertiary amine	–	–	0	–	+	0	+	+	0	+	0	0
7 Secondary amine	–	0	–	–	+	+	0	0	0	0	0	+
8 Primary amine, amide with 2H on N	–	0	–	–	+	+	0	0	+	+	+	+
9 Ether, oxide, sulfoxide	–	0	+	–	+	0	0	+	0	+	0	+
10 Ester, aldehyde, carbonate, phosphate nitrate, nitrite, nitrile	–	0	+	–	+	+	0	+	+	0	+	+
11 Aromatic, olefin, halogen aromatic multihalogen paraffin without active H monohalogen paraffin	+	+	+	0	+	0	0	+	0	+	0	0
Non-H-Bonding Groups												
12 Paraffin, carbon disulfide	+	+	+	+	+	0	+	+	+	+	0	0

Source: From Frank et al. (1999).

ethanol, water, and mixtures of these two are the solvents of choice for most nutraceutical products because of consumer acceptance. Adding some water to ethanol can enhance the rate of extraction by causing the raw material to swell, allowing the solvent to penetrate the solid particles more easily. However, even with mixtures of ethanol and water, some care in selecting the solvent must be taken. For instance, extraction of saponins or other glucosides with water can be detrimental in some cases due to hydrolysis, especially when enzymes are present that are not destroyed by drying the raw material (Bonati, 1980a). Bonati (1980b) also reports that the valepotriates in valerian are quite soluble in concentrated alcohol solutions, but the addition of appreciable water to the solvent causes an instability of these compounds. Adjustment of pH may also be necessary to extract the desired phytochemicals, as demonstrated by Metivier et al. (1980) when extracting anthocyanins from grape pomace. A low pH in the solvent can also maintain the stability of the active compounds in the extract, as reported by Crippa (1978) for alkaloids in both *Rauvolfia* and *Belladonna*.

The Robbins Chart of Solute/Solvent Group Interactions, shown in Table 11.2, is useful in selecting the appropriate solvent type for a given phytochemical. This chart helps quantify the old adage that "like dissolves like" and provides a system for choosing a solvent for a given solute, based on their respective chemical compositions. The table indicates positive, neutral, or negative deviation from an ideal solution. A negative deviation indicates an attractive interaction between groups, thus the potential for a high solubility of the solute in the solvent. Also, when a neutral attraction is indicated, solubility is likely.

Other criteria enter into the selection of the extraction solvent. These include the following:

- Selectivity—The choice of solvent can significantly affect the purity of the extract, as the ratio of phytochemical solubility to total dissolved solids solubility is greatly influenced by the solvent used.
- Chemical and thermal stability—The solvent must be stable at the operating conditions of the extraction operation as well as in downstream processing.
- Compatibility with the solute—The solvent should not react with the phytochemicals of interest.
- Low viscosity—A lower viscosity increases the diffusion coefficient, and thus increases the rate of extraction as shown in Fick's Law.
- Ease of recovery—For most nutraceutical processes, economics dictate that the solvent be recovered for reuse in the process. Solvents with a lower boiling point and heat of vaporization are more economical to recover and reuse.
- Low flammability—For safety reasons, the least flammable solvent should be selected that will meet the needs of the process.
- Low toxicity—This is important both from a product safety consideration and to minimize worker exposure.
- Regulatory issues—Environmental regulations should be considered when choosing the solvent.
- Consumer acceptance—Some solvents may not be acceptable to the consumer.
- Low cost—The cost of the solvent may also dictate solvent selection.

11.2.3.2 Temperature

Temperature affects both the equilibrium and the mass transfer rate of the extraction process. Generally, a higher temperature causes a higher solubility of compounds into the solvent; thus, a larger equilibrium constant. Additionally, temperature plays an important role in the rate of extraction, as the diffusion of the solute through the particle to the particle surface is the rate-limiting step. The higher the temperature, the larger the diffusion coefficient becomes; hence, the greater the rate of extraction. Temperature is limited by the boiling point of the solvent. In addition, many phytochemicals are thermally unstable, and the extraction temperature needs to be controlled to prevent thermal decomposition. Thus, a balancing act is needed, keeping the temperature as high as possible to provide a fast extraction rate, while not destroying the desired marker compounds.

Other criteria need to be considered in selecting the appropriate extraction temperature. These include the following:

- Selectivity—Because the extraction rates of various compounds increase differently with changes in temperature, selectivity of one compound over other compounds can sometimes be accomplished by controlling temperature.
- Viscosity—Most types of extraction equipment have liquid viscosity limitations. As the dissolved solids content in the extract increases, higher viscosity results. Higher temperatures will significantly decrease the extract viscosity.
- Deterioration of the biomass—For some extraction processes, the rigid structure of the biomass must be maintained during extraction. In some instances, this biomass structure disintegrates at elevated temperatures.

11.2.3.3 Particle Size

Fick's Second Law shows that the rate of extraction increases with decreasing particle size. If one remembers that the rate-controlling step for extraction is the migration of the solute through the pores of the particles to the particle surface, a smaller particle will have a shorter path for the solute to reach the surface. A shorter diffusion path equates to a faster extraction rate.

Gertenbach (1998) used an experimentally determined diffusion coefficient for the extraction of phenolic compounds from *Echinacea purpurea* roots and the solution of Fick's Law given by Crank (1956) to generate Figure 11.2, which shows the calculated relationship between biomass particle size and extraction time. From the graph, 80% recovery from 12.7 cm (1/2-inch) particles requires 64 hours of extraction. However, by reducing the biomass particles to 6.4 cm (1/4-inch), only 16 hours

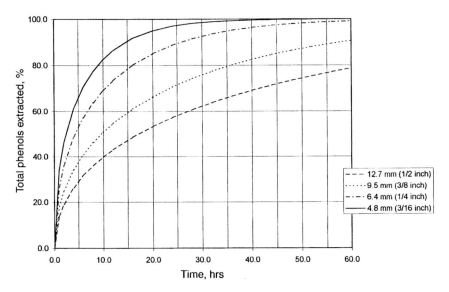

FIGURE 11.2 Calculated extraction recovery of phenolic compounds from *Echinacea purpurea* roots with ethanol-water solvent as a function of particle size. [From Gertenbach (1998).]

are required to obtain an 80% recovery. Thus, the botanical raw material should be ground to as small a particle size as the extraction equipment can handle. Limitations of particle size for each type of extractor are discussed later.

11.2.3.4   Liquid-to-Solids Ratio

A higher liquid-to-solids ratio dilutes the concentration of dissolved phytochemicals at the surface of the particle, thus providing a higher concentration gradient between the concentrations inside and at the surface of the particles. This higher concentration gradient gives a faster extraction rate. However, a more dilute extract requires much more extensive downstream processing; so, once again, a balancing act is needed in selecting the appropriate liquid-to-solids ratio.

11.3   EXTRACTION EQUIPMENT

Extraction equipment can be classified by the method used to contact the solid with the solvent. Two general extraction methods are used for extracting plant raw materials: dispersed-solids extraction and percolation extraction. During dispersed-solids extraction, enough solvent is used to suspend the solids within the liquid. A stirred tank is the simplest example of a dispersed-solids extraction. In a percolation extractor, solvent flows through a fixed bed of ground raw material.

Dispersed-solids extraction is used for finely ground plant materials, when the raw material swells considerably and prevents flow through the solids bed or when the solids disintegrate during extraction. The equipment consists of one or more tanks for mixing the solids with the liquid, then a solid-liquid separation step, such as screening, filtration, or centrifuging, to collect the extract from the extracted biomass. A typical one-stage mixed tank system is shown in Figure 11.3. Dispersed-solids extraction is most suitable for raw materials that must be finely ground to extract the phytochemicals of interest.

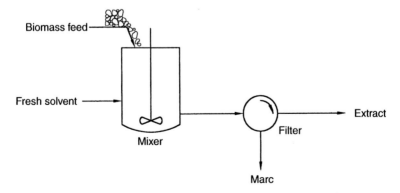

FIGURE 11.3 Dispersed solids extraction with filtration to separate the marc (extracted plant material).

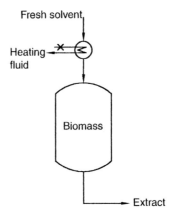

FIGURE 11.4 Percolation extraction.

A typical percolation extraction is shown in Figure 11.4. A tank is filled with ground raw material, and then the solvent is passed through the bed of solids. The flow of liquid past the particles carries extract away from the particle surface. Two types of liquid flow are used: immersed beds, in which all of the void volume between the particles is filled with solvent, and trickle flow, in which the solvent trickles over the particles without completely filling the empty spaces between particles. The advantages of percolation extraction are that a separate solid-liquid separation step is not generally needed and that raw material grinding costs are reduced due to the use of a coarser grind size.

For both dispersed solids and percolation extractors, batch or continuous equipment is used. In batch operation, solids are contacted with the liquid for sufficient time to allow an equilibrium concentration to be reached, and then the extract is collected. In continuous operation, solids and solvent are continuously fed to the extraction equipment, while extract and extracted plant material (known as marc) are removed. The most efficient method of operating continuous or batch equipment is in a countercurrent fashion, in which fresh solvent is contacted with the most extracted solid, and fresh solid is contacted with the most concentrated solvent in a series of steps. Countercurrent operation is shown schematically in Figure 11.5. The advantages of running countercurrently are that a much more concentrated extract is produced, while more complete extraction and a higher phytochemical recovery from the starting raw material are obtained. Countercurrent equipment is designed and operated such that the raw material is extracted to exhaustion, with little phytochemical remaining in the marc. Other literature sources contain information on extraction equipment suitable for plant materials, including Schwartzberg (1980, 1987), Bombardelli (1986), and List and Schmidt (1989).

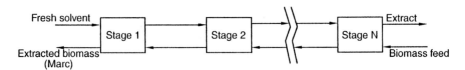

FIGURE 11.5 Countercurrent operation.

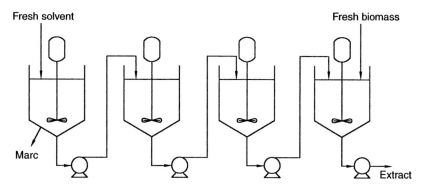

FIGURE 11.6 Countercurrent slurry tank extraction.

11.3.1 COMMERCIAL BATCH EQUIPMENT

11.3.1.1 Mixed Slurry Tank

The simplest type of extraction equipment is the stirred tank, coupled with a solid-liquid separation step such as a screen, filter, or centrifuge to separate the marc from the extract. This equipment is readily available in a wide variety of sizes and can be operated as a single-batch extractor or run in series in a countercurrent mode as shown in Figure 11.6.

This type of equipment generally is the least expensive to install but tends to produce weak extracts that require further concentration. In addition, a single extraction leaves considerable desired phytochemical behind in the marc, giving lower recoveries. To improve recovery, a second and third batch of solvent can be mixed with the partially extracted plant material, draining the tank after each extraction stage. A more concentrated extract can be generated by piping several tanks together and operating countercurrently, as shown in Figure 11.6. However, a solid-liquid separation step is needed between each tank.

If a screen is installed at the bottom of the tank, the solid-liquid separation can take place within the extraction equipment. The extract drains out of the tank, leaving the solids behind. Water can then be added to reslurry the solids, washing solvent from the marc. However, some means of removing the washed marc from a tank equipped with a screen is needed. Where external filtration or centrifuging equipment is used, solvent can be removed from the marc by water washing or conveying to an external dryer.

11.3.1.2 Tank Percolation

Percolation is another relatively inexpensive extraction system and is the technology most widely used in extracting plant materials to make herbal extracts. A wide range of tank sizes can be used, providing the capacity required by the manufacturer. A number of percolation strategies can be used, with the simplest being the pumping of solvent through the bed until an economic recovery of the desired phytochemical

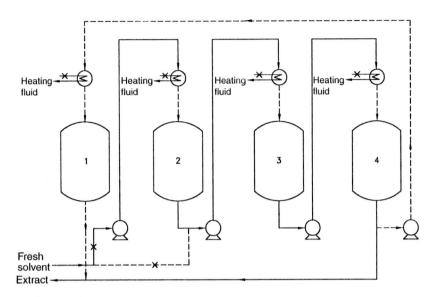

FIGURE 11.7 Shanks or cascade percolation extractors.

is achieved. The drawback of operating in this mode is the dilute extract that is produced and the large amount of solvent that is required.

To reduce the amount of solvent required, the percolator is filled with solvent, and then the extract is recirculated multiple times through the bed of raw material. Once equilibrium concentrations are approached, the extract is withdrawn, and a second charge of solvent is added to the percolator. This second charge of solvent is once again recirculated through the bed of plant material until equilibrium concentration is once again approached. After draining the extract, additional cycles using fresh solvent are repeated until the desired phytochemical recovery is achieved.

To produce even more concentrated extracts with high marker recovery, percolation tanks are hooked together in series and are operated in a countercurrent fashion. Figure 11.7 shows one countercurrent configuration, known as the Shanks Process, or cascade operation. Solvent passes through the bed of raw material from one extractor to the next, many times recirculating the solvent within one percolation tank before pumping to the next. Extract concentration is built up by contacting the most concentrated extract with the least extracted raw material. These types of extractors were used in the early 1900s for extracting sugar beets and are still used for extracting coffee in making instant coffee.

As a uniform solvent flow is required for percolation operations, care must be taken in grinding the raw material to provide a particle size that allows adequate flow at an acceptable extraction rate. When the extraction is complete, the tank is drained of liquid, eliminating a separate solid-liquid separation step. In addition, the solvent can be removed from the marc within the extractor by washing with water or heating the marc with steam. Alternatively, the solvent-soaked marc can be removed from the extractor, and solvent can be recovered in an external dryer.

Extr-o-mat

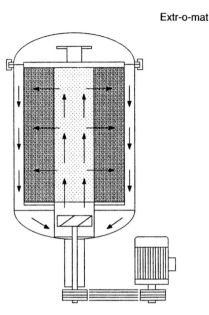

FIGURE 11.8 Acinoxa Extr-O-Mat extractor. (Courtesy of Inox, Inc. Charlotte, North Carolina.)

11.3.1.3 Acinoxa Extr-O-Mat

A schematic drawing of the Extr-O-Mat is shown in Figure 11.8. This extractor consists of a pressure-rated cylinder with the raw material enclosed in an annular basket that fits within the cylinder. An internal pump provides circulation through the bed of plant material, providing rapid percolation through the bed. Raw material is ground to 4.8–6.4 cm (3/16–1/4 inch), and the fine grind increases the extraction rate. The high recirculation also assists faster extraction, as the liquid on the particle surface is continually swept away, providing a higher concentration gradient within the raw material particles. Typically, the extractor is operated in a batch mode from one to four hours, until an equilibrium extract concentration is reached. Several batches of solvent are needed to achieve high phytochemical recovery.

Ground raw material is loaded into the baskets externally and then lowered into the extractor. When extraction is completed, solvent can be removed by washing with water, steam stripping, or using hot nitrogen. Due to the pressure rating of the cylinder, solvent removal can be operated at ambient pressure or vacuum. Once the solvent has been removed from the marc, the basket is lifted from the extractor body, and the marc is removed.

Countercurrent operation is possible with this equipment by extracting the plant material with several batches of solvent, contacting the previous partial extract with the next load of fresh raw material.

Extr-O-Mats are available in sizes capable of holding 0.5–450 kg (1–1000 lbs) of plant material per load. They are more widely used by European nutraceutical manufacturers than North American manufacturers, and a wide range of plant materials have been successfully extracted in Extr-O-Mat equipment.

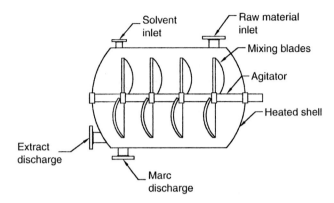

FIGURE 11.9 Cylindrical mixing extractor.

11.3.1.4 Cylindrical Mixing Extractors

Many equipment vendors manufacture cylindrical drying equipment that has been successfully used to extract plant materials. Figure 11.9 shows the configuration of this type of extractor. The unit consists of a horizontal cylinder equipped with rotating mixing internals. Dispersed solids extraction takes place, although a much higher solvent-to-solids ratio can be used, resulting in a higher concentrated extract. As the cylinder is jacketed, elevated temperatures can be reached and maintained. Dispersed solids operation allows for processing much finer plant materials, thus extraction cycles are relatively short. Multiple batches of fresh solvent must be used to achieve high extraction recoveries.

This type of equipment can be operated in a countercurrent mode, in which a load of raw material is extracted with several batches of solvent. The highest concentrated extract is used to make product. Weaker, subsequent batches of extract are saved for the next load of raw material to produce a more concentrated extract.

Solvent can be removed from the marc by heating with heat supplied by the outside shell, injecting steam, or pulling a vacuum on the extractor. Solvent-free marc is dumped out the bottom of the extractor. Extractors capable of vacuum operation are available to handle 70–450 kg (150–1000 lbs) of raw material. Much larger capacity is available for nonvacuum systems, up to 6800 kg (15,000 lb) loads.

11.3.1.5 Conical Screw Extractor

Figure 11.10 shows a conical screw extractor, which was initially designed for drying operations. Dispersed-solids mixing is provided by an internal screw, which rotates eccentrically within the cone. Extract is drained through a screen at the bottom of the cone to separate it from the extracted plant material. Solvent is removed from the marc using the same techniques as the cylindrical mixing extractor. Solvent-free marc is unloaded through a flange at the bottom of the cone.

This type of extractor is operated in a similar manner as a cylindrical mixing extractor and has similar advantages and disadvantages. Equipment is available for processing 9–2300 kg (20–5000 lbs) of raw material per batch.

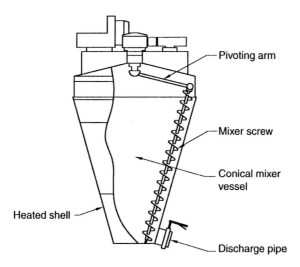

FIGURE 11.10 Conical screw extractor.

11.3.2 COMMERCIAL CONTINUOUS EQUIPMENT

11.3.2.1 Niro Screw Extractor

Used for many years in the sugar beet industry, screw extractors allow countercurrent extraction to give higher extract concentrations and better phytochemical recovery into the extract. Referring to the equipment schematic in Figure 11.11, ground raw material enters the bottom of the inclined screw and is slowly conveyed upward. Solvent enters the top of the screw and percolates down through the plant material. Extract leaves the bottom of the extractor, while solvent-soaked marc is conveyed out the top of the extractor to external solvent removal. Proper grind size of the feed material is important to achieve good contact between the solvent and the solids. Units are available that are explosion-proof for use with flammable solvents, and the equipment comes with inlet solvent heating and a heated jacket for maintaining extraction temperature. The equipment can be operated under pressure, so that extraction temperatures can exceed the normal boiling point of the solvent.

Standard sizes of explosion-proof models will handle 14–1400 kg (30–3000 lb/hr) raw material feed. Extraction times range from 20 to 90 minutes and are varied

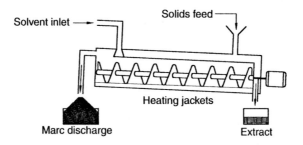

FIGURE 11.11 Niro screw extractor.

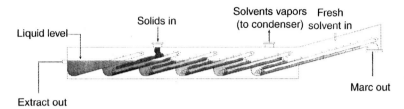

FIGURE 11.12 Crown immersion-type extractor. (Courtesy of Crown Iron Works Company.)

by changing the speed and pitch of the screw. As these extractors are highly auto-mated, capital costs are higher, but labor costs are much less. This equipment has been successfully used on a variety of plant feed materials.

11.3.2.2 Crown Iron Works Extractors

Crown Iron Works of Minneapolis, Minnesota has developed two specialty extractors that are finding use in the nutraceutical industry. Both types of extractors feature continuous, countercurrent percolation operation, requiring minimal operator time. These extractors are modifications of Crown's oilseed extractors that have been used for several decades.

The immersion-type extractor, shown in Figure 11.12, is suitable for granular solids that have a greater density than the solvent and sink when wetted. Within the extractor, solids are conveyed up a series of inclines with cleated belts, remain-ing submerged in the solvent bath. Liquid enters countercurrently, percolating past the moving bed of solids. Further contact between the solids and solvent occurs as the solids drop from one conveyor to the next. The last solids conveyor is longer, allowing the extracted solids to be washed with fresh solvent, then drained before discharging. Solvent is removed from the marc in external equipment. Concen-trated extract overflows the solids feed end of the extractor and is pumped to downstream processing.

Figure 11.13 shows the percolation type configuration, in which solvent flows through a shallow bed of solids. The shallow bed is used because the bed is subjected to less compression, and solvent channeling is less likely. This extractor is more suited for plant materials that float in the solvent. Also, the raw material must have sufficient structure to allow good distribution of the solvent through the bed of solids.

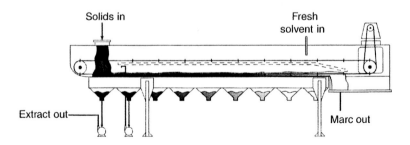

FIGURE 11.13 Crown percolation-type extractor. (Courtesy of Crown Iron Works Company.)

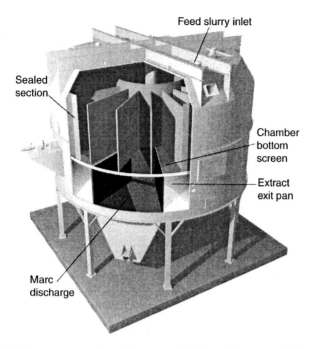

FIGURE 11.14 Carousel extraction. (Courtesy of DeSmet Process and Technology, Inc. Atlanta, Georgia.)

Feed solids are loaded on a perforated belt. Solvent is added countercurrently as a spray on the nearly spent raw material. After percolating through the bed, the extract is collected in chambers below the belt and then pumped to the next section of belt in a countercurrent mode. Extracted marc falls from the belt into an auger system that conveys the solvent-soaked marc to external solvent recovery. Concentrated extract is removed at the feed end for further processing.

Typical throughput for these types of extractors is 5–250,000 kg/hr (10 lbs/hr–275 tons/hr), with retention times ranging up to five hours. These extractors require careful grind size control when preparing the raw material for extraction to assure adequate marker recovery.

11.3.2.3 Carousel Extractor

Carousel extractors suitable for extracting phytochemicals from plant materials have also been adopted from equipment used extensively in the oilseed extraction industry. The carousel extractor consists of a sectioned cylinder, as shown in Figure 11.14. Raw material is slurried with extract and loaded into the initial cell. Solvent is added to each cell countercurrently, collecting at the bottom of each cell, and is then pumped to the next cell. Either flooded (deep bed) or trickle (shallow bed) operation is possible depending on the solvent flow rate to each cell and equipment design. After draining the solvent in the last cell, the marc is discharged from the cell and conveyed to external solvent removal equipment. Concentrated extract is collected from the raw material feed cell.

Throughputs available for carousel extractors range from 450 to 270,000 kg/hr (1–300 tons/hr). Retention times range between 45–180 minutes, again requiring proper raw material grinding to achieve percolation while having sufficiently rapid extraction. These extractors also require low labor to operate.

11.4 EXTRACTOR DESIGN CONSIDERATIONS

In selecting extraction equipment and designing a commercial extraction operation, many items need to be taken into account, including raw material issues, the extraction parameters necessary to achieve economic recovery, the size of the extraction operation, and the product mix. These are discussed below.

11.4.1 RAW MATERIAL ISSUES

Proper handling of botanical raw material can make or break an extraction operation. Each of the items discussed in the following sections must be considered in designing a commercial phytochemical extraction operation.

11.4.1.1 Raw Material Procurement and Storage

One of the more important criteria for operating a nutraceuticals extraction plant is obtaining raw material that meets the requirements of the process to produce the desired products. Many important issues need to be considered when purchasing and processing raw material.

The nutraceutical industry is moving more and more toward selling "standardized extracts," or dried extract product that is manufactured to a specified phytochemical concentration (also known as marker concentration). A mass balance, coupled with an economic evaluation of the process, will quickly show the importance of processing raw material with a high phytochemical content to keep manufacturing costs low. If the marker content is too low, it may be impossible to meet the desired marker specification for a standardized product. The marker concentration can vary widely from one lot to the next, depending on where the plant was grown, when it was harvested, and how it was dried (Bonati, 1980b). Sampling and testing incoming raw material for marker helps to ensure that high-quality raw material is used in the preparation of extracts.

Once the raw material arrives at the manufacturing facility, it needs to be inspected before use to ensure that the proper plant species has been purchased. Additional inspection is required to assure that there is no contamination from insects and rodents, mold and fungal growth, foreign plant material, and gross contamination such as stones, sticks, and plastic. Raw material also needs to be tested for heavy metals and herbicide/pesticide contamination to verify that the product is not contaminated with these chemicals. If a kosher or organic product is desired, raw material having certifications from the appropriate agencies is needed.

Adequate warehousing facilities are needed to store the raw material before extracting. Segregation of different types of raw materials is sometimes needed to prevent cross-contamination. For example, odorous botanicals such as valerian and

TABLE 11.3
Effects of Age and Location on Ginsenosides (%)
Content in *Panax ginseng*

Age of Plants	Ji-Lin Region (China)	Tong-Hua Region (China)	Japan
2	2.0	4.8	3.0
3	2.2	5.6	2.0
4	4.7	6.2	3.5
5	4.6	6.3	2.5
6	4.3	6.4	2.5

Source: From Li and Wang (1998).

peppermint need to be kept away from other types of raw materials. The storage facilities need to be operated to prevent rodent and insect infestations. Some companies treat their incoming raw material with pressurized carbon dioxide to prevent insect hatches from occurring during storage. Temperature and humidity need to be controlled to prevent phytochemical degradation and to prevent the growth of fungus and mold on the raw material.

11.4.1.2 Marker Variability in Raw Materials

The marker content in a given raw material will vary widely due to the location at which it is grown, the weather during the growing season, the time of the year that the raw material is harvested, and how it is dried, as well as genetic variation and age from one plant to the next. An example of this for *Panax ginseng* roots can be seen in data reported by Li and Wang (1998) in Table 11.3.

A proper sampling technique, such as that recommended by the American Spice Trade Association (1995) for sampling cassia and cinnamon, ensures that a representative sample is collected from a given lot. However, even with proper sampling techniques, wide variation in marker content will be found from one sample of raw material to the next, as can be seen in Table 11.4 for St. John's wort herb, valerian roots, and *Echinacea purpurea* roots. Analyzing separate 500-g samples from the same lot, where each sample was finely ground then split to the proper analytical size with a riffle sampler, generated these data. Variation such as seen in Table 11.4 can be due to the marker residing in certain portions of the raw material, such as alpha acids found in the lupulin glands of hops (Grant, 1977). And, it can be due to natural variation that is found from one plant to the next. As the economics of an extraction operation are greatly affected by the marker assay of the material extracted, proper methodology for obtaining the actual assay of a given lot must be followed. This involves using statistics to obtain a representative sample for analysis.

From knowledge of particle sampling (see Gy, 1982), it is critical that a sufficiently large sample be collected from a raw material lot. As it is generally accepted that any chemical compound of interest will be normally distributed throughout the

TABLE 11.4
Variation in Marker Assays within the Same Raw Material Lot

Sample Number	St. John's Wort % Hypericins	Valerian % Valerenic Acids	*Echinacea purpurea* % Total Phenols
1	0.109	0.262	1.95
2	0.149	0.342	3.74
3	0.109	0.396	1.47
4	0.120	0.181	2.33
5	0.118	0.219	3.75
6	0.147	0.156	3.79
Mean	0.125	0.259	2.84
Standard deviation	0.018	0.094	1.05

Data generated by Botanicals International Extracts, Boulder, Colorado, 1998.

solids within a lot, a sample size that is too small can result in an analysis that greatly differs from the average value of the lot, as Table 11.4 illustrates.

The problem of determining the sample size that will give statistically significant analytical results has also been faced by the mining industry, as most ores are quite nonhomogeneous. The method suggested by Taggart (1948) for mineral ore samples has been successfully adapted for sampling plant materials. This method uses multiple analyses of the same raw material lot, such as found in Table 11.4, and the Student's t statistic to calculate the number of samples needed to assure a given analytical error at some designated confidence interval. Based on the number of samples calculated and the amount of material that was used for each analysis, the total sample weight is calculated. These equations are given below.

$$\sigma_S = \sqrt{\frac{\sum (a_i - a)^2}{n}} \tag{11.3}$$

$$W = w \left[\frac{t\sigma_S}{za_i} \right]^2 \tag{11.4}$$

where: σ_s is the analytical standard deviation of all of the samples
a_i is the analytical average (mean) of all of the samples
a is the individual sample analysis
n is the number of samples analyzed
t is the Student's t statistic, which is listed in Table 11.5 for various confidence levels
z is the acceptable error in the analysis
w is the weight of each individual sample
W is the total sample weight

TABLE 11.5
Values for Student's _t_

Confidence Level	Student's _t_
0.90	1.645
0.95	1.960
0.98	2.326
0.99	2.576
0.999	3.291

Using the St. John's wort values in Table 11.4, the above equations can be used to calculate the sample size needed to give a 10% error ($z = 0.10$) in the analysis at a confidence level of 0.90 ($t = 1.645$ from Table 11.5). From the data in Table 11.4,

$$a_i = 0.125$$

$$\sum (a_i - a)^2 = 0.00165$$

$$n = 6$$

$$w = 500 \text{ grams}$$

Using these values, the total sample size for this raw material is 2.4 kg to be assured that the sample analysis is within 10% of the true assay of the lot 90% of the time. This entire 2.4-kg sample would need to be finely ground, and a representative sample for the marker analysis would need to be split out using a riffle splitter. These calculations also show that a sufficiently large preship sample must be obtained from the raw material vendor to accurately measure the expected marker content in a given lot.

11.4.1.3 Grinding Plant Materials

For solid-liquid extraction, the raw material needs to be ground to the proper size for the extraction equipment used. This shortens the path of diffusion, reducing the required extraction time. In grinding plant materials, the control of fines is very important for many operations. Too many fines can prevent percolation-type extractors from operating properly. In addition, fine material can be carried in the product extract, necessitating their removal by filtration or by using other solid-liquid separation equipment.

Generally, a cutting-type mill, such as that shown in Figure 11.15, is used in the comminution of the raw material to the proper size. With this type of mill, the raw material is cut between rotating and stationary blades until the particles are small enough to pass through the screen at the bottom of the mill. The screen size is chosen to control the resulting particle size. This type of mill produces coarser particle sizes with a minimum of fines, giving a particle size distribution suitable

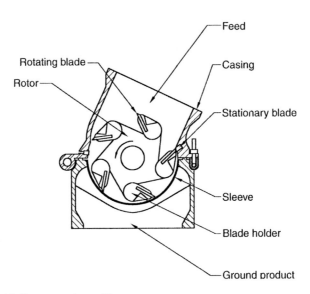

FIGURE 11.15 Rotary cutting mill.

for many types of extraction equipment. In instances when a large amount of fines are generated, even with careful grinding, these fines must be handled separately by agglomerating into pellets or extracting separately. If a very fine grind size is needed to achieve the desired extraction recovery, a hammer mill may produce a more desirable grind size. In some instances, such as extracting oilseeds, the cell walls must be ruptured in order for the solvent to penetrate the cells. In these cases, flaking equipment is used. Hard seeds, such as evening primrose, require cracking to expose the inside of the seed to the solvent.

11.4.1.4 Particle Size Distribution

Knowing and controlling the particle size distribution of the raw material is very important for the proper operation of extraction equipment. For percolation operations, too many fines can cause flow problems. For other raw materials, too coarse a grind size can result in marker recovery losses with detrimental economic consequences. One method of graphically expressing the particle size distribution of ground raw material is by using the method outlined by Falivene (1981). The particle size distribution is determined from a sieve analysis, and the results are plotted on log-probability paper as the logarithm of particle size (mesh size) vs. the cumulative weigh percent passing through that mesh size on a probability scale. In many instances, a naturally ground plant material will plot nearly linear using this method. Figure 11.16 shows a typical raw material particle size for both conventional percolators and for dispersed-solids extractors.

Plotting the particle size distribution in this manner allows one to easily determine the mean particle size, where the cumulative weight percent equals 50%. This particle size is known as the P_{50}. By determining two other points, the P_{10} and P_{90}, or the sieve size, through which 10% and 90% of the ground material will pass

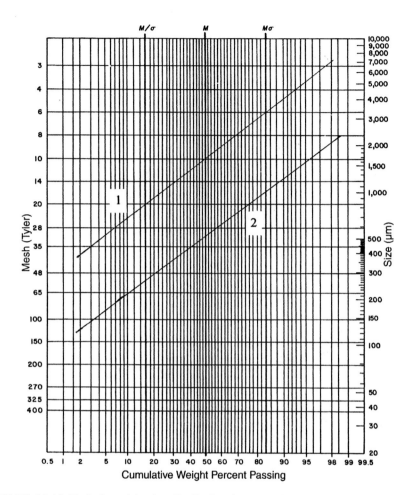

FIGURE 11.16 Typical particle size distribution for extraction: (1) percolation extraction and (2) dispersed-solids extraction.

through, respectively, the entire particle size distribution can be described. A steep slope indicates a wide particle size distribution, while a shallow slope shows a narrow particle size distribution. Extraction operation is much easier when the particle size distribution is narrow, and raw material grinding should be adjusted to achieve this goal. In addition, using log-probability paper allows one to compare multiple sieve analyses that used different mesh sizes. Also, a bimodal distribution on the log-probability graph indicates that the ground raw material is a blend of coarse and fine material, which may hamper extractor operation.

11.4.1.5 Marker Variation as a Function of Grind Size

Some raw materials show considerable marker variation with grind size, whereas others do not. Some examples of this can be found in Table 11.6. From these data, it can be seen that St. John's wort raw material shows a much higher concentration

TABLE 11.6
Marker Assay (%) of Various Screen Fractions

Screen Fraction	St. John's Wort Herb % Hypericins	*Echinacea* Aerials % Total Phenols	Kava Roots % Kavalactones
+ 16 mesh	0.054	2.41	6.15
16 × 50 mesh	0.099	2.35	7.41
– 50 mesh	0.344	2.27	6.62

Data generated by Botanicals International Extracts 1998.

of marker in the finer material, while kava and *Echinacea* aerials show less variation with grind size. For those raw materials that show significant variation with particle size, care must be taken in the conveying of the ground raw material to prevent segregation by particle size, or a large variation in marker content will result.

11.4.2 EXTRACTOR DESIGN

11.4.2.1 Preliminary Tests

Relationships between three extraction parameters—solvent composition, temperature, and particle size—can be determined from simple beaker tests. Raw material ground to the desired particle size is contacted with the chosen solvent at the chosen temperature in a beaker and mixed over time. Thief samples are removed at various time intervals and analyzed for the marker. At the completion of the test, the marc is separated from the extract, dried, and analyzed. (In instances when the phytochemical is heat sensitive, the marc can be analyzed wet and the assay calculated on a dry basis). Based on this information, the amount of marker remaining in the dry marc can be determined after correcting for the marker in the absorbed extract in the wet marc. From these data, the equilibrium relationship defined in Equation 11.1 can be determined, as well as the time it takes to reach maximum extraction. Conditions are changed to give a high equilibrium constant and a short extraction time to reach equilibrium. Based on this information, one can select the appropriate extraction equipment as well as the operating conditions to operate the equipment.

A typical extraction rate curve from Metivier et al. (1980) is seen in Figure 11.17, showing the extraction of anthocyanins from ground (–5.5 mm or –1/4 inch) wine pomace using 1% hydrochloric acid in ethanol at room temperature. The data show that approximately 11 hours are required for the anthocyanins to reach equilibrium concentrations between the anthocyanins in the liquid and those remaining in the solid. Using the data in this reference, an equilibrium constant of 0.0047, as defined in Equation 11.1, can be calculated. When extracting St. John's wort at a weighted mean particle size of 0.9 mm with 70% ethanol at 55°C, a much more rapid extraction is seen (Figure 11.18). As is typical with extracting many raw materials, the equilibrium constant for this system is fairly large at 0.26, indicating that nearly all of the marker extracts into the liquid at high liquid-to-solid ratios.

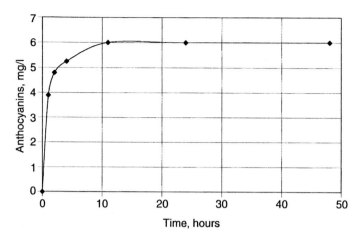

FIGURE 11.17 Extraction of anthocyanins from grape pomace at room temperature with ethanol containing 1% HCl. (From Metivier et al., 1980)

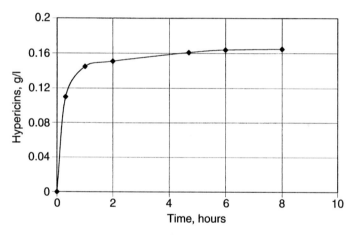

FIGURE 11.18 Extraction of hypericins from St. John's wort herb at 55°C with 70% ethanol. (Data generated by Botanicals International Extracts. Boulder, Colorado, 1999)

11.4.2.2 Design of Dispersed Solids Equipment

Dispersed solids equipment is generally operated in a batch mode, contacting solvent and the raw plant material together for sufficient time that equilibrium concentration in the extract is approached. A mass balance on the system gives the following:

$$W_r C_r + W_s C_s = W_e C_e + W_m C_m \tag{11.5}$$

where W_r, W_s, W_e, and W_m are the weights of raw material, solvent, extract, and wet marc, respectively, and C_r, C_s, C_e, and C_m are the concentrations (weight fractions) of the phytochemical in the raw material, solvent, extract, and wet marc, respectively.

When using fresh solvent, the weight fraction of the phytochemical in the solvent is zero. Complicating the use of Equation 11.5 is the fact that the marc contains unextracted phytochemical within the solids, plus phytochemical in the extract within the pore structure of the marc. Thus, the phytochemical concentration within the marc can be expressed as follows:

$$C_m = \frac{W_{dm}C_{dm} + W_{em}C_{em}}{W_m} \tag{11.6}$$

where W_{dm} is the weight of dry marc, W_{em} is the weight of extract within the marc, C_{dm} is the phytochemical concentration of the dry marc, and C_{em} is the phytochemical concentration of the extract within the marc.

At equilibrium, the concentration of the extract within the marc is the same as the concentration of the bulk extract. Also, defining X as the weight fraction of solvent in the marc, Equation 11.6 can be simplified to the the following:

$$C_m = (1 - X)C_{dm} + XC_e \tag{11.7}$$

Adding the equilibrium relationship of Equation 11.1

$$K = \frac{C_e}{C_{dm}} \tag{11.1}$$

to Equation 11.7, the concentration of the wet marc can be given by

$$C_m = C_e\left[\frac{1 - X}{K} + X\right] \tag{11.8}$$

Combining Equation 11.8 with the mass balance in Equation 11.5, the extract concentration at equilibrium for dispersed-solids extraction can be calculated from easily measured parameters:

$$C_e = \frac{W_r C_r}{W_e + W_m\left[\dfrac{1 - X}{K} + X\right]} \tag{11.9}$$

For multiple extractions using fresh solvent, Equation 11.9 can be modified to calculate the expected extract concentration from the nth wash, as follows:

$$C_{e,n} = \frac{W_{m,n-1}C_{m,n-1}}{W_{e,n} + W_{m,n}\left[\dfrac{1 - X}{K} + X\right]} \tag{11.10}$$

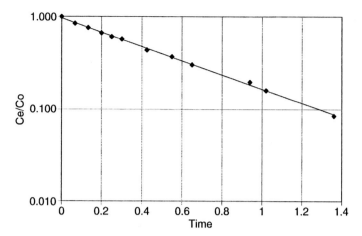

FIGURE 11.19 Percolation extraction of sucrose with water from sucrose-impregnated coffee grounds. (From Schwartzberg et al., 1982.)

Using these equations, it is possible to use the simple beaker test data to calculate expected phytochemical extract concentrations and recoveries at the temperature and particle size of the beaker test for various liquid-to-solids ratios for each successive extraction with fresh solvent.

11.4.2.3 Design of Percolation Equipment

When fresh solvent is pumped through a bed of raw plant materials, the extract leaving the percolator will initially be quite high then drop off in concentration. In many instances, the drop in extract concentration over time will display a relationship similar to that shown in Figure 11.19 that Schwartzberg et al. (1982) presented for extracting sucrose with water from coffee grounds that had been impregnated with sucrose. Schwartzberg (1987) presents a generalized expression for this percolation extract concentration over time, which can be rearranged to the following:

$$C_e = C_o \exp\left[-\frac{C_o \varepsilon}{C_r(1-\varepsilon)}t \right] \qquad (11.11)$$

where C_o is the initial concentration of the first extract leaving the percolator and t is the time that extract began to leave the extractor.

This equation can be used to scale small-scale percolation tests to larger size. The preliminary tests are used to set the solvent composition, extraction temperature, and raw material grind size. The extraction time is varied by changing the flow rate of the solvent through the bed of raw material. Using these parameters, a small diameter pilot percolator is tested to determine the parameters in Equation 11.11. From these data, commercial percolator operation can be predicted.

Percolation extractions require that the solvent flow through a bed of solids. The flow of liquid through a bed of solids can be described by the Kozeny-Carmen equation (McCabe and Smith, 1976), which relates the pressure drop through a bed of solids to the flow rate and properties of the solid and liquid as follows:

$$150 = \frac{\Delta P g_c \phi_S^2 D_{pm}^2 \varepsilon^3}{L v_o \mu (1-\varepsilon)^2} \tag{11.12}$$

where: ΔP is the pressure drop through the bed
  g_c is the gravitational constant = 32.2 $lb_m ft/lb_f$ sec^2
  ϕ_s is the sphericity or shape factor examples can be found in Perry et. al (1984), p. 5–54
  D_{pm} is the mean effective diameter, as defined below
  ε is the fractional void volume of the bed
  L is the bed height
  v_o is the superficial velocity, defined as the volumetric flow rate divided by the cross-sectional area
  μ is the fluid viscosity

Units are chosen such that each side of the equation is dimensionless. The mean particle size is defined as follows:

$$D_{pm} = \frac{1}{\sum \frac{x_i}{\phi_s D_i}} \tag{11.13}$$

where x_i is the fractional volume of particles of size D_i.

The Kozeny-Carmen equation can be rearranged to give the pressure drop through the bed as a function of liquid flow rate and the liquid flow rate as a function of pressure drop through the bed:

$$\Delta P = \frac{150 L v_o \mu (1-\varepsilon)^2}{g_c \phi_S^2 D_{pm}^2 \varepsilon^3} \tag{11.14}$$

$$v_o = \frac{\Delta P g_c \phi_S^2 D_{pm}^2 \varepsilon^3}{150 L \mu (1-\varepsilon)^2} \tag{11.15}$$

In those cases in which fluid flow is due to gravity, ΔP is related to the fluid density and the height of the column (when the column is full of liquid):

$$\Delta P = \frac{\rho g L}{g_c} \tag{11.16}$$

where ρ is the fluid density and g is the gravitational acceleration = 32.2 ft/sec^2.

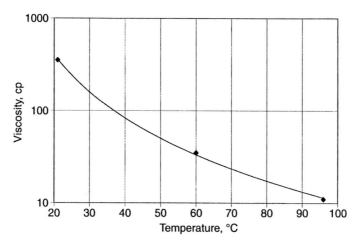

FIGURE 11.20 Viscosity of black tea extract as a function of temperature. (Data generated by Botanicals International Extracts. Boulder, Colorado, 1999)

Substituting this into the above equation, the flow rate is given by the following:

$$v_o = \frac{\rho g \phi_s^2 D_{pm}^2 \varepsilon^3}{150 \mu (1-\varepsilon)^2} \qquad (11.17)$$

It is interesting to note that the length of the bed does not affect the flow rate when the flow is due only to gravity.

Although using these equations to predict flow in percolators is impractical, as determining some of the constants is difficult, these equations show the important variables to control to provide adequate flow through a percolation bed.

- Void volume is the most critical parameter, as the flow rate is proportional to the cube of the void volume and inversely proportional to the square of one minus the void volume. As an example, if the void volume is increased from 0.4 to 0.6, the flow rate will increase by eight times.
- Particle size also greatly affects flow rate, which is proportional to the square of particle size. Larger particle sizes should be considered in cases where the flow rate is too low.
- Fluid viscosity is inversely proportional to flow rate. However, viscosity is strongly affected by temperature, as can be seen in Figure 11.20, showing the viscosity-temperature relationship of a black tea extract. Increasing temperature will greatly improve the flow through a packed bed by significantly decreasing viscosity. Offsetting this is the increased extraction rate at higher temperatures, which increases the amount of dissolved solids in the extract, increasing viscosity.
- Bed length is important when operating under pressure, and a longer bed gives lower flow. Length is directly proportional to flow rate for a fixed feed pressure. Bed length is not a factor when using gravity flow.

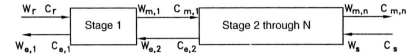

FIGURE 11.21 Strategy for determining the number of theoretical stages in countercurrent extraction.

- Fluid density becomes important when flowing by gravity. However, the variables affecting density in solid-liquid extraction—temperature and solids concentration—do not greatly affect the fluid density.

When operating percolators in recirculation mode, where the solvent is allowed to approach equilibrium concentrations, Equations 11.5 through 11.10 that were developed for dispersed solids extraction can also be used for percolator design.

11.4.2.4 Countercurrent Operation

In designing countercurrent extraction, the number of stages becomes an important design parameter. This can be determined by using techniques similar to those used for other mass transfer operations that involve equilibrium stages, such as distillation or liquid-liquid extraction. McCabe and Smith (1976) outline the methodologies that can be used. With constant underflow, where the weight of wet marc does not significantly change from one extractor to the next, McCabe-Thiele diagrams are very useful in determining the number of extractors. Experience has shown that the assumption of constant underflow is applicable to extracting many phytochemicals from plant materials, once the raw material is wetted with solvent.

As many situations start with dried raw material, the assumption of constant underflow does not hold for the first stage, when the plant material becomes soaked with extract. In this case, the first stage is handled separately as a mass balance, while the remaining stages utilize a McCabe-Theile diagram to determine the number of stages. This strategy is illustrated in Figure 11.21. McCabe-Thiele diagrams graphically determine the number of stages by stepping off the number of "steps" between an equilibrium curve and an operating line. This is illustrated in the example below.

In many cases, the equilibrium between the concentrations in the solids and liquid is constant throughout the range of interest, and the equilibrium curve is a straight line. The slope of the equilibrium line is related to the equilibrium constant defined in Equation 11.1 and the amount of liquid the marc retains by the following relationship:

$$K' = \frac{C_e}{C_m} = \frac{K}{1-(1-K)X} \tag{11.18}$$

where K' is the slope of the equilibrium line; K is the equilibrium constant, defined in Equation 11.1; and X is the weight fraction of liquid retained by the marc. With constant underflow, the slope of the operating line is the ratio of liquid-to-solids, or

$$m = \frac{W_e}{W_m} \qquad (11.19)$$

Conversely, the liquid-to-solids ratio can be determined by defining the inlet and outlet concentrations for stages 2 through n and determining the operating line between these two points.

Two constraints limit the conditions of countercurrent extraction:

(1) The marker concentration of the extract leaving the last stage cannot exceed the extract concentration that is in equilibrium with the raw material marker concentration.
(2) The liquid-to-solids ratio cannot be less than the weight fraction of liquid retained in the marc, or

$$m \geq \frac{X}{1-X} \qquad (11.20)$$

Also, for dispersed-solids extraction, the liquid-to-solids ratio will need to be higher than that calculated in Equation 11.20, as sufficient solvent is needed to keep the raw material solids suspended in the solvent.

Once the number of stages is determined, this can be related to the number of dispersed-solids mixers or recirculating percolators that would be required. For economic reasons, the stages are not allowed to come to equilibrium in many instances, and the stage efficiency is less than 100%. Thus, more stages are required than the theoretical number. The calculations also determine the liquid-to-solids ratio required. From the beaker tests, the time required to approach equilibrium is known, setting the extraction time for batch operations and relative flow rates for continuous operation. For countercurrent extraction technologies that are continuous (rather than having discrete stages), such as the Crown and Niro extractors, pilot testing is needed to determine the length of a theoretical stage. Once this is determined, then the total length of extractor can be calculated.

As an illustration, consider the extraction of hypericins from St. John's wort herb shown in Figure 11.18. This extraction test gives an equilibrium constant of 0.26 and 70% solvent remaining in the wet marc. If we assume that the raw material contains 0.15% hypericins, and the incoming solvent has no hypericins, then:

- $C_r = 0.0015$ g hypericins/g raw material
- $C_s = 0$ g hypericins/g solvent
- $X = 0.70$ g extract/g wet marc

How many stages are required to achieve 95% hypericins extraction with a liquid-to-solids ratio of 10?

From this information and assuming a basis of 100 kg/hr of raw material fed into the extractor, the following can be calculated:

- $K' = 0.589$, using the values of $K = 0.26$ and $X = 0.70$
- $W_s = (100$ kg/hr raw material$)(10$ kg solvent/kg raw material$) = 1000$ kg/hr solvent
- $W_m = (100$ kg/hr raw material$)/[(1 - 0.7)] = 333$ kg/hr marc

From an overall mass balance, the flow of extract through the extractor is as follows:

- $W_e = W_r + W_s - W_m = 100 + 1,000 - 333 = 767$ kg/hr extract

As a 95% recovery is achieved, the extract concentration out of the first stage, $C_{e,1}$, and the marc concentration out of the last or nth stage, $C_{m,n}$, is found from the following:

- $C_{e,1} = (100$ kg/hr raw material$)(0.0015$ kg hypericins/kg raw material$)(0.95)/(767$ kg/hr extract$) = 0.000186$ kg hypericins/kg extract from the extractor
- $C_{m,n} = (100$ kg/hr raw material$)(0.0015$ kg hypericins/kg raw material$)(0.05)/(333$ kg/hr marc$) = 0.0000225$ kg hypericins/kg wet marc

In the first stage, the raw material is wetted with incoming extract. The hypericins concentration in the marc leaving this stage, $C_{m,1}$, is given by Equation 11.18, and the concentration of the extract entering this stage, $C_{e,2}$, is calculated from a hypericins balance around this stage:

- $C_{m,1} = (0.000186)/(0.539) = 0.000345$ kg hypericins/kg wet marc
- $C_{e,2} = [(767)(0.000186) + (333)(0.000345) - (100)(0.0015)]/(767) = 0.000140$ kg hypericins/kg extract

For the remaining stages, the McCabe-Thiele diagram shown in Figure 11.22 graphically determines the number of stages. The slope of the equilibrium line is determined from Equation 11.18 as 0.589 and is plotted such that the line passes through the origin. The operating line is plotted where the line passes through the points $(C_{m,n}, C_s)$ or $(0.0000225, 0.00)$ and $(C_{m,1}, C_{e,2})$ or $(0.000345, 0.000140)$. Stepping off the stages between the operation and equilibrium lines, seven stages (including the first stage that was calculated separately) are required to achieve a 95% recovery of the hypericins.

11.4.2.5 Solvent Removal from the Marc

One further consideration in selecting extraction equipment is determining how the solvent is separated from the marc. Unless the solvent is water, the marc will need to be essentially solvent-free for disposal. Also, economics will dictate that solvents

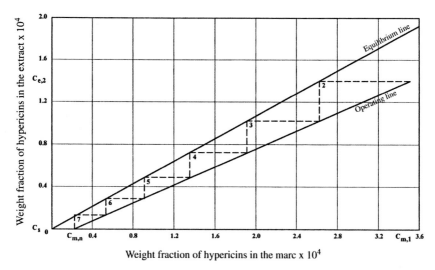

Weight fraction of hypericins in the marc x 10^4

FIGURE 11.22 McCabe-Thiele diagram for determining the number of theoretical stages for the St. John's wort example.

besides water need to be reused. For some equipment, the solvent-soaked marc is removed from the extractor, and separate equipment is used for solvent removal. In other equipment, both extraction and solvent removal take place in the same equipment. The advantage of performing solvent removal in the same equipment is that less equipment must be purchased. For larger extraction operations in which the extraction equipment cost is much higher than the solvent removal equipment, economics dictate that separate equipment for solvent removal be purchased so that the more expensive extraction equipment is used full time for extraction.

Two methods for solvent removal are used. In most instances, the marc is contacted with steam, which vaporizes the solvent from the marc. The vaporized steam is condensed, and the solvent is reclaimed from the condensate. For water-miscible solvents, the marc can also be washed with water. The solvent can again be reclaimed from the wash in thermal equipment described below. One drawback of washing with water is that a significant amount of solids will dissolve in the water fraction. This can cause problems with solvent recovery.

11.4.2.6 Solvent Recovery for Reuse

Economics dictate that the solvent must be recovered and reused in a nutraceutical extraction operation. Solvent wastes come from removing solvent from the marc and from subsequent downstream processing of the extract to a paste or solid product.

In the case in which a water-miscible solvent is used, such as ethanol, methanol, or acetone, fractional distillation equipment will be required. This type of equipment also separates any dissolved solids in the wastes, such as those generated when water washing the marc. One potential problem with significant dissolved solids in the distillation feed is that these solids can precipitate out in the distillation equipment as the organic solvent is removed, potentially plugging the distillation column. The still bottoms, containing water and solids from the solvent wastes, can usually be disposed of in the plant sewer effluent after pH adjustment and treatment for biological oxygen demand (BOD).

When the solvent is not miscible with water, such as hexane or ethyl acetate, water is easily removed from the solvent by decantation. In most instances, the decanted solvent can be directly reused in extraction. The decanted solvent will contain little dissolved solids, as this solvent stream would have been condensed from thermal processes.

The equipment for distillation and other thermal processes and information on designing this type of equipment can be found in Perry et al. (1984) and Smith (1971).

11.5 CONCLUSION

The selection of the proper extraction equipment type is governed by economics, the chemistry of the phytochemical of interest, and the physical properties of the botanical to be extracted. Each situation is different, and a careful evaluation of extraction needs should be made. Areas to consider include the following:

- Extraction parameters—From the relationship between extraction time and solvent composition, temperature, liquid-to-solids ratio, and particle size, certain types of equipment may not be suitable to meet these needs. For instance, if a fine particle size is needed, percolation equipment is not feasible.
- Desired throughput—The discussion above gives a range of capacities for each type of equipment. Equipment selection will be governed by the desired extraction capacity.
- Single product vs. multiple product operation—Equipment with more internal parts takes more time to clean when changing from one product to the next. Ease of switching from one product to the next should enter into equipment selection.
- Frequency of product change-out—If multiple product operation is planned, the frequency of changing from one product operation to the next will also dictate the appropriate equipment. Again, cleaning between runs needs to be considered.
- Labor costs vs. capital costs—Some types of equipment are very labor intensive, but the capital costs for installation are relatively cheap. Other types of equipment have low labor requirements but are more expensive. The trade-offs between labor and capital costs need to be carefully analyzed when selecting the appropriate equipment.

11.6 NOTATION

a—average analysis of multiple samples
a_i—individual sample analysis
C—concentration

C_{dm}—concentration of the dry marc, weight percent
C_e—concentration of the extract, weight percent
C_{em}—concentration of the extract within the marc, weight percent
C_m—concentration of the wet marc, weight percent
C_o—initial concentration
C_r—concentration of the botanical raw material, weight percent
C_s—concentration of the solvent, weight percent
D—diffusion coefficient, length^2/time
D_{pm}—mean effective particle diameter
g—gravitational acceleration
g_c—gravitational constant, 32.2 lb_m-ft/lb_f-sec^2
K—equilibrium constant, unitless
K′—slope of the equilibrium line in a McCabe-Thiele diagram
L—bed length
m—slope of the operating line in a McCabe-Thiele diagram
n—number of samples or number of extractors
ΔP—pressure drop
t—time or Student's *t* statistic
v_o—superficial velocity
w—individual sample weight
W—total sample weight
W_{dm}—weight of the dry marc
W_e—weight of the extract
W_{em}—weight of the extract within the marc
W_m—weight of the wet marc
W_r—weight of the botanical raw material
W_s—weight of the solvent
x—particle diameter
X—weight fraction of liquid retained in the marc
z—acceptable sample analytical error

Greek Symbols

ε − fractional void volume within a bed
φ_s—sphericity or shape factor of a particle
μ—fluid viscosity
ρ—fluid density
σ_s—analytical standard deviation of multiple samples

REFERENCES

American Spice Trade Association. 1995. "Raw Cassia and Cinnamon Sampling and Testing," Technical Bulletin 950522.
Bombardelli, E. 1986. "Technologies for the Processing of Medicinal Plants," in *The Medicinal Plant Industry*, ed. Wijesekera, R.O.B., CRC Press, Boca Raton, FL.

Bonati, A. 1980a. "Problems Relating to the Preparation and Use of Extracts from Medicinal Plants," *Fitoterapia,* 51:5–11.

Bonati, A. 1980b. "Medicinal Plants and Industry," *J. Ethnopharmacol.,* 2:167–171.

Crank, J. 1956. *The Mathematics of Diffusion,* Oxford Press, Oxford, UK.

Crippa, F. 1978. "Problems of Pharmaceutical Techniques with Plant Extracts," *Fitoterapia.* 49(6):257–263.

Falivene, P.J. 1981. "Graph Paper for Sieve Analyses," *Chem. Eng.,* February 25, 1981, p. 87.

Frank, T.C., Downey, J.R., and Gupta, S.K. 1999. "Quickly Screen Solvent for Organic Solids," *Chem. Eng. Prog.,* December 1999, pp. 41–61.

Gertenbach, D. 1998. "How to Establish a Solid-Liquid Extraction Facility for Nutraceuticals," Annual IFT Meeting, June 1998.

Grant, H.L. 1977. "Hops," in Broderick, A.M., *The Practical Brewer, A Manual for the Brewing Industry,* Master Brewers Association of the Americas, pp. 128–146.

Gy, P.M. 1982. *Sampling of Particulate Materials, Theory and Practice,* Elsevier Scientific Publishing Company, New York, NY.

Li, T.S.C. and Wang, L.C.H. 1998. "Physiological Components and Health Effects of Ginseng, *Echinacea,* and Sea Buckhorn" in Mazza, G., *Functional Foods, Biochemical and Processing Aspects,* Technomic Publishing Co., Inc., Lancaster, PA, p. 334.

List, P.H. and Schmidt, P.C. 1989. *Phytopharmaceutical Technology,* CRC Press, Boca Raton, FL.

McCabe, W.L. and Smith, J.C. 1976. *Unit Operations in Chemical Engineering,* Third Edition, McGraw-Hill Book Company, New York, NY, pp. 146, 610–613.

Metivier, R.P., Francis, F.J., and Clydesdale, F.M. 1980. "Solvent Extraction of Anthocyanins from Wine Pomace," *J. Food Sci.,* 45:1099–1100.

Perry, R.H., Green, D.W., and Maloney, J.O. 1984. *Perry's Chemical Engineers' Handbook,* Sixth Edition, McGraw Hill, Inc., New York, NY.

Plachco, F.P. and Lago, M.E. 1975. "Solid-Liquid Extraction in Countercurrent Cascade with Retention of Liquid by the Solid-Slab Geometry-Constant Diffusivity," *Ind. J. Technol.,* 13:438–445.

Plachco, F.P. and Lago, M.E. 1976. "Successive Solid-Liquid Extraction with Retention of Liquid by the Solid Constant Diffusivity-Infinite Slab Geometry," *Chem. Eng. Sci.,* 31:1085–1090.

Plachco, F.P. and Lago, M.E. 1978. "Successive Solid-Liquid Extraction with Retention of Liquid by the Solid Constant Diffusivity-Infinite Cylinder Geometry," *Ind. J. Technol.,* 16:215–222.

Schwartzberg, H.G. 1980. "Continuous Counter-Current Extraction in the Food Industry," *Chem. Eng. Prog.,* April 1980, pp. 67–85.

Schwartzberg, H.G. 1987. "Leaching—Organic Materials," in Rousseau, R.W., *Handbook of Separation Process Technology,* John Wiley and Sons, Inc., New York, NY.

Schwartzberg, H.G. and Chao, R.Y. 1982. "Solute Diffusivities in Leaching Processes," *Food Technol.,* February 1982, pp. 73–86.

Schwartzberg, H.G., Torres, A., and Zaman, S. 1982. "Mass Transfer in Solid-Liquid Extraction Batteries," Food Process Engineering AIChE Symposium Series No. 218, Vol. 78, pp. 90–101.

Smith, J.C. 1971. *Separation Processes.* McGraw-Hill Book Company, New York, NY.

Taggart, A.F. 1948. *Handbook of Mineral Dressing,* John Wiley and Sons, Inc., New York, pp. 19–12.

12 Safety of Botanical Dietary Supplements

Joseph M. Betz, Tam Garland and Samuel W. Page

CONTENTS

12.1 INTRODUCTION

Plants have been used medicinally for thousands of years, but in some parts of the world, this use has diminished as single-entity pharmaceuticals have come to dominate the marketplace (Croom and Walker, 1995). Recent disillusionment with the conventional Western medical establishment coupled with the perception that traditional and natural ingredients are inherently safer and more healthful than are synthetic ingredients has led to a resurgence in the popularity of herbs (Der Marderosian, 1977). In certain parts of the world, continued use of medicinal plants is based on expense of synthetic drugs and consequent unavailability (Vollmer et al., 1987; Bakhiet and Adam, 1995). In developed nations, many consumers look to herbs as safer, more natural alternatives to synthetic or highly purified drugs. Unfortunately, historical use does not always guarantee safety, and many of the herbs used medicinally are also mentioned in monographs on poisonous plants.

In the United States, the Dietary Supplement Health and Education Act (DSHEA) of 1994 (United States Public Law 103-417) amended the Federal Food, Drug and Cosmetic Act by defining as a dietary supplement any product (other than tobacco) that contains a vitamin, mineral, herb or other botanical, or amino acid and is intended as a supplement to the diet. Food is considered safe by definition, but new foods or food additives must be demonstrated to be generally recognized as safe (GRAS) before they are permitted onto the U.S. market. Dietary supplement constituents (dietary ingredients) that were marketed in the United States prior to 15 October 1994 may remain on the market unless proven unsafe by the U.S. Food and Drug Administration (FDA). Dietary ingredients are considered unsafe if they "present a significant or unreasonable risk of illness or injury under conditions of use recommended or suggested in labeling, or ... under ordinary conditions of use." The rationale for this "grandfathered ingredient" provision of the law was the recognition that most of the ingredients already being sold had a long history of safe use by humans. As a safeguard, the law gave the FDA authority to remove unsafe products from the market and to perform premarket safety reviews of new dietary ingredients (those introduced after 15 October 1994). Passage of the DSHEA in 1994 was interpreted as deregulation of the dietary supplement marketplace. However, companies are still prohibited from marketing unsafe products and are expected to have data to document safety, and unsafe products are subject to recall. New dietary ingredients may not be marketed until data that demonstrates the safety of the ingredient are submitted to FDA for review and FDA concurs that the ingredient is safe. Companies are prohibited from making claims that products cure or treat diseases. They are permitted to make statements about the effects of the product on the structure or function of the body, but these claims must be submitted to FDA for approval, and companies are required to have data that substantiate the claims. Finally, product labels are required to be accurate and must contain very specific information. Products with labels that do not conform to the regulations or that contain less than the declared amount of any ingredient (plant material, marker compound, vitamin, etc.) are subject to recall or seizure (Soller, 2000).

12.2 HUMAN HEALTH PROBLEMS CAUSED BY BOTANICALS

12.2.1 RECENT POISONINGS

Despite knowledge that some plants are toxic, human poisonings continue to occur. Examples include intoxications by poison hemlock (*Conium maculatum*) (Frank et al., 1995), deadly nightshade (*Atropa belladonna*) (Hartmeier and Steurer, 1996; Schneider et al., 1996; Southgate et al., 2000), Tung nut (*Aleurites fordii*) (Lin et al., 1996), tea tree (*Melaleuca alternifolia*) oil (Del Beccaro, 1995), *Zigadenus* (Heilpern, 1995), pokeweed (*Phytolacca americana*) (Hamilton et al., 1995), ackee (*Blighia sapida*) (Meda et al., 1999), coyatillo (*Karwinskia humboldtiana*) (Martinez et al., 1998), *Nicotiana glauca* (Sims et al., 1999) and others. Recent retrospectives on human intoxications caused by particular plants and their products include reviews on pennyroyal (*Mentha pulegium* or *Hedeoma pulegioides*) (Anderson et al., 1996), oleander (*Nerium oleander, Thevetia peruviana*) (Langford and Boor, 1996), angel's trumpet (*Datura* spp.) (Greene et al., 1996) and eucalyptus oil (Tibballs, 1995).

Data on human poisoning by plants are collected in a manner that is fundamentally different from that collected on animals. Whereas identification of a potential toxicological hazard to humans or animals may begin with reports of intoxication, toxicity of plants suspected of injuring animals may be confirmed in controlled feeding studies. Experiments designed solely to establish toxicity to humans are ethically problematic, and therefore, the literature remains inconsistent. Hazard to humans from a particular plant species may be predicted by extrapolation from animal data (both controlled feeding studies and poisonings in the field), by examination of the historical record for poisoning episodes in humans and from the practical experience of traditional healers. Reports of veterinary intoxications are valuable, although interspecific differences in susceptibility to particular toxins may make direct predictions of toxicity to humans difficult. Unfortunately, much of the information possessed by traditional healers is unavailable to the greater scientific community or lacks key information such as dose and duration of use. The unfortunate result of this lack of information is that while toxicity of many plants (e.g., *Conium maculatum*) to people has been firmly established, that of others remains surrounded by a fog of weak case reports and misconceptions. Reviews of human case reports such as those recently performed for mistletoe (*Phoradendron flavescens*) (Spiller et al., 1996), poinsettia (*Euphorbia pulcherrima*) (Krenzelok et al., 1996) and yew (*Taxus* spp.) (Krenzelok et al., 1998) provide occasional respite from the confusion, but such systematic reviews are rare.

12.2.2 CIRCUMSTANCES OF INTOXICATION

In addition to any medical, toxicological, or chemical system for classification, most adverse reactions to botanicals may be divided into five general categories based on the circumstances under which the reaction occurred. These categories are misidentification, misuse, deliberate adulteration, inherent toxicity and interaction with pharmaceuticals.

12.2.2.1 Misidentification

Many of the more serious cases of human poisoning associated with plants have been caused by misidentification of plant species (Huxtable, 1990), and most of these have resulted from consumption of misidentified self-collected material. A Washington State couple died after consuming a tea made from *Digitalis* that they had collected in the belief that it was comfrey (Stillman et al., 1977). A man died after he and his brother ingested "ginseng" that turned out to be water hemlock (*Cicuta maculata*) (Sweeney et al., 1994). A couple became intoxicated after consuming a pie made from deadly nightshade (*Atropa belladonna*) berries that they had collected and frozen in the belief that they were bilberries (*Vaccinium* spp.) (Southgate et al., 2000). Two people were poisoned (one fatally) after they had collected and consumed meadow saffron (*Colchicum autumnale*) in the belief that it was wild garlic (*Allium ursinum*) (Klintschar et al., 1999).

Whereas cases such as these are obviously beyond the control of any government entity, educational programs to increase public awareness and improved quality assurance requirements for commercial products may prevent intoxications caused by misidentification. Manufacturers are responsible for assuring that the contents of a package or bottle are accurately represented on the label. Failure to meet this standard causes the product to be adulterated and/or, misbranded and subject to enforcement action by the FDA. A few representative instances of intoxication caused by accidental adulteration of commercial products follow.

Atropine poisoning by "comfrey" tea was reported by Routledge and Spriggs (1989), but comfrey (*Symphytum* spp.) does not contain atropine, and it is likely that the plant material had been misidentified. In Belgium, more than 70 women were afflicted with progressive interstitial renal fibrosis after following a slimming regimen that was purported to include the Chinese herbs *Stephania tetranda* and *Magnolia officinalis* (Vanherweghem et al., 1993). The nature of the nephropathy and the similarity of the Chinese names for *Stephania* (*Fangji*) and *Aristolochia* (*Fangchi*) led investigators to suspect accidental substitution of *Aristolochia fangchi* for *Stephania*. A subsequent analysis identified the known nephrotoxin, aristolochic acid (a natural constituent of several species of *Aristolochia* but not of *Stephania*), in 12 of 13 batches of the herbal material (Vanhaelen et al., 1994). Chinese herb nephropathy has been subsequently described in Taiwan (But and Ma, 1999; Yang et al., 2000). A similar case involving substitution of *Aristolochia manschuriensis* (*Guang mu tong*) for *Akebia* spp. (*Mu tong*) in the United Kingdom has also been reported (Lord et al., 1999). In addition to the devastating rapidly progressing fibrosing interstitial nephritis caused by aristolochic acid, long-term sequelae to the Belgian cases have recently been reported in the form of urethral carcinoma in the affected women (Nortier et al., 2000).

In another highly publicized case, a young Massachusetts woman became ill with a number of symptoms, including abnormal heart rhythm, after ingesting a "cleansing regimen" that consisted of five different products. One of the products was an herbal mixture that contained 15 different species of plant. The label for this product indicated that the primary ingredient was plantain leaf (*Plantago* spp.), but the plantain had been misidentified at its source in Germany, and the product actually contained Grecian foxglove (*Digitalis lantana*). This plant contains over 60 cardiac glycosides, including digoxin and several lanatosides. A sensory evaluation intended to confirm the identity of the plant material when it arrived in the United States failed to detect the substitution. In the end, over 6000 pounds of *D. lantana* were imported over the course of two and a half years. Fifteen companies received recall notices, but the recall eventually affected several hundred manufacturers, distributors, and retailers. No deaths were reported, although there were at least two hospitalizations (Slifman et al., 1998).

The difficulty in attributing an adverse reaction to a particular plant is demonstrated by the above examples. These cases also demonstrate the absolute need for good quality assurance and especially the need to establish the identity of the raw materials used by the botanical industry. The *Aristolochia* cases also illustrate the pitfalls inherent in using common or local names for botanicals. The similarity of the pinyin transliterations of the Chinese language common names for *Stephania tetrandra* (*Fangji*), *Aristolochia fangchi* (*Fangchi*), *Aristolochia manschuriensis* (*Guang mu tong*) and *Akebia* spp. (*Mu tong*) was undoubtedly a major factor in the intoxications. Problems of this type are inherent in the use of nonstandardized common names. A cursory glance at the "Index to Colloquial Names" in the Eighth Edition of *Gray's Manual of Botany* (Fernald, 1987) yields at least eight different snakeroots in six different families. They are black snakeroot (*Zigadenus dens us*, Liliaceous), broom (*Gutierrez* spp., Asteraceae), button (*Eryngium yuccifolium*, Apiaceae), sampson's (*Psoralea psoralioides*, Fabaceae), seneca (*Polygala senega*,

Polygalaceae), virginia (*Aristolochia serpentaria*, Aristolochiaceae) and white snakeroot (*Eupatorium rugosum*, Asteraceae). In the United States, a list of standardized common names called *Herbs of Commerce* (American Herbal Products Association, 1992) was adopted by the FDA as the only permitted source for common names on product labeling. Products that contain plants not listed in *Herbs of Commerce* are required to use the correct Latin name of the plant on the label. In addition to difficulties with nomenclature, problems have arisen because of the difficulty in determining exactly which plants are present in a finished product. Capsules that contain more than one plant and/or a dried crude extract make identification of the plant(s) by traditional means such as microscopy virtually impossible (Betz et al., 1995). Determination of the identity of a misidentified plant then becomes a matter of looking for specific toxic chemical constituents, which is usually a "needle in a haystack" approach unless symptoms are characteristic of a particular compound. Good quality assurance of raw material obviates the need for this detective work.

12.2.2.2 Misuse

Misuse of botanical products has resulted in a number of adverse reaction reports. In this discussion, the word misuse will be used in its broadest sense to indicate abuse as well as incorrect or inappropriate use (including overdose). The latter definition is especially important when discussing plants that have been used as traditional medicines and are now being used for conditions and in dosages and combinations that have no historical precedent. While there is nothing inherently wrong with discovering new uses for old remedies, new combinations, dosages, and dosage forms lack the historical underpinnings of traditional use and should be scrutinized for safety using modern methodology.

Decoctions of the traditional Chinese medicine (TCM) *Má Huáng* (*Ephedra* spp.) have been used for thousands of years as a diaphoretic, stimulant and antiasthmatic (Bensky and Gamble, 1986). The plant contains (–)-ephedrine, (+)-pseudoephedrine and four other structurally related alkaloidal amines (Huang, 1993). Synthetic (–)-ephedrine, (+)-pseudoephedrine and (±)-norephedrine (phenyl-propanolamine) salts are used clinically in the United States (*United States Pharmacopeia*, 1999). There have been a number of recent reports of adverse reactions associated with weight loss products (Catlin et al., 1993; Capwell, 1995; Perrotta et al., 1996; Betz et al., 1997) and with products marketed as "safe, legal" herbal stimulants (Nightingale, 1996; Yates et al., 2000) that contain *Má Huáng* extracts and caffeine. Reported reactions range from mild irritability, nervousness and sleeplessness to death. Despite a relatively large number of reports of an association between *Ephedra* products and adverse events, causality has been difficult to establish because of serious flaws in the reporting system and in clinical follow-up. Gurley et al. (2000) hypothesize that adverse reactions to these products are caused by variability in the ratio and amount of the individual alkaloids in the products. Lee et al. (2000) performed an *in vitro* cytotoxicity study on *Má Huáng* extracts and found that cytotoxicity could not be entirely accounted for by alkaloid content, and that this toxicity was significantly reduced when the plant material was extracted in a manner similar to the traditional means of preparing *Má Huáng*. Systematic investigations of the biological activity of the whole plant (as opposed to individual alkaloids) are needed if the question of risk for nontraditional uses is to be answered. Astrup et al. (1995) and others (Toubro et al., 1993) have evaluated ephedrine and caffeine formulations for safety and efficacy, but these results shed only a little light on the ephedra situation, because the study environment and the nature of the ephedrine source are fundamentally different.

Yohimbe (*Pausinystalia yohimbe*) is a purported aphrodisiac that contains a number of alkaloids, including yohimbine. This compound has been used in prescription formulations as a treatment for certain types of impotence (Sonda et al., 1990) but has been reported to cause serious adverse reactions (Grossman et al., 1993). Despite the potency of yohimbine itself, there have been no reports of an adverse reaction to yohimbe in the peer-reviewed literature. This lack of adverse reactions to yohimbe products in the United States is probably due to the very low levels (or absence) of alkaloids in dietary supplement products on the market, and adverse reactions would be expected if alkaloid levels in products were increased by manufacturers (Betz et al., 1995).

Suicide attempts have accounted for several cases of poisoning by botanicals. For example, a very mild intoxication by the popular nighttime sleep aid Valerian (*Valeriana officianalis*) was reported after a woman ingested 40–50 capsules in an unsuccesful suicide attempt (Willey et al., 1995). An elderly woman attempted to commit suicide by ingesting a large quantity of eucalyptus oil (Anpalahan and Le Couteur, 1998), and intensive gastric irrigation was required to save the life of a middle-aged man who ingested aconite (*Aconitum* spp.) in an attempt to end his own life (Mizugaki et al., 1998).

Kava kava (*Piper methysticum*) is a plant indigenous to the islands of the South Pacific, where its rhizome has commonly been used to make a beverage to induce relaxation (Weiner, 1971; Nagata, 1971). Pharmacological activity is attributed to a series of arylethylene pyrones called kavalactones (Klohs et al., 1959). Consumption of high doses may cause drowsiness, motor impairment and changes in visual and/or auditory perception (Gajdusek, 1979; Garner and Klinger, 1985). A Utah man stopped for driving erratically was convicted of driving under the influence after ingestion of 16 cups of kava (about eight times the "usual" amount) (Swenson, 1996), while a California man faced similar charges in the year 2000 (Wilson, 2000).

Abuse of self-collected material has also resulted in intoxication. Intentional ingestion of Jimson weed (*Datura stramonium*) for recreational purposes continues to cause human injury and death (Perrotta et al., 1995; Tiongson and Salen, 1998; Francis and Clarke, 1999).

Warning labels such as those recently proposed by the FDA for *Má Huáng* (Department of Health and Human Services, 1997b; Schwartz, 1997) and by trade organizations may aid in preventing some forms of misuse.

12.2.2.3 Deliberate Adulteration

Cases of deliberate substitution of one material for another (usually cheaper) date from the dawn of commerce. Deliberate adulteration of Indian mustard (*Brassica juncea*) seed with Mexican poppy (*Argemone mexicana*) seed is a historical problem and has led to mass intoxications caused by consumption of the edible oils made from the seed (Thatte and Dahanukar, 1999). Neonatal androgenization associated with maternal ingestion of Siberian ginseng (*Eleutherococcus senticosus*) (Koren et al., 1990) was subsequently discovered to have been caused by Chinese silk vine (*Periploca sepium*) (Awang, 1991) that had been substituted for the *Eleutherococcus* in the product. In a more recent report of a substitution involving *Eleutherococcus*, a 74-year-old man who had been on digoxin therapy for a number of years developed elevated serum digoxin levels after ingesting "Siberian ginseng." When the "Siberian ginseng" was discontinued, digoxin levels dropped (McRae, 1996). The reason for this phenomenon was never determined, although the author postulated that eleutherosides (glycosidic constituents of *Eleutherococcus*) may cross-react with the serum assay used for digoxin determination or that the ginseng product contained undeclared substances. Awang (1996) prefers the second explanation and proposes that the digoxin assay may have detected the cardiac glycosides known to occur in *P. sepium* (see above).

Unfortunately, persistent misconceptions about botanical safety become part of the conventional wisdom about herbs, as the original case reports are repeatedly cited without any acknowledgement of the explicatory letters that follow in subsequent volumes of the journals in which the original reports or letters appeared. In addition to this unfortunate situation, adverse effects of particular herbs have been predicted, in the absence of case reports or even *in vitro* studies, on the basis of the chemical composition of the herb in question. For instance, Miller (1998) warned about the expected hepatotoxicity of *Echinacea* spp. based on the occurrence of the pyrrolizidine alkaloids tussilagine and isotussilagine. These compounds are indeed present (at 0.006%) in *Echinacea* root, but they are nontoxic because they lack the structural features (1,2 unsaturation in the pyrrolizidine ring) mentioned above necessary for hepatic activation into reactive pyrroles.

Deliberate adulteration of botanical products is not limited to substitution of one plant for another. Various TCMs have been found to contain undeclared synthetic medicinal ingredients. Four patients developed agranulocytosis after ingesting an herbal arthritis pain relief product called *Chui Fong*, which contained aminopyrine and phenylbutazone (Ries and Sahud, 1975). A TCM called "Black Pearl" was found to contain hydrochlorothiazide, diazepam, indomethacin and mefenamic acid (By et al., 1989). Investigators have also reported the presence of corticosteroids (including hydrocortisone) in herbal TCMs (Goldman and Myerson, 1991; Joseph et al., 1991). All of these substitutions and additions are illegal in the United States, and the FDA is charged both with removing such products from the market and with bringing criminal charges against individuals who knowingly introduce these products to the marketplace.

Substitution is not limited to herbal medicines or dietary supplements. A pink seeded cultivar of common vetch (*Vicia sativa*) called *blanche fleur* has been sold in the United States as red lentils or *masoor dahl* (*Lens culinaris*) (Tate and Enneking, 1992). The human health aspects of this substitution are unclear, but *V. sativa* is known to produce the neurotoxic amino acids β-cyanoalanine and γ-glutamyl-β-cyanoalanine (Ressler, 1962; Ressler et al., 1969), the favism factors vicine and convicine (Pitz et al., 1980; Chevion and Navok, 1983), and the cyanogenic glyco-sides isolinamarin and vicianin (Poulton, 1983).

12.2.2.4 Inherent Toxicity

A number of plants should be avoided because the seriousness of potential adverse reactions outweighs any potential benefits. All members of the Boraginaceae and several members of the Fabaceae and Asteraceae have been found to contain pyrro-lizidine alkaloids (PAs). Mass intoxications affecting up to 6000 people have been caused by bread cereals contaminated by the seeds of various species of *Heliotropium* and *Crotalaria* (Dubrovinskii, 1946; Khanin, 1948; Tandon et al., 1976a,b, 1977, 1978a,b; Krishnamachari et al., 1977).

A number of PA-containing plants (e.g., *Tussilago farfara, Cynoglossum officinale, Senecio* spp., *Symphytum* spp.) were, until recently, used medicinally in Europe (Roeder, 1995). Comfrey (*Symphytum* spp.) and its aqueous infusions contain a number of toxic PAs, including symphytine, lycopsamine and intermedine (Betz et al., 1994; Mossoba et al., 1994). Several investigators have reported human hepatic veno-occlusive disease caused by ingestion of *S. officinale* (Ridker et al., 1985; Weston et al., 1987; Bach et al., 1989), and both symphytine and *S. officinale* have been shown to be rodent carcinogens (Hirono et al., 1978, 1979). Because of these hazards, the American Herbal Products Association has placed comfrey on its restricted use list (external use only) (McGuffin et al., 1997), and the German Federal Health Department (BGA) has drastically restricted the manufacture and sale of pharmaceuticals that contain PAs with the structural features necessary to cause toxicity (Roeder, 1995).

Several cases of serious liver disease have been reported following consumption of germander (*Teucrium chamaedrys*) (Larrey et al., 1992; Mostefa-Kara et al., 1992; Laliberté and Villeneuve, 1996), and its toxicity has been established by animal studies (Kouzi et al., 1994; Loeper et al., 1994). Recognition of the inherent toxicity of this herb has led to its being banned in France (Laliberté and Villeneuve, 1996). Scullcap (*Scutellaria lateriflora*) has also been reported to cause hepatic injury (MacGregor et al., 1989). Unfortunately, the case against this herb is muddled by reports that germander is frequently substituted for scullcap (Phillipson and Anderson, 1984).

Chaparral or creosote bush (*Larrea tridentata*) is a dominant desert shrub in certain areas of the United States and Mexico. Infusions of this plant were used by native Americans for treatment of a variety of diseases (including cancer) (Hutchens, 1992). Phenolic compounds account for 83–92% of the dry weight of the plant, with nordihydroguaretic acid (NDGA) as the most abundant phenolic constituent (5–10% dry weight) (Obermeyer et al., 1995). NDGA is a potent antioxidant and was con-sidered a GRAS food additive for this purpose until animal studies revealed evidence of kidney toxicity (Grice et al., 1968). The plant has enjoyed a recent surge in popularity as an anticancer and cancer-preventing dietary supplement (Tyler, 1993).

Toxigenic potential of *Larrea* becomes apparent when animal data and a spate of reports of hepatotoxicity in humans caused by chaparral ingestion are considered together (Katz and Saibil, 1990; Clark and Reed, 1992; Smith and Desmond, 1993; Alderman et al., 1994; Gordon et al., 1995; Sheikh et al., 1997). Most of the hepatic injuries were seen in patients who had ingested chaparral capsules or tablets. Native American cultures traditionally consumed chaparral as an infusion (Mabry et al., 1977). Obermeyer et al. (1995) have demonstrated that levels of phenolic constituents (including NDGA) are much lower in such infusions than in the whole plant, and the reason for the "sudden" appearance of chaparral toxicity may be related to this departure from traditional patterns of use. Case histories of affected individuals also indicated some preexisting history of a compromised liver. Chaparral was voluntarily removed from shelves for a short time after the original reports, but further investigations failed to confirm toxicity. Products were subsequently reintroduced to the market with warning labels about preexisting liver disease. Safety of the plant to the population at large has not yet been demonstrated conclusively, but there has been no upsurge in adverse reaction reports associated with reintroduction of chaparral into the marketplace.

Despite the "history of safe use" for most of the products now in the marketplace, there remain nagging concerns about botanical dietary supplements. Shifts in traditional use patterns and introduction of new plants to the market pose new challenges to scientists, regulators and manufacturers. Although there is no good system for reporting adverse reactions to supplements in the United States, acute adverse events are often relatively easy to track. As we have seen, attribution of an event to a particular plant may be complicated by misidentification or adulteration, but a continuing pattern of adverse reactions coupled with adequate postexposure documentation will eventually lead to correct identification of acute hazards. The comfrey cases cited above illustrate the point. Other examples include adverse reactions to the undiluted juice of squirting cucumber (*Ecbalium elaterium*) in Israel (Raikhlin-Eisenkraft and Bentur, 2000) and continuing reports of bronchiolitis obliterans caused by consumption of *Sauropus androgynus* as a weight loss vegetable in Taiwan (Ger et al., 1997; Wu et al., 1997; Chang et al., 1998; Hsiue et al., 1998). More difficult to attribute are acute reactions that are rare or that require some genetic or other predisposition in the affected individual. Adverse events to chaparral (above) in individuals with compromised liver function can be predicted when the history of liver dysfunction is known. Cautions about ephedra consumption can also be provided to individuals suffering from cardiovascular disease. However, hypersensitivity to the cardiovascular effects of ephedra (even when used as directed) is difficult to predict in asymptomatic or undiagnosed individuals. Prediction of adverse events to these plants is problematic when the individuals at risk are apparently healthy but suffer from some underlying condition. Hypericism is a primary photosensitivity caused in animals by hypericin in St. John's wort (*Hypericum perforatum*). Products containing this plant are widely available in Europe and the United States, but hypericism is virtually unknown in humans, probably because humans consume considerably less of the plant than do grazing herbivores. The underlying mechanism for photosensitivity remains, however, and there has been a report of apparent hypericism in a woman who was consuming St. John's wort at recommended levels (Bove, 1998). Ginkgo is a popular herb that is considered to be relatively safe, with few documented adverse event reports. However, there have been at least two case reports associating abnormal bleeding with ginkgo (*Ginkgo biloba*) consumption (Gilbert, 1997; Vale, 1998).

A number of synthetic pharmaceuticals are known for their hepatotoxic potential. Such reactions are generally rare but represent a real risk that must be addressed by prescribing physicians. Hepatotoxic plants (comfrey, germander) notwithstanding, it is not surprising that hepatitis has been associated with some herbs. Benninger et al. (1999) report 10 cases of acute hepatitis associated with use of greater celandine (*Chelidonium majus*), and Nadir et al. (1996) report a case of acute hepatitis associated with a *Má Huáng* product. The pharmacology of the plants would not lead one to expect liver problems. The possibility that the plants were misidentified or that the hepatic involvement was due to some idiosyncratic hypersensitivity similar to the kind seen in conventional pharmaceuticals must be considered.

Chronic and delayed adverse effects of botanicals are even less well characterized than are acute reactions. In a review devoted primarily to long-term safety of phytoestrogens, Sheehan (1998) points out that virtually all of our knowledge about potential adverse effects of herbal products came from reports of acute reactions upon human exposure. As examples of the difficulty of assigning causality for a delayed or chronic illness to a particular material (botanical or otherwise), he points to the 20-year delay before clear-cell carcinoma of the cervix and vagina in young women was associated with maternal use of diethylstilbesterol (DES), the fact that lung cancer was first associated with tobacco in 1761 and that fetal alcohol syndrome was first described in the mid 1970s after millennia of human experience with alcoholic beverages.

There are a number of botanicals used as dietary supplements that have a potential for causing delayed adverse reactions. The most frequently discussed long-term effects are cancer and birth defects. Blue cohosh (*Caulophyllum thalictroides*) is known to contain the quinolizidine alkaloids N-methylcytisine, baptifoline and anagyrine as well as the aporphine alkaloid magnoflorine (Flom et al., 1967). Products containing this plant are intended for consumption by women of reproductive age, and there has been at least one report of fetal injury associated with consumption of a blue cohosh-containing product by the mother (Jones and Lawson, 1998). Anagyrine is of particular concern because of its known teratogenicity in cattle (Keeler, 1976). This and the other alkaloids have been found in dietary supplements that contain blue cohosh (Betz et al., 1998; Woldemariam et al., 1997). Bioassay-directed fractionation coupled with an *in vitro* rat embryo assay has demonstrated that N-methylcytisine is also a potential teratogen (Kennelly et al., 1999). Fully one-third of the approximately 600 plants evaluated for safety by McGuffin et al. (1997) bear warnings about use during pregnancy and lactation.

Even when delayed or chronic effects are suspected and the plants and their constituents are extensively studied, it is not always possible to conclusively demonstrate that the material in question actually causes harm in humans. Emodin, a natural anthraquinone aglycone found in anthraquinone glycoside-containing laxative plants, was reported to be mutagenic in the Ames Salmonella/mammalian microsome assay by Bosch et al. (1987). A subsequent investigation by Sandnes et al. (1992) established that senna is weakly mutagenic in the Ames test, but that all of the mutagenic activity was not attributable to the anthraquinone glycoside content of the extracts tested. Zhang et al. (1996) used computer automated structure evaluation (CASE) and multiple CASE to predict that emodin is a probable rodent carcinogen. Mueller and Stopper (1999) reported that inhibition of topoisomerase II activity by the anthraquinones emodin and aloe-emodin was likely responsible for the mutagenicity and genotoxicity of these compounds, but a review of the relevant literature led Brusick and Mengs (1997) to conclude that although individual components of senna possess demonstrable mutagenicity and genotoxicity in nonhuman models, critical evaluation of data obtained from human metabolism studies and human clinical trials do not support concerns that senna laxatives pose a genotoxic risk if used as directed. The suggestive but inconclusive *in vitro* and animal studies generated concern about possible carcinogenicity of anthraquinone glycoside-containing laxative plants such as senna (*Senna alexandrina*) and cascara (*Cascara sagrada*), and led the FDA to issue a proposed rule to remove these laxative plants from over-the-counter OTC Drug Category I (generally recognized as safe and effective and not misbranded) and place them into Category III (further testing is required) (Department of Health and Human Services, 1998). This rule, if made final, will effectively remove these plants from the market as OTC drugs in the United States, even though they will remain available as dietary supplements.

The molecular mechanism by which the PAs exert their hepatotoxicity is known. This mechanism would be expected to render the reactive pyrroles both genotoxic and mutagenic, and certain of these compounds have been shown to be rodent carcinogens (Hirono et al., 1978, 1979). Despite long-standing human dietary exposure to the PAs, no association between human cancer and ingestion of these compounds has been found, and in a recent review, Prakash et al. (1999) have concluded that PAs are not human carcinogens.

Even clear-cut cases of readily identifiable delayed reactions may take years to become apparent. The Belgian women who developed urethral cancer after *Aristolochia* ingestion were exposed almost a decade before their cancers became apparent. Had acute kidney damage not been identified at the time of the exposures, the reason for the sudden increase in this type of cancer in this population might not have been ascertained.

12.2.2.5 Herb-Drug Interactions

As noted elsewhere, herbs have been used medicinally for hundreds or thousands of years. Over the past few decades, the entire modern armamentarium of single-entity pharmaceuticals has been superimposed onto this tradition. This juxtaposition has led to a new category of safety concerns. The evolution of our current knowledge on interactions began with the realization that safety and efficacy of certain pharmaceuticals could be adversely affected by common foods. For instance, it has been known for some time that the tyramines present in red wine and certain aged cheeses can cause severe hypertensive crisis in individuals being treated with monoamine-oxidase inhibitors (MAOIs) (Asatoor et al., 1963; Blackwell, 1963), and MAOI diets limiting such common foods as pizza, soy sauce and tofu have been described (Shulman and Walker, 1999). A discussion of food-drug interactions may be found in a review by Walter-Sack and Klotz (1996), but there is one particular interaction that bears some discussion here. In 1989, a study designed to investigate the effect of alcohol on the metabolism of the prescription drug felodipine serendipitously identified grapefruit juice as a potent inhibitor of a number of important drug metabolizing enzymes

(Bailey et al., 1989). The juice was found to cause a reduction in metabolic elimination of the drug, leading to a buildup of the drug to potentially dangerous levels (Bailey et al., 1991; Fuhr et al., 1993). There followed much speculation about the mechanism for this inhibition, with most attention focused on the isoflavone content (specifically, naringin and naringenin) (Fuhr and Kummert, 1995). Eventually, Edwards et al. (1996) identified the furanocoumarin 6′,7′-dihydroxybergamottin in grapefruit juice to be a potent inhibitor of CYP3A activity in isolated rat microsomes. In the meantime, results of over 30 clinical trials have been published detailing the effects of grapefruit and other juices on the metabolism of many drugs. A thorough discussion of these studies is beyond the scope of this chapter, but Ameer and Weintraub (1997) provide an excellent review of published reports.

In the past year, several reports on the subject of herb-drug interactions have been published. These range from collations of individual case reports, to predictions of herb-drug interactions based on the hypothesized mechanisms of action of the herbs, to clinical trials in healthy individuals, to reviews of all of the above.

According to Goth (1974), a substance may influence the effect of a drug by a number of different mechanisms. These mechanisms include effects on intestinal absorption, competition for plasma protein binding, metabolism, adrenergic neuronal uptake, action at the drug receptor site, renal excretion, and alteration of electrolyte balance. Drug interaction mechanisms currently garnering the most interest are those that involve the inhibition or induction of drug biotransformation (pharmacokinetics) via hepatic enzymes such as the Cytochrome P-450 (CYP) superfamily. Six CYP isoforms (IA2, 2C9, 2C19, 2D6, 2E1, and 3A4) are important clinically (Thummel and Wilkinson, 1998). Foods and herbs generally contain a large number of pharmacologically active constituents. These compounds belong to virtually all of the classes of plant secondary metabolites, including alkaloids, terpenes, glycosides, and flavonoids. *In vitro* studies have shown saponins, including those found in ginseng, to inhibit the Cytochrome P450 system (Shi and Liu, 1996; Kim et al., 1997). Further, some coumarins may also inhibit specific CYP isoenzymes *in vitro* (Cai et al., 1997). Flavonoids are phenolic compounds that are widely distributed in plants. Dietary sources include fruits, vegetables, tea (*Camellia sinensis*), seeds, bark, leaves, nuts and herbal products (Formica and Regelson, 1995). Several decades of research into dietary causes and preventers of cancer have demonstrated that some flavonoids inhibit and some induce selected drug-metabolizing isozymes (Lasker et al., 1984; Nielson et al., 1998). *In vitro* human liver microsomal experiments have shown several bioflavonoids to inhibit the activities of CYP 3A4, 1A2, 2B, and 2E1 (Lasker et al., 1984; Li et al., 1994; Ha et al., 1995; Obermeier et al., 1995; Cai et al., 1997). Flavonoids are hydroxylated via CYP1A2 (Nielsen et al., 1998), and structure/activity relationships between hydroxylated flavonoids and modulators of CYP activity have been investigated (Tsyrlov et al., 1994). In general, hydroxylated flavonoids (quercetin, naringenin) inhibit P450 isozymes *in vitro*, and nonhydroxylated flavonoids (flavone, nobiletin, tangeretin) stimulate P450 activity *in vitro* and *in vivo* (Buening et al., 1981; Li et al., 1994; Blackman et al., 2000). Flavonoids also modulate P-glycoprotein transporter efflux (Scambia et al., 1994; Sarker, 1995) and possibly other transporters (Walgren et al., 1998).

Significantly, the flavonoid quercetin, a constituent of many herbal products including the popular herbs St. John's wort, *Ginkgo biloba* and *Echinacea purpurea*, has been shown to be a potent inhibitor of CYP3A4 and sulfotransferases (Li et al., 1994; Eaton et al., 1996). The ability of bioflavonoid constituents (naringenins and/or coumarins [6,7-dihydroxybergamotin]) in grapefruit juice to inhibit CYP 3A4 or P-glycoprotein and cause clinically significant drug interactions with dihydropyridine derivatives (i.e., the antihypertensive felodipine [Plendil®], the immune suppressant cyclosporine and the nonsedating antihistamine terfinadine), is well documented (Gruengerich and Kim, 1990; Fuhr, 1998). Conversely, the herbal medication *Eucalyptus* and possibly St. John's wort have been shown to induce hepatic isoenzymes (Liu, 1989). Recently, See et al. (1999) screened the crude extracts of 134 products that contained single herbs and 54 multiple-herb products for effects on CYP 2E1 and CYP 4A using mouse microsomes. Enzyme inhibition occurred with 28 of the single herb products and 22 of the multi-herb products, while enzyme induction was reported for 13 single-entity products and 18 combination products. CYP2E1 and CYP4A are not responsible for the metabolism of a large number of synthetic drugs, so the significance of these results in the context of herb-drug interactions is unclear.

Screening for potential drug interactions using *in vitro* models is a common and valuable approach to identifying interactions mediated by the CYP system, but the approach has inherent limitations. For instance, drug concentrations achieved in *in vitro* systems seldom accurately reflect those attained *in vivo* at the intracellular site of the enzymatic activity. Enzyme specificity for inhibition may be lost at extraphysiological concentrations (Caraco, 1998). Attempting to approximate clinically relevant concentrations of herbal constituents for *in vitro* methods is particularly problematic in that many of the compounds found within plants may not be detectable in circulating plasma in humans who ingest these herbal products. Additionally, other factors affecting adsorption, metabolism, disposition, and elimination of dietary constituents (including protein and tissue binding, hepatic blood flow, and extra-hepatic elimination) are not accounted for *in vitro* (Von Moltke et al., 1998). Information generated *in vitro*, while useful, is qualitative and not quantitative, and cannot be generalized to the clinical situation. Thus, suggestive findings of *in vitro* studies must still be confirmed with *in vivo* studies.

Mixtures of the flavonoids quercetin and apigenin inhibited the activation of zoxazolamine 6-hydroxylation by flavone and tangeretin *in vitro* but not *in vivo* (Lasker et al., 1984). Ha et al. (1995) postulated that a combination of grapefruit flavonoids may be required for full inhibitory activity, suggesting that synergistic or antagonistic effects may be seen with mixtures of phytochemicals that cannot be predicted from individually studied constituents. A further example of the limitations of *in vitro* work can be seen in two studies on St. John's wort by Carson et al. (2000) and Markowitz et al. (2000). In the former, extracts of the plant were found to potently inhibit CYP3A4 activity in human liver microsomes. In the latter, St. John's wort administered to seven healthy human volunteers at therapeutic doses in combination with the CYP3A4 probe substrate alprazolam detected no evidence of inhibitory activity. Finally, present attempts to establish chemical standards of quality for botanical products focus on identification and quantitation of "marker" compounds, substances unique to the plant in question but not necessarily biologically active. *In vitro* experiments that propose to focus on isolated "marker" constituents are hampered by the fact that the markers may not be bioactive and that the identity of the active moiety is unknown for most herbs (De Smet and Brouwers, 1997).

We have noted that foods that affect intestinal or hepatic metabolism of drugs may cause treatment failure or drug-induced toxicity. The grapefruit juice example cited above is more or less the extreme case, but it serves as a reminder that the topic of herb-drug interaction must be viewed in the context of the whole diet. The effects of flavonoids on drug metabolizing systems have also been described. This information can be combined with data on dietary flavonoid exposure obtained from attempts to determine the role of dietary flavonoids in cancer prevention to yield valuable perspective. A market-basket-type survey of Dutch consumers estimated that average daily intake of flavonoids in that country was 23 mg. Of this total, it was estimated that 16 mg/day of the total was provided by quercetin (Hertog et al., 1993).

Finally, in contrast to the extensive literature on prescription drug-drug interactions, actual case reports of herb-drug interactions are sparse. This may be because herb-drug interactions are not easily recognized by the patient, because the patient is hesitant about reporting such reactions to the health care provider, because the provider may not notify regulatory authorities or because such reactions are not occurring at any great rate. Therefore, these potential interactions do not routinely appear in the biomedical literature (D'Arcy, 1993). Formal study (as opposed to case reports) of the drug interaction potential of herbal medications is virtually nonexistent, and few pharmacokinetic studies of herbal preparations are available (DeSmet and Brouwers, 1997). As a result, the majority of reported or suspected cases of herb-drug interactions are a result of the establishment of a temporal relationship between occurrence of an adverse reaction and simultaneous use of an herbal product with a prescription drug. Utility of case reports in evaluating potential risks from herbs (see sections on Adulteration and Misidentification) and from herb-drug interactions varies. For example, kava kava (*Piper methysticum*), a popular herbal product promoted as an anxiolytic, was implicated in a case of suspected interaction with alprazolam (Almeida and Grimsley, 1996). The case illustrates the pitfalls of using isolated case reports to establish causality, because the patient was receiving a known inhibitor of alprazolam metabolism, cimetidine, in addition to the kava and alprazolam.

The most intensively studied herb-drug interactions are those involving St. John's wort (SJW). Eight case reports [reviewed by Ernst (1999) and Fugh-Berman (2000)] suggest SJW herb-drug interactions with theophylline, warfarin, ethinyl estradiol and cyclosporine. These reports indicate that SJW may induce CYP1A2, 2C9 and 3A4. Nebel et al. (1999) reported theophylline toxicity in a 42-year-old female patient when she discontinued SJW. Shortly after SJW had been initiated, her theophylline dosage required escalation from 300 mg to 800 mg BID to maintain therapeutic concentrations. When the patient discontinued the SJW, the elevated theophylline levels caused toxicity. The authors further investigated the mechanism of this herb-drug interaction utilizing *in vitro* techniques (hepG2 human hepatoma cells) and found a 1.8-fold induction of CYP1A2 with 12.5 μM of hypericin and a 5.2-fold induction with 125 μM of hypericin (Nebel et al., 1999). Case reports associate acute cardiac transplant rejection following three weeks of SJW co-administration with reduced cyclosporin concentrations caused by the herb (Ruschitzka et al., 2000) and lowered cyclosporin trough concentrations in a kidney transplant patient co-administered a *Hypericum* extract (Mai et al., 2000). In addition to these case reports, several pharmacokinetic studies suggest induction of CYP2A4 with SJW treatment longer than 8 days at therapeutic doses. Following 4 days of SJW administration (300 mg × 3) in seven healthy volunteers, Markowitz et al. (2000) reported no indication of CYP3A4 or CYP2D6 inhibition and only a trend toward CYP3A4 induction. The authors used alprazolam as a probe drug, and they found that, on average, there was the suggestion of a trend toward CYP3A4 induction after four days but little evidence of inhibition. Roby et al. (2000) found induction of CYP3A4 following 14 days of SJW (300 mg × 3) as measured by the 6β-hydroxycortisol:cortisol ratio with an average increase of 114%.

Yue et al. (2000) recently reported that SJW may reduce the effectiveness of warfarin, while a case report of breakthrough bleeding during concomitant use of oral contraceptives and SJW was reported by Bon et al. (1999). A study by Piscitelli et al. (2000) found that 14 days of SJW (300 mg × 3) reduced the AUC of indinavir by 57% ± 19% and the trough concentrations by 81% ± 16% in eight healthy volunteers, consistent with marked CYP3A4 induction. Importantly, such reductions could result in therapeutic failure and the development of viral resistance. Following the publication of the indinavir study, there was much attention given to "predicting" other CYP3A4 metabolized drugs that may interact with SJW. However, indinavir is a substrate for both CYP3A4 and P-glycoprotein transport, and the study authors noted that either (or both) mechanism may be responsible for the increased clearance of the drug. Two studies with the P-glycoprotein substrate digoxin further suggest SJW may also induce P-glycoprotein. The first was a single-blind, placebo-controlled, parallel study involving 25 healthy volunteers that looked at the potential for an herb-drug interaction between SJW and digoxin (Johne et al., 1999). The study compared the disposition of digoxin following 5 days of digoxin (0.25 mg) alone and after 1 day and 10 days of SJW (300 mg × 3, LI-160). After one day of SJW, there was a trend toward increased C_{max} for digoxin, but after 10 days of SJW, digoxin AUC significantly decreased by 28%, C_{max} significantly decreased by 26% and C_{trough} significantly decreased by 37%. In the second study, digoxin AUC was increased when co-administered with a known P-glycoprotein inhibitor, valspodar (Kovarik et al., 1999). Taken together, these two studies indicate that SJW may inhibit efflux of P-glycoprotein-dependent substrates and induce CYP3A4-dependent drug clearance.

In addition to the pharmacokinetic interactions described above, potential pharmacodynamic interactions have been described. These include case reports of abnormal bleeding events following concomitant ingestion of ginkgo and aspirin (Rosenblatt and Mindel, 1997), ginkgo and ergotamine/caffeine (Rowin and Lewis, 1996), and Dong quai (*Angelica sinensis*) and warfarin (Page and Lawrence, 1999). Three older case reports described associations between garlic consumption and prolonged postoperative bleeding (Burnham, 1995; German et al., 1995) and between excessive garlic consumption and spontaneous spinal epidural hematoma (Rose et al., 1990). St. John's wort use has been associated with mild serotonin syndrome when combined with the antidepressants trazodone (Demott, 1998), sertraline or nefazodone (Lantz et al., 1999). De Smet (1997) reviewed the literature on interactions between drugs and yohimbine (from yohimbe, *Pausinystalia yohimbe*) and noted that yohimbine should be administered with caution when used combined with tricyclic antidepressants, because the combination may cause hypertension.

Effects of botanical products on drug absorption and gastrointestinal transport have also been noted. Guar gum, the ground endosperms of *Cyamopsis tetragonolobus* is used as a stabilizer; a thickening agent for cheese, salad dressings, ice cream and soups; as a binding and disintegrating agent in tablet formulations (Merck Index, 2000); and as a dietary supplement. The substance prolongs gastrin retention and has been reported to slow absorption of digoxin, paracetemol and bumetanide, and to decrease absorption of metformin, phenoxymethylpennicillin and glibenclamide (De Smet and D'Arcy, 1996). Drug interactions due to effects of bulk forming agents (e.g., Psyllium, *Plantago ovata*) or laxative herbs such as cascara (*Rhamnus purshiana*) on gastrointestinal transit time might also be expected (Westendorf, 1993; Fugh-Berman, 2000).

There have been many more potential and reported interactions in the scientific literature. An exhaustive review is beyond the scope of this chapter. The reader is referred to recent reviews by Fugh-Berman (2000) and Miller (1998). The former places special emphasis on human case reports and highlights some of the perils and pitfalls in trying to establish causality from isolated cases. The review by Miller includes case reports but also incorporates animal and other data. The result is a catalog not only of reported events but also of potential herb-drug interactions predicted from postulated mechanisms of action of the herbs. As noted above, mechanisms and compounds responsible for biological activity are unknown for many herbs. An excellent additional resource is the three-volume series *Adverse Effects of Herbal Drugs* by De Smet et al. (1992, 1993, 1997).

12.3 CONCLUSIONS

Beyond education, public health officials can do little about cases of poisoning by self-collected material except to offer expertise and occasionally laboratory assistance after the fact. Adequate quality assurance programs are needed for commercial products, as are attempts to anticipate potential problem areas and alert the scientific community and consumers of potential problems that can be predicted by knowledge of the plants in the marketplace.

Size and complexity of the dietary supplement market are only a few of a number of difficulties that public health officials face when dealing with poisoning episodes. In North America alone, at least 700 plant species have been described as being poisonous in one way or another, and plant poisonings are often difficult to differentiate from environmental intoxications caused by pesticides and industrial chemicals and from adverse reactions to synthetic drugs (Der Marderosian and Liberti, 1988). Complicating the picture is the fact that symptoms of many intoxications mimic those of medical conditions not associated with a toxic exposure, and that establishing causality may therefore be difficult (Perrotta et al., 1996).

The occurrence of natural plant toxins in the food supply has been well documented, and several comprehensive treatises on the subject are available (National Research Council, 1973; Rechcigl, 1983; Watson, 1987; Rizk, 1990; D'Mello et al., 1991). Despite the maturity of this field, existing and potential problems persist. Methods for safe preparation of food plants such as cassava (*Manihot esculenta*) are well known in cultures that depend on it as a staple, but rapid introduction into a naive marketplace could cause cyanide poisoning. Although education and technical solutions [e.g., development of Canola varieties of low glucosinolate, low erucic acid rapeseed, *Brassica napa* (Downey, 1990)] may be enough to address certain issues, regulatory approaches may be needed for others. For example, hypoglycin A in unripe ackee (*Blighia sapida*) fruit causes a devastating illness called Jamaican vomiting sickness (Figeroa et al., 1992; McTague and Forney, 1994). Ripe, seedless pericarp of this plant is both desirable and apparently safe, and ackee is the Jamaican national fruit. Canned ackee fruit will be prohibited entry into the United States until a rigorous quality assurance system is in place to ensure that product containing toxic levels of hypoglycin does not reach the consumer.

Whereas reports of adverse reactions to botanical dietary supplements dominate this overview, changes in the legal status of these products have led to a recent explosion of the dietary supplement marketplace. The "cultural knowledge" that had historically protected the traditional consumer is often lacking in the United States. Improved quality assurance and consumer education could significantly decrease the adverse health effects associated with these products. In terms of herb-drug interactions, Treasure (2000) recommends a commonsense approach by practitioners. Much of the approach is specific to St. John's wort and is not presented here, but there are several recommendations that can be applied to all herbs. He recommends that physicians keep track of all medications and supplements that a patient is using, that they be alert when prescribing powerful medications and those with a narrow therapeutic window and that patients in vulnerable groups (i.e., those subject to polypharmacy) be closely monitored.

REFERENCES

Alderman, S., Kailas, S., Goldfarb, S., Singaram, C. and Malone, D.G. (1994) Cholestatic hepatitis after ingestion of chaparral leaf: confirmation by endoscopic retrograde cholangiopancreatography and liver biopsy. *J. Clin. Gastroenterol.* 19, 242–247.

Almeida, J.C. and Grimsley, E.W. (1996) Coma from the health food store. *Ann. Intern. Med.* 125, 940.

Ameer, M. and Weintraub, R.A. (1997) Drug interactions with grapefruit juice. *Clin Pharmacokinet.* 33, 103–121.

American Herbal Products Association (1992) Foster, S. (ed.), *Herbs of Commerce*, American Herbal Products Association, Austin, TX.

Anderson, I.B., Mullen, W.H., Meeker, J.E., Khojasteh-Bakht, S.C., Oishi, S., Nelson, S.D. and Blanc, P.D. (1996) Pennyroyal toxicity: measurement of toxic metabolite levels in two cases and review of the literature. *Ann. Intern. Med.* 124, 726–734.

Anpalahan, M. and Le Couteur, D.G. (1998) Deliberate self-poisoning with eucalyptus oil in an elderly woman. *Aust. N. Z. J. Med.* 28, 58.

Asatoor, A.M., Levi, A.J. and Milne, M.D. (1963) Tranylcypromine and cheese. *Lancet*. 2, 733

Astrup, A.V., Breum, L. and Toubro, S. (1995) Pharmacological and clinical studies of ephedrine and other thermogenic agonists. *Obes. Res.* 3, 537S–540S.

Awang, D.V.C. (1991) Maternal use of ginseng and neonatal androgenization. *JAMA*. 265, 1839.

Awang, D.V.C. (1996) Siberian ginseng toxicity may be case of mistaken identity. *Can. Med. Assoc. J.* 155, 1237.

Bach, N., Thung, S.N. and Schaffner, F. (1989) Comfrey herb tea-induced hepatic veno-occlusive disease. *Am. J. Med.* 87, 97–99.

Bailey, D.G., Spence, J.D., Edgar, B., Bayliff, C.D. and Arnold, J.M.O. (1989) Ethanol enhances the hemodynamic effects of felodipine. *Clin. Invest. Med.* 12, 357–362.

Bailey, D.G., Spence, J.D., Munoz, C. and Arnold, J.M.O. (1991) Interaction of citrus juices with felodipine and nifedipine. *Lancet*. 337, 268–269.

Bakhiet, A.O. and Adam, S.E.I. (1995) Therapeutic utility, constituents and toxicity of some medicinal plants: A review. *Vet. Human Toxicol.* 37, 255–258.

Benninger, J., Schneider, H.T., Schuppan, D., Kirchner, T. and Hahn, E.G. (1999) Acute hepatitis induced by greater celandine (*Chelidonium majus*). *Gastroenterology*. 117, 1234–1237.

Bensky, D. and Gamble, A. (1986) *Chinese Herbal Medicine-Materia Medica*. Eastland Press, Seattle, WA.

Betz, J.M., Eppley, R.M., Taylor, W.C. and Andrzejewski, D. (1994) Determination of pyrro-lizidine alkaloids in commercial comfrey products (*Symphytum* sp.). *J. Pharm. Sci.* 83, 649–653.

Betz, J.M., White, K.D. and Der Marderosian, A.H. (1995) Gas chromatographic determination of yohimbine in commercial yohimbe products. *J. A. O. A. C. Int.* 78, 1189–1194.

Betz, J.M., Gay, M.L., Mossoba, M.M., Adams, S. and Portz, B.S. (1997) Chiral gas chromato-graphic determination of ephedrine-type alkaloids in dietary supplements containing *Má Huáng*. *J. A. O. A. C. Int.* 80, 303–315.

Betz, J.M., Andrzejewski, D., Troy, A., Casey, R.E., Obermeyer, W.R., Page, S.W. and Woldemariam, T.Z. (1998) Gas chromatographic determination of toxic quinolizidine alkaloids in blue cohosh *Caulophyllum thalictroides* (L.) Michx. *Phytochem. Anal.* 9, 232–236.

Blackman, J.T., Maenpaa, J., Belle, D.J., Wrighton, S.A., Kivisto, K.T. and Neuvonen, P.J. (2000) Lack of correlation between *in vitro* and *in vivo* studies on the effects of tangeretin and tangerine juice on midazolam hydroxylation. *Clin Pharmacol. Ther.* 67, 382–390.

Blackwell, B. (1963) Hypertensive crisis due to monoamine-oxidase inhibitors. *Lancet*. 2, 849.

Bon, S., Hartmann, K. and Kuhn, M. (1999) Johanniskraut: Ein Enzyminduktor? *Schweiz. Ap. Ztg.* 16, 535–536.

Bosch, R., Friederich, U., Lutz, W.K., Brocker, E., Bachmann, M. and Schlatter, C. (1987) Investigations on DNA binding in rat liver and in Salmonella and on mutagenicity in the Ames test by emodin, a natural anthraquinone. *Mutat. Res.* 188, 161–168.

Bove, G.M. (1998) Acute neuropathy after exposure to sun in a patient treated with St. John's wort. *Lancet*. 352, 1121–1122.

Brusick, D. and Mengs, U. (1997) Assessment of the genotoxic risk from laxative senna products. *Environ. Mol. Mutagen.* 29, 1–9.

Buening, M.K., Chang, R.L., Huang, M.T., Fortner, J.G., Wood, A.W. and Conney, A.H. (1981) Activation and inhibition of benzo(a)pyrene and aflatoxin B1 meatbolism in human liver microsomes by naturally occurring flavonoids. *Cancer Res.* 41, 67–72.

Burnham, B.E. (1995) Garlic as a possible risk for postoperative bleeding [comment]. *Plast. Reconstr. Surg.* 95, 213.

But, P.P. and Ma, S.C. (1999) Chinese herb nephropathy. *Lancet.* 354, 1731–1732.

By, A., Ethier, J.C., Lauriault, G., LeBelle, M., Lodge, B.A., Savard, C., Sy, W.-W. and Wilson, W.L. (1989) Traditional oriental medicines I. Black Pearl: identification and chromatographic determination of some undeclared medicinal ingredients. *J. Chromatogr.* 469, 406–411.

Cai, Y.N., Baerdubowska, W., Ashwoodsmith, M. and Digovanni, J. (1997) Inhibitory effects of naturally occurring coumarins on the metabolic activation of benzo[A]pyrene and 7,12-dimethylbenz[A]anthracene in cultured mouse keratinocytes. *Carcinogenesis.* 18, 215.

Capwell, R.R. (1995) Ephedrine-induced mania from an herbal diet supplement. *Am. J. Psychiatry.* 152, 647.

Caraco, Y. (1998) Genetic determinants of drug responsiveness and drug interactions. *Ther. Drug Monit.* 20, 517.

Carson, S.W., Hill-Zabala, C.E., Roberts, S.H. and Hawke, R.L. (2000) Inhibitory effect of methanolic solution of St. John's wort (*Hypericum perforatum*) on cytochrome P450 3A4 activity in human liver microsomes. *Clin. Pharmacol. Ther.* 67, 99.

Catlin, D.H., Sekera, M. and Adelman, D.C. (1993) Erythroderma associated with ingestion of an herbal product. *West. J. Med.* 159, 491–493.

Chang, Y.L., Yao, Y.T., Wang, N.S. and Lee, Y.C. (1998) Segmental necrosis of small bronchi after prolonged intakes of *Sauropus androgynus* in Taiwan. *Am. J. Respir. Crit. Care. Med.* 157, 594–598.

Chevion, M. and Navok, T. (1983) A novel method for quantitation of favism-inducing agents in legumes. *Anal. Biochem.* 128, 152–158.

Clark, F. and Reed, R. (1992) Chaparral-induced toxic hepatitis-California and Texas, 1992. *Morbidity and Mortality Weekly Reports.* 41, 812–814.

Croom, Jr., E.M. and Walker, L. (1995) Botanicals in the pharmacy: New life for old remedies. *Drug Topics.* November 6, 84–93.

D'Arcy, P. (1993) Adverse reactions and interactions with herbal medicines, Part 2: Drug Interactions. *Adverse Drug React. Toxicol. Rev.* 12, 147.

Del Beccaro, M.A. (1995) Melaleuca oil poisoning in a 17-month-old. *Vet. Hum. Toxicol.* 37, 557–558.

Department of Health and Human Services. (1997a) Current Good Manufacturing Practice in Manufacturing, Packing, or Holding Dietary Supplements. *Federal Register.* 62, 5700–5709.

Department of Health and Human Services. (1997b) Dietary Supplements Containing Ephedrine Alkaloids; Proposed Rule. *Federal Register.* 62, 30678–30724.

Department of Health and Human Services. (1998) Laxative Drug Products for Over-the-Counter Human Use; Proposed Amendment to the Tentative Final Monograph. *Federal Register.* 63, 33592–33595.

De Smet, P.A.G.M. (1997) Yohimbe alkaloids—General discussion. In: De Smet, P.A.G.M., Keller, K., Hänsel, R. and Chandler, R.F. (eds.) *Adverse Effects of Herbal Drugs 3.* Springer-Verlag, Berlin, pp. 197–198.

De Smet, P.A.G.M. and D'Arcy, P.F. (1996) Drug interactions with herbals and other non-toxic remedies. In: D'Arcy, P.F., McElnay, J.C. and Welling, P.G. (eds.) *Mechanisms of Drug Interactions.* Springer-Verlag, Berlin.

De Smet, P.A.G.M. and Brouwers, J.R.B.J. (1997) Pharmacokinetic evaluation of herbal remedies: basic introduction, applicability, current status and regulatory needs. *Clin. Pharmacokinet.* 32, 427.

De Smet, P.A.G.M., Keller, K., Hänsel, R. and Chandler, R.F. (1992) *Adverse Effects of Herbal Drugs 1*, Springer-Verlag, Berlin.

De Smet, P.A.G.M., Keller, K., Hänsel, R. and Chandler, R.F. (1993) *Adverse Effects of Herbal Drugs 2*, Springer-Verlag, Berlin.

Der Marderosian, A.H. (1977) Medicinal teas—Boon or bane? *Drug Ther.* February, 178–184.

Der Marderosian, A.H. and Liberti, L. (1988) *Natural Product Medicine*. George F. Stickley Co., Philadelphia.

Demott, K. (1998) St. John's wort tied to serotonin syndrome. *Clin. Psychiatry News.* 26, 28.

D'Mello, J.P.F., Duffus, C.M. and Duffus, J.H. (eds.) (1991) *Toxic Substances in Crop Plants*. Royal Society of Chemistry, Cambridge, UK.

Downey, R.K. (1990) Canola: A quality brassica oilseed. In: Janick, J. and Simon, J.E. (eds.). *Advances in New Crops. Proceedings of the First National Symposium NEW CROPS: Research, Development, Economics*. Timber Press, Portland, OR, pp. 211–215.

Dubrovinskii, S.B. (1946) [About the alimentary toxicosis caused by heliotrope.] *J. Sov. Prot. Health* 6, 17–21 (in Russian). From: *Environmental Health Criteria 80: Pyrrolizidine Alkaloids*. World Health Organization, Geneva.

Eaton, E.A., Walle, U.K., Lewis, A.J., Hudson, T., Wilson, A.A. and Walle, T. (1996) Flavonoids, potent inhibitors of the human form phenolsulfotransferase: Potential role in drug metabolism and chemoprevention. *Drug Metab. Dispos.* 24, 232.

Edwards, D.J., Bellevue, III, F.H. and Woster, P.M. (1996) Identification of 6′,7′-dihydroxy-bergamottin, a cytochrome P450 inhibitor, in grapefruit juice. *Drug Metab. Dispos.* 24, 1287–1290.

Ernst, E. (1999) Second thoughts about safety of St. John's wort. *Lancet.* 355, 2014–2016.

Fernald, M.L. (1987) *Gray's Manual of Botany* (8th edition), Dioscarides Press, Portland, OR.

Figeroa, P., Nembhard, O., Coleman, A. and Betton, M. (1992) Toxic hypoglycemic syndrome—Jamaica, 1989–1991. *Morbidity and Mortality Weekly Reports.* 41, 53–55.

Flom, M.S., Doskotch, R.W. and Beal, J.L. (1967) Isolation and characterization of alkaloids from *Caulophyllum thalictroides*. *J. Pharm. Sci.* 56, 1515–1517.

Formica, J.V. and Regelson, W. (1995) Review of the biology of quercetin and related bioflavonoids. *Food Chem. Toxicol.* 33, 1061–1080.

Francis, P.O. and Clarke, C.F. (1999) Angel trumpet lily poisoning in five adolescents: clinical findings and management. *J. Paediatr. Child Health*, 35, 93–95.

Frank, B.S., Michelson, W.B., Panter, K.E. and Gardner, D.R. (1995) Ingestion of poison hemlock (*Conium maculatum*). *West. J. Med.* 163, 573–574.

Fugh-Berman, A. (2000) Herb-drug interactions. *Lancet.* 355, 134–138.

Fuhr, U. (1998) Drug interactions with grapefruit juice: Extent, probable mechanism and clinical relevance. *Drug Safety* 18, 251.

Fuhr, U. and Kummert, A.L. (1995) The fate of naringin in humans: A key to grapefruit juice-drug interactions? *Clin. Pharmacol. Ther.* 58, 365–373.

Fuhr, U., Klittich, K. and Staib, A.H. (1993) Inhibitory effect of grapefruit juice and its bitter principal, naringenin, on CYP1A2 dependent metabolism of caffeine in man. *Br. J. Clin. Pharmacol.* 35, 431–436.

Gajdusek, D.C. (1979) Recent observations on the use of kava in the New Hebrides. In: Efron, D.H., Holmstedt, B. and Kline, N.S. (eds.), *Ethnopharmacologic Search for Psychoactive Drugs*. Raven Press, New York, pp. 119–125.

Garner, L.F. and Klinger, J.D. (1985) Some visual effects caused by the beverage kava. *J. Ethnopharmacol.* 13, 307–311.

Ger, L.P., Chiang, A.A., Lai, R.S., Chen, S.M. and Tseng, C.J. (1997) Association of *Sauropus androgynus* and bronchiolitis obliterans syndrome: A hospital-based case-control study. *Am. J. Epidemiol.* 145, 842–849.

German, K., Kumar, U. and Blackford, H.N. (1995) Garlic and the risk of TURP bleeding. *Br. J. Urol.* 76, 518.

Gilbert, G.J. (1997) Ginkgo biloba [letter, comment]. *Neurology.* 48, 1137.

Goldman, J.A. and Myerson, G. (1991) Chinese herbal medicine: Camoflaged prescription antiinflammatory drugs, corticosteroids, and lead. *Arthritis Rheum.* 34, 1207.

Gordon, D.W., Rosenthal, G., Hart, J., Sirota, R. and Baker, A.L. (1995) Chaparral ingestion: The broadening spectrum of liver injury caused by herbal medications. *JAMA.* 273, 489–872.

Goth, A. (1974) *Medical Pharmacology*, 7th edition. C.V. Mosby Company, St. Louis, MO, p. 681.

Greene, G.S., Patterson, S.G. and Warner, E. (1996) Ingestion of angel's trumpet: An increasingly common source of toxicity. *South. Med. J.* 89, 365–369.

Grice, H.C., Becking, G. and Goodman, T. (1968) Toxic properties of nordihydroguaretic acid. *Food Cosmet. Toxicol.* 6, 155–161.

Grossman, E., Rosenthal, T., Peleg, E., Holmes, C. and Goldstein, D.S. (1993) Oral yohimbine increases blood pressure and sympathetic nervous outflow in hypertensive patients. *J. Cardiovasc. Pharmacol.* 22, 22–26.

Guengerich, F.P. and Kim, D.H. (1990) *In vitro* inhibition of dihydropyridine oxidation and aflatoxin B1 activation in human liver microsomes by naringenin and other flavonoids. *Carcinogenesis.* 11, 2275–2279.

Gurley, B.J., Gardner, S.F. and Hubbard, M.A. (2000) Content versus label claims in ephedra-containing dietary supplements. *Am. J. Health-Syst. Pharm.* 57, 963–969.

Ha, H.R., Chen, J., Leuenberger, P.M., Freiburghaus, A.U. and Follath, F. (1995) *In vitro* inhibition of midazolam and quinidine metabolism by flavonoids. *Eur. J. Clin. Pharmacol.* 48, 367.

Hamilton, R.J., Shih, R.D. and Hoffman, R.S. (1995) Mobitz Type I heart block after pokeweed ingestion. *Vet. Human Toxicol.* 37, 66–67.

Hartmeier, S.H. and Steurer, J. (1996) Mydriasis, Tachykardie. *Schweiz. Rundsch. Med. Prax.* 85, 495–498.

Heilpern, K.L. (1995) Zigadenus poisoning. *Ann. Emerg. Med.* 25, 259–262.

Hertog, M.G., Hollman, P.C., Katan, M.B. and Kromhout, D. (1993) Intake of potentially anticarcinogenic flavonoids and their determinants in adults in the Netherlands. *Nutr. Cancer.* 20, 21–29.

Hirono, L., Mori, H. and Haga, M. (1978) Carcinogenic activity of *Symphytum officinale*. *J. Natl. Cancer Inst.* 61, 865–869.

Hirono, L., Hagi, M., Fujii, M., Matsuura, S., Matsubara, N., Nakayama, M., Furuya, T., Hikichi, M., Takanashi, H., Uchida, E., Hosaka, S. and Ueno, I. (1979) Induction of hepatic tumors in rats by senkirkine and symphytine. *J. Natl. Cancer Inst.* 63, 469–472.

Hsiue, T.R., Guo, Y.L., Chen, K.W., Chen, C.W., Lee, C.H. and Chang, H.Y. (1998) Dose-response relationship and irreversible obstructive ventilatory defect in patients with consumption of *Sauropus androgynus*. *Chest.* 113, 71–76.

Huang, K.C. (1993) *The Pharmacology of Chinese Herbs*. CRC Press, Boca Raton, FL.

Hutchens, A.R. (1992) *A Handbook of Native American Herbs*. Shambala Publications, Inc., Boston, MA.

Huxtable, R.J. (1990) The harmful potential of herbal and other plant products. *Drug Safety.* 5 (Suppl 1), 126–136.

Johne, A., Brockmoller, J., Bauer, S., Maurer, A., Langheinrich, M. and Roots, I. (1999) Pharmacokinetic interaction of digoxin with an herbal extract from St. John's wort (*Hypericum perforatum*). *Clin. Pharmacol. Ther.* 66, 338–345.

Jones, T.K. and Lawson, B.M. (1998) Profound neonatal congestive heart failure caused by maternal consumption of blue cohosh herbal medication. *J. Pediatr.* 132, 550–552.

Joseph, A.M., Biggs, T., Garr, M., Singh, J. and Lederle, F.A. (1991) Stealth steroids. *N. Engl. J. Med.* 324, 62.

Katz, M. and Saibil, F. (1990) Herbal hepatitis: Subacute hepatic necrosis secondary to chaparral leaf. *J. Clin. Gastroenterol.* 12, 203–206.

Keeler, R.F. (1976) Lupin alkaloids from teratogenic and nonteratogenic lupins. III. Identification of anagyrine as a teratogen by feeding trials. *J. Toxicol. Environ. Health.* 1, 887–898.

Kennelly, E.J., Flynn, T.J., Mazzola, E.P., Roach, J.A., McCloud, T.G., Danford, D.E. and Betz, J.M. (1999) Detecting potential teratogenic alkaloids from blue cohosh rhizomes using an *in vitro* rat embryo culture. *J. Nat. Prod.* 62, 1385–1389.

Khanin, M.N. (1948) [The etiology of toxic hepatitis with ascites.] *Arch. Pathol. USSR.* 1, 42–47 (in Russian). From: *Environmental Health Criteria 80: Pyrrolizidine Alkaloids.* World Health Organization, Geneva.

Kim, H.J., Chun, Y.J., Park, J.D., Kim, S.I., Roh, J.K. and Jeong, T.C. (1997) Protection of rat liver microsomes against carbon tetrachloride-induced lipid peroxidation by red ginseng saponin through cytochrome P450 inhibition. *Planta Medica.* 63, 415.

Klintschar, M., Beham-Schmidt, C., Radner, H., Henning, G. and Roll, P. (1999) *Forensic Sci. Int.* 106, 191–200.

Klohs, M.W., Keller, F., Williams, R.E., Toekes, M.I. and Cronheim, G.E. (1959) A chemical and pharmacological investigation of *Piper methysticum* Forst. *J. Med. Pharm. Chem.* 1, 95–103.

Koren, G., Randor, S., Martin, S. and Danneman, D. (1990) Maternal ginseng use associated with neonatal androgenization. *JAMA.* 264, 2466.

Kouzi, S.A., McMurtry, R.J. and Nelson, S.D. (1994) Hepatotoxicity of germander (*Teucrium chamaedrys* L.) and one of its constituent neoclerodane diterpenes teucrin A in the mouse. *Chem. Res. Toxicol.* 7, 850–856.

Kovarik, J.M., Rigaudy, L., Guerret, M., Gerbeau, C. and Rost, K.L. (1999) Longitudinal assessment of a P-glycoprotein-mediated drug interaction of valspodar on digoxin. *Clin. Pharmacol. Ther.* 66, 391–400.

Krenzelok, E.P., Jacobsen, T.D. and Aronis, J.M. (1996) Poinsettia exposures have good outcomes … just as we thought. *Am. J. Emerg. Med.* 14, 671–674.

Krenzelok, E.P., Jacobsen, T.D. and Aronis, J.M. (1998) Is the yew really poisonous to yew? *J. Toxicol. Clin. Toxicol.* 36, 219–223.

Krishnamachari, K.A.V.R., Bhat, R.V., Krishnamurthi, D., Krishnaswamy, K. and Nagarajan, V. (1977) Aetiopathogenesis of endemic ascites in Surguja district of Madhya Pradesh. *Ind. J. Med. Res.* 65, 672–678.

Laliberté, L. and Villeneuve, J.P. (1996) Hepatitis after the use of germander, a herbal remedy. *Can. Med. Assoc. J.* 154, 1689–1692.

Langford, S.D. and Boor, P.J. (1996) Oleander toxicity: An examination of human and animal toxic exposures. *Toxicology.* 109, 1–13.

Lantz, M.S., Buchalter, E. and Giambanco, V. (1999) St. John's wort and antidepressant drug interactions in the elderly. *J. Geriatr. Psychiatr. Neurol.* 12, 7–10.

Larrey, D., Vial, T., Pauwels, A., Castot, A., Biour, M., David, M. and Michel, H. (1992) Hepatitis after germander (*Teucrium chamaedrys*) administration: Another instance of herbal medicine hepatotoxicity. *Ann. Int. Med.* 117, 129–132.

Lasker, J.M., Huang, M. and Conney, A.H. (1984) *In vitro* and *in vivo* oxidative drug metabolism by flavonoids. *J. Pharmacol. Exp. Ther.* 229, 162.

Lee, M.K., Cheng, B.W., Che, B.W. and Hsieh, D.P. (2000) Cytotoxicity assessment of *Ma-huang* (*Ephedra*) under different conditions of preparation. *Toxicol. Sci.* 56, 424–430.

Li, Y., Wang, E., Patten, C.J., Chen, L. and Yang, C.S. (1994) Effects of flavonoids on cytochrome P-450 dependent acetaminophen metabolism in rats and human microsomes. *Drug Metab. Dispos.* 22, 566–571.

Lin, T.J., Hsu, C.I., Lee, K.H., Shiu, L.L. and Deng, J.F. (1996) Two outbreaks of acute Tung Nut (*Aleurites fordii*) poisoning. *J. Toxicol. Clin. Toxicol.* 34, 87–92.

Liu, G. (1989) Pharmacological actions and clinical use of Fructus Shizandrae. *Chin. Med. J.* 102, 740.

Loeper, J., Descatoire, V., Letteron, P., Moulis, C., Degott, C., Dansette, P., Fau, D. and Pessayre, D. (1994) Hepatotoxicity of germander in mice. *Gastroenterology.* 106, 464–472.

Lord, G.M., Tagore, R., Cook, T., Gower, P. and Pusey, C.D. (1999) Nephropathy caused by Chinese herbs in the UK. *Lancet.* 354, 481–482.

Mabry, T.J., Hunziker, J.H. and DiFeo, Jr., D.R. (1977) *Creosote Bush.* Dowden, Hutchinson, and Ross, Stroudsburg, PA.

MacGregor, F.B., Abernathy, V.E., Dahabia, S., Cobden, I. and Hayes, P.C. (1989) Hepatotoxicity of herbal remedies. *Br. Med. J.* 299, 1156–1157.

Mai, I., Kruger, H., Budde, K., Johne, A., Brockmoller, J., Neumeyer, H.H., Roots, I. (2000) Hazardous pharmacokinetic interaction of St. John's wort (Hypericum perforatum) with the immunosuppressant cyclosporin. *Int. J. Clin. Pharmacol. Ther.* 38, 500–502.

Markowitz, J.S., DeVane, C.L., Boulton, D.W., Carson, S.W., Nahas, Z. and Risch, S.C. (2000) Effect of St. John's wort (*Hypericum perforatum*) on Cytochrome P-450 2D6 and 3A4 activity in healthy volunteers. *Life Sci.* 66, 133–139.

Martinez, H.R., Bermudez, M.V., Rangel-Guerra, R.A. and de Leon Flores, L. (1998) Clinical diagnosis in *Karwinskia humboldtiana* polyneuropathy. *J. Neurol. Sci.* 154, 49–54.

McGuffin, M., Hobbs, C., Upton, R. and Goldberg, A. (1997) *American Herbal Products Association's Botanical Safety Handbook.* CRC Press, Boca Raton, FL.

McRae, S. (1996) Elevated serum digoxin levels in a patient taking digoxin and Siberian ginseng. *Can. Med. Assoc. J.* 155, 293–295.

McTague, J.A. and Forney, R. (1994) Jamaican vomiting sickness in Toledo, Ohio. *Ann. Emerg. Med.* 23, 1116–1118.

Meda, H.A., Diallo, B., Buchet, J.P., Lison, D., Barennes, H., Ouangre, A., Sanou, M., Cousens, S., Tall, F. and Van de Perre, P. (1999) Epidemic of fatal encephalopathy in preschool children in Burkina Faso and consumption of unripe ackee (*Blighia sapida*) fruit. *Lancet.* 353, 536–540.

Merck Index. (2000) *Guar Gum.* 12th ed. on CD-ROM Version 12:3, Chapman & Hall/CRC-netBASE, Boca Raton, FL.

Miller, L.G. (1998) Herbal medicinals. selected clinical considerations focusing on known or potential drug-herb interactions. *Arch. Intern. Med.* 158, 2200–2211.

Mizugaki, M., Ito, K., Ohyama, Y., Konishi, Y., Tanaka, S. and Kurasawa, K. (1998) Quantitative analysis of aconitum alkaloids in the urine and serum of a male attempting suicide by oral intake of aconite extract. *J. Anal. Toxicol.* 22, 336–340.

Mossoba, M.M., Lin, H.S., Andrzejewski, D., Sphon, J.A., Betz, J.M., Miller, L.J., Eppley, R.M., Trucksess, M.W. and Page, S.W. (1994) Application of gas chromatography/matrix isolation/Fourier transform infrared spectroscopy to the identification of pyrrolizidine alkaloids from comfrey root (*Symphytum officinale* L.). *J. A. O. A. C. Int.* 77, 1167–1174.

Mostefa-Kara, N., Pauwels, A., Pines, E., Biour, M. and Levy, B.G. (1992) Fatal hepatitis after herbal tea. *Lancet.* 340, 674.

Mueller, S.O. and Stopper, H. (1999) Characterization of the genotoxicity of anthraquinones in mammalian cells. *Biochim. Biophys. Acta.* 1428, 406–414.

Nadir, A., Agrawal, S., King, P.D. and Marshall, J.B. (1996) Acute hepatitis associated with the use of a Chinese herbal product, *Ma-Huang. Am. J. Gastroenterol.* 91, 1436–1438.

Nagata, K.M. (1971) Hawaiian medicinal plants. *Econ. Bot.* 25, 245–254.

National Research Council. (1973) *Toxicants Occurring Naturally in Foods.* National Academy of Sciences, Washington, DC.

Nebel, A., Schneider, B.J., Baker, R.K. and Kroll, D.J. (1999) Potential metabolic interaction between St. John's wort and theophylline. *Ann. Pharmacother.* 33, 502.

Nielsen, S.E., Breinholt, V. and Justesen, U. (1998) *In vitro* biotransformation of flavonoids by rat liver microsomes. *Xenobiotica.* 28, 389–401.

Nightingale, S.L. (1996) Warning issued about street drugs containing botanical sources of ephedrine. *JAMA.* 275, 1534.

Nortier, J.L., Martinez, M.C., Schmeiser, H.H., Arlt, V.M., Bieler, C.A., Petein, M., Depierreux, M.F., De Pauw, L., Abramowicz, D., Vereerstraeten, P. and Vanherweghem, J.L. (2000) Urothelial carcinoma associated with the use of a Chinese herb (*Aristolochia fangchi*). *N. Engl. J. Med.* 342, 1686–1692.

Obermeier, M.T., White, R.E. and Yang, C.S. (1995) Effects of bioflavonoids on hepatic P450 activities. *Xenobiotica.* 25, 575.

Obermeyer, W.R., Musser, S.M., Betz, J.M., Casey, R.E., Pohland, A.E. and Page, S.W. (1995) Chemical studies of phytoestrogens and related compounds in dietary supplements: flax and chaparral. *Proc. Soc. Exp. Biol. Med.* 208, 6–12.

Page, R.L. and Lawrence, J.D. (1999) Potentiation of warfarin by don quai. *Pharmacotherapy.* 19, 870–876.

Perrotta, D.M., Nickey, L.N., Raid, M., Caraccio, T., Mofenson, H.C., Waters, C., Morse, D., Osorio, A.M., Hoshiko, S. and Rutherford III, G.W. (1995) Jimson weed poisoning—Texas, New York, and California, 1994. *Morbidity and Mortality Weekly Reports.* 44, 41–44.

Perrotta, P.M., Coody, G. and Culmo, C. (1996) Adverse events associated with ephedrine-containing products—Texas, December 1993–September 1995. *Morbidity and Mortality Weekly Reports.* 45, 689–693.

Phillipson, J.D. and Anderson, L.A. (1984) Herbal remedies used in sedative and antirheumatic preparations: Part 1. *Pharm. J.* 233, 80–82.

Piscitelli, S.C., Burstein, A.H., Chaitt, D., Alfaro, R.M. and Falloon, J. (2000) Marked reduction in indinavir exposure by St. John's wort. *Lancet.* 355, 547–548.

Pitz, W.J., Sosulski, F.W. and Hogge, L.R. (1980) Occurrence of vicine and convicine in seeds of some *Vicia* species and other pulses. *Can. Inst. Food Sci. Technol. J.* 13, 35–39.

Poulton, J.E. (1983) Cyanogenic compounds in plants and their toxic effects. In: Keeler, R.F. and Tu, A.T. (eds.), *Handbook of Natural Toxins. Volume 1—Plant and Fungal Toxins.* Marcel Dekker, Inc., New York, pp. 117–157.

Prakash, A.S., Pereira, T.N., Reilly, P.E. and Seawright, A.A. (1999) Pyrrolizidine alkaloids in human diet. *Mutat. Res.* 443, 53–67.

Raikhlin-Eisenkraft, B. and Bentur, Y. (2000) *Ecbalium elaterium* (squirting cucumber)—remedy or poison? *J. Toxicol. Clin. Toxicol.* 38, 305–308.

Rechcigl, Jr., M. (ed.) (1983) *Handbook of Naturally Occurring Food Toxicants.* CRC Press, Boca Raton, FL.

Ressler, C. (1962) Isolation and identification from common vetch of the neurotoxin β-cyano-L-alanine, a possible factor in neurolathyrism. *J. Biol. Chem.* 237, 733–735.

Ressler, C., Nigam, S.N. and Giza, Y.H. (1969) Toxic principle in vetch. Isolation and identification of γ-L-glutamyl-L-β-cyanoalanine from common vetch seeds. Distribution in some legumes. *J. Am. Chem. Soc.* 91, 2758–2764.

Ridker, P.M., Ohkuma, S., McDermott, W.V., Trey, C. and Huxtable, R.J. (1985) Hepatic veno-occlusive disease associated with the consumption of pyrrolizidine-containing dietary supplements. *Gastroenterology.* 88, 1050–1054.

Ries, C.A. and Sahud, M.A. (1975) Agranulocytosis caused by Chinese herbal medicines. Dangers of medications containing aminopyrine and phenylbutazone. *JAMA.* 231, 352–355.

Rizk, A.-F. M. (ed.) (1990) *Poisonous Plant Contamination of Edible Plants.* CRC Press, Boca Raton, FL.

Roby, C.A., Anderson, G.D., Kantor, E., Dryer, D.A. and Burstein, A.H. (2000) St. John's wort: effect on CYP3A4 activity. *Clin. Pharmacol. Ther.* 67, 451–457.

Roeder, E. (1995) Medicinal plants in Europe containing pyrrolizidine alkaloids. *Pharmazie.* 50, 83–98.

Rose, K.D., Croissant, P.D., Parliament, C.F. and Levin, M.B. (1990) Spontaneous spinal epidural hematoma with associated platelet dysfunction from excessive garlic consumption: A case report. *Neurosurgery.* 26, 880–882.

Rosenblatt, M. and Mindel, J. (1997) Spontaneous hyphema associated with ingestion of *Ginkgo biloba* extract [letter]. *N. Engl. J. Med.* 336, 1108.

Routledge, P.A. and Spriggs, T.L.B. (1989) Atropine as possible contaminant of comfrey tea. *Lancet.* 1 (8644), 963–964.

Rowin, J. and Lewis, S.L. (1996) Spontaneous bilateral subdural hematomas associated with chronic *Ginkgo biloba* ingestion have also occurred. *Neurology.* 46, 1775–1776.

Ruschitzka, F., Meier, P.J., Turina, M., Luscher, T.F. and Noll, G. (2000) Acute heart transplant rejection due to Saint John's wort. *Lancet.* 355, 548–549.

Sandnes, D., Johansen, T., Teien, G. and Ulsaker, G. (1992) Mutagenicity of crude senna and senna glycosides in *Salmonella typhimurium. Pharmacol. Toxicol.* 71, 165–172.

Sarkar, M.A. (1995) Quercetin not only inhibits P-glycoprotein efflux activity but also inhibits CYP3A isozymes. *Cancer Chemother. Pharmacol.* 36, 448–450.

Scambia, G., Ranelletti, F.O., Panici, P.B., DeVincenzo, R., Bonanno, G., Ferrandina, G., Piantelli, M., Bussa, S., Rumi, C., Cianfriglia, M. and Mancuso, S. (1994) Quercetin potentiates the effect of adriamycin in a multidrug resistant MCF-7 human breast cancer cell line: P-glycoprotein as a possible target. *Cancer Chemother. Pharmacol.* 34, 459–464.

Schneider, F., Lutin, P., Kintz, P., Astruc, D., Flesch, F. and Tempe, J.D. (1996) Plasma and urine concentrations of atropine after the ingestion of cooked deadly nightshade berries. *J. Toxicol. Clin. Toxicol.* 34, 113–117.

Schwartz, J. (1997) FDA proposes rules to curb risks from herbal stimulant. *Washington Post.* June 3, A02.

See, D., Gurnee, K. and LeClair, M. (1999) An *in vitro* screening study of 196 natural products for toxicity and efficacy. *J.A.N.A.* 2, 25–39.

Sheehan, D.M. (1998) Herbal medicines, phytoestrogens and toxicity: Risk:benefit considerations. *Proc. Soc. Exp. Biol. Med.* 217, 379–385.

Sheikh, N.M., Philen, R.M. and Love, L.A. (1997) Chaparral-associated hepatotoxicity. *Arch Intern. Med.* 157, 913–919.

Shi, J.Z. and Liu, G.T. (1996) Effect of alpha-hederin and sapindoside B on hepatic microsomal cytochrome P-450 in mice. *Acta Pharmacologica Sinica.* 17, 264.

Shulman, K.I. and Walker, S.E. (1999) Refining the MAOI diet: Tyramine content of pizzas and soy products. *J. Clin. Psychiatry.* 60, 191–193.

Sims, D.N., James, R. and Christensen, T. (1999) Another death due to *Nicotiana glauca. J. Forensic Sci.* 44, 447–449.

Slifman, N.R., Obermeyer, W.R., Aloi, B.K., Musser, S.M., Correll, W.A., Cichowicz, S.M., Betz, J.M. and Love, L.A. (1998) Contamination of botanical dietary supplements by *Digitalis lanata. N. Engl. J. Med.* 339, 806–811.

Smith, B.C. and Desmond, P.V. (1993) Acute hepatitis induced by ingestion of the herbal medication chaparral. *Aust. N. Z. J. Med.* 23, 526.

Soller, R.W. (2000) Regulation in the herb market: The myth of the "unregulated industry." *HerbalGram.* 49, 64–68.

Sonda, L.P., Mazo, R. and Chancellor, M.B. (1990) The role of yohimbine for the treatment of erectile impotence. *J. Sex Marital Ther.* 16, 15–21.

Southgate, H.J., Egerton, M. and Dauncey, E.A. (2000) Lessons to be learned: A case study approach. Unseasonal severe poisoning of two adults by deadly nightshade (*Atropa belladonna*). *J. R. Soc. Health.* 120, 127–130.

Spiller, H.A., Willias, D.B., Gorman, S.E. and Sanftleban, J. (1996) Retrospective study of mistletoe ingestion. *J. Toxicol. Clin. Toxicol.* 34, 405–408.

Stillman, A.E., Huxtable, R.J., Fox, D.W., Hart, M.C., Bergeson, P.S., Counts, J.M., Cooper, L., Grunenfelder, G., Blackmon, J., Fretwell, M., Raey, J., Allard, J. and Bartleson, B. (1977) Poisoning associated with herbal teas—Arizona, Washington. *Morbidity and Mortality Weekly Reports.* 26, 257–259.

Sweeney, K., Gensheimer, K.F., Knowlton-Field, J. and Smith, R.A. (1994) Water hemlock poisoning—Maine, 1992. *Morbidity and Mortality Weekly Reports.* 43, 229–231.

Swenson, J. (1996) Man convicted of driving under influence of kava. *Deseret News*, Salt Lake City, UT, 5 August.

Tandon, R.K., Tandon, B.N., Tandon, H.D., Bhatia, M.L., Bhargava, S., Lal, P. and Arora, R.R. (1976a) Study of an epidemic of veno-occlusive disease in India. *Gut.* 17, 849–855.

Tandon, B.N., Tandon, H.D., Tandon, R.K., Narendranathan, M. and Joshi, Y.K. (1976b) Epidemic of veno-occlusive disease in central India. *Lancet.* 2, 271–272.

Tandon, H.D., Tandon, B.N., Tandon, R.K. and Nayak, N.C. (1977) A pathological study of the liver in an epidemic outbreak of veno-occlusive disease. *Ind. J. Med. Res.* 65, 679–684.

Tandon, B.N., Tandon, H.D. and Mattocks, A.R. (1978a) An epidemic of veno-occlusive disease of the liver in Afghanistan. *Am. J. Gastroenterol.* 70, 607–613.

Tandon, B.N., Tandon, H.D. and Mattocks, A.R. (1978b) Study of an epidemic of veno-occlusive disease in Afghanistan. *Ind. J. Med. Res.* 68, 84–90.

Tate, M.E. and Enneking, D. (1992) A mess of red pottage. *Nature.* 359, 357–358.

Thatte, U. and Dahanukar, S. (1999) The Mexican poppy poisons the Indian mustard: Facts and figures. *J. Assoc. Physicians India.* 47, 332–335.

Thummel, K.E. and Wilkinson, G.R. (1998) *In vitro* and *in vivo* drug interactions involving human CYP3A. *Annu. Rev. Pharmacol. Toxicol.* 38, 389–430.

Tibballs, J. (1995) Clinical effects and management of eucalyptus oil ingestion in infants and young children. *Med. J. Aust.* 163, 177–180.

Tiongson, J. and Salen, P. (1998) Mass Ingestion of Jimson Weed by eleven teenagers. *Del. Med. J.* 70, 471–476.

Toubro, S., Astrup, A.V., Breum, L. and Quaade, F. (1993) Safety and Efficacy of long-term treatment with ephedrine, caffeine and an ephedrine/caffeine mixture. *Int. J. Obes. Relat. Metab. Disord.* 17, S69–S72.

Treasure, J. (2000) Herbal pharmacokinetics: A practitioner update with reference to St. John's wort (*Hypericum perforatum*) interactions. *J. Am. Herbal. Guild.* 1, 2–11.

Tsyrlov, I.B., Mikhailenko, V.M. and Gelboin, H.V. (1994) Isozyme- and species-specific susceptibility of cDNA-expressed CYP1A P-450s to different flavonoids. *Biochimica et Biophysica Acta.* 1205, 325–335.

Tyler, V.E. (1993) *The Honest Herbal.* Pharmaceutical Products Press, New York, NY.

United States Code, Title 21. (2000) *Federal Food, Drug, and Cosmetic Act of June 25, 1938, As Amended*, sec. 201 (g)(1)(B).

United States Pharmacopeial Convention. (1999) *The United States Pharmacopeia 24, The National Formulary 19.* United States Pharmacopeial Convention, Inc., Rockville, MD, pp. 642–645, 1318–1322, 1439–1441.

United States Public Law 103–417. 103rd Cong., 25 October 1994. *Dietary Supplement Health and Education Act of 1994.*

Vale, S. (1998) Subarachnoid haemorrhage associated with ginkgo biloba. *Lancet.* 352, 36.

Vanhaelen, M., Vanhaelen-Fastre, R., But, P. and Vanherweghem, J.L. (1994) Identification of aristolochic acid in Chinese herbs. *Lancet.* 343, 174.

Vanherweghem, J.L., Depierreux, M., Tielemans, C., Abramowicz, D., Dratwa, M., Jadoul, M., Richard, C., Vandervelde, D., Verbeelen, D., Vanhaelen-Fastre, R. and Vanhaelen, M. (1993) Rapidly progressive interstitial renal fibrosis in young women: Association with slimming regimen including Chinese herbs. *Lancet.* 341, 387–391.

Vollmer, J.J., Steiner, N.C., Larsen, G.Y., Muiehead, K.M. and Molyneux, R.J. (1987) Pyrrolizidine alkaloids: Testing for toxic constituents of comfrey. *J. Chem. Ed.* 64, 1027–1030.

Von Moltke, L.L., Greenblatt, D.J., Schmider, J., Wright, C.E., Harmatz, J.S. and Shader, R.I. (1998) *In vitro* approaches to predicting drug interactions *in vivo. Biochem. Pharmacol.* 55, 113–122.

Walgren, R.A., Walle, U.K. and Walle, T. (1998) Transport of quercetin and its glucosides across human intestinal epithelial Caco-2 cells. *Biochem. Pharmacol.* 55, 1721–1727.

Walter-Sack, I. and Klotz, U. (1996) Influence of diet and nutritional status on drug metabolism. *Clin. Pharmacokinet.* 31, 41–64.

Watson, D.H. (ed.) (1987) *Natural Toxicants in Food: Progress and Prospects.* VCH Publishers, New York, NY.

Weiner, M.A. (1971) Ethnomedicine in Tonga. *Econ. Bot.* 25, 423–450.

Westendorf, J. (1993) Anthranoid derivatives—general discussion. In: De Smet, P.A.G.M., Keller, K., Hänsel, R. and Chandler, R.F. (eds.), *Adverse Effects of Herbal Drugs 2,* Springer-Verlag, Berlin, pp. 105–108.

Weston, C.F.M., Cooper, B.T., Davies, J.D. and Levine, D.F. (1987) Veno-occlusive disease of the liver secondary to ingestion of comfrey. *Br. Med. J.* 295, 183.

Willey, L.B., Mady, S.P., Cobaugh, D.J. and Wax, P.M. (1995) Valerian overdose: A case report. *Vet. Hum. Toxicol.* 37, 364–365.

Wilson, M. (2000) Driver busted for kava tea—San Mateo County case first of its kind in state. *San Francisco Chronicle,* San Francisco, CA, 29 April.

Woldemariam, T.Z., Betz, J.M. and Houghton, P.J. (1997) Analysis of aporphine and quinolizidine alkaloids from *Caulophyllum thalictroides* by densitometry and HPLC. *J. Pharm. Biomed. Anal.* 15, 839–843.

Wu, C.L., Hsu, W.H., Chiang, C.D., Kao, C.H., Hung, D.Z., King, S.L. and Deng, J.F. (1997) Lung injury related to consuming *Sauropus androgynus* vegetable. *J. Toxicol. Clin. Toxicol.* 35, 241–248.

Yang, C.S., Lin, C.H., Chang, S.H. and Hsu, H.C. (2000) Rapidly progressive fibrosing interstitial nephritis associated with Chinese herbal drugs. *Am. J. Kidney Dis.* 35, 313–318.

Yates, K.M., O'Connor, A. and Horsely, C.A. (2000) "Herbal Ecstasy": A case series of adverse reactions. *N. Z. Med. J.* 113, 315–317.

Yue, Q.-Y., Bergquist, C., Gerdén, B. (2000) Seven cases of decreased effect of warfarin during concomitant treatment with St. John's wort. *Lancet.* 355, 577.

Zhang, Y.P., Sussman, N., Macina, O.T., Rosenkranz, H.S. and Klopman, G. (1996) Prediction of the carcinogenicity of a second group of organic chemicals undergoing carcinogenicity testing. *Environ. Health Perspect.* 104, 1045–1050.

Index

A

AA (Arachidonic acid), 318, 319
Abietane diterpenes, 200–206
Abortive effects, of rosemary, 190
Acetone, 100
Acetonitrile, 45, 250
Acetyl daidzin, 42, 52
Acetyl genistin, 42, 52
Acetyl glycitin, 42, 52
Acid and base treatment, 56–57
Acid hydrolysis, 45, 56, 291–292
Acinoxa Extr-O-Mat extractor, 343–344
Ackee, 383
Aconite, 372
Aconitum, 372
Acquired immune deficiency syndrome (AIDS),
      151, 203–204
Acylation, 95–98
AD (Alzheimer's Disease), 119, 319
Adenosine, 226
Adenosine diphosphate (ADP), 226
ADHD (Attention deficit hyperactive disorder),
      318
Adulteration, of botanical dietary supplements,
      372–374
Adverse Effects of Herbal Drugs (De Smet et al.),
      382
Aflurax, 271
Aged garlic extracts (AGE), 222
Aggression, effect of docosahexaenoic acid on, 322
Aging, 7, 118
Aglycones
    analysis of, 45–48
    chemical structure, 41
    conversion from glucoside to, 50, 56, 58
    effect of fermentation on, 54
    effect of processing on, 54
    effect on taste, 53
    production of aglycone-enriched protein
          extract, 58–59
    protein concentrate enriched with, 57–59
    in tempeh, 55
AIDS (Acquired immune deficiency syndrome),
      151, 203–204
Ajoene, 230
Akebia, 370
Alcohol wash, 54
Aldehydes, 246

Aliphatic acids, 76
Alkaline hydrolysis, 277
Alkaline refining, 22
Alkaline treatments, 56–57, 59
Alkamides, 243–245
    assessment, 251–252
    isolation, 250
    stability of, in echinacea, 254–255
    structure verification, 250–251
Allicin, 216, 220–221, 226, 229
Alliin, 216, 222–223
Alliinase, 216, 216–218
All-rac-a-tocopherol, 4, 6
All-rac-a-tocopheryl acetate, 6
All-rac-a-tocopheryl succinate, 6
Allyl, 215–216, 223–224
Allylmercaptan, 223–224, 226
Alpha-linolenic acid (LNA), 312, 314–316
Alpha-terpineol, 173–174
Alprazolam, 380, 381
Alzheimer's Disease (AD), 119, 319
Amaranth, 15
Ammonium oxalate, 275
Amyotrophic lateral sclerosis, 119
Anagyrine, 376
Analytical methods
    flavonoids, 99–103
        extraction and fractionation, 99–101
        HPLC analysis, 102–103
        hydrolysis of glycosides and esters, 101
        measuring total anthocyanins, 102
        measuring total flavanols, 102
        measuring total phenolics, 101–102
    limonene, 179–180
    pectin, 278–285
        analysis of neutral sugars, 280
        determination of degree of esterification,
              279
        determination of end-reducing sugars, 278
        molecular weight, 280
        quantitation in samples, 278–279
        rheological and textural characterization,
              280–285
    phenolic diterpenes, 206–207
    phytochemicals, 247–252
        alkamide assessment, 251–252
        alkamide isolation, 250
        alkamide structure verification, 250–251

395

M

N